Alternatives to Conventional Food Processing
2nd Edition

Green Chemistry Series

Editor-in-chief:
James H. Clark, *Department of Chemistry, University of York, UK*

Series editors:
George A. Kraus, *Iowa State University, USA*
Andrzej Stankiewicz, *Delft University of Technology, The Netherlands*
Peter Siedl, *Federal University of Rio de Janeiro, Brazil*

Titles in the series:
1: The Future of Glycerol: New Uses of a Versatile Raw Material
2: Alternative Solvents for Green Chemistry
3: Eco-Friendly Synthesis of Fine Chemicals
4: Sustainable Solutions for Modern Economies
5: Chemical Reactions and Processes under Flow Conditions
6: Radical Reactions in Aqueous Media
7: Aqueous Microwave Chemistry
8: The Future of Glycerol: 2nd Edition
9: Transportation Biofuels: Novel Pathways for the Production of Ethanol, Biogas and Biodiesel
10: Alternatives to Conventional Food Processing
11: Green Trends in Insect Control
12: A Handbook of Applied Biopolymer Technology: Synthesis, Degradation and Applications
13: Challenges in Green Analytical Chemistry
14: Advanced Oil Crop Biorefineries
15: Enantioselective Homogeneous Supported Catalysis
16: Natural Polymers Volume 1: Composites
17: Natural Polymers Volume 2: Nanocomposites
18: Integrated Forest Biorefineries
19: Sustainable Preparation of Metal Nanoparticles: Methods and Applications
20: Alternative Solvents for Green Chemistry: 2nd Edition
21: Natural Product Extraction: Principles and Applications
22: Element Recovery and Sustainability
23: Green Materials for Sustainable Water Remediation and Treatment
24: The Economic Utilisation of Food Co-Products
25: Biomass for Sustainable Applications: Pollution Remediation and Energy
26: From C-H to C-C Bonds: Cross-Dehydrogenative-Coupling
27: Renewable Resources for Biorefineries
28: Transition Metal Catalysis in Aerobic Alcohol Oxidation
29: Green Materials from Plant Oils
30: Polyhydroxyalkanoates (PHAs) Based Blends, Composites and Nanocomposites

31: Ball Milling Towards Green Synthesis: Applications, Projects, Challenges
32: Porous Carbon Materials from Sustainable Precursors
33: Heterogeneous Catalysis for Today's Challenges: Synthesis, Characterization and Applications
34: Chemical Biotechnology and Bioengineering
35: Microwave-Assisted Polymerization
36: Ionic Liquids in the Biorefinery Concept: Challenges and Perspectives
37: Starch-based Blends, Composites and Nanocomposites
38: Sustainable Catalysis: With Non-endangered Metals, Part 1
39: Sustainable Catalysis: With Non-endangered Metals, Part 2
40: Sustainable Catalysis: Without Metals or Other Endangered Elements, Part 1
41: Sustainable Catalysis: Without Metals or Other Endangered Elements, Part 2
42: Green Photo-active Nanomaterials
43: Commercializing Biobased Products: Opportunities, Challenges, Benefits, and Risks
44: Biomass Sugars for Non-Fuel Applications
45: White Biotechnology for Sustainable Chemistry
46: Green and Sustainable Medicinal Chemistry: Methods, Tools and Strategies for the 21st Century Pharmaceutical Industry
47: Alternative Energy Sources for Green Chemistry
48: High Pressure Technologies in Biomass Conversion
49: Sustainable Solvents: Perspectives from Research, Business and International Policy
50: Fast Pyrolysis of Biomass: Advances in Science and Technology
51: Catalyst-free Organic Synthesis
52: Hazardous Reagent Substitution: A Pharmaceutical Perspective
53: Alternatives to Conventional Food Processing: 2nd Edition

How to obtain future titles on publication:
A standing order plan is available for this series. A standing order will bring delivery of each new volume immediately on publication.

For further information please contact:
Book Sales Department, Royal Society of Chemistry, Thomas Graham House, Science Park, Milton Road, Cambridge, CB4 0WF, UK
Telephone: +44 (0)1223 420066, Fax: +44 (0)1223 420247
Email: booksales@rsc.org
Visit our website at www.rsc.org/books

Alternatives to Conventional Food Processing
2nd Edition

Edited by

Andrew Proctor
University of Arkansas, USA
Email: aproctor@uark.edu

ROYAL SOCIETY OF CHEMISTRY

THE QUEEN'S AWARDS FOR ENTERPRISE: INTERNATIONAL TRADE 2013

Green Chemistry Series No. 53

Print ISBN: 978-1-78262-518-6
PDF ISBN: 978-1-78262-659-6
EPUB ISBN: 978-1-78801-403-8
ISSN: 1757-7039

A catalogue record for this book is available from the British Library

The Royal Society of Chemistry is a charity, registered in England and Wales, Number 207890, and a company incorporated in England by Royal Charter (Registered No. RC000524), registered office: Burlington House, Piccadilly, London W1J 0BA, UK, Telephone: +44 (0) 207 4378 6556.

For further information see our web site at www.rsc.org

Printed in the United Kingdom by CPI Group (UK) Ltd, Croydon CR0 4YY, UK

Preface

The food industry is an important sector of the international business community, with food safety and food quality playing a vital role in maintaining profitability. Traditional thermal processing techniques have been used for generations to maintain a safe, nutritious food supply that is acceptable to consumers. However, increasing energy costs and the desire to purchase 'green,' environmentally responsible products have been the stimulus to develop alternative technologies. Furthermore, processing at high temperatures may cause food quality loss, which can be avoided by many alternative processing methods. This book is intended to provide food industrialists, professional academics and graduate students with a review of major alternative technology that could be used commercially to reduce energy costs while maintaining food safety and quality.

There continues to be increasing general concern regarding climate change and energy use, with additional food industry concerns regarding production costs and food safety regulation. Therefore, this second edition is a timely update of the value and commercial importance of alternative food technology research. The introductory chapter highlights the importance of evaluating new technology by sustainability assessment through life cycle analysis, which is a valuable new addition. This is followed by an update of the international legal developments that affect emerging processing methods, with subsequent chapters discussing and evaluating specific technologies. The authors have all made significant contributions to their field and are well qualified to comment on the value and significance of innovative green food processing methods. It is hoped that this book will serve as an introduction

Green Chemistry Series No. 53
Alternatives to Conventional Food Processing: 2nd Edition
Edited by Andrew Proctor

Published by the Royal Society of Chemistry, www.rsc.org

to those interested in gaining an understanding of the history and current developments in various 'green' alternative food processing methods, their potential for commercialization and role in the future of the food industry.

Andrew Proctor
University of Arkansas

Contents

Green Chemistry Series No. 53
Alternatives to Conventional Food Processing: 2nd Edition
Edited by Andrew Proctor

Published by the Royal Society of Chemistry, www.rsc.org

CHAPTER 1

Principles of Green Food Processing (Including Lifecycle Assessment and Carbon Footprint)

G. J. THOMA,*[a] S. W. ELLSWORTH[b] AND M. J. YAN[c]

[a] Ralph E. Martin Department of Chemical Engineering, University of Arkansas, Fayetteville, AR 72701, USA; [b] Department of Food Science, University of Arkansas, Fayetteville, AR 72701, USA; [c] School of Biosystems and Food Engineering, University College Dublin, Belfield, Dublin 4, Ireland
*Email: gthoma@uark.edu

1.1 Introduction

The global food and agricultural sectors are facing numerous pressures, including the burgeoning global population, the expanding middle class and the increasing desire of more people to have high-quality, low-cost food.[1] Numerous studies have shown that the main environmental hotspots within the food supply chain are associated with upstream activities (agricultural production, cultivation of crops and animal husbandry) and as a result have received the most attention from the consuming public, governmental organizations and non-governmental organizations (NGOs). These studies suggest that 70–90% of most environmental impacts in a full supply chain assessment can be attributed to the primary production phase; however,

Green Chemistry Series No. 53
Alternatives to Conventional Food Processing: 2nd Edition
Edited by Andrew Proctor

Published by the Royal Society of Chemistry, www.rsc.org

many of the same studies point to the food processing and manufacturing stage of the supply chain as being responsible for 10–20% of supply chain impact.[2–4] Also, although it is tempting to focus only on those upstream activities where the majority of impact arises, sustainability cannot be achieved by focusing on those activities alone, but must also identify opportunities and implement improvements at later stages of the supply chain. It is for that reason that this book is an especially strong addition to the literature for its focus on the food processing sector and the technologies and opportunities that exist for improvement of the environmental performance of food supply and improving food security.

The food manufacturing industry has traditionally held the role of ensuring food safety, regulatory compliance (for example, nutritional labeling), marketing and profitability. More recently, an additional layer of providing both information and documenting progress towards a sustainable food supply has been added. It should be clear that concerns over environmental sustainability of the food system will have secondary importance to the sector's traditional functions: unsafe, but environmentally friendly products will never be marketed. Hence the context of this chapter is to define the available operating space and useful techniques for understanding the role that environmental sustainability has in the food processing sector.

There is a consensus that the assessment of sustainability requires a holistic perspective of the system being evaluated. This includes the full supply chain, from cradle to grave, in addition to a full complement of environmental indicators. The cradle-to-grave perspective includes all activities necessary for the production of the item under study, extending back in the supply chain to the original extraction of resources. This means, for example, that coal mining and transport to the power plant to produce electricity for pumping or refrigeration are included. In addition, processes associated with consumption and end-of-life treatment are included. An example of the importance of including the full supply chain is in the evaluation of food packaging. One role of packaging is protection of the product, which reduces loss. Light weighting a package will make the package itself more sustainable, but if it leads to even a slight increase in food loss, the overall effect would very likely be a reduction in the overall sustainability of the system because of the relatively large impacts associated with the production of the food itself. By adopting a system perspective, tradeoffs between supply chain stages can be identified, which helps to avoid unintended consequences. In addition, a range of environmental categories should also be included in the overall assessment. Multiple categories allow the identification of potential tradeoffs between environmental impacts. For example, water use efficiency in a processing facility may be achieved at an additional energy cost and therefore the tradeoff of improved water use comes at the cost of an increased carbon footprint. This highlights the truism that "one size does not fit all." For example, in water-scarce regions a higher footprint for global warming may be a necessary and acceptable tradeoff.

1.2 Sustainability Assessment Tools

Sustainability is a complex concept with a deceptively simple definition: to meet the needs of current generations without compromising the ability of future generations to meet their needs. In general, sustainability is considered to have three pillars: social, economic and environmental. The complexity arises in attempting not only to balance environmental tradeoffs as mentioned above, but also to balance these tradeoffs with social and economic values that are deemed important. A major goal of sustainability assessment is therefore to identify the tradeoffs and tensions in the system so that fully informed decisions can be taken in an effort to maintain our collective ability to provide prosperity. Among the tools used for sustainability assessment are lifecycle assessment (LCA), lifecycle costing (LCC), social lifecycle assessment (SLCA), lifecycle sustainability assessment (LCSA), organizational lifecycle assessment (OLCA), environmental risk assessment (ERA) and, in the context of food safety, microbiological risk assessment (MRA). Some of these tools can be used in conjunction with each other or, depending on the needs of the assessment, may be used alone. An emerging paradigm in the context of systems is the so-called circular economy. In this paradigm, there is an explicit and conscious attempt to design products in a manner that makes the utilization of materials at the end of their intended life as raw materials for a subsequent use as streamlined and efficient as possible. Clearly, a fundamental principle of sustainability is resource use efficiency, and in the context of food processing this translates to minimizing energy and water use and food loss while simultaneously producing high-quality, nutritious and safe foods to enhance food security.

The most commonly used tool for system scale assessment of product systems is LCA, which is codified through a series of international standards, including general guidance in addition to specific guides for water footprint and carbon footprint.[5–8] These standards are targeted at providing guidelines for products and services and specifically require a full lifecycle perspective for the reasons outlined above. The International Organization for Standardization (ISO) has not published guidelines at the organizational level; however, the UNEP/SETAC Life Cycle Initiative and World Resource Institute have published guidelines for adapting LCA to the organizational scale.[9,10] LCC is a tool to permit the full cost of a product to be considered using the same system as used in LCA. The goal of LCC is to provide a full cost accounting of the production (including delivery and installation), operation and end-of-life costs (decommissioning and disposal) associated with a product. It may additionally include costs of externalities; for example, where environmental pollution costs that are borne by society can be quantified and verified, these externalities can also be included in the cost assessment.[11,12] Integration of LCC and LCA remains relatively uncommon, yet is an important area because all enterprises must be both economically and environmentally viable. SLCA arose from efforts within corporate social responsibility initiatives to quantify the societal metrics associated with production and consumption. SLCA is the least developed methodology, but recently guidelines have been published[13] and databases created that allow the assessment of social risks in supply chains.[14–18] SLCA attempts to evaluate and quantify socially relevant indicators, including forced child labor, excessive work time, collective bargaining rights, health and safety and human rights. The European Commission, through the European Platform on Life Cycle Assessment, initiated an effort to extend the framework of LCA to incorporate LCC and SLCA to create LCSA. There remain clear challenges associated with collecting and maintaining up-to-date data in the social databases, and for this reason there are significantly fewer publications related to SLCA and LCSA than environmental LCA. For this reason, the remainder of this chapter will focus on environmental sustainability assessment. It should be noted that despite the growing popularity and utility of LCA, there are also some significant limitations to the methodology. For example, in evaluating agricultural production systems, LCA has limited capabilities with regard to the evaluation of ecosystem services, and also has only nascent capabilities for the inclusion of health effects associated with dietary choices or food-borne pathogens.

1.3 Standards and Regulations for Assessing Sustainability

ISO requires that the selection of impact categories shall reflect a comprehensive set of environmental issues related to the product system being studied, taking the goal and scope into consideration. In other words, impact categories should be relevant to the product system under study. However, a clear definition of the criteria that define "relevant" remains elusive, but this is beginning to change. In the past decade, Product Category Rules (PCRs) have been developed to set up standardized rules for products that serve the same functions, including choice of metrics used to estimate impacts.[19] However, inconsistency in the choice of impact category has been found in a comparison of five different PCRs developed independently by different organizations. For example, The Sustainability Consortium (TSC) and the Korea Environmental Industry and Technology Institute (KEITI) share five impact categories (climate change, ozone depletion, photochemical ozone formation, acidification and eutrophication), while each requires the inclusion of another impact category: ionizing radiation by TSC and resource depletion by KEITI.[19] To address the inconsistency and duplications of PCRs, The Product Category Rule Guidance was launched in 2013 to provide more specific guidance on developing PCRs.[20–22] However, the selection of impact categories is still up to the individual PCR committee.[20]

In addition to the ISO guidelines for LCA, additional efforts in several countries have led to the development of relative guidelines. Most notable among these are the Environmental Product Declaration (EPD), based on

ISO 14025,[23] in Sweden, the Publicly Available Standard (PAS) 2050[24] from the UK and the French standard BP X30-323-0,[25] each of which provides additional guidance for the performance of LCA. Furthermore, at the level of the European Commission, through the Joint Research Council the ENVIFOOD,[26] provision of guidance specific to food and agricultural products is available. The European Commission DG Environment, in an effort to enable companies to market sustainable products without the need to perform assessments specific to each country where they may wish to market the product, established the program Single Market for Green Products Initiative. They have published the Product Environmental Footprint (PEF) and Organizational Environmental Footprint (OEF), guidance documents for developing product and sector-specific category rules defining the way in which LCA can be conducted, with the purpose of communicating to consumers regarding the environmental impact of their purchases.[27] PEF pilots were run between 2013 and 2016 to establish methods to ensure a common approach to measuring environmental performance. In doing so, the European Commission provided guidelines for the procedure for choosing impact categories in the pilot phase of conducting a PEF. It says that the identification of the most relevant impact categories shall be based on the normalized and weighted results of a screening study.[27] However, pilot studies can also choose impact categories based on the communication purpose,[27] which essentially links to the goal and target audience of the LCA.

1.3.1 The Role of Policy and Green Food Processing

In the USA, a 2010 study reported the estimated annual food waste at the retail and consumer level to be 638.7 kg per person. This food waste was valued to be $165.6 billion.[28] Similarly, another study reported that 90 million tons of food waste was generated in the European Union in 2010.[29] However, researchers looking at the methodology to calculate food waste highlighted the importance of data integrity and data collection. They suggested that government officials need to create policies that define and quantify what constitutes "food waste" due to variations among the definitions of different authors or agencies.[30] Sustainability is widely viewed as consisting of three equal pillars: social, environment and economic. However, in the *Handbook of Sustainability for the Food Sciences*, Morawicki argues that the environment is the foundational support for both social and economic sustainability.[31] In the global context, it is certainly true that without the full suite of functioning ecosystem services, society and the economy are threatened. Nonetheless, it is important that sustainability assessment considers the three pillars equally, acknowledges unavoidable tradeoffs and provides a lens for informed decision making – these two perspectives are shown in Figure 1.1. Environmental and social impacts are often external costs and are difficult to assess because there may not be an immediate and direct or obvious effect on profits.[32] Similarly to the sulfur dioxide cap and trade that played a major role in mitigating acid rain, initiatives to

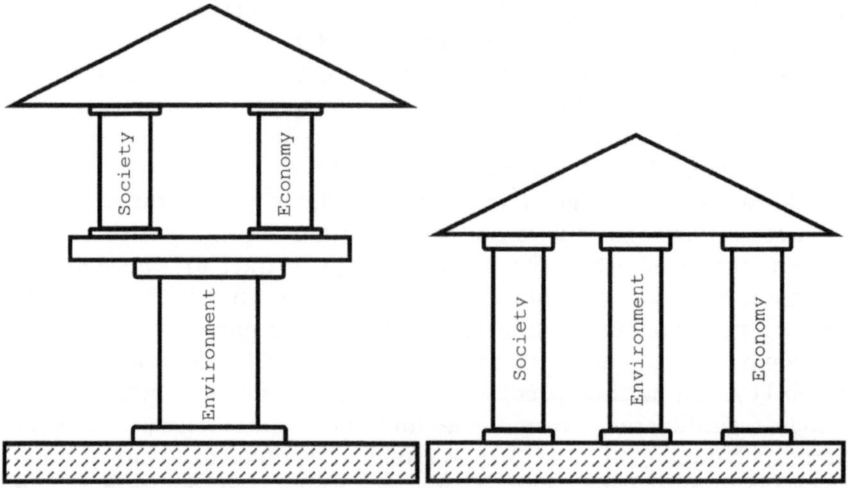

Figure 1.1 Alternative views of sustainability's three pillars. Reproduced from R. O. Morawicki, *Handbook of Sustainability for the Food Sciences*, Wiley-Blackwell, © 2012 John Wiley & Sons, Inc.

implement a carbon emission tax or excessive waste dump taxes are actions by the government intended to put a price on these externalities. Another role of government policies is to foster communication between federal, state and local levels, guiding the discussion to improve sustainable practices. These policies have the potential not only to improve food safety of the industry, but also to improve sustainable practices among food plants.[33] The impact of law and policy practices on the food industry will be further explored in a later chapter.

1.4 Introduction to LCA

In principle, LCA is simply an accounting of material and energy flows that result from each of the activities in the supply chain of a product or service. These flows are ultimately characterized and combined to provide a picture of the impact, across several environmental dimensions, of the system under study. LCA is a science-based, robust and standardized methodology for assessing the potential environmental impacts of products, services, or organizations. As described in ISO 14044, LCA consists of four stages, which will be outlined in the following subsections: Goal and Scope Definition; Life Cycle Inventory Collection; Life Cycle Impact Assessment; and Interpretation. Frequently in the process of performing an LCA it will be necessary to revise certain aspects based on new information, and thus the process becomes one of iterative refinement.

There are numerous reasons for performing an LCA, including the following: define opportunities for improvement through identifying activities that are major contributors to impacts (hotspots); develop performance benchmarks allowing the documentation of continual improvement; differentiate markets permitting the targeting of products towards consumers concerned about the impacts of products they purchase; design for environment, that is, considering the full lifecycle of a product at the initial development stages, enabling significant improvements to be made in overall lifecycle performance. With regard to marketing sustainable products, LCA is the basis for type III Environmental Product Declarations (EPDs) in which the environmental performance of a product can be communicated to consumers, following international standards.[23] An example of design for environment would be the consideration of the biodegradability or recyclability of packaging in the early design that resulted in reductions in landfill burdens for the end-of-life phase.

Broadly, there exist two modeling paradigms for performing an LCA: attributional and consequential.[34] Attributional LCA is an approach in which an average system is retrospectively evaluated and the emissions associated with the system are apportioned or allocated between multiple functions based on normative rules. Consequential LCA, on the other hand, does not consider the average system, but evaluates the anticipated change resulting from increased demand for an additional unit of product. Thus, the lifecycle inventories for the two systems are different, the first relying on averages and the second on data from the margin. Specifically, marginal processes are those processes anticipated to respond as a consequence of the additional demand. Attributional LCAs are typically used for identifying hotspots and benchmarking supply chain performance, whereas consequential LCAs are more relevant for decision support.

A relatively more recent approach for performing LCA, primarily in a business context, is known as organizational LCA.[9] Most food manufacturing businesses have multiple facilities and each facility may have multiple products produced from a potentially wide range of input materials. Frequently, attributional and consequential LCA are used to evaluate individual products in a company's portfolio. This raises several technical and methodological issues needed to perform the accounting of inputs and outputs and assign these to the individual products. This multifunctionality can be managed in different ways, all of which have advantages and disadvantages. In some cases, however, in the evaluation of an organization as an entity in which its entire product portfolio is the functional output of the system of study, the methodological issues of multifunctionality are generally less important. An organizational, or even facility, level LCA can be performed based on bills of materials, utility consumption, transportation and waste management information collected at either the organization or facility scale. The scale of analysis is valuable for benchmarking performance and documenting continual improvement when there are not specific marketing requirements for individual products.

All LCAs involve four phases as described in the international standard:[6] Goal and Scope Definition, Life Cycle Inventory, Life Cycle Impact Assessment, and Interpretation.

1.4.1 Goal and Scope Definition

Because there are a range of potential uses for the results of an LCA, a critical first step is to define the purpose of the study, as that also has important implications for the data collection requirements. In addition, the ISO standards require different levels of data quality and external review for studies that are intended to make claims of superiority among products. The Goal and Scope Definition requires the establishment of the system boundary, functional unit, impact assessment categories, data quality requirements, and other methodological choices.

1.4.1.1 System Boundary

By definition, LCA addresses environmental impacts through the whole lifecycle of products and services, that is, from raw material extraction all the way to final recycling/disposal of the products. This is called the "cradle-to-grave" approach in the system boundary, which defines the unit processes that are included in the system under study. Covering the whole lifecycle provides a comprehensive understanding about all the potential environmental impacts through all lifecycle stages. However, the amount of data to be collected to fulfill the cradle-to-grave requirement is substantially higher. Depending on the goal of the LCA, truncated system boundaries such as cradle-to-gate or gate-to-gate can also be used. ISO defines the "gate" as the production site gate.[8] Cradle-to-gate therefore means that impacts are assessed when the product leaves the production site, and further use and disposal stages are not included. Gate-to-gate therefore means that only processes at one or more single processing facilities are included. In food LCAs, cradle-to-gate often refers to cradle-to-farm or processor gate, whereas gate-to-gate often means from farm to processor or retailer, or down to a single processing stage, excluding use (*e.g.* cooking) and disposal of packaging and food waste. Note that cradle-to-gate LCA is only a partial LCA since the full lifecycle is not included and important contributors may be missed out owing to the truncation, while gate-to-gate is arguably not an LCA despite often being applied for various purposes.

System boundaries are tailored to the goal of the LCA study, which reflects the stakeholder's concerns. If the goal is to understand the impacts throughout the whole lifecycle, then obviously cradle-to-grave would be a rational choice. If the goal is to facilitate communications between a supplier and a processor, then cradle-to-gate may be sufficient for the processor to understand the "embedded" burden with its purchased products from its supplier. If the goal is to understand how consumers use the products and how to reduce food waste, then the use and disposal stages need to be considered. If the goal is for internal benchmarking (*e.g.* compare different drying methods), then gate-to-gate can also be deployed. When reading the literature, it is important to note differences in the boundaries of studies. Figure 1.2 shows an example system description for a cradle-to-grave assessment of a ready-to-eat meal.

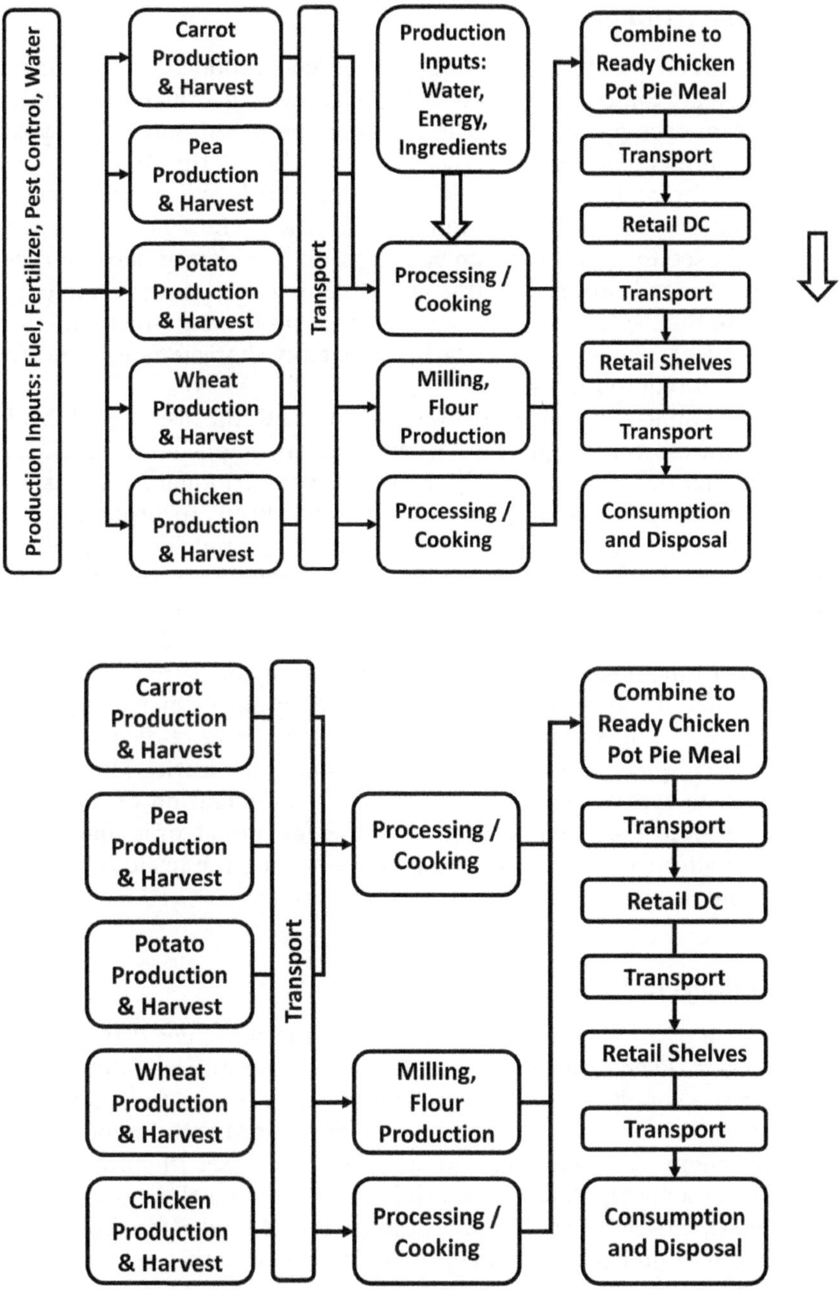

Figure 1.2 Schematic showing system boundaries for a cradle-to-grave LCA of a ready-to-heat/eat meal. Each phase of the supply chain will have inputs of fuel/energy, water, chemicals, *etc.* Each phase will also have emissions to the environment and, for SLCA or LCSA, social and economic considerations to be accounted for.

1.4.1.2 Functional Unit

LCA is intended to allow comparisons between products or production scenarios. A fair comparison can be achieved only if the functions delivered by each of the systems are identical. Hence proper definition of the system function is imperative. In principle, the definition of the functional unit should describe what function is fulfilled with which obligatory properties and for what length of time it will fulfill the function. The time characteristic is primarily used for goods that can be repeatedly used for the same purpose, *e.g.* a shirt. In food processing, this would be manifest in a specified shelf life. Because food safety is paramount, all food processing functional units must have, either explicitly or tacitly, the obligatory characteristic of meeting all food safety requirements.

The nutritional differences between foods and the associated health impacts from the consumption of different diets complicate the definition of a functional unit for food. Only recently have nutritional health aspects (non-communicable disease burden) of food been incorporated into LCA.[35–40] In the context of this book, however, the nutritional quality of foods is expected to be of relatively minor importance, and the principal use of LCA in the food processing sector will be focused on continual improvement either for individual product lines or for overall facilities. For these situations, a functional unit of a specified quantity of packaged food produced is generally adequate. However, depending on the specific goal and scope of a study, if it is known that the processing has a significant effect on the nutritional quality either through enrichment or fortification (*e.g.* vitamins added to cereals or milk, respectively),[41] the inclusion of nutritional qualities as obligatory characteristics of the functional unit may be important. An alternative approach, based on assessing the health impacts of nutrition in the diet, is discussed in Section 1.10.2.

1.4.2 Lifecycle Inventory Collection

In this phase of an LCA, a supply chain model is constructed from linked unit processes. Each unit process is a complete description of one stage or activity in the supply chain. The unit process has inputs from other unit processes or directly from nature (resource extraction) and outputs both to the environment (emissions) and to other unit processes [the product(s) of the particular activity]. Practitioners will typically try to collect primary data for processes closely related to the product under study (foreground processes) and use information from existing databases for the processes in the background system. Background processes might include electricity generation or the production of refrigerants or fuels. Frequently, the unit processes are linked using software that allows rapid calculation and analysis of the results. An important consideration in this phase of an LCA is to realize that when converting inventory into impacts, there can be mismatch of the inventory flow and impact category. This can occur, for example, if ecotoxicity is chosen as an impact category while some relevant chemicals are not included in the inventory, or a characterization factor is missing for a specific inventory flow, or even results from something as simple as a different naming convention. Some LCA software provides a tool to identify each inventory flow for which a corresponding characterization factor is missing. This highlights an important characteristic of performing LCA, namely an iterative process where information gleaned from subsequent steps may require modification of earlier steps.

1.4.3 Lifecycle Impact Assessment

LCA addresses environmental impacts through multiple metrics, *i.e.* impact categories. This step is referred as lifecycle impact assessment (LCIA). In this way, emissions and resources consumptions collected through the lifecycle inventory stage are classified into impact categories and quantified through impact indicators. This phase of an LCA is typically performed using existing frameworks and is characterized by the process in which a large number (typically thousands) of emissions to the environment are characterized for their relative contributions to a limited number of midpoint impact categories, typically between 12 and 20, depending upon the specific impact assessment framework chosen for the study. This also makes the inventory easier to interpret, as it "translates" inventory into environmental concerns. The LCIA phase of an LCA constitutes three mandatory steps:[6]

- Select impact categories and characterization models.
- Classification: assign LCI results to impact categories.
- Characterization: calculate category indicator results.

Two classifications of impacts have been defined: midpoint and endpoint categories. ISO defines an endpoint as "attribute or aspect of natural environment, human health, or resources, identifying an environmental issue giving cause for concern," and midpoints are located between the inventory and the endpoints in an environmental cause–effect chain. In other words, endpoints are the fundamental environmental damage about which we are concerned, and this currently covers three areas, namely human health, ecosystem health and resource use or depletion. Endpoint impact categories therefore signify a damage-oriented approach. Midpoint impact categories, on the other hand, reflect a problem-oriented approach, because they describe the specific problems resulting from the inventory flows (*e.g.* soil acidification). Endpoint results have a higher level of uncertainty than midpoint results owing to additional modeling down the environmental cause–effect chain, but they provide direct indicators of the environmental concerns and are therefore easier to communicate.

One of the first attempts to define best practice for impact category selection was made in 1999 during the SETAC–Europe second working group on LCIA, where 10 midpoint impact categories were proposed:[42] abiotic resource use, biotic resource use, land use, climate change, stratospheric ozone depletion, human toxicity, ecotoxicity, photooxidant formation, acidification and eutrophication. Since then, various methodologies have been developed with different packages of impact categories, but most of the baseline categories are still included as summarized in Table 1.1.

LCA researchers have developed characterization factors for a wide variety of emissions. As one example, the Intergovernmental Panel on Climate Change (IPCC) has established that methane has a radiative forcing factor that is 28 times greater than that of carbon dioxide over a 100 year time frame; thus the characterization factor for methane is 28 $kg\,CO_2e\,kg^{-1}$ methane.[43] The carbon footprint of a product is obtained by multiplying all of the greenhouse gas (GHG) emissions (across the entire supply chain) by their respective characterization factor and then summing to report the cumulative emissions as an equivalent quantity of carbon dioxide (CO_2e). Characterization factors are developed by following an environmental cause–effect chain that accounts for the cascade of events beginning with the emission of substance into the environment until it interacts with either a receptor or the environment itself to cause an impact.[6] As another example, nitrate emissions in wastewater will travel downstream, ultimately reaching the marine environment where it may cause eutrophication.

Similar, although less certain, cause–effect chains are used to develop characterization factors that combine midpoint categories into endpoint categories. For example, carcinogens, non-carcinogens and particulate matter and also ionizing radiation and climate change have direct or indirect impacts on human health. In this example, the endpoint impact is reported in units of disability-adjusted life-years (DALYs). DALYs are a measure of the decrease in life expectancy as a result of the emissions leading to ill health, disability or early death. These are the same units as used in estimating the impacts of food-borne illness from microbial risk assessment; the connection will be discussed in more detail below.

1.4.4 Interpretation

In this phase of an LCA, the implications of the results from the impact assessment are discussed and placed into context of the system under study. Hotspots and tradeoffs will be identified, and opportunities for improvement should be discussed, including the potential benefits that may be realized through their implementation. Because LCA studies complex systems and requires large amounts of data, it is also important in the interpretation phase that limitations of the conclusions be clearly articulated. Any study that reports on a comparison of two systems should also include an assessment of uncertainties that can arise from differences in data quality and potentially from methodological assumptions. Some additional,

Table 1.1 Summary of commonly used impact assessment methods for midpoint categories.[a,b]

LCIA framework	Climate change	Ozone depletion	Land use	Human toxicity	Ionizing radiation	Ecotoxicity	Ozone formation	Acidification	Terrestrial eutrophication	Aquatic eutrophication	Respiratory inorganics	Resource consumption
CML 2002	✓	✓	✓	✓	✓	N	✓	✓	✓	✓	N	✓
Eco-indicator 99	✓	✓	✓	✓	✓	N	✓	✓	✓	N	✓	✓
EDIP 2003	✓	✓	N	✓	✓	✓	✓	✓	✓	✓	✓	✓
EPS 2000	✓		✓	✓	✓	✓	✓	✓	✓	✓	✓	✓
Impact 2002+	✓	✓	✓	✓	✓	✓	✓	✓	N	✓	✓	✓
LIME	✓	✓	✓	✓	N	✓	✓	✓	✓	✓	✓	✓
LUCAS	✓	✓	✓	✓	N	✓	✓	✓	✓	✓	✓	✓
MEEuP	✓	✓	N	✓	N	✓	✓	✓	✓	✓	✓	✓
ReCiPe	✓	✓	✓	✓	✓	✓	✓	✓	✓	✓	✓	✓
Swiss Ecoscarcity 07	✓	✓	✓	✓	✓	✓	✓	✓	✓	✓	✓	✓
TRACI	✓	✓	N	✓	N	✓	✓	✓	✓	✓	✓	✓

[a] ✓ = included; N = not included.
[b] Modified from *ILCD Handbook*.[181]

optional, steps of LCIA can also be taken and can be useful for interpretation, as follows:[6]

- Normalization: calculate the magnitude of indicator results relative to reference values.
- Grouping: sort and rank the impact categories.
- Weighting: convert and aggregate indicator results across impact categories using weighting factors based on value choices.
- Data quality analysis: understand the robustness of the results.

When multiple impact categories are used, normalization can reveal the relative importance of each impact category within a certain geographical context. Normalized impact results are compared with a reference value, *e.g.* the average environmental impact per capita in one year within a certain country. Thus, normalized results are expressed as dimensionless fractions and therefore can be compared across impact categories that were previously incomparable owing to the use of different units. This gives some idea about the relative magnitude and therefore relevance of each impact category compared with the others.

1.4.4.1 Limitations and Tradeoffs in LCA

By addressing multiple impact categories, LCA avoids the narrow focus of a single impact and therefore leads to a more comprehensive understanding about the overall impacts. However, LCA addresses only the environmental issues that are specified in the goal and scope. Therefore, LCA, despite accounting for multiple metrics, may not provide a complete assessment of all environmental issues regarding the product system under study.[5] For example, noise and odor are not yet accounted for in LCIA. In addition, many impacts that are accounted for do not fully include the effect of the location of the emission in the characterization. Extending LCIA to account for geospatially explicit inventory and impact is a current area of intense investigation.[44–47]

Tradeoffs are often found between different impact categories. For example, UHT milk has a longer shelf life and potentially less spoilage, but the energy consumption for thermal treatment is higher than for pasteurized milk. The nutrient content of UHT is lower than that of pasteurized milk, if vitamins are taken into consideration. A study comparing environmental impacts of primary livestock products found similar tradeoffs (Figure 1.3). For example, pork has a relatively small land occupation but a comparatively high energy intensity.

1.5 LCIA of Food Processing

Since food processing is downstream of agricultural production, the two are often assessed together in LCA studies to meet the ISO standard requirement

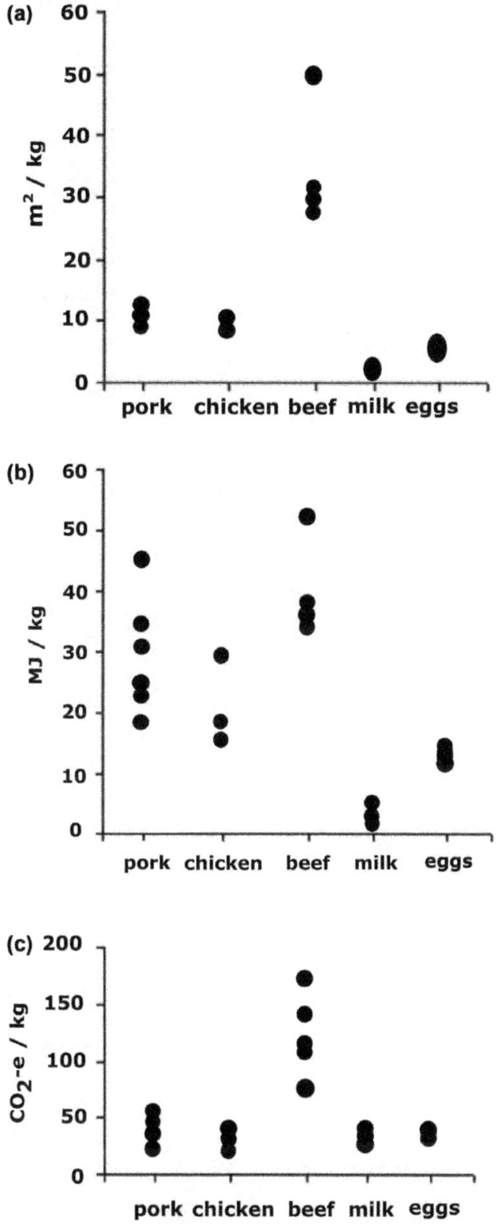

Figure 1.3 Comparison of lifecycle assessment of a few livestock products per kg wet mass basis. (a) Land use; (b) energy; (c) climate change.
Reprinted from *Livestock Science*, **128**, M. de Vries and I. J. M. de Boer, Comparing environmental impacts for livestock products: A review of life cycle assessments, 1–11, Copyright (2010), with permission from Elsevier.

of extending the system boundary to include extraction from nature. Common midpoint impact categories for agricultural production are therefore often applied to food processing, including global warming, acidification, eutrophication, water consumption, (agricultural) land occupation, abiotic resource depletion (*e.g.* use of phosphorus) and biodiversity. A recent study has found that LCAs focusing on climate-related impact categories have proportionally increased over the last 20 years.[48] This reflects the growing concerns about climate change but also narrows the environmental focus of LCA studies.

Generally, processed food has higher environmental impacts than less processed food over the lifecycle. This can be seen from the energy intensity of dairy products (Figure 1.4).[49–51] For example, liquid milk has a far lower energy intensity than powder products, which require thermal energy to remove moisture. The trend is likely to remain when making a comparison on a dry mass basis (*e.g.* protein), because further processing tends to increase the concentration of solids. For example, agricultural production tends to dominate the climate change impact for livestock-derived food products. This is because the farming stage tends to produce CH_4 and N_2O, which are GHGs with a high global warming potential (GWP) of 28 and 265 $kg\,kg^{-1}\,CO_2e$, respectively,[43] whereas inputs to industrial processing, generally related to combustion of fossil fuels, tend to be produce CO_2, which has a GWP of 1 $kg\,kg^{-1}\,CO_2e$. This may not be the case for other impact categories. An interesting study on the environmental impacts of a Spanish diet found that when human excretion is included in the whole lifecycle of food consumption, it became a significant contributor to the eutrophication owing to the emissions of nutrients in treated sewage (Figure 1.5).[52]

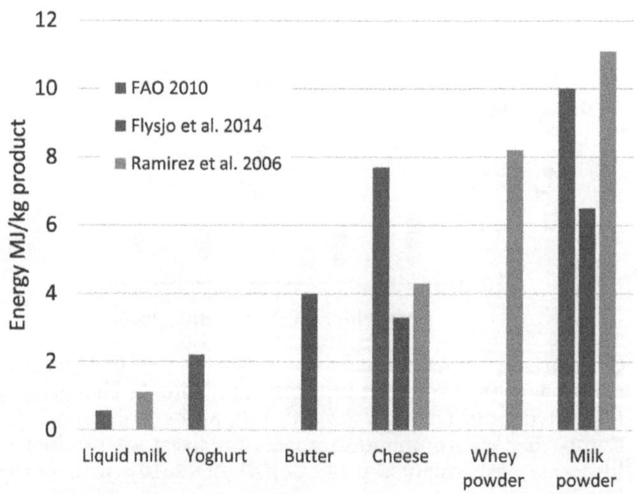

Figure 1.4 Energy intensity of dairy products. Data from ref. 49–51.

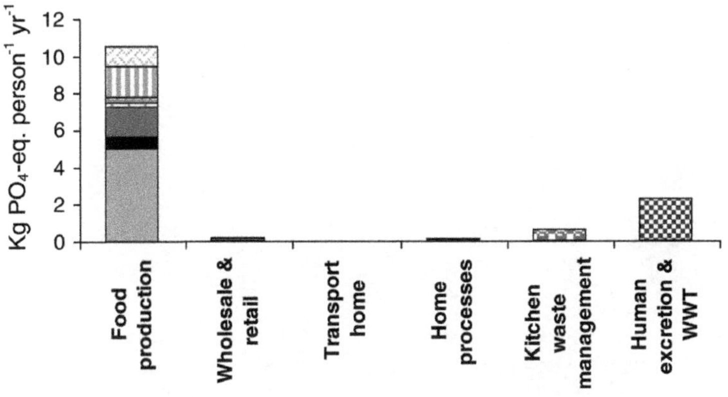

Figure 1.5 Eutrophication potential of Spanish diet.
Reproduced from *Int. J. Life Cycle Assess*, Life cycle assessment of the average Spanish diet including human excretion, **15**, 2010, 794–805, I. Muñoz, L. Milà i Canals and A. R. Fernández-Alba, © Springer-Verlag 2010, with permission of Springer.

1.6 LCA of Food Production, Processing and Consumption

Even though the majority of environmental impacts occur in the agricultural production stage, there have been a number of studies that were focused on the food processing sector. One of the first environmental LCAs to be conducted, in 1969, by the Coca-Cola Company evaluated the environmental impact of their containers; hence the food processing industry has a long history of endeavoring to understand the environmental consequences of their products. More recently, a significant number of LCAs related to dairy processing have been performed,[15,50,53–65] whereas some have taken a broader view,[66–71] and others have focused on different sectors such as ready-made meals,[72–74] vegetable oils,[75] eggs,[76] vegetables,[77,78] meat processing[79,80] and packaging.[69,81–83]

1.6.1 Cradle-to-grave Studies

Applying the cradle-to-grave approach to food LCA provides a holistic picture of the impacts throughout the system. For example, a study on phosphorus efficiency through the US food system found that only 15% of the total phosphorus extracted from nature for the provision of food was eventually ingested by humans, the remaining 75% being lost to the environment at different stages of the lifecycle (Figure 1.6).[84] The top three contributors along the food chain were identified as livestock, meat and dairy production and crop cultivation, and household food waste, mining waste and fertilizer manufacturing waste were also significant; however, the food manufacturing

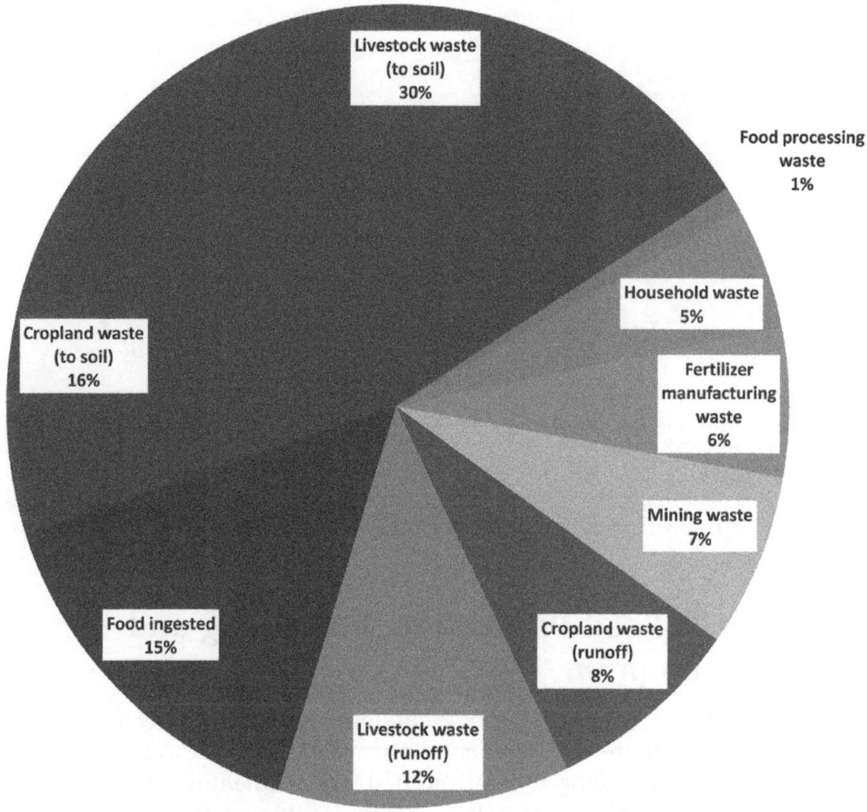

Figure 1.6 Share of total lifecycle phosphorus loss in the US food production–consumption chain.
Data from *Chemosphere*, 2011, **84**, 806–813.

sector was a very small contributor.[84] This type of analysis supports targeted efforts on those stages in the supply chain where the greatest benefits can be realized, and may provide decision support for national or international policies, where a whole-system view is necessary to manage tradeoffs. It should be noted that the 46% of emissions to soil may not be necessarily completely lost as they represent stages in phosphorus cycling, thus highlighting the importance of defining system boundaries when performing LCA.

Tracing the whole lifecycle can also reveal the locations where the impacts occur and how that affects the overall environmental impacts. A study on environmental impacts associated with Dutch private consumption showed that food production and consumption were dominant in the environmental load for pesticide use, fish extraction, water use, land use, eutrophication and acidification, most of which took place outside The Netherlands.[85] Contrary to some common beliefs, local food products may not always be an effective way to reduce the climate change impacts associated with food, since transportation contribution tends to be much smaller on a total lifecycle basis than the production stage.[86,87] This suggests that using the most efficient production technologies is frequently more important than the production location.

Full LCA can reveal the importance of food packaging and waste management beyond the farm gate. In a study of packaging and food losses of five food items, it was found that packaging was an important tool in reducing food waste, and therefore reducing the total environmental impact, even if there was an increase in impact from the packaging itself.[88] For example, it was found that reduction of losses from milk, bread, beef and cheese by 2% or more decreased the total energy use despite the increase in energy use in the packaging phase.[88] This was because all impacts are scaled against one unit of product, hence the less product is lost the lower are the overall impacts. Another study of cradle-to-grave LCA on a canned ready meal found that solid waste management was the dominant contributor to marine ecotoxicity and fresh water ecotoxicity, whereas the food ingredients (cooked pulses and pork meat cut) were the dominant contributor to land use, carcinogens, acidification/eutrophication, global warming and terrestrial ecotoxicity.[74] A further study comparing ready-made meals with homemade meals suggested that homemade meals have a lower impact, largely owing to less food waste, and also that frozen meals have a greater impact than chilled ready-made meals owing to the additional energy requirements in the supply chain. Interestingly, ingredients sourced from Brazil or Spain had a lower impact than similar ingredients sourced locally in the UK.[72] This shows the different environmental aspects to which food ingredients and solid waste contribute. A cradle-to-grave perspective and broad impact categories can ensure a comprehensive understanding of the impacts and provide insight into sourcing and design for waste minimization.

1.6.2 Cradle-to-gate

The primary production stage of food products is usually found to be a significant contributor to the overall impacts. For example, a cradle-to-grave LCA of yogurt found that dairy farming was the main contributor to acidification (91%), eutrophication (92%) and climate change (62%).[89] A cradle-to-gate LCA of quick-service restaurant chicken meat found that the poultry grow-out stage was a major contributor to climate change (41%) and non-renewable energy depletion (55%), and feed production (upstream to poultry grow-out) was a significant contributor to climate change (34%) and water depletion (66%).[90]

It is therefore not surprising to find that food LCAs are still heavily focused on the agricultural "gate." Searching for "lifecycle assessment dairy" in Thomson Reuters *Web of Science*, of the 45 journal articles published in 2016 (up to 10 December) that were focused on dairy or dairy-related systems (*e.g.* beef), 24 articles (53%) set the system boundary from the cradle-to-farm gate, meaning that environmental impact is assessed when raw milk is ready for delivery from the dairy farm.

At the farm gate, differences were found between animal- and plant-based food products. A study on 84 common foods found that animal-based foods were associated with higher energy use and GHG emissions than plant-based foods, except vegetables produced in heated greenhouses or transported *via* air freight (Figure 1.7).[91] This is mainly due to relatively low feed conversion rate (FCR) (kg feed consumed per kg product) in animal production, a measurement of an animal's efficiency in converting feed into useful products (milk, meat, eggs, *etc.*). A study on livestock systems in Europe found that FCR ranged from 19.8 : 1 for beef, 4.1 : 1 for pork, 3.3 : 1 for poultry, 2.8 : 1 for eggs, to 1.2 : 1 for cow's milk.[92]

Food-preserving technology during processing was also found to play an important role in the overall impacts of food. A study compared four thermal and non-thermal techniques (autoclave pasteurization, microwaves, high hydrostatic pressure and modified atmosphere packaging) and their effects across six impact categories (acidification, eutrophication, global warming potential, photochemical oxidation, water depletion and cumulative energy

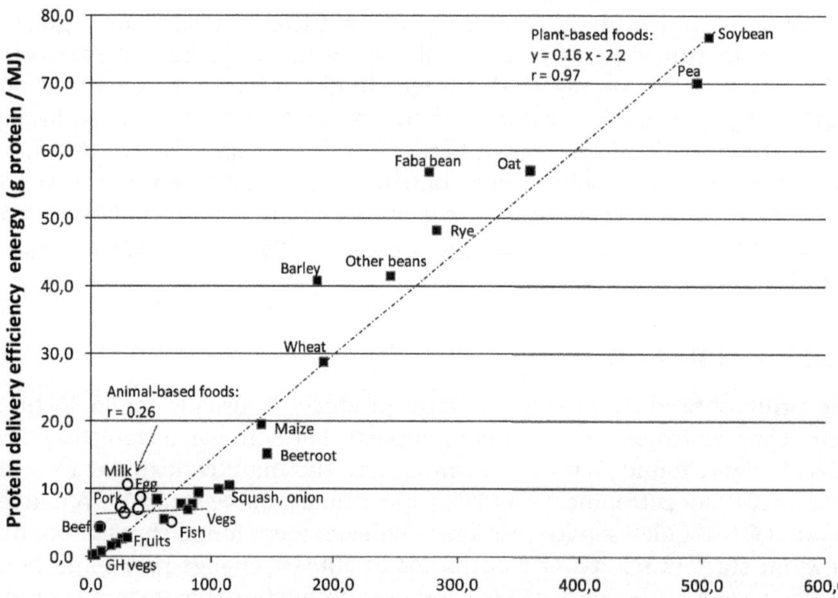

Figure 1.7 Correlation between protein delivery efficiencies for energy and GHG emissions. Correlation coefficients shown on graph.
Reprinted from *Food Policy*, **36**, A. D. González, B. Frostell and A. Carlsson-Kanyama, Protein efficiency per unit energy and per unit greenhouse gas emissions: Potential contribution of diet choices to climate change mitigation, 562–570, Copyright (2011), with permission from Elsevier.

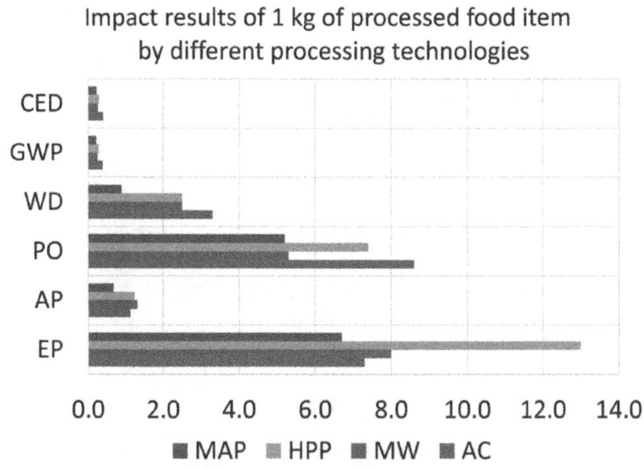

Figure 1.8 Comparative results of food preservation technologies. AP = acidification potential (g SO_2); EP = eutrophication potential (mg PO_4^{3-}); GWP = global warming potential (kg CO_2e); PO = photochemical oxidation (g C_2H_4e); WD = water depletion (L); CED = cumulative energy demand (kJ); AC = pasteurization by autoclave; MW = pasteurization by microwaves; HPP = pasteurization by high hydrostatic pressures; MAP = modified atmosphere packaging.
Data from *J. Clean. Prod.*, 2012, **28**, 198–207.

demand) and found that they each had strengths in different circumstances.[70] Microwaves and modified atmosphere packaging were found to reduce energy demand and GHG emissions, non-thermal technologies (modified atmosphere packaging, high hydrostatic pressures) were found to have lower water requirements, and modified atmosphere packaging was found to be the most sustainable option when a shelf life shorter than 30 days is required (Figure 1.8).[70]

1.6.3 Gate-to-gate

With all the merits of cradle-to-gate and cradle-to-grave LCAs, gate-to-gate studies can still provide a detailed and focused investigation on specific parts of the supply chain such as transportation. For example, a study on 50 fluid milk processing plants across the USA analyzed the farm gate to retailer loading dock GHG emissions of fluid milk processing, and found that truck fleet tailpipe emissions were the largest contributor to GHG (29%), followed by plant electricity usage (27%) (Figure 1.9).[58] The significant contribution of transport was due to the long distance from farms to processing plants, averaging 850 km per round trip.[93] The plant's annual volume of milk production (<80, 80–120, >120 million kg) was not found to be a strong indicator of the GHG total emission intensity (kg CO_2e per kg of packaged milk), despite the fact that larger plants tended to have higher distribution-related emissions. This was because the overall GHG intensity

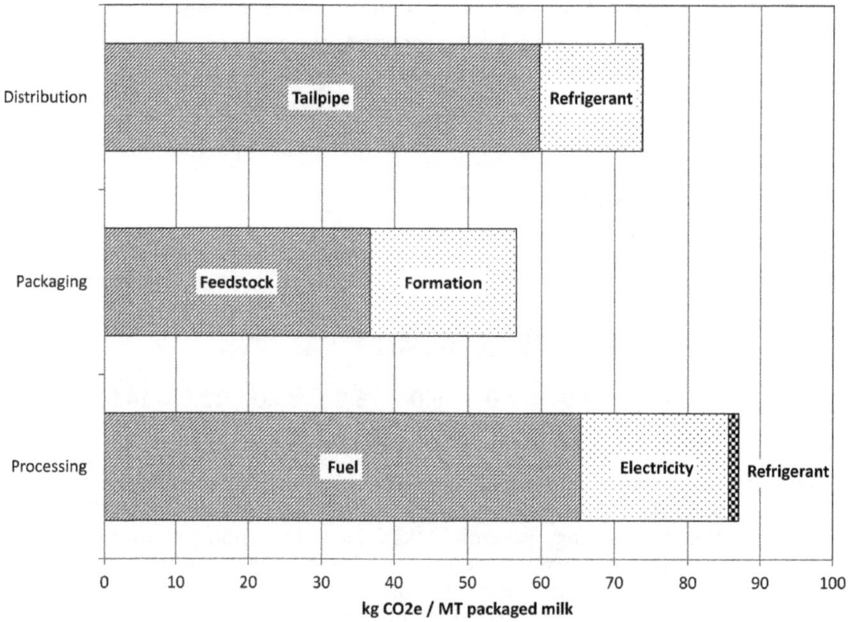

Figure 1.9 Breakdown of GHG emissions for unit processes, kg CO_2e per kg packaged milk.
Data from ref. 58.

was also dependent on factors such as regional electricity grid mix and whether or not there was onsite wastewater treatment and multiple coproducts.[58]

A gate-to-gate approach has also been applied to study specific technologies for food processing. An LCA study comparing four cleaning-in-place (CIP) methods in dairy processing found that those with small volumes and low temperatures (*i.e.* enzyme-based cleaning and one-phase alkaline cleaning) had lower impacts on energy use, global warming, acidification, eutrophication and photooxidant formation than the conventional methods (*i.e.* alkaline/acid cleaning with hot water disinfection, and with disinfection using cold nitric acid at pH 2), and phosphorus and nitrogen in the detergents influenced the eutrophication in addition to milk residue in the rinsing phase.[63] Another LCA study compared two drying methods for apple powder and found that drum drying, despite being simpler and cheaper, had higher impacts than multistage drying (ultrafiltration, crystallization and spray drying) in all selected impact categories.[94]

It may be argued that gate-to-gate studies, despite being narrowly focused on individual stages and processes, provide more insights at the operational level. It is often difficult to collect detailed data through the whole lifecycle and many approximations and assumptions must be made without possibly conducting a full validation. This is arguably an inherent weakness of the LCA methodology, namely that it may be spread too thin to reveal the details.

Gate-to-gate studies provide useful supplementary data to the whole lifecycle and therefore can make valuable contributions to understanding the bigger picture. When one or more stages of the lifecycle have been found to dominate the whole lifecycle impacts from a hot spot analysis (*e.g.* industrial stages dominate the impacts of apple powder),[94] gate-to-gate studies are particularly useful to provide in-depth knowledge by dissecting the impacts of individual stages.

1.7 Case Study: Carbon Footprint of Fluid Milk Production

This study was a LCA intended to compute the lifecycle inventories and evaluate the environmental impact in terms of global warming potential (CO_2e) for the aggregate fluid milk processing sector. In mid-February 2008, an extensive survey was completed by 50 individual processing plants reporting on their production during calendar year 2007. Information in each survey included plant major energy consumption; water consumption; truck fleet fuel consumption; refrigerant purchases for both the plant and truck fleet; on-site milk packaging production; packaged milk type and sizes; and annual production for total plant fluid, fluid milk and packaged milk. Information provided in the surveys supplemented by correspondence for clarification and verification were the primary data sources for the study.

1.7.1 Methodology

The functional unit was defined as 1 tonne of packaged milk. The environmental impact metric was chosen as kg CO_2e emitted to the atmosphere. The gate-to-gate system boundaries began with raw milk entering the plant's refrigerated storage silos and ended with the delivery of packaged milk to the customer *via* the plant's distribution truck fleet. Incidental effects such as employees' commutes and business travel for industry executives were not included. Where allocations of inputs were required, the allocation procedures followed the ISO allocation hierarchy. Primary allocations occurred for processing plant electricity and plant heating fuel; allocation was based on the fraction of total packaged milk with respect to all other total processed plant fluids. Packaging and distribution were not allocated, but were based on totals reported on the survey based on packaged milk volume. For example, diesel for the distribution of milk was specifically requested on the survey, and therefore not corrected to account for delivery of any other products.

1.7.2 Packaging

Three general types of fluid milk containers [high-density polyethylene (HDPE), polyethylene terephthalate (PET) and paperboard carton] of various sizes (gallon, half gallon, quart, pint, half pint, single serve, and other) were reported by each plant. Most plants purchased HDPE resin to blow mold their own gallon and half-gallon containers on-site. Total emissions associated with packaging included container material (and material raw-material extraction and manufacture), bottle production and caps. The majority of fuel energy in milk processing was used to produce steam for thermal processing, equipment cleaning and other plant processes.

1.7.3 Electricity and Fuel

Each facility reported overall consumption of energy in terms of the primary energy carrier (*e.g.* fuel oil or natural gas). When the electricity consumption location was known, this study used emission factors (in kg CO_2e kWh^{-1}) for the three US regional interconnection grids. National average emission factors were used when the location was unknown, specifically to account for containers that are blow molded offsite and transported preformed. The three regions were Eastern Interconnection, Western Interconnection and the Electric Reliability Council of Texas (ERCOT) Interconnection.

1.7.4 Results

The emissions profile from each of the 50 individual facilities is presented and the profile of one individual plant (the same plant in each figure) is highlighted. These results show that there is significant variability, even among modern processing facilities, with respect to environmental impacts. Furthermore, individual plants may be high performing or low performing in one or more of the operational stages. With data such as these, individual plant managers can see how they are performing with regard to a representative industry average and, if their performance is worse than average, they can target specific efforts to reduce impacts (energy and therefore costs) associated with that operational phase that is higher than average. Figure 1.9 provides a contribution analysis for each major operational component of the average emissions aggregated over all 50 plants.

Processing GHG emissions are associated with the thermal processing and standardization of raw milk, including container filling. There are three components to processing GHG emissions: plant electricity (minus any on-site blow molding, included in packaging), plant heating fuel (generally natural gas) and refrigerant leakage. The emissions for the processing operations at each of the 50 facilities are presented in Figure 1.10.

Packaging emissions are associated with the packaging materials for processed milk containers. There are two components to packaging GHG emissions: packaging materials and electricity consumption for package manufacturing. Emissions associated with packaging operations are presented in Figure 1.11.

Distribution emissions are associated with the transportation of processed and packaged fluid milk from the plant to the point of delivery to the

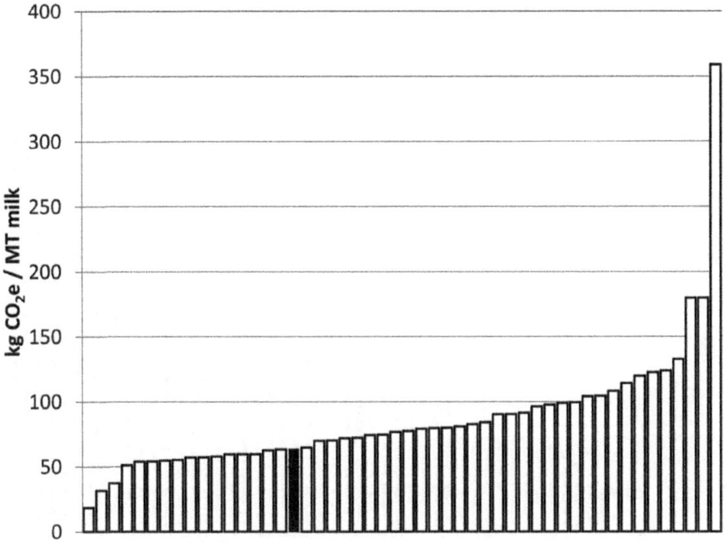

Figure 1.10 Total GHG emissions for pasteurization, standardization and filling for all 50 plants in ascending order.

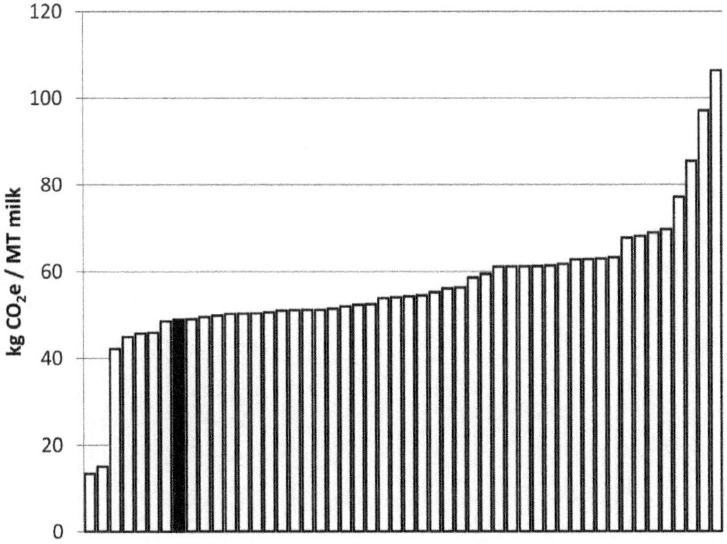

Figure 1.11 Total GHG emissions from packaging operations for all 50 plants in ascending order.

retailer. There are two components to GHG emissions: truck tailpipe emissions and truck refrigeration system refrigerant leakage. Total emissions for distribution are given in Figure 1.12.

Figure 1.13 presents the total gate-to-gate emissions profile of the 50 facilities participating in the study. The black column in each of the previous

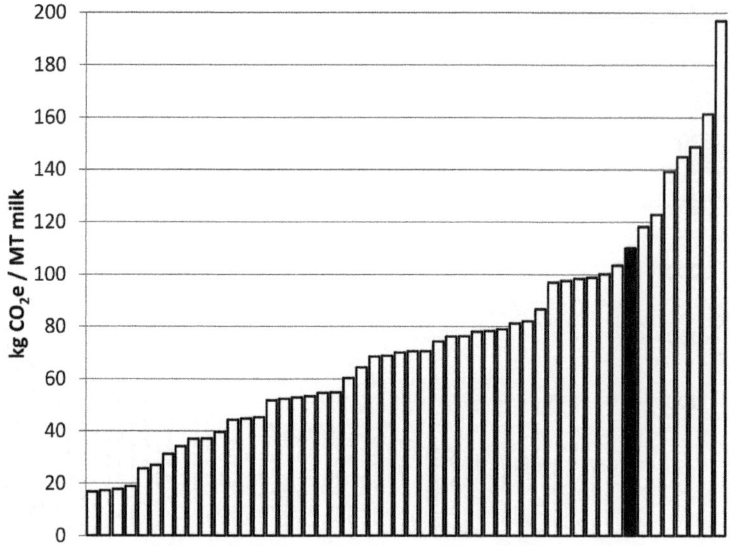

Figure 1.12 Total GHG emissions from distribution trucks for all 50 plants in ascending order.

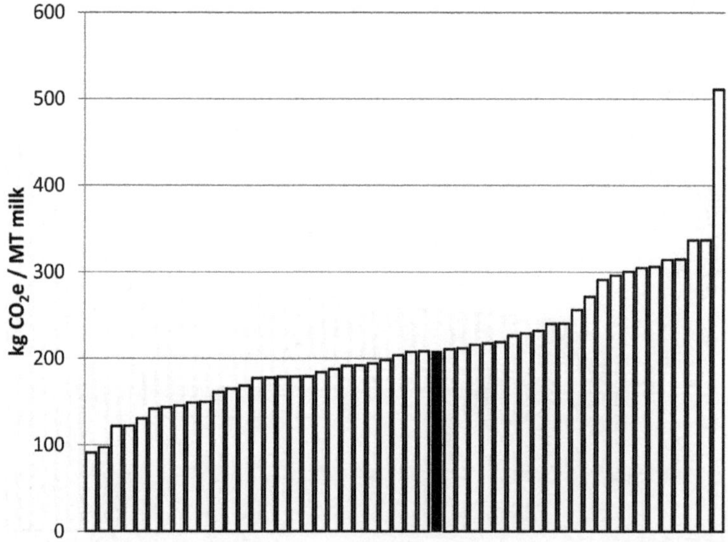

Figure 1.13 Total GHG emissions per metric ton of packaged milk in ascending order for 50 milk processing facilities.

charts represents the same facility and demonstrated the variability even among relatively modern and well-managed facilities. This suggests that there is significant potential across the food processing industry to collectively improve sustainable practices through the identification of the fundamental sources of this variability.

1.8 An Overview of Emerging Practices and Technologies for Greener Food Production

As mentioned above, a major function of the food processing sector is to provide nutritious and safe foods with a sufficiently long shelf life to minimize losses due to spoilage. Significant energy resources are expended in food processing for the control of pathogenic or toxin-producing microbes. In the following sections, we briefly introduce several emerging technologies that have potential for significant energy savings while maintaining the same food safety standards as conventional technologies. Subsequent chapters provide significant details for each of these alternative technologies.

1.8.1 High Hydrostatic Pressure Processing

High hydrostatic pressure processing (HHP), also known as high-pressure processing (HPP), improves shelf life and decreases the need for chemical preservatives. In addition, the energy requirement of HPP is expected to be lower than that of thermal processing, but more research analyzing the energy requirements of HPP needs to be carried out to validate the reduced energy claims.[95] HPP is a promising technology to combat viruses in high-risk foods because the treatment of pressure applies an equal force to all areas of the foods at the same time.[96] The pressures and temperatures generally used in industry are around 600 MPa and <45 °C. Specifically, the products are placed in an HPP chamber and the vessel is closed, filled with a pressure-transmitting medium and pressurized either by pumping into the vessel or by reducing the pressure chamber volume, for example, by using a piston. The product is then maintained at high pressure for a specified time to inactivate microbes or change the composition of the product. Several pressure-transmitting fluids are used to displace space and protect the inner vessel surface from corrosion, including water and food-grade solutions such as castor oil, silicone oil, sodium benzoate, ethanol and glycol.[95] Pressure applied to foods being processed is transmitted isostatically and instantaneously; hence the process is not dependent on the shape, size or composition of the food. Lou *et al.* reported, in a review, that HHP has a superior ability to inactive viruses compared with other novel technologies, and specifically a significant reduction (≥5 log) was shown by HPP in fresh foods and shellfish, but not by ionizing irradiation, high-intensity ultrasound or ultraviolet light processing.[96,97] They also reported that HPP has potential to inactivate both enveloped and non-enveloped virus particles, and stated that HPP is one of the most promising technologies to ensure food safety and quality.

1.8.2 Ohmic Heating of Foods

Ohmic (OH) heating of foods has been used by the food industry in applications such as blanching, drying, evaporation, dehydration and fermentation. OH is an electric resistance heating method. Electrodes are used to pass current through the food, which results in the production of heat within the food (Joule heating, due to the electrical resistance of the food) and leads to thermal inactivation of microbes. Thus, conductivity of the food is a critical factor in the effectiveness of OH processing. Other key factors are heat generation in the system, electrical field strength, residence time and how the food flows through the system.[98] In a review, Pereira and Vicente reported that OH can reduce the time required to reach temperatures for high-temperature, short-time pasteurization.[99] Moreover, OH provides uniform heating of a liquid with little or no difference in external and internal thermal penetration. Consequently, OH has a high energy efficacy with minimal heat loss during processing. Similarly to current thermal processes, OH can be used for continuous processing, but it has fewer problems due to surface overheating and fouling in comparison with conventional thermal processing. Hence OH systems are associated with low maintenance costs. Research sponsored by the US Environmental Protection Agency (EPA) highlighted potential cost savings for food manufacturers and suggested that OH can potentially reduce energy demand in the food industry. A study using sweet potatoes as the model food found that OH reduced freeze-drying time by 25%. Exploring uses of OH for food by-products may identify opportunities to reduce the environmental impact of food processing. For example, OH has recently been investigated as a means to increase the extraction yield of antioxidant bioactive compounds from winery waste products. The OH treatment increased the polyphenol extraction of grape pomace by 36%. This prevents food by-products from simply being thrown away by providing added value and reduces the overall impact of the food processing on the environment.

1.8.3 Pulsed Electric Field Processing

Pulsed electric field (PEF) processing exposes the food to an intermittent, high-intensity electric field for short periods of time. The high field intensities are achieved through storing a large amount of energy in a capacitor bank (a series of capacitors) from a DC power supply, which is then discharged in the form of high-voltage pulses.[100] Studies on energy requirements have concluded that PEF is an energy-efficient process compared with thermal pasteurization, particularly when a continuous system is used.[100] The energy savings associated with PEF result from the decreased operating temperatures, which reduce the energy needed for cooling the processed product. The potential emission savings estimated for the not-from-concentrate orange juice industry are in the range 33–66 kt CO_2e per year.[101] Laboratory- and pilot-scale treatment chambers have been designed and tested for PEF treatment; however, continuous operation is more appropriate for industry-scale applications. One limiting factor of PEF technology is that it does not have the capability to inactivate bacterial spores, most likely owing to penetration limitation through the spore cytoplasmic membrane.

In a recent review of the effectiveness of PEF processing on the deactivation of microorganisms, Saldaña *et al.* showed that PEF resistance varies greatly across different strains of bacterial species and food types (*e.g.* composition, pH).[102] Hence the potential sustainability benefits of this technology have yet to be fully demonstrated.

1.8.4 Plasma Processing

Plasma technology has been studied since the mid-1990s for food processing. It uses a mixture of electrons, ions, atomic species, UV photons, and charged particles to kill microorganisms.[103] The advantages of plasma processing, as reported by Thirumdas *et al.*, are (1) high efficiency of microbial inactivation treatment at low temperatures, (2) precise generation of plasmas tailored for processing of specific foods and (3) "just-in-time" production of the disinfection agent, (4) low impact on internal food matrix, (5) no water or solvents used and (6) no resides.[104] Plasma technology can be divided into three categories: (1) remote-treatment cold plasma systems, (2) direct-treatment cold plasma systems and (3) electrode contact systems. Niemira stressed that cold plasma antimicrobial inactivation rates require continued validation, and concurrent research should also consider the impact of cold plasma treatment on sensory attributes, *i.e.* taste, smell and flavor, because sustainable food processing must also provide desirable products.[105] As cold plasma surface and gas-phase antimicrobial treatment research continues, the economics will help guide the food industry. Li and Farid's review described recent developments in atmospheric cold plasma (ACP) technology and found that it involves lower water consumption compared with current post-wash treatment of produce.[106] Moreover, they reported that post-packaging treatment can prevent contamination associated with non-sterile packaging, which leads to improved food safety.[106] A study by Ziuzina *et al.* highlighted the effectiveness of ACP in the disinfection of *Escherichia coli*, *Salmonella enterica* serovar Typhimurium and *Listeria monocytogenes* on fresh produce.[107] They concluded that ACP treatment of food in a sealed package can significantly lower microbial activity 24 h post-packaging.[107] By reducing the microbial load on fresh produce, shelf life can be increased and thereby reduce wasted food, which has been estimated, in 2008, to be 123 kg per capita in the USA.[28] Reducing or eliminating food waste is a very effective tool for enhancing food supply sustainability because, as mentioned previously, a large share of the full supply chain environmental burden typically occurs in the primary production phase.

1.8.5 Microwave Food Processing

Microwave heating is founded on the principle of converting electromagnetic field energy into thermal energy by oscillating the polarity of molecules in materials. A frequency range of 300 MHz–300 GHz (2450 MHz is most common) is used for microwave food processing. Microwaves cause internal heating based on the food's dielectric properties. Moreover, microwaves can easily be combined with other technologies such as microwave/convection drying, microwave/vacuum drying, microwave freeze-drying, microwave-assisted fluidized bed drying, microwave drying and osmotic dehydration.[108]

A review on microwave processing in the disinfestation of cereals and pulses estimated that insects, mites, rodents and microbes in stored grains cause an estimated 10% of food waste. In developing nations, the loss is as high as 30%. Owing to differences in the dielectric properties of the grain and the insects, selective heating of the insects is possible. Thus, the dielectric based heating kills insects, but leaves the grains unaffected, reducing food waste, which represents an important sustainability gain.

Another review reported the development of "smart drying" using microwaves coupled with real-time sensors that monitor various material qualities. These sensors can measure and report parameters such as moisture content, color, shape and conditions inside the drying equipment such as pressure, velocity, temperature and humidity. If impurities or defects are detected, then automated adjustments are made to the processing line through a process control feedback loop. Thus, an optimal level of microwave power is applied during processing, which will improve energy conservation.[109]

Microwave processing has attracted renewed interest as a method to extract the essential oils and bioactive components of spices and herbs, without the use of solvents or water. In place of the solvents or water, the process uses a combination of microwave heating and dry distillation. Again, the sustainability gains associated with reduction in water and chemical solvent consumption may be noteworthy, but have yet to be fully studied.

1.8.6 High-intensity Pulsed-light Food Processing

High-intensity (20 000 times more intense than sunlight) pulsed-light (HIPL) processing is a surface decontamination technology.[110] In addition, HIPL can be used to decontaminate food contact surfaces in food production facilities. The US Food and Drug Administration (FDA) approved pulsed UV light in 1996 for the production, processing and handling of foods. Synonyms for pulsed-light technology include pulsed UV light (PUV), intense pulsed light (IPL), high-intensity broad-spectrum pulsed light (BSPL), intense light pulses (IPL) and pulsed white light (PWL).[111] HIPL uses a xenon flash lamp, converting electric pulses into short-duration (1 μs–0.1 s) and high-power pulses of radiation with a broad emission spectrum, ranging from ultraviolet (200 nm) to infrared (1100 nm). This technology targets microbes' DNA by causing irreversible damage to the DNA, leading to death because of the impaired replication. In addition, an increase in temperature from HIPL processing induces photothermal microbial inactivation, which increases the HIPL lethality.[112] The HIPL process does not produce any chemicals, carcinogenic residues or toxic by-products.[113] When HIPL is evaluated on an energy cost per microbial reduction basis, HIPL is more energy efficient than other emerging technologies; however, HIPL demands a high start-up energy input and may require cooling units to prevent overheating.[114] A study compared HIPL with near-UV–visible light (395 ± 5 nm) and continuous UV light treatment and found HIPL to inactivate more efficiently *E. coli* and *Listeria innocua* in a bench-top liquid suspension experiment.[115] Other tests have shown it to be successful in application to various food products such as produce, fruits, meats, dairy products and honey.[111,113,114] The main sustainability gains achievable using HIPL appear to be related to energy conservation.

1.8.7 Infrared Food Processing

Infrared electromagnetic waves cause radiative heating. Infrared heating can be sub-categorized into near-infrared (0.75–3 μm), mid-infrared (3–25 μm) and far-infrared (25–1000 μm). It can be used in drying, baking, roasting, grilling and reheating of food products. Owing to the variations in how infrared heating can be created, it can function either like hot air heating or like microwave heating.[116] A review summarized the advantages of infrared radiation's high thermal efficiency compared with convection as (1) high thermal and energy efficacy, (2) faster heating rate, (3) shorter response time to reach operating temperatures, (4) uniform drying temperature, (5) high degree process control, (6) cleaner working environment and (7) the possibility of selective heating. The disadvantages include (1) low penetration power depending on the electromagnetic wave source, (2) prolonged exposure may cause undesired effects on the food's structure and (3) the heating is not sensitive to the reflective properties of coatings.[117] Two compelling industry applications are in tomato peeling and corn drying operations. It was found that compared with conventional lye peeling, infrared dry peeling required a heating time of only 30 s compared with 75 s, showed a lower peeling loss (8.3–13.2% *versus* 12.9–15.8%) and reduced the energy to tear tomatoes from two different tomato cultivars.[118] When infrared heating was applied to shelled corn drying, it was found that infrared heating was most efficient for corn with a high initial moisture content (the highest used in the study was 28%).[119] Hence infrared drying could be used as an initial drying step to reduce the overall time and energy of the process.[119] Infrared food processing therefore touches on two important sustainability points: reducing food loss and energy conservation. Additional specific applications in the food industry can be found in Rastogi's review.[120]

1.8.8 Ultrasonic Food Processing

Microbial inactivation occurs in ultrasonically irradiated liquids in the frequency range 20 kHz–10 MHz. High-power ultrasound, also known as power ultrasound, in the frequency range 20–10 kHz can cause cavitation.

Cavitation, the collapse of microbubbles formed as a result of the ultrasonic sound waves, creates small regions of sufficiently high temperature and pressure to cause the formation of hydroxyl radicals. These strong oxidants provide a mechanism of microbial inactivation during ultrasonic processing. Because these temperature and pressure changes are very localized within the food matrix, there are minimal overall changes to the food.[121] One study reported that ultrasonic cleaning was superior to a cleaning wash with 0.5% KOH or water only. Ultrasonic cleaning was tested against bacterial pathogens known to be recalcitrant due to biofilm formation and also yeasts and molds commonly found in food processing plants. In addition, a case study evaluating ultrasonic application on two cheese molds and milk transportation crates showed a faster turnaround for sanitation and a reduction in the temperature needed for sanitation (below 60 °C).[122] Likewise, allergens, such as milk and wheat, were removed more efficiently from the processing line conveyor belts using ultrasonic cleaning compared with a water spray. The improved cleaning technique could reduce the amounts of energy, chemicals and water needed to clean processing lines.[123]

Moreover, ultrasound processing has been explored to create added-value products from food waste. One study assessed the ability of ultrasound-assisted extraction (UAE) to increase the high added-value compound extraction levels from olive oil waste and olive oil production by-products. Polyphenol (and antioxidant) extraction from olive mill wastewater using UAE demonstrated higher yields than conventional filtration methods.[124] Another study reviewed the potential of ultrasonic treatment for absorbent regeneration, drying and dehydration in foods. Although still inconclusive, ultrasonic treatment for adsorbent regeneration (*i.e.* active carbon, polymeric resin, *etc.*) may be more efficient than thermal methods. Pretreatment of products by ultrasonic processing resulted in greatly reduced drying and dehydration times.[125] Technologies that lead to more efficient processing or cleaning or that permit more effective creation of value-added products all support a more sustainable processing industry by reducing input requirements and potentially providing additional revenue streams.

1.8.9 Supercritical Fluid Extraction

The most common supercritical fluid used in food processing is carbon dioxide (CO_2). CO_2 has been proven to have a lower environmental impact than solvents such as chlorofluorocarbons. CO_2 is recognized as safe by the FDA and European Food Safety Authority (EFSA) for food processing. The main operating parameters of supercritical fluid extraction (SFE) are temperature, pressure, time and type and percentage of polarity modifiers. SFE of coffee grounds, a by-product of coffee production, has been studied. The oil extracted from the coffee grounds contained 44.5% linoleic acid and 37.5% palmitic acid, and it was estimated, through an economic analysis, that the extracted oil could yield a net income of €21.9 million per year.[126] Another study evaluated SFE as an alternative process for palm oil extraction to reduce the energy and water required for the intense processing of palm oil in Malaysia. Malaysia produced an estimated 17.7 million tons of palm oil in 2008 and is the second largest producer of palm oil, responsible for more than half of the world production, so even very small energy savings will have a large cumulative effect. While highlighting the benefit of SFE in the palm oil industry, the reviewers noted that there are still problems with the scalability of SFE owing to the current lack of industrial-scale SFE equipment.[127] SFE research also continues in the field of natural functional ingredients, exploring the potential of adding SFE garlic extracts to sunflower oil. The results showed that the garlic extract allicin, when evaluated against synthetic butylated hydroxytoluene (BHT), performed similarly to BHT in preventing lipid oxidation.[128]

1.8.10 Supercritical Fluid Pasteurization

Another use of supercritical fluids (SCFs) is as an alternative non-thermal method for pasteurization. Supercritical carbon dioxide ($scCO_2$) causes microbial inactivation through (1) cell wall rupturing caused by the strong interaction of the fluid with lipids, (2) enzyme deactivation, (3) disruption of the metabolic chain caused by inhibition of decarboxylases by excess CO_2 surrounding the cell and (4) intercellular electrolyte alterations. However, spore inactivation with SCFs is not fully understood, although it appears to be caused by perforation of the spores' outer layer.[129] A recent study reported 100% inactivation of both *Aspergillus niger* and *Penicillium simplicissimum* spores.[130] Parameters used in supercritical pasteurization are temperature, pressure, moisture and exposure time. Just as with thermal pasteurization, an increase in temperature results in a reduced processing time.[131] One study looked at the effectives of $scCO_2$ pasteurization for post-harvest processing (PHP) of oysters. In addition to validating that $scCO_2$ was equivalent to other FDA-approval PHP treatments of oysters, the results showed that $scCO_2$ processing was also able to remove or reduce the load of toxic metals, biotoxins and chemical environmental contaminants.[132] These contaminants are a concern in oysters along with food-borne pathogens. Other research investigated $scCO_2$ as a non-thermal alternative to pasteurizing apple cider and apple juice and found that a 5 log reduction in *E. coli* could be achieved without altering either the flavor or color.[133] Hence $scCO_2$ is a compelling non-thermal alternative.

1.8.11 Membrane Separations in Food Processing

Membrane separation has been very successful for waste water management and continues to provide innovative separation solutions.[134–136] Membranes have a wide range of applications based on the size range of the material filtered from the product. Membranes are classified based on their application as microfiltration (MF), ultrafiltration (UF), dialysis (D), nanofiltration (NF), reverse osmosis (RO), pervaporation (PV), gas separation (GS) and ion-exchange membranes (IEM). Membranes do not require additives and can perform separations at low temperatures.[137] In one study, RO separation was shown to result in approximately one-third of the total operating cost compared with evaporative separation.[138] One example of membranes supporting sustainable food processing can be found in the dairy industry. Whey, once a waste by-product, is now a functional ingredient in high demand. This process not only reduced a waste with a high biological and chemical oxygen demand produced by dairy plants, but also provided another product line for dairy processors. The filtration step also increased the food safety of these products owing to removal of bacteria and spores. Both whey protein concentrate and isolate can be used as functional ingredients, *i.e.* as emulsifying, gelling and foaming agents.[139] Other valuable individual whey proteins (immunoglobulins, lactoferrin, lactoperoxidase, bovine serum albumin, α-lactalbumin and β-lactoglobulin) and lactose can be fractionated using membrane technology.

1.9 Food Safety Surveillance Systems

The basis of food-borne illness surveillance systems follows the Food and Agricultural Organization–World Health Organization (FAO–WHO) guidelines, which describe how a surveillance system must first set a baseline of a current illness rates caused by an infection and then map how to reduce the infection rate and impact over time, as shown in Figure 1.14.

Cassini *et al.* argued that surveillance systems cannot accurately capture and measure the true impact of infectious and food-borne diseases because of two fundamental issues: (1) not everyone who becomes ill will seek care and (2) a failure of communication between healthcare providers and government agencies.[140] Therefore, risk managers must continue to find new ways to identify better both the sources and rates of infection and improve the collection and processing of the information. To address and prevent continuing food-borne illness issues effectively, epidemiological data collection and local risk assessment actives should be combined.[140] Two example programs that are improving the management of epidemiologic data are the Burden of Communicable Disease (BCoDE) database used in the European Union (EU) and the FDA-iRisk database used in the USA. The EU has developed a toolkit for the BCoDE to map the health outcomes (symptomatic levels *versus* asymptomatic) through representations of the transitions from different infection states. From a probability standpoint, each infection has a certain chance of a particular outcome which is influenced by a person's demographic risk factors. The user's specific input allows an estimation of the impact of food-borne pathogen infections as DALYs.[141,142]

The FDA-iRisk is the equivalent to EU's BCoDE food safety surveillance system. The FDA-iRisk creates models to assess, compare and rank the risks brought by multiple food hazard interventions across the entire food supply chain from "farm to fork." It addresses risk at (1) the primary production location, (2) processing, (3) retail markets and (4) the consumer. The

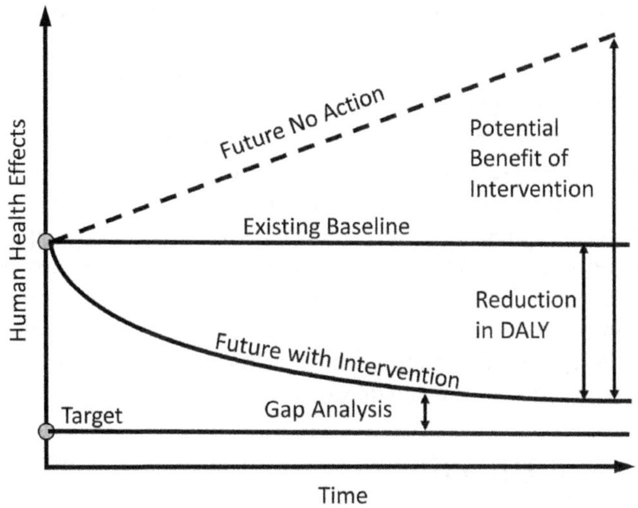

Figure 1.14 FAO–WHO conceptual model for DALY health impact. Reproduced from Food and Agriculture Organization of the United Nations, 2009, *Risk Characterization of Microbiological Hazards in Food: Guidelines*, http://www.who.int/foodsafety/publications/micro/MRA17.pdf.

structural parallel to LCA is obvious (see Section 1.10.2). The model is based on seven components: "the food, the hazard, the population of consumers, a process model (*i.e.* food production, processing and handling practices that follow the fate of the hazard through the supply chain), consumption patterns in the population, dose–response relationships and burden of disease measures associated with health effects."[33] The process models use four inputs: (1) the introduction of the hazard, (2) patters of consumption, (3) dose–response relationships and (4) health effects. Chen *et al.* reported a case study examining *L. monocytogenes* and *Salmonella* as microbial hazards using the FDA-iRisk system.[33] Both the BCoDE and FDA-iRisk are modeling programs that can effectively estimate the linked MRA and DALY impact for companies and government officials to prevent, correct and respond to food-borne illness better. This not only improves public heath safety, but also encourages management decisions based on quantitative data.[143]

Food safety surveillance systems directly impact our ability to estimate accurately the impact of food-borne illness. It is known that seven pathogens cause 90% of all food-borne illness in the USA: *Campylobacter*, *Clostridium perfringens*, *E. coli 0157*, *L. monocytogenes*, non-typhoidal *Salmonella* (NTS), *Norovirus* and *Toxoplasma gondii.*[144] The estimated total annual DALYs from these seven food-borne pathogens was 120 000 based on the health impact calculated from years of life loss (YLL), years lost due to disability (YLD) sequelae (a subsequent illness resulting from the infection) and YLD acute. Without the communication and reporting efforts orchestrated between different organizations, it would be impossible to make an accurate estimation of the impacts.

In addition to the surveillance programs growing in the USA and EU, China is continuing to improve its public health awareness and prevention of food-borne illness. In 2009, China created and enacted the Food Safety Law of the People's Republic of China, which was revised in 2015.[145] The regulations led to the National Heath and Family Planning Commission (NHFPC), which works with various government agencies within China's government. Open and fluid communication between agencies is imperative to keep the cases of food-borne illness low within a country of approximately 1.35 billion people. The NHFPC uses food surveillance to provide a foundation and technical support for MRA planning, standard setting and a promulgation of food safety laws. China has reported success of its new food surveillance system by decreasing the number of cases of *Coronobacter* spp. from 1.1 to 0.35% in infant powdered formula.[146]

1.10 Future Directions

Advances in LCA, particularly as it is implemented in food production and consumption, are ongoing. There are efforts under way to incorporate directly the epidemiology of diet on health and also the beginning of work to permit an integrated accounting of food-borne pathogen impacts on health. In addition to the continued developments in food processing technology, which are the subject of much of the remainder of this book, other technological advances involving food quality hold promise for the continual improvement of the sustainability of the global food supply chain.

1.10.1 Disability-adjusted Life-years – A Unifying Metric

The effort to evaluate the entire impact of DALYs of food-borne illness globally was launched in 2006 with the WHO initiative entitled the "Initiative to Estimate the Global Burden of Foodborne Diseases." Furthermore, the WHO created the Foodborne Disease Burden Epidemiology Reference Group (FERG) to support future DALY studies. They concluded that there are 31 significant international foodborne diseases, that is, diseases with food as the primary vector for infection. DALY is the sum of the number of healthy years of life lost (YLL) due to premature mortality and the number of years lost due to disability (YLD). YLL is calculated by multiplying the number of deaths (D) by the remaining life expectancy at the age at which death occurs in years (E). YLD is calculated by gathering the number of incidents cases, symptom duration and symptom severity, or "disability weight," ranging from 0 = perfect health to 1 = death.[147–149] This initiative built upon previous work that the FAO and WHO had conducted by helping organizations define, assess and mitigate food-borne illness risk. One method is through an MRA, which was first defined by the Codex Alimentarius Commission CAC/GL 30-1999, with amendments in 2012 and 2014, as "a scientifically based process consisting of the following steps: (i) hazard identification; (ii) hazard characterization; (iii) exposure assessment; and (iv) risk characterization."[140,150] Not only can MRA help the food industry continue to improve public heath safety from food-borne illness and disease, it also can allow the industry to allocate limited resources more effectively through the identification of hotspots of pathogenic risk in the supply chain.[143]

Figure 1.15 highlights the importance of adequately identifying the risk and shows how risk characterization, risk management and risk communication, while having separate definitions, share overlapping boundaries that must be executed well by management to mitigate a microbiological risk and reduce outbreaks of food-borne illness.[151] See Romero-Barrios *et al.* for an example of modeling quantitative MRA.[152]

The application of LCA in the context of the food system must be predicated on the understanding that the functional unit for the assessment must always meet certain obligatory characteristics: all legal and regulatory requirements (*e.g.* nutrient label requirements) and food safety requirements. Food safety is frequently modeled using quantitative MRA, where the probability of illness from exposure to a food-borne pathogen is expressed in DALYs. The observation that LCIA and quantitative MRA both report impacts in terms of DALYs suggests the possibility of combining the two assessment frameworks. Many papers have been published discussing the basis for combining risk assessment (in general) and LCA;[151,153–158] however, despite

Figure 1.15 A figure based on the Codex defining stages of effective risk management systems, showing continual improvement through iterations. Adapted from Food and Agriculture Organization of the United Nations, 2009, *Risk Characterization of Microbiological Hazards in Food: Guidelines*, http://www.who.int/foodsafety/publications/micro/MRA17.pdf.

these discussions, there have been relatively few applications in which the two methods have been combined.

There are some broad similarities between risk assessment (and MRA specifically) and LCA: both share the goal of system assessment through quantitative modeling in an effort to support decision makers. Yet there are also some differences that may partially explain the relatively few applications of the combined assessments. Risk assessment is site and scale specific and employs a conservative approach to evaluate safety in the context of an exposure or acceptable threshold level. Lifecycle assessment, although evolving, is frequently site generic, using an approach that estimates the average potential impacts frequently with the goal of identifying hotspots or comparing alternatives. Nonetheless, there have been some efforts to combine the approaches, as outlined in the following sections.

1.10.2 Food Safety and LCA

As mentioned previously, with regard to food processing, the first concerns that arise will be food safety and microbiological risk. However, it may be surprising to find that this is rarely addressed in LCA. Failure of food safety protocols ultimately leads to impacts on human health, which is an endpoint impact category in lifecycle impact assessment. For food LCAs that include accounting of human health, it is generally based on exposure to chemicals, including pesticide residues, rather than microbiological risks in the food supply chain.

Villamonte *et al.* reported that there is reduced human health impact from prepared meals treated with HPP than traditional methods and, although the DALYs from *Listeria* remain below the cumulative carcinogenic effects of the supply chain, the impact increases with the storage time.[159] This is the only report directly connecting food-borne pathogen MRA with LCA, and it provides a reasonable methodological approach. Other work connecting MRA with LCA has been published, again providing a methodological framework for future improvements.[160,161] The authors evaluated the combined environmental and pathogenic impacts on human health associated with different options for the treatment of sewage sludge and they showed that pathogen exposure can contribute up to 20% of the total human health impact measured in DALYs, with the remaining impact arising from exposure to heavy metals and other chemicals.

1.10.3 Nutrition and LCA

There is growing belief that dietary guidelines need to be based on both nutritional and environmental science.[162,163] This is likely driven by the observation that food production consumption is a significant driver of global environmental impact, and the desire to provide sound nutritional recommendations with reduced environmental impact to improve global sustainability. A growing number of studies are starting to consider the benefits and costs of differences in the nutritional qualities of food in LCA. Although the epidemiology of non-communicable diseases associated with dietary choices is relatively well established, these mechanisms have not yet been incorporated into LCA. This is not surprising, as LCA arose from industrial ecology, and only more recently has it expanded to include food production and processing and therefore generally includes only negative impacts. Thus, current LCIA categories, both midpoint and endpoint, quantify negative rather than positive impacts. To include dietary effects in LCA, a new mechanism needs to be established to connect the inventory with the impact/benefit category. A recent study proposed a combined nutritional and environmental LCA framework for diet using published epidemiologic data relating food consumption to reported nutritional health effects (both positive and negative) expressed in DALYs.[35] Dietary health impacts, such as microbiological risk assessment and chemical exposure, are all commonly quantified in units of DALY. Since DALY is the impact category metric used in LCA to assess (negative) impacts on human health, in principle it should also be applicable to account for both burdens and benefits associated with the nutritional content of foods. It can be argued that all endpoint impact categories (resource and ecosystem) could be "expanded" in the same way to include benefits derived from human activities. Benefits presented at the midpoint level may be expressed as negative values for the indicator.

1.10.3.1 Dietary Health Impact in LCA

As an example, a 2013 study conducted in France, Sweden and The Netherlands modeled the potential outcome of additional calcium intake from dairy foods to reduce the risk of osteoporotic fractures. The economic impact of supplemental calcium was quantified based the cost of additional dairy intake and the cost saving through the prevention of osteoporotic fractures. The study found that low calcium intake (<600 mg) led to a reduction of 6263 DALYs in France, 1246 DALYs in Sweden and 374 DALYs in The Netherlands. The associated total cost that could potentially be saved by the prevention of osteoporotic hip fractures was €100 311 274 for women in France, €23 912 460 for women in Sweden and €5 121 041 for women in The Netherlands.[164] Another LCA study conducted in 2015 evaluated the impact of milk on both the environment and health also measured in DALYs. The study did not focus on food-borne illness but, similarly to the previous study, focused on the health impact of non-communicable disease associated with dietary intake. The study is an excellent model for how a quantitative epidemiology-based study can estimate the synergies and tradeoffs between nutrition and environmental human health impact expressed in DALYs. The new method that the research developed is called Combined Nutritional and Environmental Life Cycle Assessment (CONE-LCA); CONE-LCA measures and compares in parallel the environmental and nutritional effects of food or diets.[35]

1.10.4 Food Waste and Sustainability

As mentioned previously, reduction of food waste is a significant opportunity for improving the sustainability of the food supply chain. Global food waste is estimated at 1.3 MT per year.[165] There are many causes for this loss. In the developing world, much of it is preharvest associated with pest damage or the inability to harvest in time, whereas in the developed world, it is more commonly caused by postproduction losses, and significantly driven by consumer behavior.[28,29,166–171] Therefore, although the efforts of the food processing sector to provide new technology that reduces avoidable food losses are important, education to change consumer behavior is equally important.

Figure 1.16 outlines the goals of the US Environmental Protection Agency (EPA) to reduce food waste and improve food recovery. Some source reduction is easily achieved through education because there is documented confusion over date labeling on foods.[29,166,172] Although the one of the goals of the EPA is to use food waste as compost material, studies must be conducted to ensure that the composted food waste does not pose a pathogen risk. For example, a study investigating the recycling of sludge and food waste to reduce gas emissions and improve phosphorus recovery highlighted

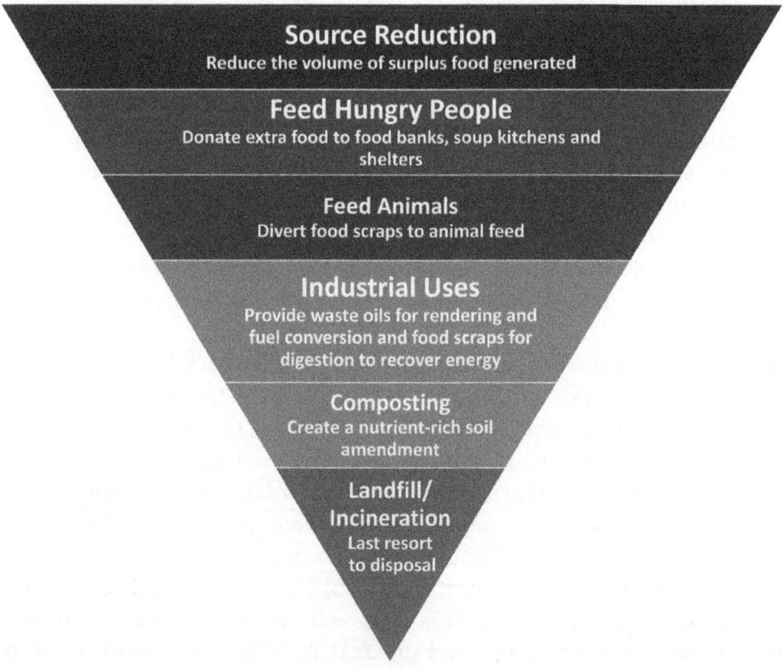

Figure 1.16 US EPA's hierarchical steps to reduce food waste. Reproduced from US EPA, *Food Recovery Hierarchy*, (2016), http://bit.ly/2hg4Fhi.

the difficulty of waste recycling programs. Food waste to be reused in the food system (*e.g.* as compost) introduces the potential risk of food-borne pathogen contamination to the food supply. All waste recycle programs must ensure that no pathogens enter the food supply during the recycling process.[173]

1.10.4.1 Food Safety and Food Waste

Another by-product of a strong food safety plan is the avoidance of large and costly food recalls, which also contribute to food waste. Table 1.2 shows the number of recalls and the associated pounds of meat recalled in the USA in 2015.

Table 1.2 Recall summary for calendar year 2015.[a]

	No. of recalls	No. of pounds recalled	Estimated carbon footprint[b]/tonne CO_2e
Total	150	21 104 848	
Class			
I	99	16 623 878	
II	39	3 176 212	
III	12	1 304 758	
Reason for recall			
STEC[c]	8	215 593	
Listeria monocytogenes	6	82 547	
Salmonella	3	4 828 874	
Undeclared allergen	58	10 268 457	
Extraneous material	11	1 104 790	
Processing defect	4	5259	
Undeclared substance	5	1 176 731	
Other[d]	55	3 422 597	
Species			
Beef	41	1 345 842	12 000
Mixed	38	10 238 498	38 800
Pork	37	1 480 768	3360
Poultry[e]	33	8 004 465	16 300
Ovine	1	35 275	370

[a]Adapted from USDA-FSIS: US Department of Agriculture Food Safety and Inspection Service, https://www.fsis.usda.gov/wps/portal/fsis/topics/recalls-and-public-health-alerts/recall-summaries (accessed December 2016).
[b]Footprints estimated from data in M. de Vries and I. J. M. de Boer, *Livest. Sci.*, 2010, **128**, 1, and *LEAP, Greenhouse Gas Emissions and Fossil Energy Use from Small Ruminant Supply Chains*, United Nations Food and Agriculture Organization, Rome, 2015.
[c]STEC includes recalls due to Shiga toxin-producing *E. coli* (STEC). STEC organisms include *E. coli* O157:H7, *E. coli* O26, *E. coli* O45, *E. coli* O103, *E. coli* O111, *E. coli* O121 and *E. coli* O145.
[d]"Other" includes producing without inspection, failure to present for import inspection and labeling issues, among others.
[e]Poultry includes egg products.

A recall of meat products is especially problematic for the environment because animal products require more water and energy input than protein pulses. At the highest end, beef requires 20 times more water than a pulse protein and accounts for one third of the world's water supply dedicated to animal production.[174] A food recall is ultimately the failure of a food safety plan being either implemented or executed. One of the easiest steps to make the food supply chain more sustainable is to reduce number of recalls due to mislabeling and microbial contamination. One innovative technology that is just emerging is blockchain technology, which is a distributed ledger that records every transaction in a completely traceable and transparent manner. Walmart is currently testing the technology, which should permit, for example, very selective rather than mass removal of contaminated food when a contamination event is identified.[175,176]

Another way to avoid food waste is through improved communication of microbial risks in food. Consumers throw away food that is still safe to eat from a microbial viewpoint over concern for food safety because of a misunderstanding between the meaning of "expiry" and "best by" date information on packaging, which is mandated by food labeling laws. An expiry date is used to describe a date when the packaged food product may begin to become unsafe by microbiological standards and should not be consumed. A "best by" date is a guarantee by the food manufacturer that all nutritive and quality aspects of the product will be maintained until at least that date. However, properly stored food that has passed its "best by" date does not automatically have an increased food-borne illness risk from a microbiological viewpoint. The food industry tries to use expiry dates primarily for more perishable goods whereas less perishable foods normally have "best by" dates. Notwithstanding the food industry efforts, consumers still discard food that is safe owing to confusion about the meaning of these two terms.[29,166,172]

1.10.5 Technological Advancement

Sensors and monitoring tools are also generating synergy for sustainability when integrated with continuing improvements in technology. A few of the benefits and potential outcomes are as follows:

- Less waste resulting from food safety inspections.
- Improved sampling techniques.
- More monitoring and measurement to improve quality and reduce waste: coupling radiofrequency identification (RFID) and blockchain technology.[176]
- Wireless sensing/remote monitoring through the Internet of things (IoT).
- Automatic lighting and temperature controls throughout the food supply chain.
- Enabling "big data" analysis to highlight risk allows a sharper focus on food safety.[177]

- Quicker response times.
- Safety and environmental best practices can be shared quickly across the industry through improvements in communication technologies.
- Focuses of sourcing energy for food processing through renewable energy sources.
- Management focus on water neutrality, in turn lessening the load on water sanitation.
- Multiple hurdle approach to extend the shelf life of foods in lieu of focusing on just one type of extended shelf life technique (*e.g.* aseptic packing).[31]

An example of how technology integration has led to the reduction of food waste and improved food safety can be seen in an experiment conducted in 2013 that mapped and identified high-risk areas for food-borne pathogens on a farm by combining GIS (geographical information systems technology) information and pathogen-specific growth patterns to predict preharvest food safety hazards. Using these predictions, the farmer has a better opportunity to mitigate the hazard before harvest, in turn reducing the risk of illness from food-borne pathogens across the entire food supply chain.[178] In addition, some databases are now using unstructured data collection methods, *e.g.* social media sites, to help track a food-borne illness.[179] Continuing innovation will improve these tools by empowering more people to report food-borne illness efficiently and ease communication across domestic and international government agencies.

1.11 Conclusion

We must encourage students to study interdisciplinary food-related programs, such as food engineering coupled with lifecycle management principles, which can provide new minds to help implement more sustainable technology and management in the food industry. In addition, it is imperative that education includes a dimension for socioeconomic innovation.[180] As disciplinary boundaries blur, our collective ability to communicate and solve more and more complex problems improves. It is important to remember that sustainability, because of its multidimensional nature and myriad tradeoffs, does not always result in immediate economic benefit to an organization. As mentioned earlier, the definition of sustainability adopted by each organization should be understood and communicated across all levels within an organization.[31]

Given the global imperative for food security under the combined pressures of climate change, resource scarcity (especially water), growing population and affluence, the opportunities and responsibilities falling on the food processing sector are significant. However, our ability to benchmark and document continuous improvement and adapt new technologies gives reason for optimism in the face of our collective challenges.

References

1. H. C. J. Godfray, J. R. Beddington, I. R. Crute, L. Haddad, D. Lawrence, J. F. Muir, J. Pretty, S. Robinson, S. M. Thomas and C. Toulmin, *Science*, 2010, **327**, 812.
2. D. Kim, G. Thoma, D. W. Nutter, F. Milani, R. Ulrich and G. A. Norris, *Int. J. Life Cycle Assess.*, 2013, **18**, 1019.
3. G. Thoma, J. Popp, D. W. Nutter, D. R. Shonnard, R. Ulrich, M. D. Matlock, D. Kim, Z. Neiderman, N. Kemper, C. East and F. Adom, *Int. Dairy J.*, 2013, **31**, S3.
4. B. Notarnicola, G. Tassielli, P. A. Renzulli, V. Castellani and S. Sala, *J. Cleaner Prod.*, 2017, **140**, 753.
5. ISO 14040, *Environmental Management–Life Cycle Assessment—Principles and Framework*, International Organization for Standardization, Geneva, Switzerland, 2006.
6. ISO 14044, *Environmental management - Life cycle assessment — Requirements and guidelines*, International Organization for Standardization, Geneva, Switzerland, 2006.
7. ISO 14046, *Water footprint - Principles, requirements and guidelines*, International Organization for Standardization, Geneva, Switzerland, 2014.
8. ISO 14067, *Greenhouse gases - Carbon footprint of products - Requirements and guidelines for quantification and communication*, International Organization for Standardization, Geneva, Switzerland, 2013.
9. J. M. Blanco, M. Finkbeiner and A. Inaba, *Guidance on Organizational Life Cycle Assessment*, UNEP/SETAC Life Cycle Initiative, 2015.
10. WBCSD and WRI, *The Greenhouse Gas Protocol: A Corporate Accounting and Reporting Standard*, World Business Council for Sustainable Development (WBCSD); World Resources Institute (WRI), 2004.
11. G. A. Norris, *Int. J. Life Cycle Assess.*, 2001, **6**, 118.
12. European Commission, *Buying green!*, European Commission, Brussels, BE, 3rd edn, 2016.
13. *Guidelines for Social Life Cycle Assessment of Products*, ed. C. Benoît and B. Mazijn, UNEP SETAC Life Cycle Initiative, 2009, vol. 15.
14. C. Benoit-Norris, *Int. J. Life Cycle Assess.*, 2014, **19**, 261.
15. C. Benoit-Norris, D. A. Cavan and G. A. Norris, *Sustainability*, 2012, **4**, 1946.
16. C. Benoit-Norris, G. A. Norris and D. Aulisio, *Sustainability*, 2014, **6**, 6973.
17. A. Ciroth and F. Eisfeldt, *Social Life Cycle Impact Assessment of the Nicaragua Canal Project*, GreenDelta, Berlin, 2016.
18. A. Ciroth and F. Eisfeldt, *PSILCA - A Product Social Impact Life Cycle Assessment database. Database version 1.0. Documentation*, GreenDelta, Berlin, 2016.
19. V. Subramanian, W. Ingwersen, C. Hensler and H. Collie, *Int. J. Life Cycle Assess.*, 2012, **17**, 892.

20. *Product Category Rule Guidance Development Initiative*, http://www.pcrguidance.org/ (accessed: December 2016).
21. W. Ingwersen, V. Subramanian, R. Schenck, L. Bushi, A. Costello, L. Draucker, C. East, C. Hensler, H. Lahd and S.-O. Ryding, *Int. J. Life Cycle Assess.*, 2012, **17**, 258.
22. W. W. Ingwersen and V. Subramanian, *Int. J. Life Cycle Assess.*, 2014, **19**, 532.
23. ISO 14025, *Environmental labels and declarations — Type III environmental declarations — Principles and procedures*, International Organization for Standardization, 2006.
24. British Standards Institution, *Publically Available Specification 2050:2011 Specification for the assessment of the life cycle greenhouse gas emissions of goods and services*, BSI, PAS 2050:2, 2011.
25. AFNOR, *General principles for an environmental communication on mass market products Part 0: General principles and methodological framework*, ADEME - The French Environment and Energy Management Agency, 2011.
26. Food SCP RT, *ENVIFOOD Protocol Environmental Assessment of Food and Drink Protocol*, Food Sustainable Consumption and Production Round Table (SCP RT), Working Group 1, Brussels, Belgium, 2013.
27. European Commission, *Environmental Footprint Pilot Guidance document, - Guidance for the implementation of the EU Product Environmental Footprint (PEF) during the Environmental Footprint (EF) pilot phase, version 5.2*, 2016.
28. J. C. Buzby and J. Hyman, *Food Policy*, 2012, **37**, 561.
29. V. Monier, S. Mudgal, V. Escalon, C. O'Connor, T. Gibon, G. Anderson, H. Montoux and H. Reisinger, *Preparatory Study on Food Waste Across Eu 27*, 2010, vol. 33.
30. K.-R. Bräutigam, J. Jörissen and C. Priefer, *Waste Manage. Res.*, 2014, **32**, 683.
31. R. O. Morawicki, *Handbook of Sustainability for the Food Sciences*, Wiley-Blackwell, 2012.
32. P. Belli, J. Anderson, H. Barnum, J. Dixon and J. Tan, *Exch. Organ. Behav. Teach. J.*, 1998, 209.
33. Y. Chen, S. B. Dennis, E. Hartnett, G. Paoli, R. Pouillot, T. Ruthman and M. Wilson, *J. Food Prot.*, 2013, **76**, 376.
34. *Global Guidance Principles for Life Cycle Assessment Databases*, ed. G. W. Sonnemann and B. Vigon, UNEP/SETAC Life Cycle Initiative, Shonan, Japan, 2011.
35. K. S. Stylianou, M. C. Heller, V. L. Fulgoni, A. S. Ernstoff, G. A. Keoleian and O. Jolliet, *Int. J. Life Cycle Assess.*, 2015, **21**, 734–746.
36. V. L. Fulgoni, D. R. Keast and A. Drewnowski, *J. Nutr.*, 2009, **139**, 1549.
37. A. Drewnowski and V. L. Fulgoni, *Nutr. Rev.*, 2008, **66**, 23.
38. A. Drewnowski, C. D. Rehm, A. Martin, E. O. Verger, M. Voinnesson and P. Imbert, *Am. J. Clin. Nutr.*, 2015, **101**, 184.

39. D. Gustafson, A. Gutman, W. Leet, A. Drewnowski, J. Fanzo and J. Ingram, *Sustainability*, 2016, **8**, 196.
40. A. Smedman, H. Lindmark-Månsson, A. Drewnowski and A.-K. M. Edman, *Food Nutr. Res.*, 2010, **54**, 1.
41. C. M. Weaver, J. Dwyer, V. L. Fulgoni, J. C. King, G. A. Leveille, R. S. MacDonald, J. Ordovas and D. Schnakenberg, *Am. J. Clin. Nutr.*, 2014, **99**, 1525.
42. H. A. Udo de Haes, O. Jolliet, G. Finnveden, M. Z. Hauschild, W. Krewit and R. Muller-Wenk, *Int. J. Life Cycle Assess.*, 1999, **4**, 167.
43. G. Myhre, D. Shindell, F.-M. Bréon, W. Collins, J. Fuglestvedt, J. Huang, D. Koch, J.-F. Lamarque, D. Lee, B. Mendoza, T. Nakajima, A. Robock, G. Stephens, T. Takemura and H. Zhan, in *Climate Change 2013: The Physical Science Basis. Contribution of Working Group I to the Fifth Assessment Report of the Intergovernmental Panel on Climate Change*, ed. T. F. Stocker, D. Qin, G.-K. Plattner, M. Tignor, S. K. Allen, J. Boschung, A. Nauels, Y. Xia, V. Bex and P. M. Midgley, Cambridge University Press, Cambridge, United Kingdom and New York, NY, USA, 2013, pp. 659–740.
44. R. Geyer, J. P. Lindner, D. M. Stoms, F. W. Davis and B. Wittstock, *Int. J. Life Cycle Assess.*, 2010, **15**, 692.
45. M. N. Pineda, PhD Thesis, Universitat Autònoma de Barcelona, 2011.
46. M. D. Matlock, G. Thoma, E. Cummings, J. Cothren, M. Leh and J. Wilson, *Int. Dairy J.*, 2013, **31**, S78.
47. S. Pfister, P. Bayer, A. Koehler and S. Hellweg, *Environ. Sci. Technol.*, 2011, **45**, 5761.
48. E. Grubert, *Int. J. Life Cycle Assess.*, 2016, 1.
49. P. J. Gerber, T. Vellinga, C. Opio, B. Henderson and H. Steinfeld, *Greenhouse Gas Emissions from the Dairy Sector: A Life Cycle Assessment*, Food and Agriculture Organization of the United Nations, Rome, IT, 2010.
50. A. Flysjö, *J. Dairy Sci.*, 2011, **94**, 5833–5841.
51. C. Ramirez, M. Patel and K. Blok, *Energy*, 2006, **31**, 1984.
52. I. Muñoz, L. Milà i Canals and A. R. Fernández-Alba, *Int. J. Life Cycle Assess.*, 2010, **15**, 794.
53. F. Milani, D. W. Nutter and G. Thoma, *J. Dairy Sci.*, 2011, **94**, 4243.
54. H. Aguirre-Villegas, F. Milani, S. Kraatz and D. Reinemann, *Trans. Am. Soc. Agric. Eng.*, 2012, **55**, 613.
55. J. Berlin, *Int. Dairy J.*, 2002, **12**, 939.
56. G. A. Keoleian, A. W. Phipps, T. Dritz and D. Brachfeld, *Packag. Technol. Sci.*, 2004, **17**, 85.
57. B. J. Singh, A. Krasowski, S. P. Singh and J. Singh, *Packag. Technol. Sci.*, 2011, **24**, 49.
58. D. W. Nutter, D. Kim, R. Ulrich and G. Thoma, *Int. Dairy J.*, 2013, **31**, S57.
59. E. Castañeda-Gutiérrez, R. Nabais, C. Pereira, A. J. D. Ferreira and M. Lopes, *The Quest for Sustainability: The Role of Environmental Management Systems and Tools*, 2006.

60. A. Daneshi, A. Esmaili-sari, M. Daneshi and H. Baumann, *J. Cleaner Prod.*, 2014, **80**, 150.
61. J. Berlin and U. Sonesson, *J. Cleaner Prod.*, 2008, **16**, 483.
62. I. Djekic, J. Miocinovic, I. Tomasevic, N. Smigic and N. Tomic, *J. Cleaner Prod.*, 2014, **68**, 64.
63. M. H. Eide, J. P. Homleid and B. Mattsson, *LWT – Food Sci. Technol.*, 2003, **36**, 303.
64. M. H. Eide, *Int. J. Life Cycle Assess.*, 2002, **7**, 115.
65. P. H. Nielsen and E. Høier, *Int. J. Life Cycle Assess.*, 2008, **14**, 137.
66. J. Sellahewa and W. Martindale, *Aspects of Applied Biology, Delivering Food Security with Supply Chain Led Innovations: Understanding Supply Chains, Providing Food Security, Delivering Choice*, 2010, vol. 102, pp. 91–98.
67. G. Egilmez, M. Kucukvar, O. Tatari and M. K. S. Bhutta, *Resour., Conserv. Recycl.*, 2014, **82**, 8.
68. P. Hansen, K. Christiansen and B. Hummelmose, *Cleaner Production Assessment in Meat Processing*, 2000.
69. L. Flanigan, R. Frischknecht and T. Montalbo, *An Analysis of Life Cycle Assessment in Packaging for Food & Beverage Applications*, UNEP/SETAC Life Cycle Initiative, 2013.
70. G. Pardo and J. Zufía, *J. Cleaner Prod.*, 2012, **28**, 198.
71. N. Sanjuán, F. Stoessel and S. Hellweg, *Environ. Sci. Technol.*, 2014, **48**, 1132.
72. X. C. Schmidt Rivera, N. Espinoza Orias and A. Azapagic, *J. Cleaner Prod.*, 2014, **73**, 294.
73. J. Davis, U. Sonesson, D. U. Baumgartner and T. Nemecek, *Food Res. Int.*, 2010, **43**, 1874.
74. L. A. Calderón, L. Iglesias, A. Laca, M. Herrero and M. Díaz, *Resour., Conserv. Recycl.*, 2010, **54**, 1196.
75. L. Schneider and P. M. Finkbeiner, *Life Cycle Assessment of EU Oilseed Crushing and Vegetable Oil Refining*, Technische Universität Berlin Chair, Berlin, 2013.
76. N. Pelletier, M. Ibarburu and H. Xin, *J. Cleaner Prod.*, 2013, **54**, 108.
77. K. Andersson, T. Ohlsson and P. Olsson, *Doktorsavhandlingar vid Chalmers Tek. Hogsk.*, 1998, **6**, 1.
78. E. Masanet, *Energy Efficiency Improvement and Cost Saving Opportunities for the Fruit and Vegetable Processing Industry*, Berkeley, 2008.
79. I. A. Gac, C. Lapasin, P. T. Laspière, S. Guardia, P. Ponchant, P. Chevillon and G. Nassy, in *9th International Conference LCA of Food*, San Francisco, 2014, pp. 4–8.
80. S. G. Wiedemann and M. Yan, in *9th International Conference LCA of Food*, San Francisco, 2014.
81. Franklin Associates, *Life Cycle Inventory of Foam Polystyrene, Paper-Based, and PLA Foodservice Products*, Kansas City, MO, 2011.
82. J. Pasqualino, M. Meneses and F. Castells, *J. Food Eng.*, 2011, **103**, 357.

83. M. Meneses, J. Pasqualino and F. Castells, *J. Environ. Manage.*, 2012, **107**, 76.
84. S. Suh and S. Yee, *Chemosphere*, 2011, **84**, 806.
85. D. S. Nijdam, H. C. Wilting, M. J. Goedkoop and J. Madsen, *J. Ind. Ecol.*, 2005, **9**, 147.
86. G. Edwards-Jones, L. Milà i Canals, N. Hounsome, M. Truninger, G. Koerber, B. Hounsome, P. Cross, E. H. York, A. Hospido, K. Plassmann, I. M. Harris, R. T. Edwards, G. A. S. Day, A. D. Tomos, S. J. Cowell and D. L. Jones, *Trends Food Sci. Technol.*, 2008, **19**, 265.
87. C. L. Weber and H. S. Matthews, *Environ. Sci. Technol.*, 2008, **42**, 3508.
88. H. Williams and F. Wikström, *J. Cleaner Prod.*, 2011, **19**, 43.
89. S. González-García, É. G. Castanheira, A. C. Dias and L. Arroja, *Int. J. Life Cycle Assess.*, 2013, **18**, 796.
90. J. Bengtsson and J. Seddon, *J. Cleaner Prod.*, 2013, **41**, 291.
91. A. D. González, B. Frostell and A. Carlsson-Kanyama, *Food Policy*, 2011, **36**, 562.
92. J. P. Lesschen, M. van den Berg, H. Westhoek, H. P. Witzke and O. Oenema, *Anim. Feed Sci. Technol.*, 2011, **166–167**, 16.
93. R. Ulrich, G. Thoma, D. W. Nutter and J. Wilson, *Int. Dairy J.*, 2013, **31**, S50.
94. I. De Marco, S. Miranda, S. Riemma and R. Iannone, *Int. J. Life Cycle Assess.*, 2015, **20**, 1659.
95. V. M. Balasubramaniam, S. I. Martínez-Monteagudo and R. Gupta, *Annu. Rev. Food Sci. Technol.*, 2015, **6**, 435.
96. F. Lou, H. Neetoo, H. Chen and J. Li, *Annu. Rev. Food Sci. Technol.*, 2015, **6**, 389.
97. K. Feng, E. Divers, Y. Ma and J. Li, *Appl. Environ. Microbiol.*, 2011, **77**, 3507.
98. K. S. Varghese, M. C. Pandey, K. Radhakrishna and A. S. Bawa, *J. Food Sci. Technol.*, 2014, **51**, 2304.
99. R. N. Pereira and A. A. Vicente, *Food Res. Int.*, 2010, **43**, 1936.
100. Q. Zhang, G. V. Barbosa-Cánovas and B. G. Swanson, *J. Food Eng.*, 1995, **25**, 261.
101. R. B. Lung, E. Masanet, L. Berkeley and A. Mckane, in *Industrial Energy Technology Conference (IETC 2006)*, New Orleans, LA, 2006.
102. G. Saldaña, I. Álvarez, S. Condón and J. Raso, *Crit. Rev. Food Sci. Nutr.*, 2014, **54**, 1415.
103. N. N. Misra, S. K. Pankaj, A. Segat and K. Ishikawa, *Trends Food Sci. Technol.*, 2016, **55**, 39.
104. R. Thirumdas, C. Sarangapani and U. S. Annapure, *Food Biophys.*, 2015, **10**, 1.
105. B. A. Niemira, *Annu. Rev. Food Sci. Technol.*, 2012, **3**, 125.
106. X. Li and M. Farid, *J. Food Eng.*, 2016, **182**, 33.
107. D. Ziuzina, S. Patil, P. J. Cullen, K. M. Keener and P. Bourke, *Food Microbiol.*, 2014, **42**, 109.
108. D. Wray and H. S. Ramaswamy, *Dry. Technol.*, 2015, **33**, 769.

109. Y. Su, M. Zhang and A. S. Mujumdar, *Dry. Technol.*, 2015, **33**, 260.
110. J. Dunn, *Poult. Sci.*, 1996, **75**, 1133.
111. V. Heinrich, M. Zunabovic, T. Varzakas, J. Bergmair and W. Kneifel, *Crit. Rev. Food Sci. Nutr.*, 2016, **56**, 591.
112. I. M. Caminiti, I. Palgan, F. Noci, A. Muñoz, P. Whyte, D. A. Cronin, D. J. Morgan and J. G. Lyng, *Innovative Food Sci. Emerging Technol.*, 2011, **12**, 118–123.
113. P. Kumar Sing, S. Kumar, P. Kumar and Z. Bhat, *Am. J. Food Technol.*, 2012, **7**, 506.
114. N. Elmnasser, S. Guillou, F. Leroi, N. Orange, A. Bakhrouf and M. Federighi, *Can. J. Microbiol.*, 2007, **53**, 813.
115. A. Birmpa, A. Vantarakis, S. Paparrodopoulos, P. Whyte and J. Lyng, *Biomed Res. Int.*, 2014, **2014**, 1.
116. A. K. Datta and V. Rakesh, *Compr. Rev. Food Sci. Food Saf.*, 2013, **12**, 24.
117. K. Krishnamurthy, H. K. Khurana, J. Soojin, J. Irudayaraj and A. Demirci, *Compr. Rev. Food Sci. Food Saf.*, 2008, **7**, 2.
118. X. Li, Z. Pan, G. G. Atungulu, X. Zheng, D. Wood, M. Delwiche and T. H. McHugh, *Innovative Food Sci. Emerging Technol.*, 2014, **21**, 123.
119. S. A. Wilson, G. G. Atungulu, A. Couch and S. Sadaka, *Appl. Eng. Agric.*, 2015, **31**, 799.
120. N. K. Rastogi, *Crit. Rev. Food Sci. Nutr.*, 2012, **52**, 737.
121. P. Piyasena, E. Mohareb and R. C. McKellar, *Int. J. Food Microbiol.*, 2003, **87**, 207.
122. S. Salo and G. Wirtanen, *Br. Food J.*, 2007, **109**, 31.
123. L. Axelsson, A. Holck, I. Rud, D. Samah, P. Tierce, M. Favre and C. F. Kure, *J. Food Prot.*, 2013, **76**, 1401.
124. E. Roselló-Soto, M. Koubaa, A. Moubarik, R. P. Lopes, J. A. Saraiva, N. Boussetta, N. Grimi and F. J. Barba, *Trends Food Sci. Technol.*, 2015, **45**, 296.
125. Y. Yao, *Ultrason. Sonochem.*, 2016, **31**, 512.
126. M. M. R. de Melo, H. M. A. Barbosa, C. P. Passos and C. M. Silva, *J. Supercrit. Fluids*, 2014, **86**, 150.
127. M. J. H. Akanda, M. Z. I. Sarker, S. Ferdosh, M. Y. A. Manap, N. N. N. Ab Rahman and M. O. Ab Kadir, *Molecules*, 2012, **17**, 1764.
128. E. Bravi, G. Perretti, C. Falconi, O. Marconi and P. Fantozzi, *J. Sci. Food Agric.*, 2017, **97**, 102.
129. M. Perrut, *J. Supercrit. Fluids*, 2012, **66**, 359.
130. E. A. Noman, N. N. N. A. Rahman, M. Shahadat, H. Nagao, A. F. M. Al-Karkhi, A. Al-Gheethi, T. N. T. Lah and A. K. M. Omar, *CLEAN – Soil, Air, Water*, 2016, **44**, 1700.
131. K. Khosravi-Darani, *Crit. Rev. Food Sci. Nutr.*, 2010, **50**, 479.
132. D. A. F. Meujo, D. A. Kevin, J. Peng, J. J. Bowling, J. Liu and M. T. Hamann, *Int. J. Food Microbiol.*, 2010, **138**, 63.
133. H.-G. Yuk, D. J. Geveke and H. Q. Zhang, *Int. J. Food Microbiol.*, 2010, **138**, 91.
134. G. Turner, *Desalin. Water Treat.*, 2013, **51**, 4817.

135. L. Suárez, M. A. Díez, R. García and F. A. Riera, *J. Ind. Eng. Chem.*, 2012, **18**, 1859.
136. R. Y. Avula, H. M. Nelson and R. K. Singh, *Innov. Food Sci. Emerg. Technol.*, 2009, **10**, 1.
137. M. Kumar, M. A. Khan, Z. A. Al-Othman and T. S. Y. Choong, *Sep. Purif. Rev.*, 2013, **42**, 187.
138. K. Jevons and M. Awe, *Desalination*, 2010, **250**, 961.
139. P. Kumar, N. Sharma, R. Ranjan, S. Kumar, Z. F. Bhat and D. K. Jeong, *Asian-Australas. J. Anim. Sci.*, 2013, **26**, 1347.
140. A. Cassini, S. Hathaway, A. Havelaar, M. Koopmans, K. Koutsoumanis, W. Messens, G. Müller-Seitz, B. Nørrung, V. Rizzi and F. Scheutz, *EFSA J.*, 2016, **14**, 1.
141. *European Center for Disease Prevention and Control*, http://bit.ly/2hfzvrh (accessed: December 2016).
142. FDA/CFSAN, JIFSAN, RSI and FDA, *Food and Drug Administration Center for Food Safety and Applied Nutrition (FDA/CFSAN), Joint Institute for Food Safety and Applied Nutrition (JIFSAN), Risk Sciences International (RSI)*, https://irisk.foodrisk.org/ (accessed: December 2016).
143. A. H. Havelaar, J. Bräunig, K. Christiansen, M. Cornu, T. Hald, M.-J. J. Mangen, K. Mølbak, A. Pielaat, E. Snary, W. Van Pelt, A. Velthuis and H. Wahlström, *Zoonoses Public Health*, 2007, **54**, 103.
144. *USDA Food Safety and Inspection Service*, https://goo.gl/ttksk2 (accessed: December 2016).
145. J. Clever, *People's Republic of China's Food Safety Law*, USDA, Foreign Agriculture Service, 2015.
146. X. Pei, N. Li, Y. Guo, X. Liu, L. Yan, Y. Li, S. Yang, J. Hu, J. Zhu and D. Yang, *Int. J. Environ. Res. Public Health*, 2015, **12**, 10662.
147. K. Gibney, M. Sinclair, J. O'Toole and K. Leder, *J. Public Health Policy*, 2013, **34**, 439.
148. E. Scallan, R. M. Hoekstra, B. E. Mahon, T. F. Jones and P. M. Griffin, *Epidemiol. Infect.*, 2015, **143**, 2795.
149. A. H. Havelaar, M. D. Kirk, P. R. Torgerson, H. J. Gibb, T. Hald, R. J. Lake, N. Praet, D. C. Bellinger, N. R. de Silva, N. Gargouri, N. Speybroeck, A. Cawthorne, C. Mathers, C. Stein, F. J. Angulo and B. Devleesschauwer, *PLoS Med.*, 2015, **12**, e1001923.
150. WHO and FAO, *Risk Characterization of Microbiological Hazards in Food: Guidelines*, World Health Organization, Food and Agriculture Organization of the United Nations, 2009, vol. 17.
151. S. J. Cowell, R. Fairman and R. E. Lofstedt, *Risk Anal.*, 2002, **22**, 879.
152. P. Romero-Barrios, M. Hempen, W. Messens, P. Stella and M. Hugas, *Food Control*, 2013, **29**, 343.
153. L. Breedveld, *J. Risk Res.*, 2013, **16**, 459.
154. J. W. Owens, *Risk Anal.*, 1997, **17**, 359.
155. D. W. Pennington, M. Margni, J. Payet and O. Jolliet, *Hum. Ecol. Risk Assess. An Int. J.*, 2006, **12**, 450.

156. I. Linkov and T. P. Seager, *Environ. Sci. Technol.*, 2011, **45**, 5068.
157. G. Sonnemann, F. Castells and G. Schuhmacher, *Integrated Life–Cycle and Risk Assessment for Industrial Processes. Advanced Methods in Resource and Waste Management Series*, 2003.
158. T. E. McKone, A. D. Kyle, O. Jolliet, S. Irving Olsen and M. Hauschild, *Int. J. Life Cycle Assess.*, 2006, **11**, 137.
159. G. Villamonte, M. De Lamballerie and V. Jury, in *Proceedings of the 9th International Conference on Life Cycle Assessment in the Agri-Food Sector Consideration*, 2014.
160. R. Harder, S. Heimersson, M. Svanström and G. M. Peters, *Environ. Sci. Technol.*, 2014, **48**, 9438.
161. S. Heimersson, R. Harder, G. M. Peters and M. Svanström, *Environ. Sci. Technol.*, 2014, **48**, 9446.
162. C. van Dooren, M. Marinussen, H. Blonk, H. Aiking and P. Vellinga, *Food Policy*, 2014, **44**, 36.
163. Ministry of Health of Brazil, *Dietary Guidelines for the Brazilian Population*, Secretariat of Health Care, Primary Health Care Department, 2014.
164. F. J. B. Lötters, I. Lenoir-Wijnkoop, P. Fardellone, R. Rizzoli, E. Rocher and M. J. Poley, *Osteoporos. Int.*, 2013, **24**, 139.
165. J. Gustavsson, C. Cederberg, U. Sonesson, R. van Otterdijk and A. Meybeck, *Global food losses and food waste: extent, causes and prevention*, Food and Agriculture Organization of the United Nations, Rome, Italy, 2011.
166. N. Lucifero, *Agric. Agric. Sci. Procedia*, 2016, **8**, 282.
167. J. C. Buzby, J. Bentley, B. Padera, C. Ammon and J. Campuzano, *Agriculture*, 2015, **5**, 626.
168. M. S. Tom, P. S. Fischbeck and C. T. Hendrickson, *Environ. Syst. Decis.*, 2016, **36**, 92.
169. M. C. Heller and G. A. Keoleian, *J. Ind. Ecol.*, 2015, **19**, 391.
170. FAO, *Food Wastage Footprint. Impacts on Natural Resources. Summary Report*, Food and Agriculture Organization of the United Nations, Rome, Italy, 2013.
171. C. Schader, U. Schmidt and J. Brei, *Food Wastage Footprint: Full Cost-accounting*, Food and Agriculture Organization of the United Nations, Rome, Italy, 2014.
172. B. Kim, L. P. Houser, A. Rosenthal and R. Neff, *Literature Review of Methods and Tools for Quantifying the Indirect Environmental Impacts of Food Procurement*, Baltimore, MD, 2008.
173. T. Nakakubo, A. Tokai and K. Ohno, *J. Cleaner Prod.*, 2012, **32**, 157.
174. A. K. Chapagain and A. Y. Hoekstra, *Unesco*, 2010, **1**, 80.
175. O. Kharif, *Bloomberg Online*, http://bloom.bg/2i1×44q (accessed: December 2016).
176. F. Tian, in *13th International Conference on Service Systems and Service Management*, 2016.

177. L. Strawn, E. Brown, J. David, H. den Bakker, P. Vangay, F. Yiannas and M. Wiedmann, *Food Technol.*, 2015, **69**, 42.
178. L. K. Strawn, E. D. Fortes, E. A. Bihn, K. K. Nightingale, Y. T. Grohn, R. W. Worobo, M. Wiedmann and P. W. Bergholz, *Appl. Environ. Microbiol.*, 2013, **79**, 588.
179. M. Keller, M. Blench, H. Tolentino, C. C. Freifeld, K. D. Mandl, A. Mawudeku, G. Eysenbach and J. S. Brownstein, *Emerging Infect. Dis.*, 2009, **15**, 689.
180. Y. H. Roos, P. J. Fryer, D. Knorr, H. P. Schuchmann, K. Schroën, M. A. I. Schutyser, G. Trystram and E. J. Windhab, *Food Eng. Rev.*, 2016, **8**, 91.
181. European Commission – Joint Research Centre, International Reference Life Cycle Data System (ILCD) Handbook: Analysis of existing Environmental Impact Assessment methodologies for use in Life Cycle Assessment, 2010.

CHAPTER 2

Food Law and Sustainable Food Processing: A Comparison of the EU and the USA

NEAL D. FORTIN

Institute for Food Laws and Regulation, Michigan State University, 469 Wilson Road, 139 Trout Food Science, East Lansing, MI 48824, USA
Email: fortinne@msu.edu

2.1 Introduction[†,1]

Agriculture has achieved impressive advances over the past 50 years: fewer farmers produce more food on the same acreage while resource input and energy use per output have decreased. Notwithstanding these achievements, the world faces daunting challenges ahead in meeting the food needs of a growing global population while facing rising production costs, increasingly scarce natural resources, and climate change.

These constraining factors have made sustainability an important driver in the food industry. Food and agricultural sustainability generally encompasses four concerns: satisfying human food needs, preserving environmental quality, maintaining the economic viability of agriculture, and supporting shared social goals, such as fostering quality of life.[2]

[†]Some ideas and parts of this chapter are derived with permission from M. T. Roberts and E. H. Leibovitch, Comparison of EU and US Law on Sustainable Food Processing, in *Alternatives to Conventional Food Processing*, 1st edn, ed. A. Proctor, Royal Society of Chemistry, Cambridge, 2011, Ch. 2, pp. 11–92.

Green Chemistry Series No. 53
Alternatives to Conventional Food Processing: 2nd Edition
Edited by Andrew Proctor

Published by the Royal Society of Chemistry, www.rsc.org

A primary environmental concern is that food production is approaching a tipping point. Limitations in natural resources may directly result in a catastrophic failure to satisfy human needs, or the disruption of the environment and ecosystems may indirectly result in a catastrophic failure to satisfy human needs. Adequate nutrition and a hospitable environment are critical to the survival of nation states, our civilization, and even our species. Thus, food and agricultural sustainability ranks alongside pandemics, world war, and climate change in the magnitude of risk to humanity.[3]

A significant economic force behind sustainability is the rise of the "power of green." Consumers are motivated to purchase products that are perceived to be delivered by a more sustainable food supply chain. Sustainability also has a self-rewarding economic aspect in that a supply chain or processing method that consumes less resources or produces less waste can result in financial benefits in production. A significant social concern behind sustainability is that food and agricultural production supports quality of life for farmers and farm workers.

Today's sense of sustainability represents a pivotal shift in society's food production paradigm. Starvation has stalked humankind throughout our history. Through prehistory and nearly all recorded history, securing sufficient supply has been the dominant factor in food production. Because hunger haunted humanity through history, "more" was an overriding good embedded in our food production systems.

In this sense, food processing was one of humankind's greatest achievements. Simple processing, such as drying and fermentation, permitted preservation of food and increased food security. The mastery of fire was perhaps the crowning achievement in the history of food processing because it gave us cooking. Some anthropologists assert that cooking is what made us human; without the extra calories and nutrients released by cooking, the human brain could not have developed into what it is today. Brain is "expensive tissue" to maintain, and in evolution hominid brains were increasing in size at the same time as energy-intensive gut tissue was decreasing.[4]

It is not a coincidence that the fermentation of beer and the making of bread are as old as recorded history. Such food processing techniques were necessary antecedents to making recorded history possible. Food processing and agriculture, which began roughly 10 000 years ago,[5] brought additional food security, an important measure of sustainability, which allowed recording of history.

However, at the same time, agriculture carried seeds of its own unsustainability because agriculture intrinsically departs from natural ecosystems. Agriculture can never fully be in harmony with natural systems. Agricultural cultivation means modification of animals, plants, and ecosystems away from the natural ecosystem.

It is easy for our minds to conflate early agriculture with natural ecosystems because of the limited impact compared with today's industrial agriculture. During most of the earliest agricultural era, food systems and food processing centered around individual farms and homes. Nonetheless, we should not mistake the smaller scale and the reference to the past as the measure of sustainability. We must deal with the scale of the present population, and the methods of the past are not necessarily sustainable. Moreover, sustainability cannot just seek continuity with a static environment and ecology. Sustainability also means adapting to changing environment and ecology, that is, sustainability must include a measure of resilience.[6]

As modern food processing evolved, scientific methods and techniques not only converted raw food into edible, convenient forms but also provided benefits in preservation, nutrition, quality, and safety. A primary focus of modern food processing remained producing more. More food was needed to feed growing populations, and more food was needed in preserved forms suitable for shipping to growing urban populations.

However, when the concern over adequate supply faded in memory within developed countries, concern over sustainability grew. The recent generation has also seen a major focus on producing a greater variety of food to provide more options to consumers. At the same time, modern food processing has faced increasing scrutiny. For example, concerns about nutrition have led to advocacy for less processing of food.[7,8] Owing to the negative environmental consequences associated with food processing, some have called for a reduction in the inputs of raw materials, water, and energy into food processing to reduce its total environmental impact.[9] One aim is to use renewable energy (or process waste to produce energy) and to process food with zero waste.[9]

A fundamental question is whether achieving these or other sustainability objectives in food processing can be assisted by government regulation. The inherent limits to what can be achieved with regulation are further complicated by the diverse array of agricultural products and production methods. Moreover, the nature of the global food supply chain, stretching across ecosystems and national borders, and also the multitude of national political environments, all make it difficult to regulate for sustainability objectives.

The nature of political systems everywhere is such that they are rarely conducive to proactive changes in food regulation. Most significant changes in national food laws result in response to serious problems. In addition, each governing system has its own history, politics, traditions, institutions, and resources that affect its priorities and strategies for the regulation of food production and processing. Any new legislation must be developed with these factors in mind to ensure that the law works within the context of the governing circumstances. Moreover, national regulation faces challenges of budget limitations and cuts, new emerging food safety concerns, and increasing pressure to respond to consumer interest in non-science traits of food products.

In addition to national systems, there also is an overarching international food law system. This system is based on the World Trade Organization (WTO) agreements. The two most important WTO agreements related to sustainability are the WTO Agreement on the Applications of Sanitary and Phytosanitary Measures (SPS Agreement) and the WTO Technical Barriers to Trade Agreement (TBT Agreement). The SPS Agreement adopts as default reference standards the food standards set in the three international bodies, known as the "three sisters:" Codex Alimentarius, the World Organisation for Animal Health (OIE), and the International Plant Protection Convention (IPPC). Thus, the SPS Agreement, TBT Agreement, and Codex Alimentarius are important considerations for food processing sustainability.

Another control mechanism in the international food supply chain is private standards, or non-governmental standards. Although not directly part of the regulatory system, private standards play an increasing role in the global food system. Private standards are incredibly diverse and numerous, such as religious (halal, kosher), ethical concern (animal welfare, free-range chickens), social issue (fair trade), conservation (dolphin friendly, Rainforest Alliance), food safety (KenyaGAP, ChileGAP), and production method standards (genetically modified organism (GMO) free, safe quality food (SQF)).

This chapter offers an analysis of the European Union (EU) and US food systems and compares their respective regulation of sustainability as it relates to "green" or "environmentally friendly" processing. Although sustainability also includes economic and social values, this comparative analysis largely confines itself to environmental values.

The EU and US food regulatory systems offer both similarities and differences in their regulatory approaches. Examination of these similarities and differences through a methodic comparative approach of the two systems can reveal insights into both and broader understanding generally.[10] Comparative law can also help to provide insights into the transplanting of law and legal institutions from one system to another. Notwithstanding the differences in regulatory theories, attitudes, and approaches, the EU and the USA may be able to borrow regulatory approaches from each other. Comparative law can also provide useful guideposts in shared regulatory successes and regulatory dead ends.

2.1.1 Roadmap for This Chapter

This chapter begins with the history, development, and general food law provisions of the respective EU and US food regulatory systems. This overview is not meant to be a complete summary of the systems but rather an overview of aspects pertinent to sustainability. Certain regulatory patterns, tendencies, and perceptions that are similar and different between the two systems are explored.

Next, the chapter introduces the concept of sustainability and the emerging "green processing" in the food law of the EU and USA. These two regulatory regimes are analyzed and compared. In addition to government measures, this chapter also considers private standards that have emerged both to promote and to regulate sustainability in food processing. Concerns in the EU and USA over the use of private standards are evaluated, including the international trade implications.

The chapter concludes with observations from this comparison of EU and US regulatory approaches to sustainability in food processing.

2.2 EU and US Law and Policy

Rarely throughout the history of all regions has the development of food law sprung from proactive, prudent planning. Almost always, significant new food law was passed only after a crisis or serious problem had served as a catalyst. We see this pattern repeated in the history of food law development in both the EU and the USA.

2.2.1 History and Development of Food Law in the EU

The EU is a unique political and economic community of currently 28 Member States that features a supra-national legal system. It traces its origins to the European Economic Community (EEC) formed in 1957 by the Treaty of Rome among six European countries.[11] This union grew through the accession of new Member States, and the purely economic union evolved to span policy areas, from climate, environment, and health to security, justice, and migration.[‡] The European Economic Community name change to the European Union in 1993 reflected this evolution.

The main objective of the establishment of a European Economic Community was to create a common market without internal borders and with freedom of movement of goods, services, persons, and capital. Agriculture was one of three common policies established under the initial Treaty of Rome (the others being trade and transport). Consumer protection was expressly mentioned in European legislation only later, in the mid-1980s.

Only after major food safety scares occurred were significant changes made in EU food law. Foremost was the bovine spongiform encephalopathy (BSE) crisis of the mid-1990s, which may have been the starting point of a reform of the whole EU food law system. BSE, also called mad cow disease, is a fatal, neurodegenerative disease that spread among cattle through contaminated animal feed. In 1996, it became clear that a link existed between BSE in UK cattle and Creutzfeldt–Jakob disease (CJD), a fatal human disease.[12] The European Commission banned all British beef exports, and the UK slaughtered and destroyed more than a million UK cattle.[13] The European Parliament established an inquiry into BSE to investigate the BSE crisis, which found that the crisis had been mismanaged both nationally and by the EU.[14]

The Treaty of Amsterdam of 1997 recognized the need for public health and consumer protection by adding duties for the EU institutions,[15] and today's consolidated version of the Treaty of Rome still includes these provisions.[16] In 1997, the Commission separated risk management from risk assessment in part to keep scientific advice independent from the politics of risk management.[17]

Another crisis hit EU food safety regulation in 1999. In May of that year, the Belgian authorities found that farm animals had been fed

[‡]The motivation behind the Community's formation was mainly economic, but politics was in the mind of the founders.

dioxin-contaminated feed.[13] European authorities had Belgian chickens, eggs, pork, and beef removed from the European market. In the aftermath, the European Commission issued a White Paper on Food Safety in 2000, in which it declared that food safety would be best ensured through an independent EU food authority responsible for independent scientific advice, operation of rapid alert systems, and risk communication "from farm to table."[18]

In 2002, the main EU food law, Regulation[§] (EC) 178/2002, laid down general principles and requirements of food law, establishing the European Food Safety Authority (EFSA), and set procedures in matters of food safety.[19] The EFSA is an independent agency with the role of providing the European Commission and Member States with independent and transparent scientific advice [ref. 20, Art. 22(5)]. The EFSA has no enforcement authority. The regulation of food is handled by the European Commission and Member States. The European Commission is split into Directorate Generals (DGs), which are departments with specific areas of responsibility for EU policy and law. Each DG is led by one of the Commissioners.

In 2015, the Commission reorganized the DGs, and health and food safety matters were put in a new Directorate General for Health and Food Safety (DG SANTE). The EFSA, the Commission, and Member States are expected to cooperate to ensure that risk assessment, risk management, and risk communication are consistent [ref. 20, Art. 22(8)]. Although food law is increasingly created at the EU level, the primary responsibility for implementing EU law lies with the EU Member States.

The Commission is also assisted by standing committees to develop food safety measures. These committees are composed of representatives of the Member States.[20] Regulation 178/2002 recommended the creation of a Standing Committee on the Food Chain and Animal Health (SCFCAH).[21] The SCFCAH is composed of representatives of Member States and is chaired by a Commission representative and examines the Commission's proposals under the detailed procedures that are called comitology.[22]

2.2.2 History and Development of Food Law in the USA

The US government is a system of federalism, where the 50 US states maintain some sovereign authority and have their own independent state laws. The federal government has plenary power over commerce and other

[§]The difference between Regulations and Directives lies in the implementation. Article 189 of the Treaty of Rome provides that a Regulation "shall have general application … [and] shall be binding in its entirety and directly applicable in all Member States." The same Article provides that a Directive "shall be binding, as to the result to be achieved, upon each Member State to which it is addressed, but shall leave to the national authorities the choice of form and methods." Treaty of Rome, Art. 189. On the other hand, in the USA "Regulations" is a synonym for administrative rules, which are laws promulgated by administrative agencies, such as the FDA and USDA, under the authority granted by the US Congress through an enabling statute, also called a "Public Act."

areas. Although the USA inherited the British common law system, today's food law is substantially based on statutory law and administrative law.

The history of the development food law in the USA follows a pattern similar to that seen in the EU, evolving in reaction to crises. New legislation of a significant nature was usually passed after a food safety crisis that was combined with serious public concern. The significant first such crisis in the history of US federal food law arrived with the beginning of the modern era of chemistry. Chemistry brought new food additives and new means of adulteration of food. The lack of government regulation allowed the addition of untested chemicals such as formaldehyde to food and substitution of cheap ingredients for those represented on labels.

In 1883, Harvey Wiley became the chief chemist of the US Bureau of Chemistry.[23] He expanded research and testing of food and documented widespread adulteration of foods. He helped to spur public concern over food safety and quality by his publications and by campaigning for a national food and drug law. Wiley galvanized public attention with his well-publicized "poison squad." In a special dining hall, Wiley fed the poison squad volunteers food containing various chemicals used in food manufacturing to document the ill effects.[24] Although shockingly crude by today's standards, Wiley's methods were instrumental in the passage of the Pure Food and Drug Act of 1906.[24]

In that same era, Upton Sinclair published *The Jungle*. This novel portrayed nauseating practices and unsanitary conditions in Chicago's meat packing plants. For example, Sinclair described diseased meat and rat dung being processed into sausages. The novel galvanized the public's attention and put the focus squarely on the conditions of food processing. The book was instrumental in the passage of the Federal Meat Inspection Act of 1906.[25]

The Pure Food and Drug Act of 1906 did not require drugs be proven safe or effective by the manufacturer. Again, it took a tragedy to propel change. In 1937, a company manufactured elixir of sulfanilamide (an antibiotic), using diethylene glycol to sweeten the liquid, without ever testing the safety. The elixir killed at least 107 people, many of whom were children, who suffered agonizing deaths. Congress responded by strengthening federal food safety regulatory authority with the passage of the 1938 Federal Food, Drug and Cosmetic Act (FDCA).

The food law continued to evolve based upon the concerns and issues of the times. The following are some key examples. In response to concerns over synthetic food additives, the Food Additives Amendment to the FDCA was enacted in 1958, mandating the proof of safety of food additives before entering commerce.[26] Following well-publicized outbreaks of botulism food poisoning from canned foods in 1973, the Food and Drug Administration (FDA) issued Low-Acid Food Processing Regulations.[27] In 1970, the Environmental Protection Agency (EPA) entered the food regulatory system when President Richard M. Nixon delegated pesticide regulation to the EPA to oversee governmental pesticide research and regulate residues in food.[28]

In 1993, an outbreak of *Escherichia coli* O157:H7 associated with fast-food hamburgers caused over 400 illnesses and four deaths.[29] In response to a public outcry for safer ground beef, in 1996 the US Department of Agriculture (USDA) issued a rule implementing Pathogen Reduction/Hazard Analysis and Critical Control Point (HACCP) Systems.[30,31]

The above are just examples to illustrate the reactive nature of food law. Unfortunately, there are many more examples in both the EU and the USA. However, the nature of food law evolution tending towards being reactive rather than proactive is seen in both regions. This tendency towards reactiveness foretells challenges in passing proactive sustainability laws. Thus, proposed legislation that is crafted to sugar-coat prospective benefits with some immediate benefits will have an improved likelihood of being enacted.

2.2.3 General Food Law Provisions in the EU and the USA

There are profound differences in US and EU policies regarding biotechnology, growth hormones in beef, and antibiotics in animal feed.[32] Because these controversial differences in food and agriculture policy command our attention, they can obscure the fact that both regions have developed similar food and agricultural policies and apply similar scientific evidence of safety and risk. Moreover, both the USA and EU have similar or identical objectives for a safe and plentiful food supply and economic protection of consumers.

The EU and USA have general food laws applicable to all food. In addition to these fundamental provisions, specialized requirements apply to particular foods. The general food law provisions in the EU are in Regulation (EC) 178/2002,[33] and in the USA in the Federal Food, Drug, and Cosmetic Act[34] and the Federal Meat Inspection Act.[35] The overriding objectives of all these laws is food and feed safety and honest trade.

Regulation (EC) 178/2002 aims at ensuring "a high level of protection of human life and health and the protection of consumers' interests, including fair practices in food trade, taking account of, where appropriate, the protection of animal health and welfare, plant health and the environment" and it "aim[s] to achieve the free movement in the Community of food and feed manufactured or marketed."[36] The Federal Food, Drug, and Cosmetic Act aims at ensuring safe and wholesome food and promoting honesty and fair dealing in the interests of consumers.[37]

Both the EU and the USA use similar approaches of risk assessment and risk management. Risk assessment is to be based on the available scientific evidence and undertaken in an independent and objective manner. In the EU, risk assessment is carried out by the European Food Safety Authority (EFSA).[38] In the USA, risk assessment is carried out by scientific committees within federal agencies, such as the FDA, Centers for Disease Control and Prevention (CDC), or USDA, or by independent scientific bodies, such as the Institute of Medicine and the National Academy of Sciences.

In both the EU and the USA, the risk assessment from the scientists is supposed to be kept separate from the risk management. In the EU, the EFSA passes the risk assessment over to DG SANTE for the risk management. In the USA, the risk assessment is passed either to risk managers within a federal agency, such as the FDA, or to the US Congress.

In the EU, DG SANTE is expected to take the risk assessment into account along with the precautionary principle and other legitimate factors when they are relevant to achieve the law's general objectives.[39] These factors are listed in Recital 19 as societal, economic, traditional, ethical, and environmental factors, and the feasibility of controls.[40] The precautionary principle permits government action when a scientific assessment of available information indicates possible harmful effects even when scientific uncertainty exists.[41] Measures taken under the precautionary principle should be "proportionate and no more restrictive of trade than is required to achieve the high level of health protection chosen in the Community, regard being had to technical and economic feasibility and other factors regarded as legitimate in the matter under consideration."[42,43]

In the USA, the federal agencies are expected to take the risk assessment into account along with the governing laws and their risk management principles and any other factors specified in the law. Thus, the US Congress, which writes the laws, is the top-level risk management. The resulting law on each matter can be distinct. On some matters, the law is more stringent than the precautionary principle. On other matters, the precautionary principle applies only indirectly. For instance, the US law on food additives is more precautionary regarding potential carcinogens than necessary even under strict application of the precautionary principle – food additives that are safe under conditions of use are still banned if the substance can cause cancer in laboratory animals at large doses.[44] On the other hand, the law regulating naturally occurring toxins in food permits only limited application of the precautionary principle – the FDA must have scientific evidence that the toxin is ordinarily injurious.[45]

It is commonly espoused that the precautionary principle is the major point of divergence between EU and US food law. However, close examination reveals that, while the EU adopts the precautionary principle by name and practice in the region's food law, the US exhibits many examples where the precautionary principle is adopted into the nation's food law in everything but name.

2.2.4 Development of the International Concept of Sustainability

The international concerns over sustainability arose from a number of environmental, economic, and social realizations, all of which relate back to recognition of the limitations of the planet on which we depend. Sustainability not only means environmental durability, but also refers to the idea that one must produce and use resources in a way that does not jeopardize another's present and future access to these resources.

The global actions taken to increase sustainability have resulted from a series of international meetings where nations came together and recognized how collectively current policies and practices jeopardize the future and demand action. In 1972, the preservation of the environment was addressed at the Declaration of the United Nations Conference on the Human Environment, where nations recognized the need to safeguard natural resources such as the air, water, land, flora, and fauna and the need to protect non-renewable resources and prevent pollution.[46]

One definition of sustainability or sustainable development that is often referred to is the one mentioned in the 1987 Report of the World Commission on Environment and Development entitled *Our Common Future.* This World Commission was created by the United Nations in 1983 through General Assembly Resolution 38/161 to recommend long-term environmental strategies to achieve sustainable development in 2000 and thereafter and to make recommendations on ways to translate environmental concerns into cooperation activities between countries at varying stages of economic and social development, in order to achieve "common and mutually supportive" goals that take into account the interactions "between people, resources, environment and development."[47] The report defines sustainable development as follows:

> [development that] meets the needs of the present without compromising the ability of future generations to meet their own needs ... [Such development implies] limitations imposed by the present state of technology and social organization on environmental resources and by the ability of the biosphere to absorb the effects of human activities ... Sustainable development requires meeting the basic needs of all and extending to all the opportunity to fulfill their aspirations for a better life. A world in which poverty is endemic will always be prone to ecological and other catastrophes ... In the end, sustainable development is ... a process of change in which the exploitation of resources, the direction of investments, the orientation of technological development and institutional change are made consistent with future as well as present needs ... Painful choices have to be made ... Sustainable development must rest on political will.[48]

In 1992, the United Nations Conference on Environment and Development issued the Rio Declaration on Environment and Development, which called for countries' recognition that environmental protection is an integral part of the development process, for the development of sustainable patterns of production and consumption, for the cooperation between nations to improve the scientific understanding and the technology transfer for sustainable development, and for the creation of environmental policies and objectives at national levels.[49] This conference established the United Nations Framework Convention on Climate Change, which sets a general framework for intergovernmental efforts to address the challenges posed by climate change. It acknowledges that the climate system is a shared resource and that emission of carbon dioxide and other greenhouse gases can affect its stability.[50] The signatories committed to sketch out strategies for sustainable development in time for the 2002 World Summit on Sustainable Development. In 1997, an addition to this treaty, the Kyoto Protocol set binding targets forcing member countries to reduce their greenhouse gas emissions.[51] Although the EU ratified the Protocol, the USA did not.

In 2002, the international community reaffirmed its commitment to sustainable development in the Johannesburg Declaration on Sustainable Development. The nations recognized their responsibility towards sustainable development by strengthening economic and social development and environmental protection at international, regional, national, and local levels, and the need to improve access to "clean water, sanitation, adequate shelter, energy, healthcare, food security, and the protection of biodiversity."[52]

2.2.5 History of Sustainability Approach in the USA and EU

The USA and the EU have both established sustainable agriculture and food production policies. The following compares some of the government approaches taken to encourage sustainability in agriculture and food production processes.

2.2.5.1 History of Sustainability Approach in the USA

The USA set up several voluntary conservation programs starting around the time of the Great Depression. The Soil Conservation Act of 1935 had a main objective of the protection of land resources from soil erosion and established the Soil Conservation Service to administer the Conservation Options Program to assist producers in planning soil management programs. A year later, the Soil Conservation and Domestic Allotment Act amended the Soil Conservation Act to allow the voluntary Agricultural Conservation Program, which gave financial assistance to producers who followed approved soil conservation practices. In 1956, the Great Plains Conservation Program was established and focused on wind and water erosion and anti-pollution practices designed to persuade producers to apply soil conservation practices in the Great Plains states in the USA.¶

¶The "Great Plains" is an extensive prairie and grassland that lies west of the Mississippi River and east of the Rocky Mountains in the USA and Canada. The Canadian portion of the plains is known as the Prairies.

A significant step towards environmental sustainability in the USA was the enactment of the National Environmental Policy Act (NEPA) in 1969. NEPA's purpose is to

> declare a national policy which [would] encourage productive and enjoyable harmony between man and his environment; to promote efforts which will prevent or eliminate damage to the environment and biosphere and stimulate the health and welfare of man; to enrich the understanding of the ecological systems and natural resources important to the Nation; and to establish a Council on Environmental Quality.[53]

Following this, in 1970, the US Environmental Protection Agency (EPA) was established as an independent agency in the federal government's executive branch. Its mission is to protect human health and the environment.[54] This mission recognizes that environmental protection is an integral consideration nationally and globally in policies concerning natural resources, human health, economic growth, energy, transportation, agriculture, industry, and international trade.[54] The EPA recognizes that "environmental protection contributes to making our communities and ecosystems diverse, sustainable and economically productive."[54]

Many other initiatives and programs have been created in the decades since the creation of the EPA. The following are some examples. In 1993, President Clinton signed Executive Order 12852 to establish the President's Council on Sustainable Development, to "forge consensus on policy by bringing together diverse interests to identify and develop innovative economic, environmental and social policies and strategies; demonstrate implementation of policy that fosters sustainable development by working with diverse interests to identify and demonstrate implementation of sustainable development; get the word out about sustainable development; and evaluate and report on progress by recommending national, community and enterprise level frameworks for tracking sustainable development."[55] In 1996, the Council issued a report entitled *Sustainable America: A New Consensus for Prosperity, Opportunity and a Healthy Environment for the Future.* With respect to agriculture-related issues, the document calls for renewable resources to be used and for public land to be looked after.

In March 1997, the Council published a new report, entitled *The Road to Sustainable Development: A Snapshot of Activities in the United States*. The report lays out actions under the Extended Product Responsibility strategy, which is based on the principle of shared responsibility between suppliers, manufacturers, and consumers for reducing the environmental impacts of products throughout their lifecycles. The report lists initiatives including recycling programs, enhancing environmental performance in chemical manufacturing, energy efficiency, the creation of eco-industrial parks, educational initiatives, the creation of federal working groups focusing on sustainable development, and natural resource management programs.[56]

The EPA regularly issues strategy documents for a certain period, as required under the Government Performance and Results Act of 1993. These strategies list the goals that the EPA wants to achieve in the timeframe established. These strategic plans often have mentioned the need to improve efforts towards sustainability.[57] In 2006, the EPA announced the *EPA Strategy for Agriculture*, in which it indicated its commitment to protect the country's food, water, land, and air for future generations. The *Strategy* listed the framework that the agency planned to follow in order to achieve these objectives: "increasing the awareness understanding within EPA of agricultural impacts and benefits to human health and the environment and working across media in a more systems approach to environmental protection," "working with the agricultural sector – including production, processing and distribution – in developing and demonstrating environmental protection solutions that express the value of farmland environmental stewardship activities to the public," "coordinating research and technology development and transfer so the needs of agriculture and EPA can be more efficiently met," and "identifying existing environmental measures and developing new ones, where appropriate, to demonstrate environmental improvement related to agriculture."[58]

In January 2007, President Bush signed Executive Order 13423 called Strengthening Federal Environmental, Energy and Transportation Management. This Order also set goals on issues of renewable energy, energy efficiency, recycling, and water conservation.

2.2.5.2 History of Sustainability Approach in the EU

Following the commitments that the Rio Declaration signatories agreed to in 1992, the 1997 Treaty of Amsterdam added sustainable development in the list of EU commitments.[59] The European strategy for sustainable development was published in 2001. In 1998, the European Council met in Cardiff, UK, and launched what is known as the Cardiff Process, which required the integration of environmental considerations in all the European Community's activities.

At the Berlin European Council of 1999, the Heads of States agreed on an action program for the period from 2000 to 2006 called Agenda 2000.[60] Agenda 2000 contained two priorities related to sustainability, concerning the environment and the reform of the Common Agricultural Policy (CAP). The document called for a "greener CAP," with agri-environmental measures to "support the sustainable development of rural areas" and to encourage farmers to use farming practices that are "compatible with environmental protection and natural resources conservation."[60] The document also introduced the idea of making direct payments to farmers conditional on the observance of environmental requirements and called for Member States to set out codes of good farming practices in their rural development plans that should represent the type of farming that a reasonable farmer would follow in the region concerned.[60,61]

In 1999, in Helsinki, the European Council asked the European Commission to issue a proposal for a strategy laying out economic, social, and ecological policies for sustainable development for the 2001 European Council. At the 2001 European Council in Gothenburg, the Commission presented *A Sustainable Europe for a Better World: A European Union Strategy for Sustainable Development*.[62] This document was the first step in developing the strategy and lays out both the challenges and opportunities brought on by sustainable development, to generate discourse between European institutions, Member States, and society.

The Commission recognized the EU's responsibility in setting a European sustainable development strategy to limit and eventually eliminate threats to sustainable development. The Commission also welcomed the ideas that sustainable development can be achieved only if a balance is found among its economic, social, and environmental dimensions, and that today's generation has a duty to leave sufficient economic, social, and environmental resources for future generations. The Commission listed public health along with climate change and clean energy, management of natural resources, poverty, and social exclusion, aging and demography and mobility, land use, and territorial development as the topics to be addressed with priority within the European sustainable development strategy.

The Commission listed environmental pollution from agricultural and industrial processes as a source of public health concerns. The Commission recognized that although technology can bring risks to sustainable development, it can also bring opportunities for a more efficient use of resources. It acknowledged that without investment in technological progress and scientific research, climate change could only be addressed through difficult changes in production and consumption patterns. The Commission also recommended that each EU policy be assessed based on its impact on sustainable development and that subsidies that encourage wasteful use of natural resources be eliminated.

In February 2002, the Commission issued *Towards a Global Partnership for Sustainable Development*, which focused more on Europe's external policy. The Commission recognized the EU's duty to ensure that globalization contributes to sustainable development in part by providing incentives for environmentally and socially sustainable production and trade and the need to develop intermediate objectives in key sectors such as land and soil, water, energy, and biodiversity. It recommended continuing to implement the international and external objectives of sustainable development to key policies, such as the CAP and the Common Fisheries Policy.[63]

Since then, the EU has implemented several initiatives and actions to achieve progress towards sustainable development. These measures include the creation of the European Climate Change Program, an EU Emission Trading Scheme, and a renewable energy coalition. In 2002, the European Parliament and the European Council launched the Sixth Community Environment Action Program for 2002 until 2012, which set concrete targets for the stabilization of greenhouse gas emissions, which aimed at "protecting, conserving, restoring and developing the functioning of natural systems, natural habitats, wild flora and fauna with the aim of halting desertification and the loss of biodiversity, including diversity of genetic resources, both in the European Union and on a global scale," at "contributing to a high level of quality of life and social well-being for citizens by providing an environment where the level of pollution does not give rise to harmful effects on human health and the environment and by encouraging a sustainable urban development," and at "better resource efficiency and resource and waste management to bring about more sustainable production and consumption patterns."[64] The EU also included a long-term perspective for sustainable agriculture in its reform of the CAP in 2003, by creating a rural development policy composed of new measures to promote the environment, quality, and animal welfare and to assist farmers in meeting EU production standards.

In 2005, the Commission issued a *Review of the EU Sustainable Development Strategy*.[65] The European Council adopted a renewed sustainable development strategy in 2006.[66] The strategy repeats the seven priority challenges (climate change and clean energy; sustainable transport; sustainable production and consumption; public health threats; better management of natural resources; social inclusion, demography and migration; fighting global poverty), addresses how the EU will fulfill its sustainable development commitments, and stresses the need to change the current unsustainable consumption and production patterns and create communities that are able to use and manage the resources efficiently. The EU Sustainable Development Strategy requires the Commission to submit every 2 years a progress report on its EU-wide implementation.[67]

2.2.6 Sustainable Agriculture in the USA and EU

The causes of the current trend towards sustainability in the context of agriculture and food production can be traced back to the beginning of the industrialization period in both the USA and the EU. The modernization of agriculture allowed farmers to produce more food, which is sold at lower prices, and government subsidies have encouraged abundant production.

However, agricultural practices have been costly to the environment. For example, intensive farming has led to issues such as soil erosion, compaction, and salinization. Fertilizer and pesticide use led to water pollution, and nutrient runoff led to eutrophication. Subsequent water pollution affected agricultural production, fisheries, and drinking water supplies. Agriculture also impacted biodiversity and reduced wetlands and wildlife habitats. Deforestation for agricultural production played a role in elevated levels of emissions of greenhouse gases such as carbon dioxide and methane (from livestock digestion processes and stored animal manure) and nitrous oxide (from organic and mineral nitrogen fertilizers).[68]

Certain farming practices are deemed more sustainable.[69] For example, integrated pest management practices can minimize pesticide use, manure and nitrogen-rich crops can maintain soil fertility without synthetic fertilizer, and winter crop management can reduce soil erosion.[69] Both the USA and EU have taken measures to make their agricultural practices more sustainable.

2.2.6.1 Sustainable Agriculture in the USA

As we saw previously, in the USA environmental laws and national programs were created over the years at federal, state, and local levels, to guide the agricultural community towards more sustainable agricultural practices. Sustainable agriculture is defined in the Food, Conservation and Energy Act of 2008 as

> an integrated system of plant and animal production practices having a site-specific application that will, over the long-term:
>
> (A) satisfy human food and fiber needs;
> (B) enhance environmental quality and the natural resource base upon which the agriculture economy depends;
> (C) make the most efficient use of non-renewable resources and on-farm resources and integrate, where appropriate, biological cycles and controls;
> (D) sustain the economic viability of farm operations; and
> (E) enhance the quality of life for farmers and society as a whole.[70]

The Act describes the research projects that should be conducted to learn and inform about ways to reduce chemicals use, promote crop diversification, and make farming practices more sustainable.[71] Rather than dictate any particular farming practice, the USDA National Institute of Food and Agriculture (NIFA) promotes sustainable agriculture through national program leadership and funding for research and extension education.

Farmers and ranchers run diverse operations with diverse needs and ways to improve their sustainability. These vary from region to region, state to state, and farm to farm. However, some common practices have emerged, such as integrated pest management to minimize health and environmental risks; soil conservation methods, such as strip cropping, reduced tillage, and no-till; water conservation and protection, such as planting riparian buffer strips; proper management of manure, nitrogen, and other plant nutrients; and agroforestry, such as inter-planting trees with crops or pasture, better managing woodlots and windbreaks, and using trees and shrubs along streams as buffer strips.

Over the years, the USDA has supported dozens of activities and programs intended to improve agricultural sustainability. Many of these are programs to encourage agricultural producers and landowners to follow conservation practices on their land. Mostly these are voluntary programs. The goals include reducing soil erosion, improving water quality, increasing wildlife habitat, and reducing damage caused by floods. The USDA Natural Resources Conservation Service (NRCS) and the Farm Service Agency (FSA) are the main bodies in charge of administering and implementing the programs.

An example of one such program is the Conservation Stewardship Program, which provides financial and technical assistance to producers "to promote the conservation and cropland, grassland, prairie land, improve pasture and range land, and forested land that is an incidental part of an agriculture operation." To be eligible, the producer must have already addressed one or two resource concerns on their own. Once enrolled, the producer receives payments to adopt new conservation activities or retain the existing ones[72] and to adopt resource-based crop rotation.

Similarly, the Conservation Innovation Grants Program is a voluntary program meant to "stimulate the development and adoption of innovative conservation approaches and technologies while leveraging Federal investment in environmental enhancement and protection, in conjunction with agricultural production."[73]

In addition to these programs, the USDA has developed research programs such as the Sustainable Agriculture Research and Education (SARE) program, which helps "advance farming systems that are profitable, environmentally sound and good for communities through a nationwide research and education grants program."[74] It administers a competitive grant program for producers to fund on-farm research or demonstration projects. The USDA also funds information and research centers focusing on sustainable and environmentally friendly agricultural production practices (*e.g.* The National Sustainable Agriculture Information Service). The NIFA also invests in and supports initiatives that ensure the long-term viability of agriculture and sustain natural resources and the environment.[75]

The regulation of agricultural practices regarding environmental concerns is in part the role of the EPA under various statutes. Although the EPA has standards for sustainable practices, its focus is more on limiting the impact of unsustainable practices on the environment and health. For example, water pollution on or off farms is regulated under the Clean Water Act,[76] which is intended to "restore and maintain the chemical, physical and biological integrity of the Nation's waters."[77] The Safe Drinking Water Act requires that the EPA set national drinking water standards and establish a compliance system.[78] The Clean Air Act[79] addresses air pollution and aims to "protect and enhance the quality of the Nation's air resources so as to promote the public health and welfare and the productive capacity of its population," "to initiate and accelerate a national research and development program to achieve the prevention and control of air pollution," "to provide technical and financial assistance to state and local governments in connection with the development and execution of their air pollution prevention and control programs," and "to encourage and assist the development and operation of regional air pollution prevention and control programs."[80] Pesticide uses are also regulated under the EPA, through the Federal

Insecticide, Fungicide and Rodenticide Act, which mandates the EPA to regulate the use and sale of pesticides to protect human health and preserve the environment.[81] Through its National Pollutant Discharge Elimination System, the EPA oversees the discharge of pollutants from point sources such as concentrated animal feeding operations.

Individual US states also have enacted their own laws to encourage sustainable practices. For instance, California passed the Sustainable Agriculture Research and Education Act in 1986 to encourage research and education on sustainable agriculture practices.[82] The Act founded the Sustainable Agriculture Research and Education Program at the University of California, which supports competitive grants. Iowa enacted the Iowa Groundwater Protection Act of 1987, which was at the origin of the sustainable agriculture program at the Leopold Center for Sustainable Agriculture, a recognized research center in the USA for issues pertaining to sustainable agriculture.

2.2.6.2 *Sustainable Agriculture in the EU*

In the EU, the CAP has been amended several times since 2000 to take into account the environmental concerns surrounding farming and food production. The CAP is now aimed at addressing the risk of environmental degradation while continuing to encourage farmers to maintain the countryside and the environment through rural development measures and through measures that ensure farming profitability throughout the EU.

The 2003 CAP reform decoupled most of the direct aid payments from production of specific crops or breeding of animals, which reduced the incentives to resort to intensive farming practices that damage the environment. It established a single payment scheme based on historical reference levels of direct support received by farmers. The reform has also made cross-compliance and modulation mandatory. Cross-compliance refers to the principle that farmers should comply with environmental protection, animal health and welfare, and food safety requirements and should maintain their land in good agricultural and environmental condition in order to benefit from market support.[83] Modulation consists in shifting support from market policy to measures contributing to practices favorable to the environment and thereby reducing direct payments by a certain percentage every year and allocating the sums saved among the Member States for rural development measures supported by the European Agricultural Fund for Rural Development.[84]

Farmers who adopt environmentally friendly farming techniques that go beyond usual good farming practice (such as managing low-intensity pasture systems, resorting to integrated farm management and organic agriculture, preserving the landscape and historical features, ensuring the conservation of high-value habitats and their biodiversity, *etc.*) for a minimum of 5 years receive payments meant to compensate for the additional costs and loss of income that resulted from the altered farming practices.

In addition, Regulation 1782/2003 used its cross-compliance requirements to incorporate the requirements pertaining to soil conservation, water quality, pesticide use, and biodiversity. For instance, it incorporated the provisions laid out in the Nitrates Directive, those laid out in Council Directive 91/414/EEC concerning the placing of plant protection products on the market, and also those regarding groundwater found in Council Directive 80/68/EEC on the protection of groundwater against pollution caused by certain dangerous substances.[85] The same regulation incorporated the practices aimed at protecting the soil from erosion and maintaining the soil's organic matter and structure, in its definition of "good agricultural and environmental conditions" referred to in its cross-compliance section.[86] It also integrated Council Directive 79/409/EEC on the conservation of wild birds and Council Directive 92/43/EEC on the conservation of natural habitats and of wild flora and fauna.[87]

Moreover, the EU, through Council Regulation (EC) 1698/2005, supports rural development by the European Agricultural Fund for Rural Development.[88] This is designed to help the rural areas of the EU to meet a wide range of economic, environmental, and social challenges. These include "restoring, preserving and enhancing ecosystems related to agriculture and forestry; promoting resource efficiency; and supporting the shift toward a low-carbon and climate-resilient economy in the agriculture, food and forestry sectors."[89]

The general EU environmental regulations, although not specifically targeted at agriculture, also positively impact on the sustainability of agricultural practices. This is the case, for instance, with the legislation on waste management such as the Waste Framework Directive,[90] the Landfill Directive,[91] the Directive on the incineration of waste,[92] and the Directive on urban wastewater treatment.[93] The Integrated Pollution Prevention and Control Directive requires industrial and agricultural activities with a high pollution potential to have a permit.[94] Moreover, the Strategic Environmental Assessment Directive[95] and the Environmental Impact Assessment Directive require environmental consequences to be assessed before the implementation of certain projects.[96]

To ensure the safe use of chemicals, the EU enacted Council Regulation (EC) 1907/2006, introducing the Registration, Evaluation, Authorization and Restriction of Chemicals (REACH) and establishing the European Chemicals Agency (ECHA).[97] In brief, under the REACH, the industry (manufacturers and importers) bears the responsibility for managing the risks associated with its chemicals and for gathering information on their chemicals' properties. The industry is also expected to register this information in a central database operated by the ECHA, which coordinates evaluations of suspicious chemicals and runs a public database where consumers can find information on potential hazards.

2.2.6.2.1 Organic Food Programs. In addition to the programs, regulations, incentives, and research initiatives related to sustainable agriculture, both the USA and EU have developed an organic program. The organic approach to farming is of interest because this alternative to conventional farming is a systems approach that recognizes the importance of interconnections and functional relationships among components of the farm (for example, plants, soils, animals, and water). Production and handling requirements are related to sustainability, such as the focus on ensuring soil fertility and crop nutrient management, and crop rotation. The organic philosophy emphasizes the use of biological processes (as opposed to synthetic chemicals) to achieve soil quality, control pests, and provide favorable growing environments. The philosophy prohibits use of most synthetically produced inputs.[98]

The regulation of organic production is based on certification of production methods more than on regulation of product. Without process certification, there is limited ability to verify that food presented as organic was truly a result of organic production. The premium price for organic products and the multiple hand offs in the system create a motive and opportunity for fraud, so certification helps to ensure fair practices.

Both the US and EU organic programs are fairly comprehensive. Both aim at producing food in an environmentally and animal-friendly fashion, and programs combine a marketing priority with the goal of sustainability. Regarding organic farming regulations, both the USA and the EU are strict about the standards that producers and handlers must follow in order to represent and sell their products as organic. The standards cover the production of processed food and mandate specific agricultural practices for plant production, such as crop rotation, tillage, and cultivation practices that maintain or increase soil organic matter, enhance soil stability and soil biodiversity, and prevent soil compaction and soil erosion.[99] In 2012, the USA and the EU agreed to treat each other's regulatory system of organic certification as equivalent to its own.[100]

2.2.6.2.2 Organic Farming in the USA. By the 1990s, the practice of producing organically grown agricultural products had existed for several decades in the USA and had evolved from a small-scale and localized system into a highly organized and global production and marketing system. These changes motivated the organic industry to establish uniform standards for organic production and marketing. Congress responded to these issues by enacting the Organic Foods Production Act (OFPA).[101] In 2000, the Agricultural Marketing Service, an agency within USDA, published a final rule that implemented OFPA.[102] The combination of OFPA and the final rule created the National Organic Program (NOP). A producer or handler that sells, labels, or represents agricultural products as organic must comply with the NOP. The policy goals of the OFPA are to establish national standards governing marketing as organically produced, assuring consumers that organically produced products meet a consistent standard; and facilitating interstate commerce food that is organically produced.[103]

The organic standards center around reducing synthetic inputs into agriculture and following an organic plan.[104]

2.2.6.2.3 Organic Farming in the EU. The EU defines organic production as "the use of the production method compliant with the rules established in Regulation (EC) 834/2007, at all stages of production, preparation and distribution."[105] The goal is to establish a sustainable management system for agriculture in order to produce foods of a high quality and of a wide variety that "respond to consumers' demand for goods produced by the use of processes that do not harm the environment, human health, plant health or animal health and welfare."[106]

The first EU organic regulation was enacted in 1991. It initially covered plant products, but was amended several times to regulate other issues such as animal protection, animal feed, the use of livestock manure, and prevention of illness.[107] The EU regulation sets minimum standards, but Member States were free to set their own, stricter standards.

2.2.6.2.4 Agrarian Philosophy *Versus* Sustainability. Although organic standards aspire to address environmental issues and create sustainable agricultural practices, whether organic agriculture is sustainable or not is a debated topic.[108,109] Some organic practices and standards are designed to be more sustainable, such as ensuring soil fertility, crop nutrient management, and crop rotation. Nonetheless, organic farming is not synonymous with sustainability, and some critics find aspects of organic methods to be unsustainable.[113,114] For example, it is commonly presumed that organic agriculture causes only minimal pollution.[110] However, intensive organic agriculture can create the same types of problems as seen with intensive conventional methods, such as intensive use of composted manure resulting in nitrate pollution of groundwater.[111]

Organic programs depart from sustainability when the organic ideals arise from a philosophical or aesthetic belief beyond the ability of science to quantify.[111] For instance, the belief that organic pesticides or fertilizers are categorically better than synthetic chemicals is not aligned with scientific understanding. For instance, a meta-analysis of EU organic farming found that "Organic farms tend to have higher soil organic matter content and lower nutrient losses" but "ammonia emissions, nitrogen leaching, and nitrous oxide emissions per product unit were higher from organic systems. Organic systems had lower energy requirements, but higher land use, eutrophication potential, and acidification potential per product unit."[115]

Similarly, both the US and EU organic programs strictly forbid the use of GMOs and irradiation. These prohibitions are a valid societal choice, an expression of biophilia, the innate human love of nature and the living world.[112] Nonetheless, the prohibitions are disconnected from any assessment of sustainability. The use of technologies in a prudent way can improve sustainability. For example, irradiation can help reduce food waste due to pests and spoilage and reduce the number of food-borne illnesses. The use of recombinant bovine somatotropin (rBST) produced from GMOs can result in a significant improvement in the efficiency of milk production and mitigate the environmental impact, including reduced greenhouse gas emissions, fossil fuel, water, and feed use, and cropland area.[113]

The organic prohibition of technology *per se* represents philosophical beauty striking a victory over dismal science.[114] Certainly, society has a choice about which agricultural system and what type of agriculture are desirable. This debate over what the future of agriculture ought to be transcends science. Science cannot with validity tell us what ought to be.

The two major conflicting perspectives of what agriculture ought to be have been described as the "industrial philosophy" *versus* the "agrarian philosophy."[115] The agrarian philosophy frequently focuses on social sustainability. This can be seen in the concerns for "fair trade," local foods, and "humanely produced" animal products.[116] In contrast, the industrial agriculture view focuses on producing at the lowest cost and in a manner that provides sufficient food for society. The industrial agriculture view is illustrated by the infamous Earl Butz statement, "Get big or get out!."[117] This "more is better" paradigm is discordant to the agrarian view. Where the industrial view sees fewer and larger commercial farms as a way to secure economies of scale and lower the costs of food, the agrarian view sees a problem in the loss of small farms.[118]

However, it is essential that we disentangle the agrarian ideology from the examination of the environmental and natural resource impact from the use of technologies. The coupling of the marketing of organic foods with the mantle of sustainability rewards the pecuniary interests of organic producers over the scientific determination of sustainability. This can even be viewed as a form of political capture of sustainability efforts by agrarian fundamentalists.[119]

In addition, this coupling of organic foods with the mantle of sustainability hinders thoughtful analysis of organic standards. In the end, this poses a risk to achieving more sustainable production.[119] Only by distinguishing agricultural regulation's economic and environmental objectives from the social objectives can we hope to discern whether the law is succeeding.[119]

Although science cannot with validity tell us what ought to be, it can provide an objective analysis of our options and make predictions about potential outcomes from the use of different approaches. Science can be an honest broker to help us identify and clarify issues surrounding the choices available about food and agriculture systems.[120] For example, the Clean Water Act requires use of best management practices, and guidelines provide acceptable soil phosphorus upper limits, soil erosion rates, water conservation, tillage practices, and pesticide use. Science can sort through the wide array of practices and determine the effectiveness of each approach.

2.2.7 Sustainable Food Processing

The general environmental standards mentioned previously in the context of sustainable agriculture, whether in the USA or the EU, apply to all sectors, including those involved in food production and processing. However, food processing is not sustainable merely because the agricultural practices followed for the primary ingredients were sustainable. The whole production process must be sustainable.

The food supply system must implement environmentally friendly measures across the food chain, from production to delivery. At the international level, ISO 14001 gives the generic requirements for an environmental management system and allows a company to "identify and control the environmental impact of its activities, products or services," to "improve its environmental performance continually," and to "implement a systematic approach to setting environmental objectives and targets, to achieving these and to demonstrating that they have been achieved."[121] Many companies in the USA and EU follow the ISO 14001 standard. However, countries have also set up their own environmental management systems for food production and processing.

Both the USA and the EU set up guidance (through the EPA in the USA and through programs such as the eco-management and audit scheme in the EU) for companies to follow when developing environmental management systems. These systems include an analysis of the present conditions of the company, an agreement on environment-related goals, identification of the failures of the current system and ways to prevent pollution or minimize waste, and implementation of an environmental management program.

2.2.7.1 *Market Benefits*

The market can reward processors that adopt sustainable methods by providing a premium price. Sustainability can also be internally self-rewarding; for example, the reduction in energy input and waste output can lower the costs of production.[122] Of important sustainability concerns in food processing, at the top is the handling of waste. Sustainable methods also can provide positive business opportunities.[123]

Moreover, the regulation of the *safety* of food processing provides a measure of improvement in sustainability. The prevention of food-borne illness is a form of improved sustainability because it reduces recalls, reduces food waste, and reduces expenses for healthcare and loss of productivity. In addition, science-based systems of food safety preventive control, such as the HACCP system, increase process efficiency. The HACCP system's precision in the prevention of food safety problems also provides greater precision in the control of materials and other inputs, which can reduce losses and waste. Thus, the firms that implement scientific safety controls generally are more profitable.[124]

2.2.7.2 Sustainable Food Production/Processing in the USA

The USA is the largest consumer and producer of processed food products in the world. The US food-processing industry accounts for about 26% of global food processing production.[125] The general environmental regulations mentioned previously in the context of sustainable agriculture are also applicable to the food processing sector. Like agriculture, the primary resources used by the food processing industry are water, raw materials, and energy.

Sustainability in food processing centers around efforts to reduce use of these resources, while also reducing pollution or waste. Manufacturing processes or technologies that reduce pollution, waste, energy use, or other resource use in comparison with the technologies that they replace are sometimes referred to "clean technologies."[125] These include techniques such as advanced wastewater treatment practices, water and wastewater reduction, improved packaging, improved sensors, and process control.

The EPA is the main federal regulatory agency regarding environmental compliance for food processors. The EPA enforces the air pollution regulations, hazardous waste management regulations, hazardous materials management, solid waste management, water pollution control requirements, drinking water regulations, toxic substances, pesticide regulations, environmental clean-up, restoration, and corrective action.

The EPA has also issued a number of different guidance documents to assist food processors with environmental compliance. The EPA defines pollution prevention as "the use of materials, processes or practices that reduce or eliminate the generation of pollutants or waste at the source," encompassing both "source reduction and in-process recycling."[126] The EPA advises companies to identify the environmental aspects and impacts through a number or methods of analysis, such as the HACCP system, process hazard analysis (to identify and evaluate potential impacts associated with unexpected releases of hazardous materials), failure mode and effects analyses (used in the quality field to identify and prioritize potential equipment and process failures and to identify potential corrective actions), process mapping (mapping out the various processes that are the core functions of an organization, indicating inflows and outflows of materials, products, and by-products), lifecycle assessments (to assess the full range of impacts from products, from raw material procurement through product disposal), risk assessments (to assess potential health and environmental risks typically associated with chemical exposure), emission inventories (to quantify emissions of pollutants to the air), pollution prevention or waste minimization audits (to identify opportunities to reduce or eliminate pollution at the source and to identify recycling options), and environmental property assessments (to assess potential environmental liabilities associated with plant or business acquisitions or divestitures).[126]

The EPA has also developed a number of voluntary programs to encourage clean technologies. For example, in 1992, the EPA with the Department of

Energy created Energy Star, a rating system to encourage energy-efficient products to reduce greenhouse gas emissions and to allow consumers to identify them. Over time, the program has expanded in partnership with food processing companies to improve energy efficiency in the food processing business and create financial value by improving energy efficiency.[127]

2.2.7.2.1 Food Packaging Materials. In the late 1980s and early 1990s, environmental concerns grew over food packaging materials. With the rise of these concerns there was a rise in marketing claims related to solid waste, recycling, and the ozone layer. In 1992, the Federal Trade Commission (FTC) adopted voluntary *Guides for the Use of Environmental Marketing Claims* (the "Green Guides").[128] The Green Guides provide guidance on preventing false or misleading use of environmental advertising claims. Four general principles constitute the foundation for environmental claims: qualifications and disclosures should be sufficiently clear and conspicuous to prevent deception; claims should make clear whether they apply to the product, packaging, or just a component of either; claims should not overstate environmental benefits; and comparative claims should be presented in such a way that the basis for comparison is clear. The Green Guides also address claims concerning environmental friendliness, degradability, compostability, recyclability, recycled content, source reduction, refill ability, and ozone friendliness.

2.2.7.3 Sustainable Food Production/Processing in the EU

The EU's general environmental regulations also have an impact on the sustainability of food processing activities. Like the USA, the EU has been considering the idea of waste management in order to avoid environmental impact. In 2005, the Commission proposed a new strategy on the prevention and recycling of waste, where, as in the USA, it lists waste prevention as the main objective and recycling and recovery as the next most important options.[129]

One of the seven priorities cited in the EU Sustainable Development Strategy, for which it set environmental objectives and called for concrete actions, was sustainable consumption and production. In December 2008, the European Council adopted the Sustainable Consumption and Production Action Plan initially proposed by the Commission, which aims at updating the current legislation and enacting new measures to decrease the environmental impact of production mechanisms and consumption trends.[130]

The EU established the European Ecolabel (represented by a flower logo) in 1992, which is a voluntary scheme to encourage businesses to market environmentally friendly products and services. It covers a wide range of products and services ranging from cleaning products, appliances, paper products, textile and home and garden products and lubricants, to services such as tourist accommodation. The Ecolabel does not apply to food products, but it has been proposed by the Commission to extend it to some food products, such as processed foods, fishing, and aquaculture. Council Regulation (EC) 1980/2000 set the criteria that products must meet in order to bear the Ecolabel logo.[131] However, the European Parliament issued a legislative resolution on the Commission proposal and Article 6 makes it clear that "before developing criteria for food and feed products ... the Commission [must] undertake a study ... exploring the feasibility" and "the study should pay particular attention to the impact of any Ecolabel criteria on food and feed products, as well as unprocessed agricultural products that lie within the scope of [the Organic] Regulation (EC) No. 834/2007."[132] In other words, the fear is that an Ecolabel for food products would conflict with the organic logo.

The EU has also created environmental management systems in order for companies to manage their environmental impacts. The Community EcoManagement and Audit Scheme (EMAS) was initially created in 1995 to provide manufacturing companies with a voluntary tool to improve their environmental performance.[133] The standards are close to the requirements of ISO 14001. Many food and drink industry stakeholders implement environmental management systems, either through the EMAS or through ISO 14001. In April 2009, the European Parliament adopted a position in first reading on the 2008 Commission for a regulation on the voluntary participation by organizations in a Community eco-management and audit scheme,[134] where it suggests reducing the system's costs and administration in order to make it more appealing too small- to medium-sized businesses.[135]

Furthermore, the Commission adopted the Environmental Technologies Action Plan (ETAP) in January 2004 to promote innovation, growth, and sustainable development in Europe.[136] The Action Plan aims in part at boosting eco-innovation and the use of environmental technologies to improve the EU's competitiveness and sustainability. The plan encompasses a broad array of actions that the European Commission and national and regional governments should undertake, such as research, environmental technology verification, performance targets, green public procurement, awareness raising and training, supporting eco-technologies in developing countries, and promoting foreign investment. The collaborative efforts of Member States help promote the use of clean technologies, such as water treatment and waste management systems, recycling, wind turbines, solar panels, wave power, and clean transport.

Some stakeholders in the food processing industry also participate in the European Emission Trading Scheme (ETS), the largest multicountry, multisector greenhouse gas emission trading system, which came into force in October 2003.[137] Given the fact that the food processing industry discharges greenhouse gas owing to its use of cooking, refrigeration, freezing, and air compressor systems, some food processing companies are subject to this requirement.

In addition, the Directorate General on Research funds research projects focusing on waste management techniques. For example, the Reducing Food Processing Waste Specific Targeted Research Project (REPRO) gathered food research organizations from Europe, Turkey, and South Africa to find methods that allow the transformation of plant-based organic waste co-products derived from food processing into added-value food, feed, and pharmaceutical products, such as biopolymers, phytochemicals, nutrients, and micronutrients.[138]

Other projects included the SAFEWASTES Project and the PROTECTOR Project. SAFEWASTES was set up to develop innovative biotechnological processes for food wastes to recover valuable compounds. One key goal was the development of organic feed additives to replace in-feed antibiotics. The PROTECTOR Project aimed at developing a technique to recycle calcium- and phosphorus-rich animal bone meal into safe and sterile mineral char, using an advanced high-temperature carbonization and surface-modification device developed by the consortium.

In addition to these governmentally established measures, a new trend towards voluntary sustainability has been emerging within the private sector. Sustainable food production and processing are also regulated through private standards between processors, producers, and retailers, which use the basic principles set by law and expand on them to set stricter production and processing requirements.

2.2.8 Consideration of Trade Agreements

The World Trade Organization (WTO) is the institutional foundation of international trade. The WTO agreements provide the legal ground rules for international commerce.[‖] One of those foundational principles is the Principle of Nondiscrimination in Trade.[139] Among members, imported goods must be treated equally with domestic goods.

Those sustainability requirements in law that apply to imported foods fall under the provisions of international free trade agreements because these new requirements are barriers to trade. Therefore, depending on how new sustainability requirements are written and implemented, they could violate WTO agreements. If the requirements place more restrictive requirements on foreign goods than domestic goods, then the requirements could violate obligations under the WTO, specifically the GATT, the SPS Agreement, and the TBT Agreement. However, additional requirements on foreign producers are permitted if based on sound scientific reasons for health or safety.[140]

[‖]GATT 1947 was established on a provisional basis after World War II in the wake of other new multilateral institutions dedicated to international economic cooperation. Despite its provisional nature, the GATT 1947 remained the only multilateral instrument governing international trade from 1948 until the establishment of the WTO in 1995. Annex 1A of the WTO Agreement contains the GATT 1994, which incorporates by reference (and with a few adjustments) the GATT 1947.

Most sustainability requirements in law fall under the TBT Agreement, which prohibits imported products being treated less favorably than like domestic products.[141] Technical regulations cannot be more trade-restrictive than necessary to fulfill a "legitimate objective." Legitimate objectives are defined "national security requirements; the prevention of deceptive practices; [and] protection of human health or safety, animal or plant life or health, or the environment."[141] Thus, the validity of any sustainability law will hinge on the soundness of the scientific assessments used in writing the rules and implementing the law.

A few sustainability requirements in law may fall under the SPS Agreement, under which a country that adopts a higher level of sanitary or phytosanitary protection must conduct a risk assessment.[142] In the risk assessment, the country must consider the available scientific evidence among other factors.[142] Therefore, the validity of sanitary or phytosanitary requirements will hinge on the soundness of the scientific risk assessments applied. The risk assessment must identify the potential adverse effects and, if any adverse effects exist, the country must evaluate the potential of occurrence of those adverse effects.[143]

2.3 Private Standards

In addition to government regulations, sustainability is promoted through private standards in both the USA and the EU. The reasons for the development of private standards in the EU and the USA are similar. One is the interest in doing what is right for society. Another is consumer demand for food products that are produced in "ethical" ways or in a way that does not adversely affect the environment, animal welfare, or labor conditions. Moreover, globalization of the food supply system encourages the creation of private standards to facilitate global transactions.

Private standards have emerged both to promote and to standardize aspects of sustainability in food and agriculture. Profit is certainly an important driver of the use of these private standards. Another contributing factor is the interest in "doing the right thing." As noted in an FAO report:

> For businesses, making a profit and 'doing the right thing' are not mutually exclusive. Contributing to food sustainability . . . not only helps create a safer and more prosperous world. It also helps secure future commercial success in new markets and build a more profitable business environment. In fact, business simply cannot thrive if society fails.[144]

The quest after this objective has resulted in ambitious endeavors. For instance, the United Nations Global Compact is an initiative that draws together businesses, non-governmental organizations (NGOs), the United Nations, and other international organizations to create a sustainable and inclusive economy.[145]

Some commentators herald the benefits attributed to private sustainability standards.[146,147] On the other hand, others have criticized corporate social responsibility as gloss rather than constructive responsibility.[148] Regardless of which side of the public debate one falls, social responsibility has become an integral part of doing business in the food sector. Social responsibility for sustainability often translates into private standards. Consumers have become interested in non-safety traits of food products, including the impact of food production on the environment (other traits include animal welfare and labor conditions). These concerns are couched in terms of "ethical consumerism" and resonate especially in developed areas, such as the EU and USA.[149]

Ethical consumerism has driven a demand for information that is not necessarily science based. It also drives a demand for information that cannot be determined from the characteristics of the product. Therefore, consumers look to food companies to provide this information and ensure that it is accurate. Private standards help food firms to provide this information about food product, and third-party certification helps to ensure that the information conforms with the standard.[150]

The emergence of food retailers as gatekeepers and as power brokers has spurred the development of private standards. Retailers prefer to minimize price competition and compete based on qualities not related to safety and that are not necessarily science based. Retailer market concentration increases the bargaining power of retailers, which allows them to impose conditions they want on suppliers and processors via private standards.[151]

In addition, food companies have produced sustainability standards to demonstrate performance to consumers and shareholders. For example, Nestlé, a multinational food company with headquarters in Vevey, Switzerland, publishes a policy on environmental sustainability that relies on specific performance indicators, including production tonnage, water withdrawal, energy consumption, water discharge, greenhouse gas emissions, by-products, and waste.[152]

Other examples include Tyson Foods, headquartered in Springdale, Arkansas, USA, the world's largest provider of chicken, beef, and pork products, which publishes its sustainable values in an annual report that focuses on performance.[153] PepsiCo, headquartered in Purchase, New York, USA, a world supplier of snacks, foods, and beverages, produces an annual sustainability report that includes environmental sustainability along with "human sustainability" and "talent sustainability."[154] Danone, based in Paris, France, a worldwide producer of dairy products, bottled water, and other products, publishes an annual sustainability report.[155]

2.3.1 Special Challenges of Private Standards

Although private sustainability standards can be a positive driving force towards the improvement of sustainability aspects of the food system, they also present special challenges to the respective EU and US food systems and to food processors. These challenges derive from the practical concerns that relate to capacity, contract compliance, and the risks of careless standard setting.

2.3.1.1 Capacity Pressure

New standards can be a catalyst for progressive change. Stricter standards can provide a stimulus for investments in supply-chain modernization, provide incentives for the adoption of better safety and quality control practices, and help clarify the appropriate and necessary roles of government in food safety and agriculture health management. Rather than degrading the comparative advantage of developing countries, the compliance process can result in new forms of competitive advantage and contribute to more sustainable and profitable trade over the long term.[156]

However, although private standards may provide a stimulus to improved processing practices and give a competitive advantage to complying processors, they may also be a proportionately greater disadvantage to smaller scale processors. The up-front capital costs for new buildings and facilities, the initial and ongoing costs of third-party certification, higher operating costs, and reduced profit margins present hurdles. Smaller processors, whether in developing countries or in the EU or the USA, may find the scope and content of standards difficult and compliance costs prohibitive.

2.3.1.2 Contract Compliance

Private standards are generally imposed on suppliers by way of contract. The contract terms may place a higher burden of performance than what is imposed by government regulation. Smaller processors may find themselves at a disadvantage due to a weaker bargaining power.

Private standards may act as barriers to trade, but being private rather than government sanctions puts them largely outside the fair-trade agreements (see Section 2.3.1.3). Fairness in contract compliance falls within national commercial law. Both the EU and the USA have modern commercial codes governing the sale of goods. The Uniform Commercial Code (UCC) is the primary body of commercial law for domestic transactions in the USA and has been adopted in all 50 states. For international transactions involving US companies, the UCC is gradually being supplanted by the United Nations Convention on Contracts for the International Sale of Goods (CISG).[157] The CISG has been adopted by more than 50 countries, including most European countries. The contract law of all countries requires that the parties mutually reach agreement and understanding about the essential terms of a contract.[158]

2.3.1.3 Speech and Public Scrutiny

Where sustainability standards are used in marketing, there are legal risks for the food firms, whether in the USA or the EU. Firms must ensure that claims about sustainability are neither false nor misleading. Both US and EU law prohibits false or misleading claims, and the standard can be stringently enforced. Being factually accurate is not sufficient. Statements must not mislead by implication or by omission of facts. One must strive to avoid ambiguity so that no group of ordinary consumers is misled.

In addition to government regulators, private lawsuits from competitors are not uncommon. Law requiring honest dealing in trade is as old as the history of trade.[159] The long history of law requiring honest trade is due in part to the support from leaders among traders, who recognized the need for a level playing field. Firms are also susceptible to public criticism and negative media attention if the processor's performance falls short of the standard.[160] Failure to live up to self-imposed norms may be used to shame publicly and stigmatize the processor.

To stave off these threats, the processor should assess the need for outside help when drafting sustainability standards. Inflated, unrealistic expectations can create liability or vulnerability problems. Using auditors, third-party certifiers, and outside legal counsel can help guide formulation and enforcement and avoid liability and vulnerability. There should be a periodic review of sustainability standards given the changing dynamic of sustainability and changing global food production conditions.[160]

Care should be taken to ensure that company sustainability standards in the business strategy match what realistically can be delivered. A processor's goals for sustainability standards should be weighed against economic considerations. Also, limitations exist in controlling upstream suppliers. With the difficulty in controlling all of the upstream activity, a processor should be cautious about promising sustainable practices that involve activity over which it has little control. Common sense should be applied to ensure that expectations are legally and operationally reasonable and yet also appropriately capture the market value of sustainability.[160]

2.3.2 International Trade Implications of Private Standards

In reaction to strict private standards, some developing countries have turned to the World Trade Organization (WTO) to explore whether relief may be available under international trade agreements.[161] In particular, the SPS Agreement[162] and the TBT Agreement[163] would be pertinent if the standards were unfairly restrictive of trade *and* adopted by national governments.

Generally, government environmental and sustainable standards fall within the scope of the TBT Agreement, which is limited to technical regulations procedures that are not covered by the SPS Agreement, such as worker health, safety, and welfare and waste and pollution management. A basic principle of the TBT Agreement is that technical requirements may not be more trade restrictive than necessary to achieve a legitimate policy objective. Legitimate objectives include the protecting the environment, national security requirements, prevention of deceptive practices, and protection of human health or safety and animal and plant life.

However, private standards may be largely outside the scope of the TBT agreement. TBT Article 4.1 states:

> Members shall ensure that their central government standardizing bodies accept and comply with the Code of Good Practice for the Preparation, Adoption and Application of Standards in Annex 3 to this Agreement (referred to in this Agreement as the 'Code of Good Practice'). They shall take such reasonable measures as may be available to them to ensure that local government and *non-governmental standardizing bodies* within their territories, as well as regional standardizing bodies of which they or one or more bodies within their territories are members, accept and comply with this Code of Good Practice. In addition, Members shall not take measures which have the effect of, directly or indirectly, requiring or encouraging such standardizing bodies to act in a manner inconsistent with the Code of Good Practice. The obligations of Members with respect to compliance of standardizing bodies with the provisions of the Code of Good Practice shall apply irrespective of whether or not a standardizing body has accepted the Code of Good Practice. [*emphasis added*]

Hence the application of the TBT Agreement to private standards hinges on the interpretation of "non-governmental standardizing bodies." The TBT Agreement does not define "non-governmental standardizing bodies," but TBT Annex 1 defines "non-governmental body" as a "body other than a central government body or a local government body, including a non-governmental body which has *legal power to enforce a technical regulation*" [*emphasis added*]. A "technical regulation" is one for which compliance is mandatory.[164] Thus, a non-governmental body is one that has legal power that is derived from a Member country to enforce some aspect of a law. In other words, the non-governmental body is delegated certain government tasks regarding government regulations. Consequently, provided that private standard-setting bodies are not entangled with government enforcement, private standards fall outside the control of the WTO TBT Agreement.

Notwithstanding private sustainability standards being outside the TBT's authority, some developing countries have challenged the legitimacy of private standards through the WTO.[165] Without the WTO as a resolution option, the best, although meager, alternative is to monitor private standards and standard setters for fairness in trade. Given the important role that private standards have in the food sector, both the EU and the USA will continue to have a stake in the fairness and suitability of private sustainability standards.

2.4 Conclusion

Food processing law through history and into the present era has centered on two goals: a requirement for the production of safe and wholesome food and honest presentation. In short, do not harm and do not lie or mislead.

Rarely has the law mandated measures to increase sustainability in food processing.

Historically, sustainability in food processing was limited to ensuring sufficient quantity and safety. This aspect of sustainability has deep roots in human history because shortage and hunger were constant threats through most of history. This quantity focus carried through to the modern era. Other aspects of sustainability were not considered or were left to market forces.

Ensuring a sufficient quantity of food remains an important goal of sustainability; however, today we face the question of how to prioritize a variety of sustainability goals. Sustainability in food processing is a complex mix of a wide range of environmental, resource, economic, and social concerns. Just as important, who will decide how to prioritize? The values in the industrial farm and the agrarian views clash over what matters.

At best, the enshrining the agrarian philosophy as the mantle of sustainability hinders thoughtful analysis of agriculture and food policy. At worst, this coupling poses a risk to achieving more sustainable food production.[166] Only by distinguishing agricultural regulations' economic and environmental objectives from the social objectives can we hope to discern whether the law is succeeding.[166]

Science cannot tell us what ought to be, and is best suited to providing an objective analysis of our options. Sustainable food production is a complex and wide-ranging set of concerns. There is no black and white division of sustainable and unsustainable systems. All types of agricultural and food production systems potentially can contribute to achieving aspects of sustainability. Sustainability objectives should be subjected to the same type of "hard-look" environmental assessment as required by the US National Environmental Policy Act (NEPA).[167] Thus applied, science can serve as an honest broker regarding potential outcomes from different approaches available in our food and agriculture systems.

2.4.1 Food Law in the USA and EU

Throughout this comparison, hypotheses may be drawn from the similarities and differences in procedural and substantive approaches taken by the USA and the EU. Both recognize that the concept of sustainability must be included as a fundamental principle in any new legal, political, economic, or social undertaking. However, the approaches adopted are not always identical owing to differences in historical and cultural backgrounds, political environments, and institutional organizations.

The USA lacks a comprehensive policy that broadly promotes sustainable food and agricultural systems. Sustainability is identified and addressed in various laws and initiatives but in pieces rather than holistically. For instance, the Clean Air Act, the Clean Water Act, and the Federal Insecticide, Fungicide and Rodenticide Act address certain parts of environmental sustainability. The US government funds various research incentives on sustainability, which is supplemented by use law, consumer activism, and public and private standards.

On the other hand, the EU has enacted broad regulations making the Member State governments take measures towards agriculture-related environmental sustainability. The cross-compliance requirement of the CAP is a good example of such measures. The EU has taken a more proactive stance to encourage and sometimes oblige private parties to perform their duties in sustainable ways or in ways that do not adversely affect the environment. The production and processing of food are increasingly regulated to promote sustainability, generally following the guidance given by the international standard ISO 14001.

2.4.2 Private Standards and Actions

In both regions, food regulation is generally reactive, not proactive. Typically, major new food safety laws were enacted following a food scare, such as BSE in the case of the EU and the unsanitary conditions of the meat packing plants at the beginning of the twentieth century in the USA. Initially in the EU, food regulation was an economic measure to ensure free movement of goods within the Community; the regulation of food safety remained with the EU Member States. Nonetheless, EU-wide food safety law with the specific goal of consumer protection came later. Both the USA and the EU set consumer protection as the main food safety regulation objective.

Owing to the reactive tendency in food legislation, food law should not be expected to lead the vanguard in creating food processing sustainability. Thus, the private actions of food companies rise in importance. Private actions include decisions for greater sustainability, where it rewards the company by increasing company efficiency, such as reduction of waste and use of energy. Private standards may be derived from a desire to do good or to demonstrate performance to shareholders. In addition, private standards are a means to improve profitability through "green" marketing. In the end, it is consumer demand for more information and the innovation of food companies that will serve as the driving force for sustainability.

Early food processing was perhaps humankind's greatest achievement because it allowed the development of the human brain. Early food processing may also have been humankind's greatest advancement in sustainability. Once again, food processors have a pivotal role to play in humankind's necessary future movement towards sustainability.

References

1. M. T. Roberts and E. H. Leibovitch, in *Alternatives to Conventional Food Processing*, ed. A. Proctor, Royal Society of Chemistry, Cambridge, 1st edn, 2011, ch. 2, pp. 11–92.

2. *Sustainability in the Food Industry*, ed. C. J. Baldwin, Wiley-Blackwell, 2009.
3. E. O. Wilson, A biologist's manifesto for preserving life on Earth, *Sierra*, 2017, January–February, 28–33.
4. J. Adler, The mind on fire: a million or so years ago, the amazing evolution of the human brain was sparked by a breakthrough that people everywhere still enjoy every day: cooking, *Smithson. Mag.*, 2013, **44**(3), 42.
5. R. R. Sokal, N. L. Oden and C. Wilson, Genetic evidence for the spread of agriculture in europe by demic diffusion, *Nature*, 1991, **351**, 143.
6. R. K. Craig and M. H. Benson, Replacing sustainability, *Akron Law Rev.*, 2013, **46**, 841.
7. E. N. McIntosh, *American Food Habits in Historical Perspective*, Praeger Publishers, Westport, CT, 1995, pp. 129–131.
8. S. Daniells, Food industry dooms children to obesity, says scientist, Food Navigator-usa.com, 14 August 2006.
9. T. Lang, in *Food Policy Old and New*, ed. S. Maxwell and R. Slater, Blackwell, Oxford, 2004, p. 27.
10. R. B. Schlesinger, *et al.*, *Comparative Law*, Foundation Press, New York, 1998, p. 2.
11. Treaty Establishing the European Economic Community, Rome, 25 March 1957.
12. W. S. Collins, The Commission's delegation dilemma: is the European Food Safety Authority an independent or an accountable agency? *Univ. Calif., Davis, J. Int. Law Policy*, 2004, **10**, 279–280.
13. A. J. Valleron, Estimation of epidemic size and incubation time based on age characteristics of vCJD in the United Kingdom, *Science*, 2001, **294**, 1726–1728.
14. Report on Alleged Contraventions of Maladministration in the Implementation of Community Law in Relation to BSE, European Parliament Document (A4-0020/97) 1 (7 Feb 1997).
15. Treaty of Amsterdam, 1997, art. 26–27 amending art. 129 and 129a.
16. Consolidated version of the Treaty Establishing the European Community, art. 152–3, *Off. J. Eur. Commun.*, 2002, L325, 33, 100–101.
17. A. Alemanno, in *What's the Beef? The Contested Governance of European Food Safety*, ed. C. Ansell and D. Vogel, MIT Press, Cambridge, MA, 2006, pp. 240–241.
18. Commission White Paper on Food Safety, p. 3, COM (1999)719 final (12 Jan 2000).
19. Regulation (EC) No. 178/2002.
20. European Commission, Standing committees, available at https://ec.europa.eu/food/plant/standing_committees_en (accessed 5 Mar 2017).
21. Regulation (EC) No. 178/2002, recital 62.
22. Regulation (EC) No. 178/2002, art. 58.
23. W. Harvey Wiley: pioneer consumer activist, FDA Consumer Mag., 2006, January–February.

24. Pure Food and Drug Act, Pub. L. No. 59-384, 34 Stat. 768 (1906), repealed by Federal Food, Drug and Cosmetic Act, Pub. L. No. 75-717, 52 Stat. 1040 (1938) (codified as amended at 21 U.S.C.).
25. Federal Meat Inspection Act of 1906, P.L. 59-252, 34 Stat. 2156 (30 Jun 1906).
26. Food Additives Amendment of 1958, 72 Stat. 1784 (1958) (codified in various sections of 21 U.S.C.).
27. 21 C.F.R. § 108 (Emergency Permit Control); § 113 (Thermally Processed Low-Acid Foods); § 114 (Acidified Foods).
28. U.S. Environmental Protection Agency (hereafter 'EPA'), Reorganization Plan No. 3 of 1970, https://www.epa.gov/history/origins-epa (accessed 5 Mar 2017).
29. J. N. Sofos, ASAS Centennial Paper, Developments and future outlook for posts laughter food safety, *J. Anim. Sci.*, 2009, **87**, 2448.
30. N. D. Fortin, The hang-up with HACCP: the resistance to translating science into food safety law, *Food Drug Law J.*, 2003, **58**, 565.
31. USDA FSIS, Pathogen Reduction: Hazard Analysis and Critical Control Point (HACCP) Systems; Final Rule. Federal Register, 25 Jul 1996, pp. 38806–38989.
32. N. Kalaitzandonakes, Why Does Biotech Regulation Differ So Much Between the EU and the US? *Agbioforum*, 2000, **3**, 75–76.
33. Regulation (EC) No. 178/2002.
34. Federal Food, Drug, and Cosmetic Act, Pub. L. No. 75-717, 52 Stat. 1040 (1938) (codified as amended at 21 U.S.C.) (hereafter 'FDCA').
35. Federal Meat Inspection Act, 21 U.S.C., § 601 et seq.
36. Federal Meat Inspection Act, 21 U.S.C., art. 5(1) & (2).
37. FDCA § 401, 402, and 403.
38. Regulation (EC) No. 178/2002, art. 10.
39. Regulation (EC) No. 178/2002, art. 6(3).
40. Regulation (EC) No. 178/2002, recital (19).
41. Regulation (EC) No. 178/2002, art. 7(1).
42. Regulation (EC) No. 178/2002, art. 7.
43. European Commission, Communication from the Commission on the Precautionary Principle, COM (2000) 1FINAL, 2000.
44. FDCA § 409(c)(3).
45. FDCA § 403(a)(1).
46. United Nations, *Declaration of the United Nations Conference on the Human Environment*, 1972.
47. General Assembly Resolution, 38/161 (1983).
48. United Nations, Our Common Future: Report of the World Commission on Environment and Development, U.N. Doc. A/42/427 (1987), pp. 24–25.
49. United Nations Environment Program, Rio Declaration on Environment and Development, principles 4, 8, 9, 11 (1992).
50. United Nations Framework Convention on Climate Change, 1992, available at http://unfccc.int/resource/docs/convkp/conveng.pdf (accessed 11 Mar 2017).

51. United Nations Framework Convention on Climate Change, Kyoto Protocol, 1997.
52. United Nations, Johannesburg Declaration on Sustainable Development, principle 5, 2002.
53. 42 U.S.C. § 4321.
54. EPA Web site, https://www.epa.gov/aboutepa/our-mission-and-what-we-do (accessed 12 Mar 2017).
55. President's Council on Sustainable Development website, https://clinton2.nara.gov/PCSD/Publications/Snapshot.html# (accessed 12 Mar 2017).
56. President's Council on Sustainable Development, *The Road to Sustainable Development: A Snapshot of Activities in the United States*, Mar 1997.
57. EPA, *2006–2011 EPA Strategic Plan: Charting Our Course*, 2006.
58. EPA, *EPA National Strategy for Agriculture*, 25 Apr 2006.
59. European Union, Treaty of Amsterdam, 1997, art. 1–2.
60. European Union, *Europe's Agenda 2000, Strengthening and Widening the European Union*, available at http://ec.europa.eu/agenda2000/public_en.pdf (last accessed 17 Jul 2009).
61. Commission Regulation (EC) No. 445/2002, art. 29, *Off. J. Eur. Commun.*, 2002, L 74, 1, 9.
62. European Commission, *Commission Communication, a Sustainable Europe for a Better World: a European Union Strategy for Sustainable Development*, COM (2001) 264 final, 15 May 2001.
63. European Commission, *Commission Declaration, Towards a Global Partnership for Sustainable Development*, COM (2002) 82 final, 13 Feb 2002.
64. European Parliament and Council Decision No. 1600/2002/EC, art. 2, *Off. J. Eur. Commun.*, 2002, L 242, 1, 3.
65. European Commission, *Commission Communication, The 2005 Review of the EU Sustainable Development Strategy: Initial Stocktaking and Future Orientations*, COM (2005) 37 final, 9 Feb 2005.
66. European Union, *Review of the EU Sustainable Development Strategy (EU SDS) – Renewed Strategy*, 10917/06, 26 Jun 2006.
67. European Union, Progress on EU Sustainable Development Strategy, Final Report, p. 18.
68. C. B. Connard, Sustaining Agriculture: An Examination of Current Legislation Promoting Sustainable Agriculture as an Alternative to Conventional Farming Practices, *Penn State Environ. Law Rev.*, 2004, **13**, 125.
69. C. B. Connard, Sustaining Agriculture: An Examination of Current Legislation Promoting Sustainable Agriculture as an Alternative to Conventional Farming Practices, *Penn State Environ. Law Rev.*, 2004, **13**, 136.
70. Food, Conservation and Energy Act of 2008, § 2709 (7 U.S.C. § 3103).
71. USDA, Sustainable Agriculture Program, available at https://nifa.usda.gov/program/sustainable-agriculture-program (accessed 5 Feb 2017).

72. USDA, Conservation Innovation Grants Program, https://www.nrcs.usda.gov/wps/portal/nrcs/main/national/programs/financial/csp/ (accessed 13 Mar 2017).
73. USDA, Conservation Innovation Grants Program, available at https://www.nrcs.usda.gov/wps/portal/nrcs/main/national/programs/financial/cig/ (accessed 13 Mar 2017).
74. SARE, Sustainable Agriculture Research and Education Program website, http://www.sare.org (accessed 15 Mar 2017).
75. National Institute of Food and Agriculture, About NIFA, available at https://nifa.usda.gov/about-nifa (accessed 15 Mar 2017).
76. Federal Water Pollution Control Act, 33 USC, §§ 1251–1387.
77. 33 U.S.C. § 1251.
78. 42 U.S.C. §§ 300f–300j-26.
79. Federal Air Pollution and Control Act, 42 U.S.C. §§ 7401–7671q.
80. 2 U.S.C. § 7401(b).
81. 7 U.S.C. §§ 136–136y.
82. California Food and Agriculture Code, § 550–554.
83. Council Regulation (EC) No. 1782/2003, Chapter 1, *Off. J. Eur. Commun.*, 2003, L270, 1, 8–9.
84. Council Regulation (EC) No. 1782/2003, Chapter 1, *Off. J. Eur. Commun.*, 2003, art. 10, p. 9.
85. Council Regulation (EC) No. 1782/2003, Chapter 1, *Off. J. Eur. Commun.*, 2003, Annex III, p. 56.
86. Council Regulation (EC) No. 1782/2003, Chapter 1, *Off. J. Eur. Commun.*, 2003, Annex IV, p. 58.
87. Council Regulation (EC) No. 1782/2003, Chapter 1, *Off. J. Eur. Commun.*, 2003, Annex III, p. 56.
88. Council Regulation (EC) No. 1698/2005, *Off. J. Eur. Commun.*, 2005, L 277, 1.
89. European Commission, *Rural development 2014–2020*, available at https://ec.europa.eu/agriculture/rural-development-2014-2020_en (accessed 19 Mar 2017).
90. Council Regulation (EC) No. 1698/2005, *Off. J. Eur. Commun.*, 2005, L 277, 1.
91. Council Directive 1999/31/EC, *Off. J. Eur. Commun.*, 1999, L 182, 1.
92. Council Directive 2000/76/EC, *Off. J. Eur. Commun.*, 2000, L 332, 91.
93. Council Directive 91/271/EEC, *Off. J. Eur. Commun.*, 1991, L 135, 40.
94. Council Directive 2008/1/EC, *Off. J. Eur. Commun.*, 2008, L 24, 8.
95. Council Directive 2001/42/EC, *Off. J. Eur. Commun.*, 2001, L 197, 30.
96. Council Directive 85/337/EEC, 1985, L 175, 40, amended by Council Directive 97/11/EC and by Council Directive 2003/35/EC.
97. Council Regulation (EC) No. 1907/2006, *Off. J. Eur. Commun.*, 2006, L 396, 1.
98. National Research Council *et al.*, *Toward Sustainable Agricultural Systems in the 21st Century*, Washington, DC, National Academies Press, 2010, at 221.

99. Council Regulation 834/2007, art. 12.
100. Press Release, European Comm'n, European Union and United States Agree to Historic New Partnership on Organic Trade, available at http://europa.eu/rapid/press-release_IP-12-138_en.htm (accessed 12 Feb 2017).
101. Title XXI of the Food, Agriculture, Conservation, and Trade Act of 1990 (Public Law 101–624); codified as the Organic Foods Production Act, 7 U.S.C. §§ 6501–6522.
102. 7 CFR Part 205.
103. OFPA § 2101; 7 U.S.C. § 6501.
104. OFPA § 2105; 7 U.S.C. § 6504.
105. Regulation (EC) No 834/2007, art. 2, *Off. J. Eur. Commun.*, 2007, L 189, 1–4.
106. Regulation (EC) No 834/2007, art. 3, p. 6, amended by Regulation (EC) No 967/2008, *Off. J. Eur. Commun.*, 2008, L 264, 1.
107. Regulation (EEC) No. 2092/91. Later, Council Regulation (EC) 834/2007 of 28 June 2007 repealed Regulation (EEC) 2092/91.
108. N. Bazilchuk, How sustainable is organic food? *ScienceNordic* (18 Oct 2016) available at http://sciencenordic.com/how-sustainable-organic-food (accessed 2 Jan 2017).
109. H. Miller and R. Cornett, Why organic isn't 'sustainable', *Forbes* (19 Nov 2014) http://onforb.es/1wX7XTP.
110. H. L. Tuomisto *et al.*, Does organic farming reduce environmental impacts? – A meta-analysis of European research, *J. Environ. Manag.*, 2012, **112**, 309.
111. J. M. Chen, Food and superfood: organic labeling and the triumph of gay science over dismal and natural science in agricultural policy, *Idaho Law Rev.*, 2012, **48**, 213–224.
112. E. O. Wilson, *Biophilia*, Harvard University Press, 1984.
113. J. L. Capper *et al.*, The environmental impact of recombinant bovine somatotropin (rbST) use in dairy production, *Proc. Natl. Acad. Sci. U. S. A.*, 2008, **105**(28), 9668–9673.
114. Chen *supra* note 118 at 214.
115. P. B. Thompson, *The Agrarian Vision: Sustainability and Environmental Philosophy*, University Press of Kentucky, 2010.
116. National Research Council *et al.*, *Toward Sustainable Agricultural Systems in the 21st Century*, National Academies Press, Washington, DC, 2010, at 60.
117. J. Chen, *Get green or get out: decoupling environmental from economic objectives in agricultural regulation*, Oklahoma Law Rev. **48**, 333–352, 1995 at 335.
118. Thompson, *supra* note 122.
119. Chen, *supra* note 124 at 338.
120. National Research Council, *supra* note 123, at 31.
121. International Organization for Standardization, ISO 14001:2015, available at http://www.iso.org/iso/iso_catalogue/management_stand-%20ards/iso_9000_iso_14000/iso14000 (accessed 19 Feb 2017).

122. D. Esty and A. Winston, *Green to Gold*, John Wiley & Sons, 2009.
123. C. Laszlo and N. Zhexembayeva, *Embedded Sustainability*, Greenleaf Publishing, 2011.
124. M. E. Porter, *The Competitive Advantage of Nations*, Palgrave Macmillan, UK, 1998.
125. U.S.-Asia Environmental Partnership, *Clean Technologies in US Industries: Focus on Food Processing*, http://www.p2pays.org/ref/09/08853.htm#Table%201 (accessed 19 Feb 2017).
126. EPA, *Multimedia Environmental Compliance Guide for Food Processors*, 1999.
127. EPA, Energy Star, www.energystar.gov (accessed 19 Feb 2017).
128. Federal Trade Commission, *Guides for the Use of Environmental Marketing Claims*, 16 C.F.R. § 260 (1992). The guides were revised in 1996 and 1998. 61 Fed. Reg. 53,311 (11 October 1996); 63 Fed. Reg. 24,240 (1 May 1998).
129. European Commission, Commission Communication, *Taking Sustainable Use of Resources Forward: a Thematic Strategy on the Prevention and Recycling of Waste*, COM(2005) 666 final, 21 December 2005, available at http://eur-lex.europa.eu/legal-content/EN/TXT/?uri=celex:52005DC0666 (accessed 19 Feb 2017).
130. Council of the European Union, *Sustainable Consumption and Production and Sustainable Industrial Policy Action Plan: Council Conclusions*, 16914/ 08, 5 December 2008, available at http://register.consilium.europa.eu/pdf/en/08/st16/st16914.en08. pdf (accessed 19 Feb 2017).
131. Council Regulation (EC) No. 1980/2000, *Off. J. Eur. Commun.*, 2000, L 237, 1.
132. European Parliament, *European Parliament Legislative Resolution of 2 April 2009 on the Proposal for a Regulation of the European Parliament and of the Council on a Community Ecolabel Scheme*, COM(2008)0401 – C6- 0279/2008 – 2008/0152(COD).
133. Council Regulation (EEC) No. 1836/93, *Off. J. Eur. Commun.*, 1993, L 168, 1.
134. European Commission, *Commission Proposal for a Regulation on the Voluntary Participation by Organizations in a Community Eco-management and Audit Scheme*, COM (2008) 402/2, 16 July 2008.
135. European Parliament, *Position of the European Parliament Adopted at First Reading on 2 April 2009 with a View to the Adoption of Regulation (EC)No … /2009 of the European Parliament and of the Council on the Voluntary Participation by Organizations in a Community Eco-management and Audit Scheme (EMAS)*, P6_TC1-COD(2008)0154.
136. European Commission, *Commission Communication, Stimulating Technologies for Sustainable Development: an Environmental Technologies Action Plan for the European Union*, COM(2004) 38 final, 28 Jan 2004.
137. Directive 2003/87/EC, *Off. J. Eur. Commun.*, 2003, L 275, 32.
138. Reducing Food Processing Waste Specific Targeted Research Project.

139. General Agreement on Tariffs and Trade, art. III, Oct. 30, 1947, 61 Stat. A-11, 55 U.N.T.S. 194.
140. GATT article XX(b).
141. The Agreement on Technical Barriers to Trade art. 2.2.
142. The Sanitary and Phytosanitary Agreement art. 2.2.
143. World Trade Organization, *EC Measures Concerning Meat and Meat Products (Hormones)* 1–105, 1998.
144. Food and Agriculture Organization (FAO), *Food Sustainability: A Guide to Private Sector Action*, 2008, p. 6.
145. United Nations Global Compact, available at www.unglobalcompact.org (accessed 2 Jan 2017).
146. E. Assadourian, The state of corporate responsibility and the environment, *Georgetown Int. Environ. Law Rev.*, 2006, **18**, 571.
147. S. Deva, Sustainable good governance and corporations: an analysis of asymmetries, *Georgetown Int. Environ. Law Rev.*, 2006, **18**, 707.
148. V. Besmer, The legal character of private codes of conduct: more than just a pseudo-formal gloss on corporate social responsibility, *Hastings Bus. Law J.*, 2006, **2**, 279; H. I. Miller, For Business, 'Socially Responsible' Should Mean Earning a Big Profit, *Investors Business Daily*; and A. Karnani, The Case Against Corporate Social Responsibility, *Wall Street Journal*, 22 Aug 2010.
149. T. Lang and M. Heasman, *Food Wars: The Global Battle for Mouths, Minds and Markets*, Earthscan, London, 2004.
150. M. Hatanaka, C. Bain and L. Busch, Third-party certification in the global agrifood system, *Food Policy*, 2005, **30**, 354.
151. L. Busch, D. Thiagariajan, M Hatanaka, C. Bain, L. Flores and M. Frahm, *Relationship of Third-Party Certification (TPC) to Sanitary/Phytosanitary (SPS) Measures and the International Agri-Food Trade: Final Report*, USAID, Washington, DC, December 2005; and The Next Step in the Ethical Consumerism Revolution, *Datamonitor*, London, February 2008.
152. Nestlé, Environmental Performance Indicators http://www.nestle.com/csv/performance/environmental-indicators (accessed 2 Jan 2017).
153. Tyson, 2015 Sustainability Report, available at http://www.tysonsustainability.com (accessed 2 Jan 2017).
154. PepsiCo, 2015 Sustainability Report, available at http://www.pepsico.com/Purpose/Sustainability-Reporting (accessed 2 Jan 2017).
155. Danone, 2015 Annual Report, available at http://www.danone.com/en/# (accessed 2 Jan 2017).
156. World Bank, *Food Safety and Agricultural Health Standards: Challenges and Opportunities for Developing Country Exports*, World Bank, Washington, DC, 2005, p. xii.
157. UN convention on contracts for the international sale of goods, 1980.
158. R. Schaer, B. Earle and F. Agusti, *International Business Law and Its Environment,* West Publishing, New York, 2002, p. 116.

159. N. D. Fortin, The History of Food Safety Regulation in the United States and Canada, Chapter in, *Ensuring Global Food Safety: Exploring Global Harmonization*, Academic Press/Elsevier, 2009, 32–36.
160. M. T. Roberts, *No Good Deed Goes Unpunished: A Legal Guide to Avoiding the Perils of Publishing Agriculture Sustainability Standards in the Agri-Food Sector*, Venable LLP White Paper, February 2007.
161. WTO, G/SPS/R/43, para. 4, 11 October 2006.
162. SPS Agreement available at https://www.wto.org/english/tratop_e/sps_e/spsagr_e.htm (accessed 14 Feb 2017).
163. TBT Agreement, available at https://www.wto.org/english/tratop_e/tbt_e/tbt_e.htm (accessed 13 Feb 2017).
164. TBT Annex 1.
165. WTO, G/SPS/W/230, 25 September 2008.
166. WTO, G/SPS/W/338.
167. 42 U.S.C. §§ 4321-4370d.

CHAPTER 3

Ohmic Heating of Foods

JAMES G. LYNG,* BRIAN M. McKENNA AND CRISTINA ARROYO

UCD Institute of Food and Health, UCD School of Agriculture and Food Science, University College Dublin, Dublin, Ireland,
Email: b.mckenna@ucd.ie; cristina.arroyocasabona@ucd.ie
*Email: james.lyng@ucd.ie

3.1 Introduction

Most food heating processes would not normally rank highly in any listing of green processes since the amount of energy needed to raise a food through a given temperature range is the same no matter what process is employed. However, the 'greenness' of ohmic heating comes from two sources. First, it is a very rapid process and therefore the time available for heat losses from the product being heated is small and environmental losses are minimised. Second, and more importantly, it is a direct application of electrical energy to the product and consequently the significant energy loss implicit at each energy transformation or exchange step in the process is significantly reduced, if not entirely eliminated. Consequently, it can justifiably be regarded as a green process and within heating processes will rank as one of the greenest.

Although ohmic heating as applied to foods has developed significantly over the past three decades, it has been known for significantly longer than that. In the 1950s and 1960s, significant research was undertaken into the process, largely at the request of electrical utility companies rather than the food industry, but those efforts did not result in a significant industrial response. This is primarily because of problems of electrode design, electrode polarisation and fouling, difficulties in applying the electric current

Green Chemistry Series No. 53
Alternatives to Conventional Food Processing: 2nd Edition
Edited by Andrew Proctor

Published by the Royal Society of Chemistry, www.rsc.org

directly to the food and difficulties with food particles of differing conductivity to the main food matrix.

While the basic physics and mechanisms of ohmic heating are dealt with in more detail in Section 3.2, it is worth noting here that the process involves the direct passage of an electric current through the food and the consequential heating of the food due to its electrical resistance. Simple in concept, it is anything but simple in application, as is apparent from the later sections of this chapter. However, the absence of an intermediate step between the energy source (electricity) and the food such as is common in other processes (*e.g.* using the electricity first to heat a resistive element, which in turn heats air or water with subsequent heat transfer to the food) brings with it the energy savings and efficiencies noted above. Finally, the advances in materials forming the electrical electrodes used to conduct the electrical energy to the food, together with the deeper understanding of the heating process of complex foods due to improved modelling and computational techniques, give this 'green' process a very bright future.

3.2 Basic Principle of Ohmic Heating

3.2.1 The Electrical Circuit

Voltage, current and resistance are the primary characteristics of any electrical circuit. Voltage, measured in volts (V), is the electrical driving force and can be supplied from a variety of sources such as the ac mains supply, a battery or a generator. This driving force causes a flow of electric current measured in amperes (A) and the physical makeup of the circuit (wires, *etc.*) contributes a resistance that opposes the flow and is measured in ohms (Ω). In ohmic heating, this resistance is provided by the food material through which the current is passed. To provide a physical or visual analogy to this abstract process, the concept of water flowing from a tap and garden hose is often used. When the tap is opened fully, there is a good water pressure (electrical pressure or voltage) and a substantial water flow (current flow or amperage) through the hose (wires, electrical components or food). Squeezing the hose (increasing resistance) will decrease the water flow (decrease the electrical current) but will not change water pressure (electrical pressure or voltage). However, if the water is allowed to flow through the hose normally but the tap is closed to the half-way position, the water pressure (electrical pressure or voltage) in the hose will decrease, as will the water flow (current flow or amperage) though no change in the resistance will occur. Another good analogy is that of a shower with an inadequate flow of water. Adding a pump to the system will increase the pressure (voltage) but all the other fixtures remain the same so there is no change in the resistance. However, the flow rate of water from the shower (the current) is greatly increased. For most systems involving flow, there is a fundamental relationship that states that the pressure or driving force is equal to the product of the flow rate and the resistance. For electrical systems, this relationship is known as Ohm's law and is given by the equation

$$V = I \times R \tag{3.1}$$

where V is the voltage (V), I is the current (A) and R is the resistance (Ω). It is also interesting that in the conventional heating of food a similar law exists, where the driving force for heat transfer to the food (temperature difference) equals the product of the flow rate of thermal energy and the resistance to heat flow.

3.2.2 Mechanism of Ohmic Heating

It follows from the above section that for a material to be ohmically heated, it must be physically capable of conducting electricity. For a material to be classified as a conductor, electrical charges must be able to move from one point to another within it to complete an electrical circuit.[1] While we are well used to the concept of metals being the best conductors of electricity (wires, *etc.*) and display metallic conduction due to the relatively free movement of electrons through metallic lattices, even solid foods are vastly different from metals. However, most foods contain high levels of water and dissolved salts and these solutions can conduct electricity through electrolytic conduction.

A description of ion flow in an electrolytic cell connected to a dc source can be found in any basic electrochemistry textbook. When an electrolytic solution *e.g.* NaCl solution is placed in contact with a pair of electrodes to which a dc source is connected, mobile positive ions present in the product (*e.g.* Na^+) migrate towards the negative electrode (cathode) and mobile negative ions (*e.g.* Cl^-) move towards the positive electrode (anode). At the negative electrode (which has an excess of electrons), the positive ions pick up electrons and are reduced whereas at the positive electrode (which has a deficiency of electrons) the negative ions give up electrons and are oxidised. This accumulation of reduced or oxidised ions at the electrodes would cause a concept known as electrode polarisation and the flow of current would stop owing to the increased resistance of these accumulated materials. If the current supply used in ohmic heating was dc, electrode polarisation would rapidly occur. Therefore, to prevent this, ac current is used as the cyclic change (generally low frequency, 50 Hz in Europe and 60 Hz in the USA) in the current direction allows insufficient time for appreciable polarisation to occur as the electrochemical reactions described above are considered undesirable in ohmic heating. Coating of the electrodes with titanium or titanium coated with colloidal deposits of platinum black to facilitate adsorption of the extremely small quantities of electrode reaction products can be an additional preventive measure against polarisation. A work by Samaranayake *et al.*[2] showed that changing the pulse characteristics from the traditional 60 Hz, sine wave pulses to bipolar pulses of 10 or 4 kHz or changing the pulse width and delay times can further reduce electrochemical reactions in stainless-steel, titanium and platinised titanium electrodes and thereby reduce polarisation.

When a food product conducts electricity by electrolytic conduction, the moving ions within it collide with other molecules and these collisions lead to momentum transfer to these molecules, which in turn increases their kinetic energy, thereby heating the product. De Alwis and Fryer[3] stated that the interaction between the local field strength and local electrical conductivity will govern the local heat generation rate:

$$Q = E^2 k = \lambda J^2 \tag{3.2}$$

where Q is the heat generation rate per unit volume (W m^{-3}), E is the electrical field strength (V m^{-1}), k is the electrical conductivity (S m^{-1}), λ is the resistivity (Ω m) and J is the current density (A m^{-2}). The actual heating rate for the substance can then be calculated from the equation

$$\frac{dT}{dt} = \frac{Q}{\rho C} \tag{3.3}$$

where ρ is the density (kg m^{-3}) and C is the specific heat capacity (J kg^{-1} °C^{-1}).[3] The term ρC is often referred to as the volumetric heat capacity (*i.e.* the amount of heat energy that is required to raise the temperature of 1 m^3 of the substance by 1 °C). Eqn (3.3) shows that a high Q does not guarantee a rapid rate of temperature rise, which is also dependent on ρC.

3.2.3 Factors Influencing Heat Generation Rate

As can be seen from eqn (3.2), the rate of heat generation during ohmic heating is influenced by both the electrical field strength E and the electrical conductivity k.

3.2.3.1 Electrical Field Strength

The main method of adjusting the electrical field strength is to change the applied voltage. Additionally, it can be varied by adjusting the gap between the electrodes.

3.2.3.2 Electrical Conductivity

Basic physics dictates that the electrical conductivity of a product determines its suitability for ohmic heating.[4] Although it has been stated that it is theoretically possible to provide any food with enough ohmic power to induce a target temperature rise,[5] this could require the use of increasingly large current densities or increasingly large electrical field strengths for foods where the electrical conductivity values become very large or very small, respectively (eqn (3.2)). However, practical limits to electrical field strength and current density will be dictated by safety, cost and product quality considerations. Piette *et al.*[5] also stated that ohmic heating is only practically possible between a range of electrical conductivity values (0.01–10 $S\,m^{-1}$) and that it works optimally in the range 0.1–5 $S\,m^{-1}$.

3.2.3.3 *Temperature* Versus *Electrical Conductivity*

Parrott[6] stated that, in general, the electrical conductivity of food products increases with increase in temperature and it is believed that this increase is mainly due to increased ionic mobility. This agrees with the work of Shirsat *et al.*,[7] who showed that the conductivity of model and commercial meat batters increased with increase in temperature across the range 15–80 °C. A similar effect of temperature on electrical conductivity was also found by Piette *et al.*[8] This phenomenon is often factored into the design of continuous ohmic heaters for pumpable fluid foods. These heaters have multiple heating sections and the spacing between the electrodes is often increased towards the outlet to ensure that each heating section has the same electrical impedance.

3.3 Electrical Conductivity of Foods

It is understandable that the electrical conductivity of foods will vary significantly with composition, given the expected effect of electrically conducting components within the food matrix. Water content, salt content and content of other ionic and electrolytic components will normally increase the conductivity and consequently the heating rate during ohmic heating. However, apart from the preceding comments, it is impossible to generalize across foodstuffs.

A study of the literature over the past 20 years reveals a wealth of data on conductivity measurement. Fruits and their juices[9–15] and comminuted meats[7,8,10,16–18] have been extensively studied and reported. However, fish (especially surimi) has also seen significant study (*e.g.* by Pongviratchai and Park[19]).

One common trend that has been identified is that the conductivity increased (normally linearly) with increase in temperature,[10–13,19] a phenomenon that is generally attributed to increased ionic mobility induced by changes in tissue structure (*e.g.* reduced viscosity and softening). This would imply that the rate of heating will not remain uniform throughout the heating process. Deviations from linearity are often attributed to phase changes and their impact on current flow *e.g.* boiling (and associated vapour development within the liquid),[20] starch gelatinisation (with the solution changing from an electrolyte pre-gelatinization to a polyelectrolyte post-gelatinisation[21]) or structural changes induced by protein denaturation and fat melting.[22]

For food liquids, it has also been shown that viscosity will affect the electrical conductivity,[9] as will the concentration of the liquid.[12] Specifically, Castro *et al.*[14] showed that electrical conductivity decreased with increase in solids and sugar contents, suggesting that at higher solids content (>20% w/w) and sugar contents of >40 degrees Brix (°Bx) a different design of ohmic heater may be necessary. However, when dealing with salt concentration, the effects of the salt rather than the other components becomes dominant once its concentration exceeds about 1%.[23]

With the emergence of high-frequency power supplies designed to reduce electrolysis, recent work[24] has explored the impact of frequency 50 Hz *versus* 20 kHz on electrical conductivity and showed that an increase in frequency led to a decrease in resistance.

3.4 Physical and Chemical Changes to Foods During Ohmic Heating

3.4.1 Nutritional Effects

The limited literature on the nutritional impact of ohmic heating was reviewed by Ruan *et al.*[25] in 2002. They categorised ohmic heating effects on nutrient losses into thermal destruction and diffusion but also mentioned the possibility of electrolysis at the electrodes, which leads to product contamination. In relation to thermal destruction, Lima *et al.*[26] found no significant difference when they compared the impact of electrical *versus* conventional heating on ascorbic acid (vitamin C) degradation in orange juice.

Ruan *et al.*[25] described how ohmically enhanced diffusion of charged components from solid particles to the surrounding liquid could lead to a loss of nutrients from solid particles to carrier fluid. This could be problematic in situations where the carrier liquid is not consumed in conjunction with the solid particles. However, as shown in vegetables by Mizrahi,[27] the converse is also possible. Mizrahi[27] found that the key factors determining solute loss were surface-to-volume ratio and process time. As ohmic heating allows the processing of larger particles (and hence smaller surface-to-volume ratios) in shorter times than conventional methods, under the appropriate conditions it can actually lead to a reduction in solute losses.

3.4.2 Protein Coagulation/Denaturation

In the area of protein coagulation/denaturation under ohmic heating, much of the work has been in the area of surimi production. Wastewater from this process can contain relatively high levels of protein, which contribute to a high biological oxygen demand (BOD) of the water that requires a long treatment time and large storage capacity when treated by traditional methods. Kanjanapongkul *et al.*[28] described the construction of a laboratory-scale ohmic heating system, which was capable of heating the wastewater to a sufficient temperature (60–70 °C) to coagulate the protein (which could then be removed by centrifugation), thus reducing the BOD of the wastewater.

In general, high molecular weight proteins are more susceptible to heating.[29] Ohmic heating has been used to coagulate fish proteins and partially purify proteolytic enzymes in surimi wash water prior to membrane filtration. Fish muscle is a rich source of proteinases such as cathepsins, which have many industrial applications. Benjakul and co-workers[30,31] found that in a processing sequence that included ultrafiltration and freeze-drying, a mild ohmic heating protocol (55 °C for 3 min at 90 V) was the most efficient step for concentrating the proteinase.

3.5 Non-preserving Ohmic Heating Processes

3.5.1 Parboiling

Ohmic heating has been used for parboiling of rice. As with many foods, rice is cooked as thermal processing induces starch gelatinisation and makes it more palatable and digestible. For many products, the electrical conductivity increases with increase in temperature. However, when the electrical conductivity of native starches was measured across temperature ranges encompassing their gelatinisation temperatures, linear relationships were noted before and after gelatinisation, although a decline was noted in the gelatinisation range.[32] This decline in conductivity in the gelatinisation range was attributed to the swelling of the starch granules and the increase in viscosity. An and King[33] provided a detailed overview of the thermal properties and behaviour of rice starch and rice flours during ohmic heating.

Dried rice can be sold in either raw or parboiled forms.[34] In Europe, the two most popular methods for cooking rice are the excess water method and the optimum water level method. The former method consists of boiling rice in a large volume of water and, once cooking is complete, the excess is simply drained away, whereas in the latter the rice is cooked in a precisely measured volume of water (*e.g.* in an electronically controlled rice cooker), so that all the water has been absorbed once cooking is complete.[35] Cooking in water leads to leaching of water-soluble nutrients, including starch. In the excess water method these nutrients are lost when the cooking liquor is discarded, whereas in the optimum water level method the leached nutrients stick to the surface of the cooked rice as the water becomes absorbed.[36] Both cooking methods involve bringing the rice to boiling point with a subsequent holding time. The exact cooking length varies with individual tastes. Industrially, rice is usually cooked in large vats using the excess water method. One major advantage of industrially cooking rice by ohmic means is the potential for continuous processing and a possible increase in output owing to the faster heating rates achievable with ohmic technology.

Uemura *et al.*[37] compared four rice varieties cooked under high-pressure ohmic conditions to rice cooked in a conventional rice cooker. They found that the ohmic processing enhanced the texture of high-amylase Indica rice. In recent years there have been a number of publications in the area of rice cooking using ohmic technology. Sivashanmugam and Arivazhagan[34] found that the milling performance of ohmically parboiled rice was superior and its cooking qualities were comparable to those of conventionally parboiled rice.

Yang *et al.*[38] reported differences in the microstructure of ohmically *versus* traditionally cooked rice and also found that the ohmically cooked rice had superior flavour to and a greater level of gelatinisation than conventionally cooked rice.

An and King[39] studied the impact of ohmic heating on different rice starch and rice flours and found that ohmic heating made commercial rice starch swell faster and produced different pasting characteristics (*e.g.* one of the white rice flours that they examined had better cooking stability, less retrogradation potential and lower pasting temperature than conventionally heated rice flour). They suggested that starches and flours could be produced by ohmic heating that have different cooking characteristics and stabilities to retrogradation.

Kanjanapongkul[40] reported that the energy consumption during ohmic cooking of rice was about 25% of the total energy consumed by an electric rice cooker, with reductions in cooking times also evident and no fouling following the ohmic process.

3.5.2 Blanching

Blanching has a range of functions, one of the principle ones being as a pretreatment aimed at destroying enzymatic activity (which often leads to discoloration) in fruit and vegetables prior to further processing (*e.g.* canning, dehydration or freezing).[41] It is also used for preshrinking, air removal (*versus* oxidative changes) and microbial load and cooking time reduction and also making products more pliable to facilitate filling.[42] In relation to ohmic heating, the technology itself has been used as a blanching method, but blanching by conventional means has also been employed as a pretreatment prior to ohmic heating. These two applications will be discussed separately.

3.5.2.1 Ohmic Blanching

Oliveira *et al.*[43] reviewed the potential for ohmic blanching and other drying pretreatments on the nutritional and bioactive characteristics of dried vegetables. In relation to the inactivation of enzymes responsible for discoloration, Icier *et al.*[44] compared peroxidase inactivation and instrumental colour of pea purée blanched by conventional (100 °C in a water-bath) and ohmic methods (voltage gradients from 20 to 50 V cm^{-1}). They found that ohmic blanching (50 V cm^{-1}) gave the shortest critical inactivation time (54 s) while also producing the product with the best colour quality.

Sensoy and Sastry[45] assessed the impact of ohmic blanching on the preshrinkage of mushrooms in sodium chloride solutions. They found a greater weight loss than could be attributed to the heat treatment alone, and attributed this to some combination of thermal and non-thermal permeabilisation mechanisms. Ohmic processing was also found to have the advantage of allowing blanching at higher solid : liquid ratios (50%) than conventional methods while also being more rapid and heating in a volumetric fashion.

Salengke and Sastry[46] noted that ohmic pretreatment (either by placing in direct contact with electrodes or by heating in a 0.11% salt solution) of potato slices prior to frying led to a subsequent reduction in oil uptake in the direct-contact pretreated samples only. They attributed this sample behaviour difference between the two ohmic application methods to the uptake of water and the extensive swelling of the potato tissues (which increased sample porosity) in the immersion-treated sample. They concluded that these phenomena increased oil uptake during frying and subsequent cooling and suggested that this ohmic treatment may find application in the development of less fatty fried products.

Ohmic blanching prior to osmotic dehydration has been shown to enhance the drying kinetics of strawberries and is another area of great potential, as shown by the work of Allali *et al.*[47] Over a 4 h osmotic dehydration cycle, the inclusion of a 3 min ohmic heating step at 85 °C increased the strawberry dry matter from 20.3% (in the absence of ohmic heating) to 68% (where the ohmic heating step was included).

In order to avoid undesirable product softening, which is a common side effect in the blanching of large vegetables, size-reducing pretreatments such as slicing and dicing are often used.[27] However, these size-reducing operations increase the product surface area, which in turn can have the adverse effect of increasing solute losses, thereby reducing product quality and yield and increasing BOD in the plant wastewater. Mizrahi[27] compared conventionally blanched (sliced or diced in boiling water) with ohmically blanched beetroot (whole, sliced or diced in saline with an ac voltage). They concluded that by removing the need for dicing and reducing the blanching process time, solute losses could be reduced under ohmic conditions.

Guida *et al.*[48] reported that ohmic blanching of artichoke heads inactivates the relevant target enzymes, retains the colour of the raw product and preserves the textural integrity of the product to a greater extent than conventional heating. Further work on artichoke by-products by Icier[49] also showed comparable enzyme inactivation and a higher retention of vitamin C and total phenolics in ohmically blanched products.

3.5.2.2 Blanching Prior to Ohmic Heating

Conventional blanching pretreatments prior to ohmic heating are generally aimed at reducing conductivity differences between food components, thereby improving the uniformity of heating. Sarang *et al.*[50] studied a food system (chicken chow mein) in which the sauce was more conductive than the solid components; when pretreatments in which the solid components were blanched in highly conductive sauces at 100 °C, this conductivity difference could be reduced and ohmic heating (up to 140 °C) uniformity improved compared with untreated controls. Sensory analysis showed that the blanched product had good quality and overall acceptability, suggesting the practical feasibility of this pretreatment prior to ohmic heating.

Eliot *et al.*[51] studied the effect of blanching (90–95 °C in boiling saline (5% NaCl) for 1, 2 or 4 min) of potato cubes (1.6×1.6 cm) followed by ohmic heating (135 °C for 2 or 4 min) on texture. Blanching increased the electrical conductivity and heating rates of the cubes while reducing quality loss and texture degradation.

3.5.3 Thawing

Defrosting involves increasing the temperature of a product from frozen storage temperatures to just below (*i.e.* tempering) or above (*i.e.* thawing) its freezing point. Two categories of thawing system exist. The first are conduction-based, where heat is supplied to the outer product surfaces by means of air, water, contact with a heated surface or condensing steam (vacuum thawing) and this heat is then transferred from these outer surfaces to the centre of the product by conduction (hence the name conduction-based thawing). The second category is electrical or volumetric systems where heat is generated within the product, by passing an electrical current directly through the product (ohmic) or by converting electrical energy to electromagnetic energy, which subsequently generates heat within the product as in dielectric heating (*i.e.* microwave or radiofrequency heating).

Roberts *et al.*[52] designed a prototype ohmic thawing unit with surface temperature sensing and automated control for thawing two shrimp blocks simultaneously. The substantial amount of wastewater generated with conventional warm water immersion thawing is avoided using ohmic heating, which is also more energy efficient. The design features of the system prevented runaway heating. A subsequent comparison of the sensory, microbial and quality attributes of shrimp thawed by conventional water immersion *versus* ohmic methods[53] revealed no significant differences in sensory or total aerobic microbial counts whereas small shrimp had a higher moisture content when conventionally thawed. Overall, it was concluded that ohmic thawing is effective for shrimp blocks.

Yun *et al.*[54] integrated two electrodes into an immersion vessel with the aim of combining ohmic (0–210 V at frequencies of 60 Hz–60 kHz) with conventional water immersion (10 or 20 °C) for thawing frozen 10 cm^3 chunks of meat. Under these conditions, there was no direct contact between the meat and the electrodes. Thawing time was reduced with increase in voltage, although frequency changes had no significant effect. Samples treated at lower voltages showed reduced drip loss and improved water-holding capacity.

Miao *et al.*[55] compared the thawing rate and gel strength in frozen surimi in saline cubes thawed by conventional and ohmic methods. They observed faster thawing rates and stronger gels in the ohmically thawed product. Homogeneous temperature distributions were obtained when the saline solution concentration was <4% with a 20 V applied voltage and a frequency of 60 Hz. Hong *et al.*[56] compared two ohmic thawing methods: brine immersion and electrode contact, and found more rapid thawing with the electrode contact method whereas the total colour difference was slightly higher with the ohmic immersion thawing. Increasing voltage increased the water-holding capacity and generally tended to decrease levels of thiobarbituric acid-reactive substances (TBARS). Overall the immersion thawing proved more suitable at high voltage levels whereas electrode contact was more suited to low voltage levels. Kim *et al.*[57] ohmically thawed frozen pork patties formulated with 0% and 0.5% added salt. The thawing rate increased with increase in power intensity whereas patties with added salt thawed faster than those with no added salt. Increasing the power intensity tended to increase the TBARS levels and it was suggested that thawing at 50 V induced lipid oxidation in pork patties.

Seyhun *et al.*[58] assessed the impact of product salt content (0.5–1.0%) and process frequency (10–30 kHz) on the tempering of frozen potato purée and noted that increases in both of these factors increased the electrical conductivity and heating rate, thereby reducing tempering time, although more uniform temperature distributions were noted at the lowest salt content and frequency. A more recent study by Liu *et al.*[24] noted that increasing the frequency from 50 Hz to 20 kHz reduced the resistance of frozen tuna and increased the heating rate, which increased the thawing rate. Bozkurt and Icier[59] reported on the impact of beef cut size on ohmic thawing (three voltage gradients) compared with conventional methods and found that as the voltage gradient increased the thawing time decreased but thawing loss was unaffected.

3.5.4 Dough Proofing

In more recent years, ohmic heating has been shown to have potential in the proofing of bread dough.[60] The more rapid elevation of the dough temperature to 35 °C under ohmic heating conditions ($\leq$10 °C min^{-1}), due to the reduction in the lag phase, significantly reduced (53–57% reduction) the time to an expansion ratio of 3.

3.5.5 Dehydration and Evaporation

It is widely accepted that conventional air drying processes consume substantial amounts of energy[61] and processors are being forced to investigate alternative techniques that could help to reduce this energy consumption. Ohmic heating is a technique with potential in this regard as it has been shown to induce tissue disintegration and increase drying rates while avoiding reductions in product quality.[62] The mechanism by which it enhances moisture expression and drying in plant tissue is not fully elucidated. However, it is assumed that a key factor is electroporation under mild thermal conditions (<50 °C). In their paper on ohmic dehydration of potato tissue, Lebovka *et al.*[62] also briefly reviewed emerging techniques such as electrohydrodynamic drying and the application of pulsed electric fields (PEF), which can be used to accelerate dehydration. Electroporation is also believed to occur under PEF conditions although ohmic heating differs from PEF in that the electrical field strengths are lower (<100 *versus* 500–100 V cm^{-1}, respectively) and treatment times longer (*e.g.* 10–30 min *versus* 10^{-4}–10^{-2} s, respectively). The use of ohmic heating in the dehydration of foods is a relatively recent phenomenon, as evidenced by the fact that it was not included in two prominent reviews of drying technology in 1995 and 2001.[63,64] For instance, Simpson *et al.*[65] investigated the potential of ohmic heating in accelerating the dehydration of raspberries. When ohmic heating (variable voltage (<100 V cm^{-1}) with the aim of maintaining a temperature of 45–50 °C) in a 57% glucose solution was compared with control samples in 62% glucose solution (50 °C), processing times were reduced by up to 50% compared with control samples.

Lebovka *et al.*[62] investigated the effect of ohmic pretreatment prior to convective drying of potatoes. The influence of electric field strength (<100 V cm^{-1}), total energy input and ohmic heating temperature on the degree of material damage (which was assumed to be induced by electroporation) was assessed. Material with a high degree of disintegration was obtained at electric field strengths of >70 V cm^{-1} with both electric field strength and total energy input having a considerable influence on drying rate. They concluded that ohmic pretreatments showed promise for enhancing air drying. Salengke and Sastry[66] assessed the impact on seedless red grapes of ohmic pretreatments (14 V cm^{-1} at 30, 50 or 7500 Hz) in 2% citric acid medium to a final temperature of 60 °C. They found that the ohmic pretreatment significantly increased the drying rates of the grapes. Zhong and Lima[67] found ohmic pretreatments (50, 70 or 90 V cm^{-1}) to three endpoint temperatures (45, 60 or 80 °C) of sweet potato significantly increased the vacuum drying rate and led to an overall 24% reduction in drying time. Moreno *et al.*[68] found that ohmic heating in conjunction with vacuum impregnation could be used to enhance the uptake of folic acid into apple slices successfully. Other studies by this group show an ohmically induced enhancement of mass transfer in strawberries[69,70] and pears,[71] which was attributed to enhanced cell rupture induced by ohmic heating. Ohmic heating has also been demonstrated to be a successful predrying technique for tomato paste,[72] reducing the moisture content from 90% to 70% and leading to an overall 80–97% reduction in the drying time compared with conventional hot air drying.

Ohmic heating has also shown potential in the evaporation and concentration of foods (*e.g.* pomegranate juice[73]), raising the solids content to 40% while reducing the evaporation time by about 56%.

3.5.6 Ohmically Assisted Peeling

Peeling is a key unit operation in tomato processing. Lye peeling is a commonly used method but alternative chemical-free peeling methods are preferred. Wongsa-Ngasri and Sastry[74] reported that ohmic heating has potential in the peeling of tomatoes without lye, but also showed potential for this technology in combination with lye for enhancing the rate of lye diffusion and reducing the peeling time.

3.5.7 Ohmically Assisted Extraction of Compounds from Vegetable Tissues

Work by Pereira *et al.*,[75] El Darra *et al.*,[76] and Khuenpet *et al.*[77] examined the relevant ohmic heating parameters and reported that low-energy ohmic heating can be used successfully to extract phytochemical compounds from coloured potato (*Solanum tuberosum* L. var. Vitelotte), polyphenols from red grape pomace and inulin from Jerusalem artichoke tuber powder in a shorter time and more sustainable 'green' fashion than using conventional methods. In a review, Vorobiev and Lebovka[78] provided an overview of pulsed ohmic heating to enhance the extraction of compounds from vegetable tissues.

3.5.8 Ohmically Assisted Vegetable Softening

Farahnaky *et al.*[79] reported that ohmic heating resulted in greater softening rates for cylindrical pieces of the root vegetables carrot, red beetroot and golden carrot and also the final hardness of the samples was significantly lower than those of other samples treated by either conventional or microwave methods.

3.5.9 Ohmically Assisted Hydrodistillation

Ohmic heating has been shown to reduce energy costs and shorten ethanol distillation times during bioethanol production while also offering the potential to stop distillation in seconds (avoiding unwanted distillate).[80] Similar observations were made regarding the potential for ohmic heating in the distillation of essential oils.[81,82]

3.6 Microbial Inactivation During Ohmic Heating

From the published literature, it is generally accepted that microbial inactivation during ohmic heating is mainly due to thermal effects.[83,84] However, two other non-thermal inactivation mechanisms (chemical and mechanical) have also been proposed. Chemical inactivation has been attributed to the formation of free oxygen and hydrogen, hydroxyl and hydroperoxyl radicals and metal ions that cause bacterial death.[85–87] Mechanical inactivation has been ascribed to disruption of the microbial cell membrane leading to leakage of the cell contents[88,89] *via* mild electroporation,[90–92] similar to that observed in microbes exposed to high-voltage pulses (PEF).[93,94] Early investigations of non-thermal inactivation mechanisms proved inconclusive owing to the difficulty in eliminating thermal effects.[15]

Hyung-Yong *et al.*[95] showed that ohmic heating was more effective at killing *Bacillus subtilis* spores than conventional heating and attributed this mainly to a thermal effect, although there was a suggestion that the electrical current also exerts an additional effect on the germination of spores. Table 3.1 summarises some recent research studies on microbial inactivation by ohmic heating in different food matrices and buffers. Some include combined treatments with pressure or essential oils. Methods for microbial validation of ohmic processes have been described in the literature.[6,90,96,97]

3.7 Ohmic Sterilization

It is widely agreed that a commercially sterile product can be defined as a heat-processed food that, under ordinary storage conditions, will neither spoil nor endanger the health of the consumer. From the late 1980s to the mid- to late 1990s there was considerable commercial interest in ohmic heating as an alternative method for heat sterilization. Frampton[107] described APV Barker's early ohmic sterilization systems and Sperber,[108] Rice[109] and Giese[110] gave overviews of early commercial applications. Other scientific reviews from around that time[111–114] discussed the potential for ohmic heating as an alternative method for sterilization.

3.7.1 Technological Challenges in Validating Ohmic Sterilization Procedures

As with all thermal sterilization procedures, an ohmic heating protocol must destroy pathogenic spore formers in the 'coldest spot' in the product. This generally requires validation of a heat process.

3.7.2 Temperature Measurement

Temperature measurement in static ohmic heating systems can be achieved using electrically insulated thermocouples[115] or fibre-optic probes. However, for continuous-flow systems (which would be the most common configuration proposed for sterilization), difficulties arise in foods containing suspended particles as it is not possible to measure non-invasively the temperature within the moving solids.[116] Ruan *et al.*[117] described a FLASH MRI technique for temperature mapping in particles during ohmic heating and noted very different heating patterns within ohmically *versus* conventionally heated particles. However, although this may be a valid technique, such systems are not commonly used in commercial validation procedures. Ruan *et al.*[118] stated that the lack of suitable temperature monitoring techniques (for heat profiling distribution and hot and cold spot location) is a major barrier to commercial uptake of ohmic heating. Other work[119] has shown potential for a combination of microwave and ohmic technologies for heat treating particulates leading to a reduction in temperature differences between solid and liquid phases.

Table 3.1 Recent research studies on the microbial inactivation effects of Ohmic heating.[a]

Microorganism	Medium	*F*	*V*	*T*/°C	Time	Results	Ref.
Escherichia coli O157:H7 and *Salmonella* Typhimurium	Tomato-based salsa (pH 4.16)	60 Hz–20 kHz	12.5 V cm^{-1}	90	0–80 s	The higher the frequency, the faster the inactivation. Undetectable levels of microorganisms after 71 and 82 s at 100 and 60 Hz, respectively	98
Bacillus amyloliquefaciens and *Geobacillus stearothermophilus* spores	0.1% NaCl solution (pH 5.0), green pea purée, carrot purée (pH 5.0) and tomato juice	60 Hz	50 V cm^{-1} + 600 MPa	105	0–30 min	For all the food matrices tested, *B. amyloliquefaciens* was most resistant. Approximately 2.80 and 4.11 $\log_{10}$ cycles of inactivation for *B. amyloliquefaciens* and *G. stearothermophilus* in carrot purée and 2.95 and 4.06 $\log_{10}$ cycles in 0.1% NaCl solution after 10 min of a combined treatment	99
Bacillus coagulans spores	Tomato juice (pH 4.4)	60 Hz–10 kHz	13 V cm^{-1}	95–110	0–30 min	At 95 °C, *D* values of 7.96, 8.81 and 10.1 min for 60 Hz and 10 kHz ohmic and conventional treatments, respectively	100

Table 3.1 (*Continued*)

Microorganism	Medium	*F*	*V*	*T*/°C	Time	Results	Ref.
Mesophilic aerobic and *Enterobacteriaceae*	Chilean blue mussel in 0.03% saline solution	60 Hz	9.15 V cm^{-1}	50–90	4 min	Mesophilic aerobic count was reduced by 1.7 $\log_{10}$ (initial count: 3.8 $\log_{10}$ cfu g^{-1}), while *Enterobacteriaceae* (initial count: 2.5 $\log_{10}$ cfu g^{-1}) were reduced to undetectable levels by OH at 90 °C for 4 min	101
E. coli O157:H7, *S.* Typhimurium and *Listeria monocytogenes*	Orange juice (pH 2.5, 3.5 and 4.5)	20 kHz	16 V cm^{-1}	50–60	0–60 s	Greater reductions at pH 2.5. After 60 s at 60 °C and pH 3.5, 1.84, 3.18 and 1.88 $\log_{10}$ cycles of *E. coli* O157:H7, *S.* Typhimurium and *L. monocytogenes* were inactivated, respectively	102
Total bacteria	Gochujang (fermented red pepper paste)	60 Hz	30 V	100	0–10 min	OH was effective in inactivating vegetative cells as indicated by a 2.5 $\log_{10}$ reduction compared with conduction heating at 100 °C producing a <1 $\log_{10}$ reduction after a 10 min treatment	103

Bacillus cereus spores	Doenjang (thick soy bean paste)	25 kHz	26.7 V cm^{-1}	95–125	30–90 s	OH at 115 °C for 90 s reduced the spore count by 6 $\log_{10}$ but only <2 $\log_{10}$ for thermal-only treatment. SEM revealed shrinkage and a distorted outer spore coat	104
Alicyclobacillus acidoterrestris spores	Orange juice	NR	30–50 V cm^{-1}	70–90	0–30 min	For OH at 30 V cm^{-1}, *D* values at 70, 80 and 90 °C were 58.48, 12.24 and 5.97 min, respectively. *D* values for conventional heat were 83.33, 15.11 and 7.84 min, respectively	105
E. coli O157:H7, *S.* Typhimurium, *L. monocytogenes* and MS-2 bacteriophage	Salsa (pH 3.7) + carvacrol (1.3 mM)	60 Hz	12.1 V cm^{-1}	22–90	0–100 s	OH + carvacrol exhibited a synergistic effect for microbial inactivation. Approximately 5 $\log_{10}$ bacterial reduction after 50 s and 6.2 $\log_{10}$ viral reduction after 100 s of combined treatment	106

[a]*F*, frequency; *V*, voltage, *T*, temperature; OH, ohmic heating; SEM, scanning electron microscopy; NR, not reported.

3.7.3 Modelling of Ohmic Sterilization

If suitable models were available, they could help to avoid some of the difficulties in temperature measurement. Sastry's overview[116] referred to the static models of Fryer and co-workers[4,120] and models for flow systems.[121–123] Vicente *et al.*[124] provided a more detailed review of ohmic modelling. Ruan *et al.*[118] stated that one of the greatest challenges affecting the commercial uptake of ohmic heating is difficulties in modelling or predicting the heating characteristics of the particulates and liquid. Kamonpatana *et al.* performed various modelling and microbial verification studies with electric fields parallel[125] and perpendicular[126] to the product flow with thermal validation showing good agreement between calculated and measured temperatures. Other work accounted for heat loss to the surroundings.[127]

3.7.4 Markers

Kim *et al.*[97] evaluated lethality in food particle suspensions (meatballs in starch solution) undergoing ohmic heating using microbiological (*Bacillus stearothermophilus* spores) and chemical (D-glucose and D-ribose) markers. They found good correlations between chemical marker yield and bacterial survivor population, suggesting that chemical markers can be used to evaluate lethality distribution among particles.

3.7.5 Conductivity Differences

In mixed-phase systems, one factor influencing the rate of ohmic heating of the two phases is their relative conductivities. Wang and Wu[128] evaluated the impact of vacuum infiltration of particulates with isotonic sucrose and salt addition to the carrier fluid in a model system that simulated aseptic canning of apple particulates. They found that it was possible to manipulate the conductivity so that the two phases reached their target temperatures (85 °C) simultaneously.

3.7.6 Solid–Liquid Flow

Another factor governing temperature differentials between the phases is the extent of fluid motion, as this affects the fluid–solid convective heat transfer coefficient.[116] Zitoun and Sastry[129] employed particle tracking velocimetry (a technique that can be used to measure the flow behaviour of solid–liquid mixtures) to study the flow of a mixture during ohmic heating. They found that the relative velocity between the two phases decreased when the solid fraction was increased. As mentioned in Section 3.7.3, Kamonpatana *et al.*[125,126] used a radiofrequency identification technique to estimate residence time distributions in continuous-flow ohmic heaters, providing an estimate of the fastest moving particle, which was subsequently used in a mathematical model.

3.7.7 Commercial Uptake

Parrott,[6] Ruan *et al.*[130] and Vicente *et al.*[124] overviewed the work of a US-based consortium of 25 partners from industry, academia and government who developed products and evaluated the suitability of ohmic heating for commercial application using a 5 kW pilot-scale continuous ohmic heater. The range of products tested included shelf-stable (*i.e.* commercially sterile) low- and high-acid products but also included refrigerated extended shelf-life products. In their review, Ruan *et al.*[130] described how the quality (texture, colour and flavour) of these products was comparable to or exceeded that of traditionally processed frozen, retorted and aseptically processed products. They concluded that the technology is viable and in economic terms had operational costs similar to freezing and retorting of low-acid products. However, although these aspects appear favourable, there still remains the business risk associated with start-up costs and uncertainties in terms of market potential.[130] Tucker[131] stated that ohmic heating is no longer used for ready-meal sterilization in the UK and attributed this to consumer concerns over differences in texture and flavour (which seems to be in contrast with the findings of the aforementioned US consortium) that prevented the technology from taking off for this application. Another explanation for the lack of uptake by industry was given by Ruan *et al.*,[118] who stated that one of the key stumbling blocks has been the greater difficulty in identifying, controlling and validating the critical control points necessary to render a product commercially sterile under ohmic compared with conventional (*e.g.* canning) conditions. This causes problems with regulatory agencies and will not entice the food industry to consider the uptake of ohmic heating until suitable solutions are found.

3.8 Specific Food Products

3.8.1 Meat

Meat is a complex food material consisting of muscle fibres and connective tissue with a certain amount of intramuscular fat and its ohmic heating behaviour will approach some degree of uniformity only in a finely comminuted structure.

The lean component of meat generally has a higher electrical conductivity than fat. Shirsat *et al.*[17] showed that lean pork had a conductivity range of 0.64–0.86 $S\,m^{-1}$ and also, in the same and other papers, showed the electrical conductivity of pork fat to be very low (0.01–0.09 $S\,m^{-1}$). They also presented a graph showing the effect of added fat (0–100%) on the conductivity of a lean meat–fat blend. This showed the poorly conductive nature of fat compared with lean minced muscle. However, the relationship between fat content and conductivity appeared to be asymptotic rather than linear. As the percentage of fat in lean meat was reduced, the conductivity of the lean meat increased until the level of fat (<10%) had a limited effect on the conductivity of the mixture. They also suggested that the addition of fat to lean meat could possibly influence the conductivity of the mixture in two ways: (a) a reduction due to the fat itself and its poor conductivity and (b) coating of lean particles with fat thereby creating a barrier for the passage of electrical current. This research alone is evidence of the complex nature of meat and the difficulty that will be encountered in subjecting it to ohmic heating.

Of course, formulated meat products will also include other non-meat ingredients such as salt (for flavour, water binding and preservation), seasoning (flavour), nitrates and nitrites (colour and antimicrobial), proteins (increase nutritional value and gelation aid), bulking agents such as starch and artificial colourings, all of which will exhibit different conductivities and consequent differences in ohmic heating performance. Lyng *et al.*[16] presented comparative data on many such ingredients. The distribution of any ionic components within a formulation is also important as an uneven distribution can lead to uneven heating.

Shirsat *et al.*[7] conducted further studies on the effects of salt and fat on conductivity and ohmic heating times by treating model meat emulsions with different salt levels and also an emulsion with no added fat. They showed that salt or other electrolytic ingredients were essential for ohmic heating and, for a given emulsion formulation, an increase in salt content increased both electrical conductivity and ohmic heating rate. They also noted that an increase in fat content caused a reduction in conductivity with an increase in ohmic heating time. Piette *et al.*[8] also reported that an increase in salt content (1–2.5%) in pseudo ham resulted in a linear increase in electrical conductivity while an increase in fat content (10–30%) in Bologna sausages led to a linear decrease in electrical conductivity.

As already stated, the structure of entire meat is considerably more complex as the muscle fibre and connective tissue structures are still intact and there will normally be a certain amount of intramuscular fat (marbling) within the tissue. A further complication is that many muscle fibres change direction along the length of the muscle and the potential pathways for the passage of electrical current will be considerably more complex, and non-conductive fat deposits are potentially larger and lead to greater obstacles to current flow. Also, although directionality is not really a major issue in emulsions owing to the homogeneity of the structure, this is not so in entire meats where the relative positioning of electrodes parallel or perpendicular to myofibrillar and connective tissue structure could provide very different pathways for the movement of ions within the product. A review on the effect of ohmic heating on different quality characteristics of meat was presented by Yildiz-Turp *et al.*[132] Some of the recent applications of ohmic heating to meat products are given in Table 3.2.

Table 3.2 Selected recent studies on the application of ohmic heating in meat.[a]

Muscle/meat sample	F/Hz	V/V cm^{-1}	Final T/°C	Results	Ref.
Beef *longissimus dorsi* (cylindrical: ∅ 4.2, L 10)	—	12	72	Cook loss and shear force value of the OH-cooked meat were lower and redness was higher than that of the WB-cooked samples. Proteomics analysis showed that OH-cooked meat had less protein damage than WB-cooked meat	133
Beef *semitendinosus* (cylindrical: ∅ 7.1, L 12)	50	8.3	72 and 95	Two OH cooking protocols *versus* conventional cooking in a steam oven. OH provided a 7–15-fold reduction in cooking time. OH-heated meat had lighter surface-colours than and comparable texture characteristics to conventionally cooked samples	134
Beef–fat blends (2%, 9% and 15% fat) (cylindrical: ∅ 2.5, L 5)	50	20, 30 and 40	70	OH-cooked samples were firmer than those cooked conventionally in an electric grill but yield and fat retention were similar	135
Chicken breast (cylindrical: ∅ 3, L 10)	20 000	10	121	Water content, inosine monophosphate, glutamic acid and sensory tests suggested that the quality of the OH-heated samples was similar to that of the retort-heated samples	136
Turkey crowns (cylindrical: ∅ 7.1, L 12)	50	8.3	72 and 95	Two OH cooking protocols *versus* conventional cooking in a steam oven. OH-heated meat had lighter surface colours than and comparable texture profile analysis attributes to conventionally cooked samples. Lipid oxidation and sulfur flavour compound were higher for conventional than for OH-treated samples	137
Pork *longissimus dorsi* (cylindrical: ∅ 4.2, L 10)	50	10	10–80	Colour parameters, deoxymyoglobin and metmyoglobin were lower for the OH-cooked meat than those obtained by WB cooking. WB-cooked samples had a paler colour and lower water-holding capacity, especially above 40 °C	138
Pork *longissimus dorsi* (rectangular: 3×4×6 cm)	60	2.5 and 3.8	4	Different results on the thawing characteristics and NaCl diffusion of pork depending on the voltage and medium salinity. The NaCl gain during OH could be regarded as an advantage in the meat-salting process, if the individual thawing and salting processes of meat were combined	139
Pork *longissimus dorsi* (cylindrical: ∅ 7.1, L 12)	50	4.16	70 and 95	Two OH cooking protocols *versus* conventional cooking in a steam oven. Although the OH cooking protocols reduced cook loss and yielded products with acceptable eating qualities, sensory evaluation found the overall quality of the conventionally cooked ham to be significantly superior	140

[a] F, frequency; V, voltage; T, temperature; OH: ohmic heating; WB, water-bath; ∅, diameter (cm); L, length (cm).

3.8.2 Fish

Fish has attracted limited research as far as ohmic heating is concerned, although it would be expected that many of the conclusions for meat and meat product heating would show similar outcomes. Some published work are reviews that mention ohmic heating (among other methods) as potential heating methods for fish and fish products.[141] However, Matsubara *et al.*[142] studied ohmic heating as a method for improving the quality of salt-dried salmon, where ohmic heating was added to the traditional salting and drying process. They found that heating at 45 °C for 5 min before salting had a significant lowering effect on texture, as measured by the texture profile analysis characteristics, hardness and cohesiveness. This made the salted and dried meat more fragile. Myotomial blocks in the finished salmon muscle were detached at the connective tissue regions after ohmic heating. This, together with a decrease in the amount of salt-soluble protein with no change in the water-soluble protein, substantially increased the acceptability of the product. Similar effects were found when a number of salmon meat slices were ohmically heated in a carrier solution containing 0.5% NaCl.

Abu Dagga and Kolbe[143] studied and modelled the ohmic heating of surimi paste produced from Pacific whiting. Ohmic heating at a high voltage gradient (1200 $V m^{-1}$) not unexpectedly resulted in a heating rate 11 times higher than that of conventional heating for a cylindrical sample of 19 mm diameter. More importantly, the time during which the centre of the sample remained within the critical enzyme activity range was significantly reduced. In addition, the effects of a voltage gradient across the sample, moisture content, electrode design and heat exchange with the surroundings were also investigated. The heat leakage to the surroundings significantly lowered the heating rate near the sample surface. The same group[144] found that ohmic heating also improved the gel functionality of this surimi. Yongsawatdigul *et al.*[145] investigated the electrical conductivity of the surimi paste. Electrical conductivity increased with increase in temperature and salt content and slightly increased with increase in moisture content while it was well correlated linearly with temperature. The voltage gradient did not affect conductivity whereas variations with voltage gradient observed in surimi containing 3–4% salt were probably caused by electrochemical reactions at the electrode surfaces. Finally, Zhao *et al.*[146] studied the corrosion of stainless-steel electrodes during heating of surimi. As outlined previously, electrode polarization and corrosion are serious inhibitors to the wide-scale application of ohmic heating. It was reported that the electrolytic reaction observed at the electrodes below 500 Hz and 3000 $A m^{-2}$ had disappeared when the frequency was increased to 5000 Hz and higher at a current density of 2300 $A m^{-2}$. Several studies looked at the quality of the surimi paste thus processed. Chai *et al.*[147] studied the texture, colour and microstructure of ohmically heated surimi seafood gels. The gel properties were affected by the type of additive, concentration and cooking method. Starch had a more pronounced effect on gel texture at low concentrations, with potato starch giving an improved gel strength compared with wheat starch. However, potato starch decreased the gel whiteness. Non-fish protein additives gave equal or improved texture to the gels, as did fish protein additives, but there was a deterioration in gel colour. In general, ohmic cooking with its rapid heating gave improved gel strength over conventional methods. A similar exercise was conducted by Park *et al.*[148] on kamaboko (fish protein gel) made from minced Alaska pollack. Samples containing additional NaCl (0–1% w/w) and/or starch (0–10% w/w) were heated to 90 °C with an alternating current (0–20 V) at various frequencies (50 Hz–10 kHz). Breaking strength, colour and structure were examined. The samples, which had heating rates of 7.5 times that of conventional boiled samples, were similar in colour but had breaking strengths 1.6 times that of the boiled samples. The results suggest that water is individually separated in the protein gel matrix, and that the gel has a maximum dielectric constant at a frequency dictated by the gel structure and component distribution. Recently, Lascorz *et al.*[149] reported the potential of ohmic heating for faster and more uniform cooking of shrimps (*Pandalus borealis*) while providing a similar texture, colour and yield compared with the traditional steam cooking process. Thawing of frozen shrimp blocks[52,53] is another application, discussed in Section 3.5.3.

3.8.3 Milk

Several groups have investigated the effects of ohmic heating on milk. Sun *et al.*[150] compared the effects of heating milk using ohmic heating with conventional indirect heating using hot water. In particular, they compared the effects of two identical temperature conditions on microbial activity (viable aerobes and *Streptococcus thermophilus*) and on protein denaturation. Microbial counts and decimal reduction times (*D* values) were significantly lower for ohmic heating, leading to the presumption that there was an electrical effect in addition to the conventional thermal effect. In addition, there was no significant difference in protein denaturation from the two treatments. It was concluded that ohmic heating can be effective for milk pasteurization with no additional protein destruction. Ansari and Datta[151] reached similar conclusions in a review of milk heating using emerging technologies. Working with liquid infant formula, Roux *et al.*[152] demonstrated that ohmic heating is comparable to steam injection for UHT sterilization. The nutritional quality (soluble protein denaturation and vitamin C content) and chemical markers (Maillard products) were comparable for the two techniques.

Of course, ohmic heating of milk is not without its problems. The well-known fouling and deposit formation on conventional plate heat exchangers will also occur on the electrodes in ohmic heating and will lead to a severe increase in electrical resistance and a decrease in current. However, as determined by Bansal and Chen,[153] the electrode surface temperature can be significantly lower than the plate temperature in conventional heating.

However, as soon as fouling starts to occur, the electrode temperature starts to increase and can, over time, become higher than the bulk liquid temperature. Needless to say, this will, in turn, increase the fouling and deposit formation. They examined the heating process in a cylindrical ohmic heater and found that frequency was a critical factor in the development of fouling. At a frequency of 10 kHz, electrode corrosion and fouling were lower than at the 50 Hz European mains supply frequency. More recently, Crattelet *et al.*[154] developed a new fouling sensor based on the hot-wire technique and validated it as an accurate device to monitor and quantify the fouling phenomena online during continuous ohmic heating of a fluid jet.

Pereira *et al.*[155] investigated the effects of ohmic heating on the free fatty acids in goat's milk. The primary objective was to determine the effects of conventional and ohmic heating on the disruption of the milk fat globule membrane. Such disruption can lead to excessive free fatty acids in milk, with its associated production of rancid flavours. They examined the short- and medium-chain fatty acid profile in raw goat's milk that passed through pumps and heating units (but unheated) and was subjected to conventional and ohmic pasteurization. Their primary finding was that these treatments were not responsible for the variation in the measured free fatty acids and that ohmic heating at 72 °C for 15 s did not promote more free fatty acid release than conventional heating and should therefore be a viable process. With a different approach, the same group demonstrated that ohmic heating technology can be used to tailor the denaturation and aggregation behaviour of whey proteins due to the presence of a constant electric field together with the ability to provide very fast heating, thus overcoming heat transfer limitations that naturally occur during conventional thermal treatments.[156]

Soymilk was studied by two groups. Liu and Chang[157] looked at the effects of various heating methods on soymilk viscosity. This is important for particle sedimentation after processing, in sedimentation prevention in soymilk-shake products and in coagulation during tofu making. Water-bath heating, ohmic heating at 1.5 $^\circ C\,s^{-1}$ and steam heating (steam injection) were compared. One- and two-step heating were compared (heating directly to 97 °C and holding for 10 min *versus* heating to a lower temperature of 55–80 °C for 10 min followed by heating to 97 °C with a second 10 min hold). The use of the one- or two-step method had a more significant effect that any of the heating methods (including ohmic). This suggests that ohmic heating could be used as an alternative 'green' heating method for such milk. Lei *et al.*[158] looked at the production of protein–lipid films using ohmic and conventional heating. They found that both the yield and protein incorporation efficiency were improved by ohmic heating, as was the rate of film formation. In addition, the rehydration capacity was increased without any significant loss of whiteness. Therefore, for this special application of soymilk heating, it was concluded that ohmic heating offered considerable advantages.

3.8.4 Fruit and Fruit Juices

In an article entitled 'The fruit juice factory of the future', Worrall[159] included ohmic heating along with a number of other emerging technologies that were considered to have potential in this area. Research in the use of ohmic heating in this context has generally either focused on its use in assisting juice extraction or investigated its potential as an alternative to conventional heat processes (specifically blanching, pasteurization or sterilization). Two recent reviews by Jiménez-Sánchez *et al.*[160,161] include ohmic heating as an alternative to conventional thermal treatments (microbial and enzyme inactivation and yield improvement) in fruit-juice processing. Sections 3.5.5 and 3.5.6 include ohmic heating for the dehydration and assisted peeling of fruits.

3.8.4.1 *Juice Extraction*

A number of studies have shown that ohmic heating helps to maximise the mass transfer properties of fruit and vegetable tissue. This effect can be used to accelerate drying but also has potential as an operation that could enhance juice yields during extraction. Much of the work investigating ohmic technology for juice extraction has focused on vegetable tissues (*e.g.* potato slices,[162] sugar beet,[163] soybeans,[164] sweet potato tissue[67] and beetroot[165]). However, it has been shown[166,167] that ohmic treatment of apple tissue prior to juice extraction significantly improved the extraction yields compared with non-treated apple juice.

3.8.4.2 *Juice Preservation*

To meet with commercial success as an alternative juice preservation technique, ohmic heating must be able to inactivate key enzymes and microorganisms in fruit juices while producing a product economically with at least comparable or preferably superior quality and nutritional attributes. Castro *et al.*[168] compared the inactivation kinetics of selected enzymes (from a variety of sources, including soybeans, fruit and milk) of key importance in a food processing context under conventional and ohmic heating. For enzymes where inactivation is the main objective (*e.g.* lipoxygenase and polyphenoloxidase), they found that ohmic heating is a beneficial alternative to conventional heating as the electrical field reduces their D values compared with conventional heating (which means that different kinetic parameters are required under ohmic heating conditions). In contrast, pectinase, β-galactosidase and alkaline phosphatase gave inactivation comparable to that obtained conventionally. The fact that these enzymes are not inactivated to a greater extent than with conventional heating bodes well for ohmic heating. Similarly, the fact that alkaline phosphatase was inactivated to a comparable extent as in conventional heating means that it may still be used as a time–temperature indicator in milk processing. Leizerson and

Shimoni[169] found that ohmic heating reduced pectin esterase activity in orange juice by 98% but found no significant difference in inactivation effect between conventional and ohmic heat processes. Icier *et al.*[170] also evaluated the impact of ohmic heating conditions on the inactivation kinetics of polyphenoloxidase (an enzyme often used as an indicator of blanching efficiency) in grape juice. Two recent studies tested ohmic heating as an alternative to thermal pasteurization for grapefruit juice and blood orange juice[171] and apricot syrup.[172] The latter study reported that the quality characteristics of the fresh apricots in syrup were not adversely affected by ohmic processing and a high retention of the quality attributes was observed during the storage time (52 weeks).

The impact of ohmic heating on the nutritional quality of fruit juices has not been extensively investigated. Much of the work has focused on orange juice, most likely because of its large market share. Owing to its heat sensitivity, vitamin C is often used as an indicator of heat damage in heat processing. Vikram *et al.*[173] compared the effects of ohmic and other thermal treatment methods (infrared, microwave and conventional) on vitamin C levels in orange juices. They observed the highest degradation during microwave heating (which they attributed to uncontrolled temperature) whereas ohmic heating facilitated better vitamin retention at all temperatures. However, Leizerson and Shimoni[174] found comparable decreases in vitamin C in both ohmic and conventionally heated orange juice. Similarly, Lima *et al.*[26] found no significant effect on vitamin C degradation when ohmic and conventional heating methods were compared.

In terms of flavour, Leizerson and Shimoni[169,174] analysed five representative flavour compounds in orange juice and found higher concentrations during storage in the ohmically heat processed than the conventionally heated juice. The sensory shelf life of the ohmically treated juice (100 days) was nearly double that of conventionally pasteurized juice.

3.8.5 Egg

Liquid egg is a commonly used ingredient in the food industry owing to its high protein content, relatively low cost and also the fact that it provides functionality in terms of texture, mouth-feel and flavour. However, because *Salmonella* spp. are commonly found in egg and egg products,[175] it is mandatory in the USA and other jurisdictions to pasteurise (*e.g.* at 61.1 °C for 3.5 min)[176] these products prior to use. Heat treatments can have a negative impact on both physical and functional properties of egg products,[177] such as protein coagulation, which is often evident as fouling on the heat transfer surface,[178] and can affect the subsequent foaming and emulsifying properties of the egg. Despite the availability of commercial-scale ohmic heating systems for pasteurizing liquid eggs (*e.g.* the USDA-approved Raztec electroheating system, which is a form of ohmic heating in which frequencies in the range 100–450 kHz are used[179]) and the existence of many patents in this area (*e.g.* by Reznik and Knipper,[180] Polny,[181,182] Papetti and Polny,[183] and Knipper and Polny[184,185]), very little has been published in the scientific literature on the use of ohmic heating in this context. Imai *et al.*[186,187] evaluated changes in the heating rate of egg albumin solution during ohmic heating and noted a change in heating rate in 10% egg albumen solutions above the gel formation temperature (75 °C), but at lower albumen concentrations (2%) no gel was formed and a constant heating rate was obtained. In fresh egg white, the change in heating rate did not occur until the temperature reached 90 °C. More recently, Izier and Bozkurt[188] provided information on the rheological behaviour and fluid dynamics of liquid whole egg for the design of continuous ohmic heating systems for its pasteurization. The liquid whole egg exhibited the characteristics of a non-Newtonian shear thinning fluid in the temperature range of 4–60 °C while the flow was predicted as laminar for flow rates in the range 5.6–16.6 $g\,s^{-1}$.

3.8.6 Vegetables

In the area of vegetable processing, a number of studies have evaluated the potential for ohmic heating as a method for blanching vegetable purées and pieces or as an alternative method for heat processing vegetable pieces. Other applications of ohmic heating of vegetables have been described in Sections 3.5.7 and 3.5.8.

3.8.6.1 Ohmic Blanching of Vegetables

Pea purée is an ingredient used in the manufacture of baby foods but requires blanching prior to sterilization in order to inactivate enzymes and improve colour. Icier *et al.*[44] found that ohmic blanching could be used as an alternative method to inactivate peroxidases in pea purée with a much shorter processing time than in conventional methods with voltage gradients above 20 $V\,cm^{-1}$, giving better colour values than conventional water blanching. These enzymes are known to be the most heat-stable enzymes in vegetables and their inactivation is regularly used as an indicator of the efficiency of blanching. In a study using carrot pieces, Lemmens *et al.*[189] compared the effects of different thermal pretreatments (conventional heating, microwave heating and ohmic heating) used for blanching (high-temperature blanching, low-temperature blanching and low-temperature blanching in combination with calcium soaking) on peroxidase, pectin-methylesterase, degree of methoxylation, texture and β-carotene content.

3.8.6.2 Ohmic Heat Processing of Vegetables

In relatively recent years, there has been an increase in demand for ready meals and prepared consumer foods in which particles are suspended in a liquid matrix. Eliot *et al.*[51] evaluated the impact of ohmic sterilization (at 135 °C for 0, 2 or 4 min) on the texture of potato cubes (1.6 × 1.6 cm).

They found that conventional blanching pretreatments (at 90–95 °C for 1, 2 or 4 min) in salted water (5% NaCl) prior to ohmic sterilization increased the electrical conductivity of potato cubes, which in turn increased their ohmic heating rates. Blanching pretreatments also decreased texture degradation during ohmic sterilization with potato cubes that were exposed to blanching pretreatments, showing better residual firmness. They also assessed the impact of ohmic heating on the texture of cauliflower,[190] which is a brittle product that does not resist conventional heat processing methods. They further studied the impact of conventional mild precooking in unsalted water (at 40–70 °C for 15–60 min) and salted water (at 40–95 °C for 5–60 min) prior to batch ohmic heating (holding time 30 s at 135 °C). The results showed that low-temperature precooking prior to ohmic heating increased the firmness of cauliflower and offers a viable method for the sterilization of cauliflower florets. In a subsequent study, Eliot-Godéreaux *et al.*[191] ohmically processed cauliflower florets under continuous-flow conditions. However, limitations on the diameter of the tubing restricted the size of floret that could be treated to 2 cm whereas their earlier work on this topic[190] suggested that larger florets could withstand ohmic heating better. Furthermore, the available power in the continuous ohmic system (10 kW) required preheating to >60 °C prior to ohmic processing to ensure that the target sterilization temperatures were met, which had a damaging effect on the cauliflower during passage from the pump to the ohmic system. Overall, however, although further optimization of the system is required, they found the ohmic heating resulted in a product with an attractive appearance and good firmness.

3.9 Economics of Ohmic Processing

Increasing competition dictates that the profitability of the food industry is very dependent on its ability to innovate, adapt and develop new processing technologies and products that are both economically attractive and sustainable. Ohmic heating technologies offer potential economic and practical benefits to all sectors of the food industry, which could possibly include increased energy efficiencies, increased throughput, and potential for continuous *versus* batch processing. Our group has shown that in the meat industry in particular, such opportunities are attractive.[192]

The basic energy balance equation governing pasteurization is

$$P\psi = \frac{mC(\Delta T)}{t_{\rm p}} = \frac{H}{t_{\rm p}} \tag{3.4}$$

where P = power consumption of cooker (kW), ψ = efficiency of cooker, m = mass of meat (kg), ΔT = temperature rise in meat during cooking (°C), H = energy required to cook the meat (kJ), C is the specific heat capacity (kJ kg^{-1} °C^{-1}) and $t_{\rm p}$ is the duration of the heating process (s). Both conventional and ohmic pasteurization require the same energy input to heat a certain mass of product. However, ohmic heating is generally regarded as being a more efficient power source than conventional methods with energy efficiencies of >90% being reported.[193] Therefore, all things being equal, one can suggest that ohmic heating will require less power input than conventional cooking for a given temperature rise. De Halleux *et al.*[193] reported that ohmic cooking can reduce energy consumption by 82–97% compared with conventional smokehouse cooking while meeting the minimum specifications necessary for pasteurization. In addition, the substantially shorter time taken to reach pasteurization temperatures in meat emulsions demonstrates the potential for improved energy efficiencies and increased throughput.

In addition, Allen *et al.*[194] compared the costs of two ohmic heating systems with those of conventional retorting and heating in a tubular heat exchanger using both low- and high-acid foods. The results showed that the process was potentially profitable and was a viable technology for low-acid foods. This was not the case for some high-acid foods, presumably because of the milder heat treatments required for such foods.

3.10 Ohmic Heater Control Options

3.10.1 Control of Electricity Supply During Ohmic Heating

Ohmic heaters can be controlled using several different strategies:[8] constant voltage, constant current intensity and constant power. These are briefly outlined below.

3.10.1.1 Constant Voltage

If the voltage is kept constant in an ohmic heater, then there will be an increase in current intensity as the temperature and the conductivity of the product rise. This will then manifest itself as an increase in heating rate as the heating cycle progresses. Although this is a simple control strategy, it does not necessarily lead to optimal heating since the requirement to limit the voltage towards the end of the cycle so as to avoid excessive current densities will result in most of the cooking being carried out at below optimal conditions.

3.10.1.2 Constant Current Intensity

Controlling the ohmic heater in this manner means that less voltage is required as cooking progresses. This then results in a decrease in power supply and heating rate with time.

3.10.1.3 Constant Power

Using constant power during ohmic heating should result in an almost constant heating rate. However, there will be a slight increase in the specific heat capacity with increase in temperature and this may lead to a slight decrease in heating rate as the temperature rises. This form of control is more difficult to achieve and requires more sophisticated feedback control because the voltage and current intensity will vary with time.

3.10.2 Control of the Extent of Pasteurization/Cooking

For adequate safety and shelf life, a sufficient number of pasteurization units must be delivered to the ohmically heated product. On the other hand, quality will depend on the extent of heat-induced chemical changes quantified by the cook value. Both pasteurization units and cook values increase cumulatively as the heating and cooling phases progress. However, as previously outlined, the shape of the time–temperature profile during heating will differ substantially between conventional and ohmically heated products. Conventional cooking normally exhibits a sigmoid-shaped temperature profile where there is an initial settling down period followed by a period of almost linear temperature rise and finally a final phase where the rate of rate of temperature increase declines.

In the commercial pasteurization of cooked meats, processors generally heat the slowest heating part of the product in excess of a target temperature for a specified time (*e.g.* ≥70 °C for 2 min or equivalent). Ohmic heating is different because of its volumetric nature and is characterized by an almost linear increase in product temperature during cooking. Of course, the ohmic process must result in a fully pasteurized product. This would be facilitated by the use of semi-continuous systems since they would lend themselves to very precise control of pasteurization by using a continuous feedback system whereby cook value and/or pasteurization units could be sent to a programmable logic controller (PLC), programmed to release only products reaching a target pasteurization unit/cook value. Alternatively, ohmic heating could be used to preheat the product to a target temperature, before releasing the product into a chamber that would hold the product at this temperature for a sufficient time to pasteurize and cook the product.

3.10.3 Packaging for Ohmic Processing

Meat products provide good examples of packaging systems used with ohmic heating. In the commercial production of cooked meats, the majority of products are traditionally packaged in sealed casings or vacuum package bags prior to cooking. Packaging products in this way has the advantage of preventing post-process contamination of the pasteurised product following cooking. However, because ohmic heating is a direct electroheating method, then if a product is to be packaged before processing, conductive regions (which could be placed in contact with a series of electrodes) will be required in the packaging to allow current to flow through the product. Nowadays, with the advances in packaging including the development of conductive polymers, there is the possibility of low-cost in-container ohmic processing systems, which would not have a requirement for aseptic conditions. Such packaging would substantially reduce the initial capital outlay associated with these systems. Jun and Sastry[195] modelled the temperature distribution in chicken noodle soup and black beans reheated within flexible packaging with integrated foil electrodes. An improved version of this packaging, which facilitated more uniform ohmic heating, was described in a subsequent paper.[196] However, further feasibility and development work is needed before this approach could be considered for commercial application. More recently, Kanogchaipramot *et al.*[197] described a polymer film integrated with an electrically conductive film to form a conductive package that was successfully used to contain orange juice during ohmic heating.

3.11 Modelling

Unlike many other aspects of ohmic heating, the modelling of the heating system has developed a substantial literature. Basically, the studies can be divided into the following six sections.

3.11.1 General Heating Theory

Although their work was not specific to ohmic heating alone, Balsa-Canto *et al.*[198] specifically recommended its use for more complex thermal processes such as microwave and ohmic heating. A method was presented for deriving simple (reduced order) models for the rapid design and optimization of thermal processing of foods by using orthogonal decomposition and projecting original partial differential equations from thermal processing models onto a low-dimensional subspace. This was claimed to retain the relevant features of the original system. The resulting models were considered to be as accurate as the widely used finite difference techniques and facilitate faster solution for both simple and more complex geometries and heating systems.

3.11.2 Model Development

Although most work reported in this section contains elements of model development, a few authors have published specific work related to ohmic heating. Jun and Sastry[199] studied a pulsed ohmic heating system and flexible packaging for food reheating and sterilization, primarily for the International Space Shuttle programme. Preliminary tests on a flexible pouch package between metal foil electrodes showed that foods such as chicken noodle soup and black beans could be heated using pulsed ohmic heating. The electrical conductivities of selected samples ranged between 0.01 and 0.03 $S\,cm^{-1}$. A two-dimensional (2D) thermal–electric model was developed using commercial computational fluid dynamics (CFD) software to optimize the design and layout of electrodes to ensure uniform heating of the material. A static system under the assumption of there being no fluid motion in a microgravity environment was implemented. The effect of field overshoot near the electrode edge is expected to be crucial for uniformity of heating. In an earlier exercise, Sastry and Li[200] used mathematical models to aid in the qualitative prediction of some ohmic heating phenomena and to identify some of the problems inherent in the process. They considered the development of models to describe the physical phenomena during ohmic heating, the problems of model verification and the problems with both high and low solids concentrations.

3.11.3 Prediction of Temperature Profiles in Liquid Foods

Several groups have published work on the prediction of temperature during ohmic heating of liquid foods. Some of this work[201] was not specific to temperature prediction during ohmic heating but was aimed at predicting the related ohmic heating effect during PEF treatments. While this is regarded as a non-thermal process, a slight temperature rise does occur due to ohmic heating effects. Computer models were used to simulate and compare output temperature increases in different PEF continuous-flow treatments.

More specific to the ohmic heating of liquid foods is the work of Fu and Hsieh[202] on modelling a 2D heating system using commercial software. They minimized the vertical dimension in their system to approximate the 2D model and minimize convection during the ohmic heating. Heat loss to the environment and physical property values used in the simulation were major factors affecting differences between experimental and simulation results. Their work was extended to include a model rectangular particle. A more recent study[203] modelled the heating of reconstituted potato flakes in a batch ohmic heater and showed slightly cold regions and heat losses due to the electrodes and cell surfaces. Fu and Hsieh extended their modelling into 3D systems with and with particles.[204] This work is reviewed in the next section.

Finally, Uemura *et al.*[205] used finite element analysis to predict the temperature profiles of liquid foods when the liquids were heated in a ring electrode heater. An equation for electrical conduction was used to calculate the potential gradient and assumed that the liquid travelled as a plug across this gradient. The experimental and predicted results agreed well.

3.11.4 Prediction of Temperature Profiles in Liquid Foods Containing Particulates

Not surprisingly, given the nature of most foods, the greatest number of publications fall into this category. One of the earliest modelling exercises[121,122] showed that for constant-voltage simulations, large populations of low electrical conductivity particles results in slow heating of the entire mixture compared with the single phase alone, with particles tending to heat faster than liquid. However, the presence of low-conductivity particles can give rise to the danger of under-processing. Consequently, this first of many studies identified particle conductivity and the relationship between liquid and particle conductivity as critical processing parameters. Obviously, heat transfer between the two parts of the food is also important, with conventional heat transfer coefficients playing a crucial role.

When a food contains particles suspended in a liquid medium, the particle can often be of lower conductivity than the liquid. This can lead to a shadowing effect behind the low-conductivity particle and give rise to temperature inhomogeneities.[206] The authors conducted experiments on three model foods containing Perspex blocks to simulate low-conductivity particles. Model simulations were carried out using and compared with experimental data. The models were solved using the Newton–Raphson method and good predictions were obtained. It appeared from the results that internal convection within the liquid is crucial, with good mixing minimizing the shadow effects. With higher conductivity particles, ohmic heating of the particles can occur and the temperature changes in the particles can influence the temperature of the surrounding liquid.

Salengke and Sastry[207] studied ohmic heating for inactivating pathogenic spore formers. Liquids containing particles that differed significantly in electrical conductivity from the carrier liquid were studied, with two cases where the particle was either less or more conductive than the liquid. When the particle conductivity was lower than that of the liquid, the particle temperature lagged behind that of the carrier liquid. When the conductivity relationship was reversed (particle more conductive), the opposite was true, with the particle temperature being higher than that of the liquid throughout the process. Particle size was also significant, with the possibility of a temperature crossover occurring between the particle and the liquid for large particles.

In another simulation, Hsieh and Fu[204] studied sterilization under ohmic heating conditions (also studied by Zaror *et al.*[208]). In particular, they studied voltage and temperature distributions in particle-containing food liquids. The effects of ohmic heating during aseptic processing of particulate foods were simulated. Heating was shown to be a function of particle location when there was a major difference in conductivity between particle and liquid. Potato was used as a model particle in a sodium sulfate medium. A steep temperature gradient was found in the outer regions of the particles and in the liquid between particles. The spatial density of the particles within the food suspension influenced this. Sastry and Salengke[209] used models involving circuit theory and the solution of Laplace equations to investigate the reliability of modelling heating of liquid foods containing particles. Their models gave identical results when the conductivities of both parts of the suspension were equal. With solids of lower conductivity than the liquid, the circuit theory solution was more conservative than the Laplace equation solution, a prediction verified experimentally. Reversing the conductivity ratio reverses the degree of conservatism of the solution. Li *et al.*[120] conducted a similar study on model comparisons and showed that circuit analogies may not predict the heating rate correctly, especially at low solids fraction. Finally, Uemura *et al.*[210] incorporated conventional heat-transfer regimes (conduction and convection) in addition to resistive heating to explain some inconsistencies in the heating of a traditional particle-containing food (miso).

3.11.5 Modelling the Fouling Behaviour of Ohmic Heaters

As with conventional heating equipment, fouling, especially of the electrodes, has become a major problem in ohmic heating of liquid foods, semi-liquid foods, and liquid food containing particles. Several groups have studied this problem with different foods. Ayadi *et al.*[211] used a model liquid food in a flat, rectangular, ohmic heating cell. They found that the pattern of deposits on the electrodes was influenced by flow patterns in the cell, with most fouling at locations of lowest temperature and non-uniform flow velocity. In studies on another food liquid (milk), Fillaudeau *et al.*[212] used simulation models to predict the temperature profile of the fluid and the wall temperature of the electrode. An empirical correlation was established to estimate the temperature gradient, $T_w - T_b$ (where T_w is the wall temperature and T_b is the product temperature) under clean conditions (without fouling) and was used to define the operating conditions for pure-volume and direct-resistance heating. In a second exercise, the ability of ohmic heating to sterilize milk by ultra-high-temperature treatment was investigated and compared with a plate heat exchanger. In particular, the temperature range most critical to fouling (105–138 °C) was examined. The study showed that an increasing heating effect in any wall deposit formed would cause fouling to increase in an uncontrolled manner. Sulc and Novy[213] also investigated ohmic heating of milk using semi-empirical generalized correlations. The effects of flow rate on fouling were investigated. It was found that fouling layer growth was not reduced by using the higher wall shear stress associated with higher flow rates. However, such higher wall shear stress increased the time during which fouling layer growth was negligible.

3.11.6 Other Factors

Various other factors that affecting ohmic heating have been the subject of individual modelling exercises. Quarini[214] modelled the thermal and hydraulic behaviour of a model ohmic heater to predict the temperature distribution and possible overheating of the food being processed. The model showed that even though heating rates may be uniform, overheating can occur because of significant variations in the temperature distribution. The highest temperatures were found in regions of lowest fluid velocity.

Electrical conductivity is obviously a major factor in the effectiveness of ohmic heating. Palaniappan and Sastry[215] developed models to determine the effective electrical conductivity of liquid–particle mixtures and Wang and Sastry[216,217] modelled the diffusion of salt into a food as a means of increasing conductivity.

References

1. J. E. Brady and G. E. Humiston, Electrochemistry, in *General Chemistry: Principles and Structure*, John Wiley and Sons, New York, 1986, pp. 596–639.
2. C. P. Samaranayake, S. K. Sastry and H. Zhang, *J. Food Sci.*, 2005, **70**, E460–E465.
3. A. A. P. De Alwis and P. J. Fryer, *Chem. Eng. Sci.*, 1990, **45**(6), 1547–1599.
4. A. A. P. De Alwis, K. Halden and P. J. Fryer, *Chem. Eng. Results Dissemination*, 1989, **67**, 159–168.
5. G. Piette, M. Dostie and H. Ramaswamy, *47th International Congress of Meat Science and Technology*, 62–67, 2001.
6. D. L. Parrott, *Food Technol.*, 1992, **46**(12), 68–72.
7. N. Shirsat, J. G. Lyng, N. P. Brunton and B. McKenna, *J. Muscle Foods*, 2004, **15**, 121–137.
8. G. Piette, M. L. Buteau, D. de Halleux, L. Chiu, Y. Raymond, H. S. Ramaswamy and M. Dostie, *J. Food Sci.*, 2004, **69**(2), FEP71–FEP78.
9. S. Pal Singh, P. S. Tarsikka and H. Singh, *J. Food Sci. Technol.*, 2008, **45**(4), 371–372.
10. S. Sarang, S. K. Sastry and L. Knipe, *J. Food Eng.*, 2008, **87**(3), 351–356.
11. F. Icier and C. Ilicali, *Food Res. Int.*, 2005, **38**(10), 1135–1142.
12. F. Icier and C. Ilicali, *Eur. Food Res. Technol.*, 2005, **220**(3–4), 406–414.
13. I. Castro, J. A. Teixeira, S. Salengke, S. K. Sastry and A. A. Vicente, *Innov. Food Sci. Emerging Technol.*, 2004, **5**(1), 27–36.
14. I. Castro, J. A. Teixeira, S. Salengke, S. K. Sastry and A. A. Vincente, *J. Food Process Eng.*, 2003, **26**(1), 17–29.
15. S. Palaniappan and S. K. Sastry, *J. Food Process Eng.*, 1991, **14**(4), 247–260.
16. J. G. Lyng, L. Zhang and N. P. Brunton, *Meat Sci.*, 2005, **69**, 589–602.
17. N. Shirsat, J. G. Lyng, N. P. Brunton and B. McKenna, *Meat Sci.*, 2004, **67**, 507–514.
18. W. Engchuan and W. Jittanit, *Int. J. Food Prop.*, 2013, **16**(8), 1676–1692.
19. P. Pongviratchai and J. W. Park, *J. Food Sci.*, 2007, **72**(9), E503–E507.
20. J. R. Sarkis, G. D. Mercali, I. C. Tessaro and L. D. F. Marczak, *Innov. Food Sci. Emerging Technol.*, 2013, **18**, 145–154.
21. L. L. Wong, Y. W. Xu, Z. H. Lu and L. T. Li, *J. Food Eng.*, 2011, **7**(5), 15.
22. H. Bozkurt and F. Icier, *J. Food Eng.*, 2010, **96**(1), 86–92.
23. M. Marcotte, M. Trigui and H. S. Ramaswamy, *J. Food Process. Preserv.*, 2000, **24**(5), 389–406.
24. L. Liu, Y. Llave, Y. Jin, D. Y. Zheng, M. Fukuoka and N. Sakai, *J. Food Eng.*, 2017, **197**, 68–77.
25. R. Ruan, X. Ye, P. Chen, C. Doona and I. Taub, Ohmic heating, in *The Nutrition Handbook for Food Processors*, ed. C. J. K. Henry and C. Chapman, Woodhead Publishing Ltd., Cambridge, 2002, pp. 241–265.
26. M. Lima, B. F. Heskitt, L. L. Burianek, S. E. Nokes and S. K. Sastry, *J. Food Process. Preserv.*, 1999, **23**(5), 421–434.

27. S. Mizrahi, *J. Food Eng.*, 1996, **29**(2), 153–166.
28. K. Kanjanapongkul, S. Tia, P. Wongsa-Ngasri and T. Yoovidhya, *J. Food Eng.*, 2009, **91**(2), 341–346.
29. C. S. Cheng and J. R. Parrish, *J. Food Sci.*, 1979, **54**, 1116–1119.
30. S. Benjakul, M. T. Morrissey, T. A. Seymour and H. An, *J. Food Biochem.*, 1997, **21**(6), 431–443.
31. S. Benjakul, T. A. Seymour, M. T. Morrissey and H. An, *J. Food Biochem.*, 1998, **22**(1), 1–16.
32. F. D. Li, L. T. Li, Z. Li and E. Tatsumi, *J. Food Eng.*, 2004, **62**(2), 113–120.
33. H. J. An and J. M. King, *J. Food Sci.*, 2007, **72**(1), C84–C88.
34. P. Sivashanmugam and M. Arivazhagan, *Int. J. Food Eng.*, 2008, **4**(1). Article 6.
35. B. O. Juliano and C. M. Perez, *J. Texture Stud.*, 1983, **14**, 235–243.
36. B. O. Juliano, *Rice in Human Nutrition, Food and Agriculture of the United Nations, Series Title FAO Food and Nutrition Series-26*, 1993, ISBN 9251031495.
37. K. Uemura, H. Toyoshima and H. Okadome, *J. Jpn. Soc. Food Sci. Technol.*, 1998, **45**(9), 533–538.
38. M.-D. Yang, X. Chen, Z.-Y. Sun, G. Li and Z.-D. Liu, *Food Sci. Technol.*, 2006, **8**, 74–76.
39. H. J. An and J. M. King, *J. Food Sci.*, 2006, **71**(7), C437–C441.
40. K. Kanjanapongkul, *J. Food Eng.*, 2017, **192**, 1–10.
41. P. Fellows, Food Processing Technology – Principles and Practice, Woodhead Publishing Limited, Cambridge, England, 2000.
42. K. P. Poulson, *Food Technol.*, 1986, 122–129.
43. S. M. Oliveira, T. R. Brandão and C. L. Silva, *Food Eng. Rev.*, 2016, **8**(2), 134–163.
44. F. Icier, H. Yildiz and T. Baysal, *J. Food Eng.*, 2006, **74**(3), 424–429.
45. I. Sensoy and S. K. Sastry, *J. Food Process Eng.*, 2004, **27**(1), 1–15.
46. S. Salengke and S. K. Sastry, *J. Food Process Eng.*, 2007, **30**(1), 1–12.
47. H. Allali, L. Marchal and E. Vorobiev, *Food Bioprocess Technol.*, 2010, **3**(3), 406–414.
48. V. Guida, G. Ferrari, G. Pataro, A. Chambery, A. Di Maro and A. Parente, *LWT – Food Sci. Technol.*, 2013, **53**(2), 569–579.
49. F. Icier, *J. Food Process Eng.*, 2010, **33**(4), 661–683.
50. S. Sarang, S. K. Sastry, J. Gaines, T. C. S. Yang and P. Dunne, *J. Food Sci.*, 2007, **72**(5), E227–E234.
51. S. C. Eliot, A. Goullieux and J. P. Pain, *Sci. Aliments*, 1999, **19**(1), 111–117.
52. J. S. Roberts, M. O. Balaban, R. Zimmerman and D. Luzuriaga, *Comput. Electron. Agric.*, 1998, **19**(2), 211–222.
53. J. S. Roberts, M. O. Balaban and D. A. Luzuriaga, *J. Aquat. Food Product Technol.*, 2002, **11**(2), 3–11.
54. C. G. Yun, D. H. Lee and J. Park, *Korean J. Food Sci. Technol.*, 1998, **30**(4), 842–847.

55. Y. Miao, J. Y. Chen and A. Noguchi, *Food Sci. Technol. Res.*, 2007, **13**(4), 296–300.
56. G. P. Hong, S. G. Min, S. H. Ko, K. B. Shim, E. J. Seo and M. J. Choi, *Korean J. Food Sci. Anim. Resour.*, 2007, **27**(4), 416–423.
57. J. Y. Kim, G. P. Hong, S. H. Park, W. E. L. Spiess and S. G. Min, *Korean J. Food Sci. Anim. Resour.*, 2006, **26**(2), 223–228.
58. N. Seyhun, H. S. Ramaswamy, S. Zhu, G. Sumnu and S. Sahin, *Food Bioprocess Technol.*, 2013, **6**(11), 3200–3205.
59. H. Bozkurt and F. Icier, *J. Food Process Eng.*, 2012, **35**(1), 16–36.
60. T. Gally, O. Rouaud, V. Jury, M. Havet, A. Ogé and A. Le-Bail, *Innov. Food Sci. Emerg. Technol.*, 2017, **39**, 55–62.
61. S. A. Ould Ahmedou, O. Olivier Rouaud and M. Havet, *Food Bioprocess Technol.*, 2009, **2**(3), 240–247.
62. N. I. Lebovka, M. V. Shynkaryk and E. Vorobiev, *Dry. Technol.*, 2006, **24**(5), 601–608.
63. J. S. Cohen and T. C. S. Yang, *Trends Food Sci. Technol.*, 1995, **6**(1), 20–25.
64. H. M. Vega-Mercadoa, M. Góngora-Nietob and G. V. Barbosa-Cánovas, *J. Food Eng.*, 2001, **49**, 271–289.
65. R. R. Simpson, P.M. Jimenez, G. E. Carevic and M. R. Grancelli, *Arch. Latinoam. Nutr.*, 2007, **57**(2), 192–195.
66. S. Salengke and S. K. Sastry, *Dry. Technol.*, 2005, **23**(3), 551–564.
67. T. Zhong and M. Lima, *Bioresour. Technol.*, 2003, **87**(3), 215–220.
68. J. Moreno, C. Espinoza, R. Simpson, G. Petzold, H. Nuñez and M. P. Gianelli, *Innov. Food Sci. Emerg. Technol.*, 2016, **33**, 381–386.
69. J. Moreno, R. Simpson, N. Pizarro, K. Parada, N. Pinilla, J. E. Reyes and S. Almonacid, *J. Food Eng.*, 2012, **110**(2), 310–316.
70. J. Moreno, R. Simpson, A. Baeza, J. Morales, C. Muñoz, S. Sastry and S. Almonacid, *LWT – Food Sci. Technol.*, 2012, **45**(2), 148–154.
71. J. Moreno, R. Simpson, M. Sayas, I. Segura, O. Aldana and S. Almonacid, *J. Food Eng.*, 2011, **104**(4), 621–627.
72. A. Hosainpour, H. Darvishi, F. Nargesi and A. Fadavi, *Food Sci. Technol. Int.*, 2014, **20**(3), 193–204.
73. F. Icier, H. Yildiz, S. Sabanci, M. Cevik and O. F. Cokgezme, *Innov. Food Sci. Emerg. Technol.*, 2017, **39**, 241–246.
74. P. Wongsa-Ngasri and S. K. Sastry, *J. Food Eng.*, 2016, **186**, 10–16.
75. R. N. Pereira, R. M. Rodrigues, Z. Genisheva, H. Oliveira, V. de Freitas, J. A. Teixeira and A. A. Vicente, *LWT – Food Sci. Technol.*, 2016, **74**, 493–503.
76. N. El Darra, N. Grimi, E. Vorobiev, N. Louka and R. Maroun, *Food Bioprocess Technol.*, 2013, **6**(5), 1281–1289.
77. K. Khuenpet, M. Fukuoka, W. Jittanit and S. Sirisansaneeyakul, *J. Food Eng.*, 2017, **194**, 67–78.
78. E. Vorobiev and N. Lebovka, *Food Eng. Rev.*, 2010, **2**(2), 95–108.
79. A. Farahnaky, R. Azizi and M. Gavahian, *J. Food Eng.*, 2012, **113**(2), 275–280.

80. M. Gavahian, A. Farahnaky and S. Sastry, *Food Bioprod. Process.*, 2016, **98**, 44–49.
81. M. S. Damyeh and M. Niakousari, *Innov. Food Sci. Emerging Technol.*, 2016, **33**, 387–396.
82. N. El Darra, N. Grimi, E. Vorobiev, N. Louka and R. Maroun, *Food Bioprocess Technol.*, 2013, **6**(5), 1281–1289.
83. C. H. Bis, S. A. Coombes and P. J. Skudder, in *Process Engineering in the Food Industry*, ed. R. W. Field and J. A. Howell, Elsevier, London, 1989, pp. 17–25.
84. P. J. Skudder, *Asian Food J.*, 1989, **4**(4), 10–11.
85. E. A. Martin, *J. Appl. Phys.*, 1960, **31**, 255–267.
86. S. E. Gilliland and M. L. Speck, *Appl. Microbiol.*, 1967, **15**(5), 1031–1037.
87. S. E. Gilliland and M. L. Speck, *Appl. Microbiol.*, 1967, **15**(5), 1038–1044.
88. K. Shimida and K. Shimahara, *Agric. Biol. Chem.*, 1985, **49**, 3605–3607.
89. S. Palaniappan and S. K. Sastry, *Biotechnol. Bioeng.*, 1992, **39**(2), 225–232.
90. H. Y. Cho, S. K. Sastry and A. E. Yousef, *Biotechnol. Bioeng.*, 1999, **62**(3), 368–372.
91. M. B. Fox, D. C. Esveld, H. Mastwijk and R. M. Boom, *Innov. Food Sci. Emerging Technol.*, 2008, **9**(1), 101–108.
92. I. K. Park and D. H. Kang, *Appl. Environ. Microbiol.*, 2013, **9**(23), 7122–7129.
93. T. Grahl and H. Maerkl, *Appl. Microbiol. Biotechnol.*, 1996, **45**(1–2), 148–157.
94. S. K. Sastry and J. T. Barach, *J. Food Sci. Suppl.*, 2000, **65**, 42–46.
95. C. Hyung-Yong, A. E. Yousef and S. K. Sastry, *Biotechnol. Bioeng.*, 1999, **62**(3), 368–372.
96. G. D. Betts and J. E. Gaze, *Food Pasteurisation Treatments, Technical Manual 27*. Campden Food and Drink Research Association, Chipping, Campden, Gloucestershire, UK, 1992.
97. H. J. Kim, Y. M. Choi, A. P. P. Yang, T. C. S. Yang, I. A. Taub, J. Giles, C. Ditusa, S. Chall and P. Zoltai, *J. Food Process. Preserv.*, 1996, **20**(1), 41–58.
98. S. Y. Lee, S. Ryu and D. H. Kang, *Appl. Environ. Microbiol.*, 2013, **79**(1), 10–17.
99. S. H. Park, V. M. Balasubramaniam, S. K. Sastry and J. Lee, *Innov. Food Sci. Emerg. Technol.*, 2013, **19**, 115–123.
100. R. Somavat, H. M. H. Mohamed and S. K. Sastry, LWT – Food Sci. Technol., 2013, **54**, 194–198.
101. J. M. Bastías, J. Moreno, J. Pia, J. Reyes, R. Quevedo and O. Muñoz, *Innov. Food Sci. Emerging Technol.*, 2015, **30**, 98–102.
102. J. Y. Lee, S. S. Kim and D. H. Kang, *LWT – Food Sci. Technol.*, 2015, **62**, 83–88.
103. W.-I. Cho, J. Y. Yi and M. S. Chung, *Innov. Food Sci. Emerging Technol.*, 2016, **34**, 180–186.
104. J. H. Ryang, N. H. Kim, B. S. Lee, C. T. Kim and M. S. Rhee, *Lett. Appl. Microbiol.*, 2016, **63**(1), 66–73.

105. A. H. Baysal and F. Icier, *J. Food Prot.*, 2010, **73**(2), 299–304.
106. S.-S. Kim and D. H. Kang, *Food Control*, , **73**, 300–305.
107. R. Frampton, *Food Manuf.*, 1988, **6**, 61–63.
108. B. Sperber, *Food Process.*, 1992, **11**, 97–99.
109. J. Rice, *Food Process.*, 1995, **3**, 87–91.
110. J. Giese, *Food Technol.*, 1996, **20**(10), 114–115.
111. S. K. Sastry, Ohmic heating, in *Minimal Processing of Foods and Process Optimisation*, ed. R. P. Singh and F. A. R. Oliveria, CRC Press, London, 1994, pp. 17–32.
112. J. W. Larkin and S. H. Spinak, *Food Technol.*, 1996, **50**(5), 242–245.
113. D. Reznik, *Food Technol.*, 1996, **50**(5), 250–251.
114. P. Zoltai and P. Swearingen, *Food Technol.*, 1996, **50**(5), 263–266.
115. M. Zell, J. G. Lyng, D. J. Morgan and D. A. Cronin, *J. Food Eng.*, 2009, **91**, 56–63.
116. S. K. Sastry, *Novel Food Processing Technologies*, ed. G. V. Barbosa-Cánovas, N. S. Tapia and M. P. Cano, CRC Press, New York, 2005, pp. 491–500.
117. R. Ruan, P. Chen, K. Chang, H.-J. Kim and I. A. Taub, *Food Technol.*, 1999, **64**(6), 1024–1026.
118. R. Ruan, X. Ye, P. Chen, C. J. Doona and T. Yang, *Improving the Thermal Processing of Foods*, ed. P. Richardson, Woodhead Publishing Limited, Cambridge, 2004, pp. 224–252.
119. L. T. Nguyen, W. Choi, S. H. Lee and S. Jun, *J. Food Eng.*, 2013, **116**(1), 65–71.
120. L. Zhang and P. J. Fryer, *Chem. Eng. Sci.*, 1993, **48**(4), 633–642.
121. S. K. Sastry and S. Palaniappan, *J. Food Process Eng.*, 1992, **15**(4), 241–261.
122. S. K. Sastry, *J. Food Process Eng.*, 1992, **15**(4), 263–278.
123. S. Orangi, S. K. Sastry and L. Qiong, *Int. J. Heat Mass Transf.*, 1998, **41**(14), 2211–2220.
124. A. A. Vicente, I. de Castro and J. A. Teixeira, in *Thermal Food Processing*, ed. D. W. Sun, CRC Press, Boca Raton, FL, 2006, pp. 425–468.
125. P. Kamonpatana, H. M. Mohamed, M. Shynkaryk, B. Heskitt, A. E. Yousef and S. K. Sastry, *J. Food Sci.*, 2013, **78**(11), E1721–E1734.
126. P. Kamonpatana, H. M. Mohamed, M. Shynkaryk, B. Heskitt, A. E. Yousef and S. K. Sastry, *J. Food Eng.*, 2013, **118**(3), 312–325.
127. T. Tumpanuvatr and W. Jittanit, *J. Food Eng.*, 2012, **113**(2), 226–233.
128. C.-S. Wang and J. Swi-Bea Wu, *Food Sci. Agric. Chem.*, 1999, **1**(2), 154–161.
129. K. B. Zitoun and S. K. Sastry, *IFT Annual Meeting: Book of Abstracts*, IFT, Chicago, 1996.
130. R. Ruan, X. Ye, P. Chen, C. J. Doona and I. Taub, in *Thermal Technologies in Food Processing*, ed. P. Richardson, Woodhead Publishing Limited, Cambridge, 2000, pp. 241–265.
131. G. Tucker, in *Thermal Food Processing*, ed. D. W. Sun, 363–385, CRC Press, Boca Raton, FL, 2006.

132. G. Yildiz-Turp, I. Y. Sengun, P. Kendirci and F. Icier, *Meat Sci.*, 2013, **93**, 441–448.
133. X. Tian, W. Wu, Q. Yu, M. Hou, F. Jia, X. Li and R. Dai, *Innov. Food Sci. Emerging Technol.*, 2016, **34**, 259–266.
134. M. Zell, J. G. Lyng, D. A. Cronin and D. J. Morgan, *Meat Sci.*, 2010, **86**(2), 258–263.
135. H. Bozkurt and F. Icier, *J. Food Eng.*, 2010, **96**(4), 481–490.
136. R. Ito, M. Fukuoka and N. Hamada-Sato, *Meat Sci.*, 2014, **96**(2), 675–681.
137. M. Zell, J. G. Lyng, D. A. Cronin and D. J. Morgan, *Food Chem.*, 2010, **120**(3), 724–729.
138. Y. Dai, J. Miao, S. Z. Yuan, Y. Liu, X. M. Li and R. T. Dai, *Meat Sci.*, 2013, **93**(4), 898–905.
139. G.-P. Hong, J.-Y. Chun and M. J. Choi, *Korean J. Food Sci. Anim. Resour.*, 2012, **32**(3), 293–300.
140. M. Zell, J. G. Lyng, D. A. Cronin and D. J. Morgan, *Food Bioprocess Technol.*, 2012, **5**(1), 265–272.
141. M. Perez-Mateos and A. J. Borderias, *Aliment., Equipos Tecnol.*, 1997, **16**(1), 79–84.
142. H. Matsubara, J. Tanaka, S. Narita and N. Seki, *Nippon Suisan Gakkaishi*, 2007, **73**(3), 470–477.
143. Y. Abu Dagga and E. Kolbe, *J. Aquat. Food Product Technol.*, 2000, **9**(2), 43–54.
144. J. Yongsawatdigul, J. W. Park, E. Kolbe, Y. Abu Dagga and M. T. Morrissey, *J. Food Sci.*, 1995, **60**(1), 10–14.
145. J. Yongsawatdigul, J. W. Park and E. Kolbe, *J. Food Sci.*, 1995, **60**(5), 922–925.
146. Y. Zhao, E. Kolbe and B. Flugstad, *J. Food Process Eng.*, 1999, **22**, 81–99.
147. P. Chai, P. Pongvirat and W. Jae, *J. Food Qual.*, 2007, **30**(5), 783–796.
148. S. J. Park, D. Kim, K. Uemura and A. Noguchi, *J. Jpn. Soc. Food Sci. Technol.*, 1995, **42**(8), 569–574.
149. D. Lascorz, E. Torella, J. G. Lyng and C. Arroyo, *Innov. Food Sci. Emerging Technol.*, 2016, **37**, 329–335.
150. H. Sun, S. Kawamura, J. Himoto, K. Itoh, T. Wada and T. Kimura, *Food Sci. Technol. Res.*, 2008, **14**(2), 117–123.
151. M. I. A. Ansari and A. K. Datta, *Indian Dairyman*, 2007, **59**(11), 59–65.
152. S. Roux, M. Courel, I. Birlouez-Aragon, F. Municino, M. Massa and J. P. Pain, *J. Food Eng.*, 2016, **179**, 36–43.
153. B. Bansal and X. D. Chen, *Food Bioproducts Process.*, 2006, **84**(C4), 286–291.
154. J. Crattelet, S. Ghnimi, P. Debreyne, I. Zaid, A. Boukabache, D. Esteve, L. Auret and L. Fillaudeau, *J. Food Eng.*, 2013, **119**(1), 72–83.
155. R. N. Pereira, R. C. Martins and A. A. Vicente, *J. Dairy Sci.*, 2008, **91**(8), 2925–2937.
156. R. N. Pereira, R. M. Rodrigues, O. L. Ramos, F. X. Malcata, J. A. Teixeira and A. A. Vicente, *Food Bioprocess Technol.*, 2016, **9**, 576–587.

157. Z.-S. Liu and S. K. C. Chang, *J. Food Process. Preserv.*, 2007, **31**(3), 320–333.
158. L. Lei, H. Zhi, Z. Xiujin, I. Takasuke and L. Zaigui, *J. Food Eng.*, 2007, **82**(3), 292–297.
159. P. Worrall, *Fluessiges Obst.*, 1994, **61**(8), 220–227.
160. C. Jiménez-Sánchez, J. Lozano-Sánchez, A. Segura-Carretero and A. Fernández-Gutiérrez, *Crit. Rev. Food Sci. Nutr.*, 2017, **57**(3), 501–523.
161. C. Jiménez-Sánchez, J. Lozano-Sánchez, A. Segura-Carretero and A. Fernández-Gutiérrez, *Crit. Rev. Food Sci. Nutr.*, 2017, **57**(3), 637–652.
162. K. Wigerstrom, *U. S. Pat.* 3997678, 1976.
163. I. Katrokha, A. Matvienko, I. Vorona, M. Kupchik and V. Zaets, *Sakh. Prom-st*, 1984, **7**, 28–31.
164. J. Kim and Y. Pyun, *Extraction of Soy Milk Using Ohmic Heating. Abstract, 9th Congress of Food Sci. Technol.*, Budapest, Hungary, 1995, P125.
165. M. Lima, B. F. Heskiti and S. K. Sastry, *J. Food Process Eng.*, 2001, **24**, 331–340.
166. M. Lima and S. Sastry, *J. Food Eng.*, 1999, **41**(2), 115–119.
167. W. Wang and S. Sastry, *J. Food Process Eng.*, 2000, **23**, 403–418.
168. I. Castro, B. Macded, J. A. Teixeira and A. A. Vicente, *J. Food Sci.*, 2004, **69**(9), C696–C701.
169. S. Leizerson and E. Shimoni, *J. Agric. Food Chem.*, 2005, **53**(9), 3519–3524.
170. F. Icier, H. Yildiz and T. Baysal, *J. Food Eng.*, 2008, **85**, 410–417.
171. N. Achir, C. Dhuique-Mayer, T. Hadjal, K. Madani, J.-P. Pain and M. Dornier, *Innov. Food Sci. Emerging Technol.*, 2016, **33**, 397–404.
172. G. Pataro, G. Donsi and G. Ferrari, *LWT – Food Sci. Technol.*, 2016, **44**(6), 1546–1554.
173. V. B. Vikram, M. N. Ramesh and S. G. Prapulla, *J. Food Eng.*, 2005, **69**, 31–40.
174. S. Leizerson and E. Shimoni, *J. Agric. Food Chem.*, 2005, **53**(10), 4012–4018.
175. F. E. Cunningham, in *Egg Science and Technology*, ed. W. J. Stadelman and O. J. Cotterill, Avi Publishing Corporation, Westport, CT, 1986, pp. 294–295.
176. M. M. Góngora-Nieto, P. D. Pedrow, B. G. Swanson and G. V. Barbosa-Cánovas, *J. Food Eng.*, 2003, **57**, 209–216.
177. N. Hermawan, G. A. Akdemir Evrendilek, W. R. Dantxer, Q. H. Zhang and E. R. Richter, *J. Food Saf.*, 2004, **24**, 71–85.
178. W. D. Powrie and S. Nakai, *in Food Chemistry*, ed. O. R. Fennema, Marcel Dekker Inc., New York, 1985, pp. 829–856.
179. L. E. McBee, *Poult. Sci.*, 1996, **75**, 1137–1140.
180. D. Reznik and A. Knipper, *UK Pat.*, GB2265538 (A), 1993.
181. T. J. Polny, *US Pat.*, US5571550 (A), 1996.
182. T. J. Polny, *UK Pat.*, GB2301271 (A), 1996.
183. S. T. Papetti and T. J. Polny, *DE Pat.*, DE4413500, 1994.
184. A. J. Knipper and T. J. Polny, *WO Pat.*, WO9638045 (A1), 1996.
185. A. J. Knipper and T. J. Polny, *US Pat.*, US5741539 (A), 1998.

186. T. Imai, K. Uemura, S. Yoshizaki and A. Noguchi, *J. Jpn. Soc. Food Sci. Technol.*, 1996, **43**(12), 1249–1255.
187. T. Imai, K. Uemura and A. Noguchi, *Adv. Exp. Med. Biol.*, 1998, **434**, 101–108.
188. F. Icier and H. Bozkurt, *Food Bioprocess Technol.*, 2011, **4**(7), 1253–1263.
189. L. Lemmens, E. Tibäck, C. Svelander, C. Smout, L. Ahrné, M. Langton, M. Alminger, A. Van Loey and M. Hendrickx, *Innov. Food Sci. Emerg. Technol.*, 2009, **10**(4), 522–529.
190. S. C. Eliot, A. Goullieux and J. P. Pain, *J. Sci. Food Agric.*, 1999, **79**(11), 1406–1412.
191. S. C. Eliot-Godéreaux, F. Zuber and A. Goullieux, *Innov. Food Sci. Emerging Technol.*, 2001, **2**(4), 279–287.
192. J. G. Lyng and B. M. McKenna, in *Handbook of Farm, Dairy and Food Machinery*, ed. M. Kutz, William Andrew Inc., Norwich, NY, USA, 2007, 553–577.
193. D. De Halleux, G. Piette, M.-L. Buteau and M. Dostie, *Can. Biosyst. Eng.*, 2005, **47**, 341–347.
194. K. Allen, V. Eidman and J. Kinsey, *Food Technol.*, 1996, **50**(5), 269–273.
195. S. Jun and S. Sastry, *J. Food Process Eng.*, 2005, **28**, 417–436.
196. R. Somavat, P. Kamonpatana, H. M. Mohamed and S. K. Sastry, *J. Food Eng.*, 2012, **112**(3), 134–143.
197. K. Kanogchaipramot, K. Tongkhao, T. Sajjaanantakul and P. Kamonpatana, *J. Food Sci.*, 2016, **81**, 12.
198. E. Balsa-Canto, A. A. Alonso and J. R. Banga, *J. Food Eng.*, 2002, **52**(3), 227–234.
199. S. Jun and S. Sastry, *J. Food Eng.*, 2007, **80**(4), 1199–1205.
200. S. K. Sastry and Q. Li, *Food Technol.*, 1996, **50**(5), 246–248.
201. M. Lindgren, K. Aronsson, S. Galt and T. Ohlsson, *Innov. Food Sci. Emerging Technol.*, 2002, **3**(3), 233–245.
202. W. R. Fu and C. C. Hsieh, *J. Food Sci.*, 1999, **64**(6), 946–949.
203. F. Marra, M. Zell, J. G. Lyng, D. J. Morgan and D. A. Cronin, *J. Food Eng.*, 2009, **91**, 56–63.
204. C. C. Hsieh and F. Wen-Rong, *Food Sci., Taiwan*, 1999, **26**(4), 404–410.
205. K. Uemura, S. Isobe and A. Noguchi, *J. Jpn. Soc. Food Sci. Technol. (Nippon Shokuhin Kagaku Kogaku Kaishi)*, 1996, **43**(11), 1190–1196.
206. L. J. Davies, M. R. Kemp and P. J. Fryer, *J. Food Eng.*, 1999, **40**(4), 245–258.
207. S. Salengke and S. K. Sastry, *J. Food Eng.*, 2007, **83**(3), 324–336.
208. C. A. Zaror, D. L. Pyle and G. Molnar, *J. Food Eng.*, 1993, **19**(1), 33–53.
209. S. K. Sastry and S. Salengke, *J. Food Process Eng.*, 1998, **21**(6), 441–458.
210. K. Uemura, S. Isobe, T. Imai and A. Noguchi, *J. Jpn. Soc. Food Sci. Technol. (Nippon Shokuhin Kagaku Kogaku Kaishi)*, 1996, **43**(5), 510–519.
211. M. A. Ayadi, J. C. Leuliet, F. Chopard, M. Berthou and M. Lebouche, *J. Food Eng.*, 2005, **70**(4), 489–498.

212. L. Fillaudeau, P. Winterton, J. C. Leuliet, J. P. Tissier, V. Maury, F. Semet, P. Debreyne, M. Berthou and F. Chopard, *J. Dairy Sci.*, 2006, **89**(12), 4475–4489.
213. R. Šulc and M. Nový, *Int. J. Food Eng.*, 2007, **3**(1), DOI: 10.2202/1556-3758.1214.
214. G. L. Quarini, *J. Food Eng.*, 1995, **24**(4), 561–574.
215. S. Palaniappan and S. K. Sastry, *Food Bioprod. Process.*, 1991, **69**(C4), 167–174.
216. W.-C. Wang and S. K. Sastry, *J. Food Eng.*, 1993, **20**(4), 311–323.
217. W.-C. Wang and S. K. Sastry, *J. Food Eng.*, 1993, **20**(4), 299–309.

CHAPTER 4

Cold Plasma Processing to Improve Food Safety

BRENDAN A. NIEMIRA

United States Department of Agriculture, Eastern Regional Research Center, 600 Mermaid Lane, Wyndmoor, PA 19038, USA
Email: brendan.niemira@ars.usda.gov

4.1 Introduction

The *per capita* consumption of fresh fruits and vegetables in the USA is increasing, with fresh and fresh-cut products such as bagged salads being one of the fastest growing market segments.[1,2] Unfortunately, with this increase in fruit and vegetable consumption has come a persistent increase in the number of fresh fruit- and vegetable-related illness outbreaks, particularly with fresh products and prepackaged ready-to-eat produce products.[3,4] These outbreaks accounted for 6% of all reported food-borne outbreaks in the 1990s compared with only 0.7% in the 1970s,[5] but doubled between the periods 1973–1987 and 1988–1992.[6] Fresh fruits and vegetables accounted for 713 of 3204 outbreaks from 1990 to 2005, or 22% of the most common outbreaks.[7,8] Raw sprouts, leafy greens and prepared salads, juices, melons, and berries were the primary vectors.[1,5,9] The economic impact of food-borne illness is felt in costs of medical care, lost productivity, and fatalities, and amount to ~$15 500 000 000 annually for US consumers.[10,11] Conventional postharvest washing and sanitizing treatments are incompletely effective for the decontamination of food-borne pathogens, often resulting in only 2–3 log unit reductions of pathogens on fresh and fresh-cut

Green Chemistry Series No. 53
Alternatives to Conventional Food Processing: 2nd Edition
Edited by Andrew Proctor

Published by the Royal Society of Chemistry, www.rsc.org

fruits and vegetables.[12–14] There remains a critical need for new, safe, affordable, and more effective interventions.[15,16]

In recent years, cold plasma has emerged as a promising technology to improve the safety and quality of a variety of foods.[17–20] Plasma is the fourth state of matter, achieved by providing a gas with sufficient energy to induce partial or complete ionization. The resulting mixture of ionized molecules and atoms, free radicals, and free electrons is extremely reactive. As a food safety intervention, the reactive species in the plasma are directed against contaminating bacteria and viruses, disrupting membranes and cell structures, to inactivate the pathogen. With conventional plasmas, the free energy of the particles and the density of the chemical species within the typically dense plasma are characterized by distinct thermal effects. Thus, the more familiar forms of plasma are hot, whether initiated by chemical reactions (candle flames, blowtorches, *etc.*) or high voltages (welding arcs, lightning discharges). The challenge of using plasma as a food processing technology is, first and foremost, a technological challenge of creating a plasma with a sufficient density of reactive products to be useful, but in a temperature range that will not cause damage to the treated food. In the context of food processing and food microbiology, these technologies may contribute thermal energy to treated commodities, but are commonly referred to as "cold," "cool," or "non-thermal" plasma because of the non-thermal antimicrobial modes of action. This is in contrast to the use of the term "non-thermal" in plasma physics, where the term would refer to non-equilibrium or non-Maxwellian plasmas.[17] Also, some systems are described by the technology used to generate the plasma, as in "dielectric barrier discharge cold plasma"[21,22] and "cold atmospheric gas plasma,"[23] *etc.* As will be discussed, this rich and varied terminology for the various types of systems under development and evaluation can create difficulties in effectively collating the literature. This chapter refers to these technologies as "cold plasma," with additional clarification presented as needed.

4.2 Antimicrobial Modes of Action

Parallel to the development of cold plasma as a food processing technology, the field of plasma medicine research has produced a body of literature on the application of cold plasma in medical devices, with associated inquiry into the modes of action in mammalian tissues.[24,25] Cold plasma inactivates microbes *via* three general mechanisms:[17,26,27] (1) direct chemical interaction of cells with reactive species and charged particles; (2) UV damage of cellular components and membranes; and (3) UV-mediated DNA strand breakage. For example, for systems using oxygen-containing feed gases, the cold plasma ionization creates reactive species such as O_2^+, O_2^-, O_3, O, O^+, O^-, ionized ozone, metastable excited oxygen, and free electrons. These can react with organic compounds to break the single and double bonds of carbon with nitrogen, hydrogen, oxygen, *etc.*, that comprise cellular materials.[28] For any given cold plasma system, one mode of action may predominate over another; the greatest antimicrobial efficacy will be obtained in optimized systems that balance each of these mechanisms.[17,20,26,29]

Cold plasma led to membrane lipid peroxidation and cell damage, caused by reactive oxygen species such as singlet oxygen and hydrogen peroxide-like species.[30] Chemical scavengers of reactive oxygen species served to inhibit membrane lipid peroxidation and reduce oxidative DNA damage. It was concluded that these membrane effects are a major mechanism in cold plasma's inactivation of bacteria. From the body of literature relating to plasma medicine, Bekeschus *et al.*[25] noted that hydrogen peroxide and induced oxidative stress are key mechanisms in human blood exposed to cold plasma, a finding that will inform investigations of mechanisms in metabolically active food commodities (fruits and vegetables, nuts, *etc.*).

4.3 Cold Plasma Feed Gases

Different gases and gas mixtures ionize at different voltages.[31] Noble gases, such as helium, argon, and neon, ionize more readily than oxygen, nitrogen, and air. This characteristic has led to the long-standing use of these types of gases in familiar high-voltage plasma applications such as neon lights. Although cold plasma-based medical devices are relatively insensitive to the costs of the feed gases,[24] the much greater volume of material to be treated and the much lower per-unit value of food commodities are guiding principles in the evaluation of cold plasma technologies being developed for food processing. An economic analysis of helium, nitrogen, oxygen, and air found that the cost of the feed gases becomes a significant factor in implementation on a commercial scale.[17] As helium or neon cold plasma can be created with less electricity than air plasma, these gases have been the subject of much research in antimicrobial applications.[20] However, an emerging consensus has identified oxygen ions and monoatomic oxygen singlets as the primary reaction elements for the antimicrobial efficacy of cold plasma systems that use air or defined gas mixtures containing oxygen.[32]

Oxidative stress leads to cellular damage *via* lipid peroxidation and enzyme inactivation, in addition to DNA cleavage. Localization of the generation of the reactive species will determine the precise nature of the damage done to cells. A challenge for the full elucidation of the antimicrobial modes of action of cold plasma treatment is the identification of the complex mixtures of reactive species produced. Using an 80 kV cold plasma system,[32] *Escherichia coli* was inactivated mainly by cell leakage and low-level DNA damage, whereas in contrast, *Staphylococcus aureus* was mainly inactivated by intracellular damage. For *S. aureus*, significantly higher levels of intracellular reactive oxygen species were observed, along with comparatively little envelope damage. A time-dependent increase in reactive species meant that both organisms demonstrated the effects of increased reactivity damage with increased time of cold plasma treatment and exposure.[32] An interesting goal for future research is to optimize simultaneously (1) the production of reactive oxygen species, (2) the delivery and/or localization of these species to maximal antimicrobial effect, and (3) in a species-specific manner that accounts for differential cell wall, cell membrane, and intercellular features and functions (*e.g.* DNA repair capacity).

In general, therefore, although oxygen-containing gas mixtures are typically more difficult to ionize, their increased efficacy and lower cost per unit commodity treated are sufficient justification. In the rapidly expanding field of cold plasma inactivation of pathogenic bacteria and viruses, the specific gas mixtures used in one study or another can vary widely. The resulting plasma chemistries can be notably dissimilar, with associated dissimilarities in antimicrobial efficacy, sensory impact, rapidity of response, and other factors. The variation in cold plasma equipment and differences in cold plasma feed gases make it difficult to compare data from different published studies. These issues should prompt caution and awareness when evaluating and comparing results among studies.[34]

4.4 Cold Plasma Equipment

A wide variety of equipment can be used to create cold plasma. This multiplicity of elements in cold plasma equipment is one of the greatest challenges in comparing the results obtained in the rapidly expanding literature relating to this class of technologies as a food safety and food processing intervention. Generally, energy is applied to a gas in order to ionize it to plasma, which serves to sanitize the target. Cold plasma technologies can be grouped into those which rely on applying a partial vacuum and those which operate at atmospheric pressure.

4.4.1 Vacuum and Partial-pressure Cold Plasma Systems

Gases at lower pressures are more readily ionized, which allows the plasma to be generated at a lower voltage.[31] Although this serves to reduce the energy requirement, the reduced pressure means that a lower absolute volume of reactive species is produced. Also, specialized equipment is required for partial-pressure operation: vacuum chambers, seals, pumps, pass-through doors, *etc.* An example of a partial vacuum-type cold plasma discharge is shown in Figure 4.1. This system uses two spherical electrodes, with a cold plasma discharge maintained between them. The shape, configuration, material, and separation of the electrodes will have a strong influence on the initiation of the ionization process. In the system shown, the cold plasma discharge assumes a natural arc shape, rising due to the increased temperature of the ionized gas within the plasma plume. In this type of system, the vacuum chamber will fill with the secondary reaction products of the plasma ionization. Products to be treated will be positioned within the vacuum chamber, in a defined geometry in relation to this plasma discharge: proximate, above, or below, depending on the commodity and the application.

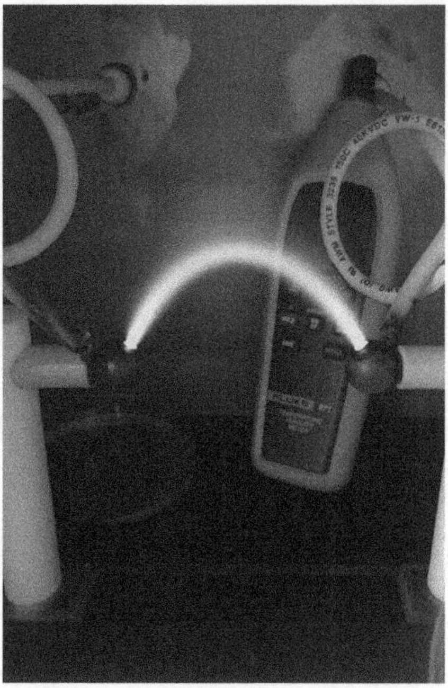

Figure 4.1 Vacuum chamber cold plasma treatment system, using spherical electrodes charged to a voltage of 5 kV. In this type of cold plasma system, the product to be treated would be placed above, proximate to, or below the cold plasma arc.

The system shown in Figure 4.1 is based on evacuated air, which can be modified with additional gases at partial pressures. Using this system, *E. coli* inoculated onto the calyx of blueberries treated with partial vacuum cold plasma (84 kPa) augmented with water vapor. Treatment in this plasma chamber significantly reduced the *E. coli* recovered from samples compared with non-treated controls, with the greatest reduction after 10 min, 1.84 log cfu g^{-1} (Hertrich *et al.*, unpublished data). Other augmentation gases (ethanol, CO_2) did not provide comparable effects. Thus, in a partial vacuum, a high-humidity environment in the cold plasma chamber may significantly enhance the effects of cold plasma, either through facilitated generation of reactive products such as H_2O_2, or through a prolonged half-life of free radicals. It is further expected that the indirect application of cold plasma to fragile commodities will enhance the preservation of sensory qualities.

Augmentation of partial vacuum cold plasma systems with a variety of gases has been explored as a means to enhance antimicrobial efficacy. Some of these are common feedstock gases for modified atmosphere processing in the food technology arena, such as pure N_2, CO_2, or O_2. Using a more exotic gas mixture, Selcuk *et al.*[35] used a vacuum-based plasma system based on evacuated air or sulfur hexafluoride as the plasma feed gas. The goal of the work was to reduce *Aspergillus* spp. and *Penicillum* spp. artificially contaminated on the surfaces of different seeds. Using either gas, treatments for up of to 20 min were able to achieve 3 log reductions of the fungal contaminants. Both gas systems gave comparable results.

Won *et al.*[36] treated whole mandarin fruits with a microwave-pumped cold plasma at 0.7 kPa. Various gases were used for the process. Nitrogen plasma at 900 W for 10 min gave the highest inhibition of the mold *Penicillium italicum* and an 84% reduction in disease incidence. The treatment also significantly increased the total phenolic content and antioxidant activity of mandarin peels, but physiological factors of the intact fruits (CO_2 generation, weight loss, content of soluble solids, titratable acidity, pH, ascorbic acid concentration, surface color) were not affected during storage at 4 and 25 °C. It was concluded that this vacuum cold plasma process had potential for improving (1) the storability of mandarins (reduced spoilage organisms) and (2) the total phenolic content and antioxidant activity.[36]

4.4.2 Atmospheric Cold Plasma Systems

Since gases are typically more difficult to ionize at atmospheric pressure,[31] such systems require higher voltages and more energy, with concomitantly more stringent equipment needs.[37] Nevertheless, because atmospheric operation is more amenable to adaptation to conveyor belt-type operations, a great deal of research effort has been directed in this area.[23,29,32,38] Reactive species in the plasma plume will recombine with each other if they do not interact with a product to be treated. Therefore, in systems with a long time of flight between plasma generation and food commodity contact, only the most long-lived reaction products (*e.g.* ozone, NO_x) will react. Cold plasma systems can therefore be understood based on proximity of cold plasma electrodes and the food target.[37]

Systems in which the electrodes are relatively distant from the target must move the cold plasma to where it is desired, using airflow, magnetic fields, *etc.* An example of this type of system is the plasma jet, shown in Figure 4.2. The electrodes are physically separate from the commodity (in the case of Figure 4.2, apple sections), with pressurized air expanding the cold plasma discharge and directing it towards the food. Systems of this design increase the flexibility of the shape of products to be treated, as the plasma plume will extend outwards to wrap around irregular objects. It should be noted, however, that the most reactive chemical species are lost to recombination in systems with a greater physical separation. This type of cold plasma system is relatively flexible, as it allows the positioning of the food target within the plasma plume (with a higher concentration of active plasma species, but a greater risk of negative sensory impact), or just beyond the perimeter of the plume (with lower reactivity, but greater control for sensitive commodities).

Mature *Salmonella* biofilms (either 24, 48, or 72 h old) were grown on food-contact surfaces and treated with short exposures to an atmospheric pressure plasma jet similar to that shown in Figure 4.2.[39] The cold plasma treatment was carried out on a variable-speed, laboratory-scale conveyor belt, to assess

Figure 4.2 Atmospheric plasma jet, applied to apple slices. This system uses an operating voltage of 30 kV. This image shows the area of active plasma (the visible plume) and the zone of "quenched" plasma, the region where the reactive species are more dispersed.

the potential of this technology for surface decontamination in a food processing environment. Test plates received cold plasma (with filtered air as the feed gas) at a distance of either 5 or 7.5 cm for 5, 10, or 15 s. At 5 cm, *Salmonella* biofilms were reduced by up to 1.57 log cfu mL^{-1} (5 s), 1.82 log cfu mL^{-1} (10 s), and 2.13 log cfu mL^{-1} (15 s). For the 15 s treatment, increasing distance generally reduced efficacy, but had a variable effect on the shorter 5 and 10 s treatments. Biofilms responded comparably to cold plasma, regardless of the age or maturity; even the oldest, most durable biofilms were readily inactivated by cold plasma.[39] In a companion study, biofilms of *E. coli* O157:H7 were even more sensitive, yielding a maximum reduction of 3.29 log for optimized cold plasma treatment (B. A. Niemira *et al.*, unpublished data). In that study, the frequency at which the cold plasma was generated had a greater influence on overall antimicrobial activity than for treated *Salmonella* biofilms, although the effects were variable. A summary of these results is presented in Figure 4.3.

A nitrogen cold plasma jet system was applied to lettuce and strawberry surfaces and potato tissue inoculated with *Salmonella enterica* serovar Typhimurium (*S.* Typhimurium).[23] The inactivation rate for this pathogen was independent of growth phase or growth temperature. A 2 min treatment with the nitrogen cold plasma reduced *Salmonella* by 2.71 log on membrane filters,

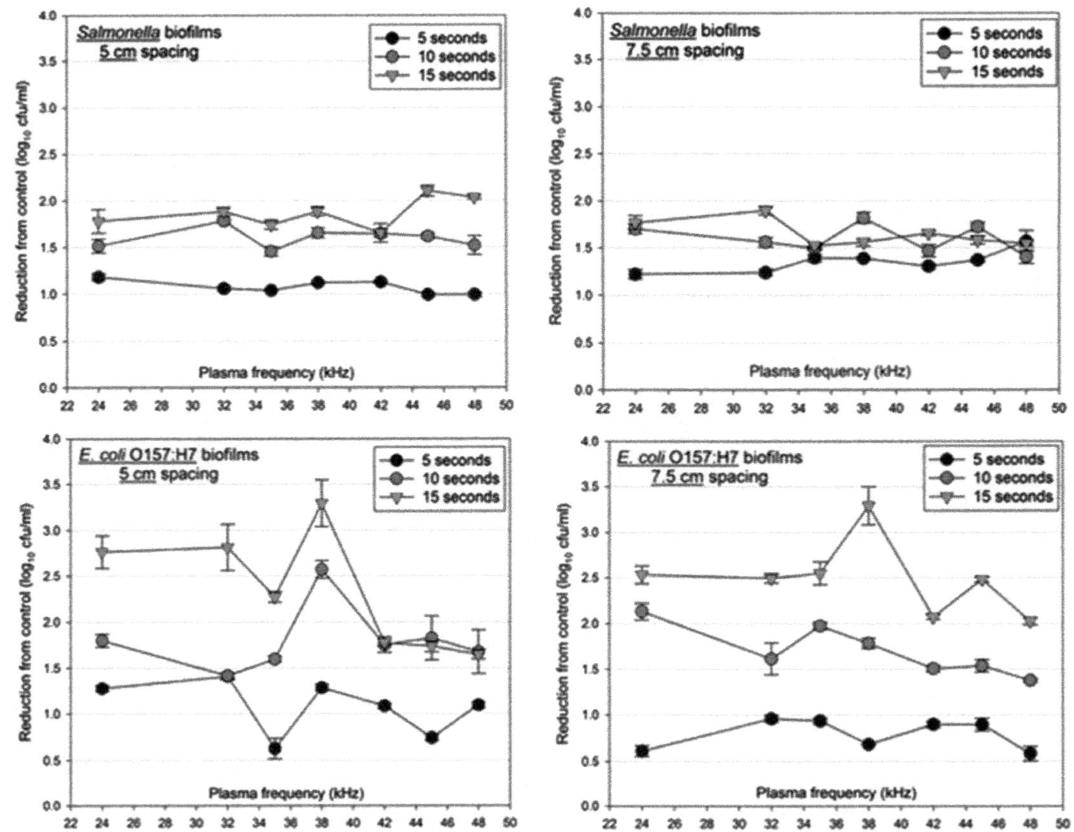

Figure 4.3 Reductions of *Salmonella* and *E. coli* O157:H7 biofilms following rapid cold plasma treatments of 5 s (black circles), 10 s (gray circles), or 15 s (triangles). Biofilms on inert substrates were 5 or 7.5 cm from cold plasma emitter head. Electricity used to generate cold plasma varies from 24 to 48 kHz. Bars = standard error.
Adapted from ref. 39 and B. A. Niemira, unpublished data.

but a much longer treatment of 15 min was necessary to achieve 2.72, 1.76, and 0.94 log reductions on lettuce, strawberry, and potato, respectively. These results are consistent with a body of literature for cold plasma and other antimicrobial interventions where sanitizing efficacy on a defined media *in vitro* is greater than that observed on actual food commodities. Fernández *et al.*[23] identified surface features of various commodities as key factors in the differential inactivation of *Salmonella*. Scanning electron microscopy of the lettuce, strawberry, and potato revealed topographical features that may shield pathogen cells from the cold plasma active species. Such protective features on leaves, melons, *etc.* have been noted in the response of pathogens to chemical and other non-thermal sanitizing processes.[15,40]

Blueberries were treated with atmospheric cold plasma and sampled for total aerobic plate count (APC) and yeasts/molds, both immediately after treatment and after cold storage for 1, 2, and 7 days.[41] Cold plasma treatments of up to 120 s significantly ($p < 0.05$) reduced APC immediately after exposure, with reductions of 0.8–1.6 log cfu g^{-1}, depending on the treatment time. After 7 days of storage at 4 °C, APC counts were 1.5–2.0 log cfu g^{-1} less than for the control. Firmness and total anthocyanins were reduced by treatments longer than 60 s.[41] In a companion study, the same cold plasma system was used to inactivate norovirus surrogates on fresh fruit.[38] Viruses are the leading cause of food-borne outbreaks, so a non-thermal intervention with efficacy against viruses is highly desirable. Two surrogates for human norovirus, Tulane virus and murine norovirus, were treated with this same atmospheric cold plasma jet for 0, 15, 30, 45, and 60 s. The Tulane virus was significantly reduced by 1.5 pfu g^{-1} after 45 s, with a maximum log reduction of 3.5 log pfu g^{-1} after 120 s of treatment. The murine norovirus was more sensitive, with a significant reduction (0.5 log pfu g^{-1}) after only 15 s, and a maximal reduction (>5 log pfu g) after 90 s. Temperature measurements showed that these reductions were produced by non-thermal means.[38] It was concluded that, with further optimization, this type of cold plasma system may be an effective intervention against food-borne viruses, such as human norovirus.

Two different atmospheric pressure cold plasma systems (plasma jet and a microwave-driven remote plasma) were compared as a means of decontamination of whole black pepper.[42] After a 30-min treatment with remote plasma, *Bacillus subtilis* spores, *B. atrophaeus* spores, and *Salmonella enterica* on naturally contaminated peppercorns counts were reduced by 2.4 log, 2.8 log, and 4.1 log, respectively. Direct plasma jet treatment was less effective on both the spores and the vegetative cells. Measurement of surface color and essential oils indicated minimal impact of the cold plasma treatment on the sensory quality of the peppercorns.

4.4.3 Dielectric Barrier Discharges

Some cold plasma systems position the food between the plasma generation electrodes rather than generating the cold plasma remotely from the food. This exposes food to a very high density of free electrons, radicals, ions, and

UV radiation. An example of this type of system is the Dielectric Barrier Discharge (DBD). In these systems, instead of an electric arc being established in the air gap between electrodes, an electrically resistive material (a dielectric) is placed between the electrodes (*i.e.* a barrier). If a sufficiently high voltage is generated, electricity flows through the dielectric, creating a uniform plasma field, with reduced risks of arcs, sparks, or other dense, inhomogeneous discharges that could damage foods. DBD systems can, with proper equipment design, create a stable, uniform field.[17,43,44] These systems can achieve significant reductions of food-borne pathogens, but the spacing between the electrodes means that the plasma discharge is sensitive to perturbation. The electrical conductivity of the commodity within the plasma field can facilitate stray discharges. As these discharges move across (or through) the product, unwanted spot heating can lead to damage. Therefore, the design of electrodes in DBD systems is critical. This includes their shaping, composition, spacing, operational voltage and frequency, and other factors.

Using a DBD system operated at 1.0–2.5 kHz, almonds inoculated with *E. coli* were placed between the electrodes and treated with cold plasma using air as the feed gas.[45] A DBD treatment of 30 s at 25 kV, 2.0 kHz or 2.5 kHz reduced the *E. coli* by 4 and 5 logs, respectively. Butscher *et al.*[46] examined sprouting seeds (alfalfa, onion, radish, and cress) that were artificially inoculated with *E. coli* and treated in a DBD system using pulsed electricity and argon gas flushing. The plasma treatment resulted in reductions of up to 3.4 log on cress seeds. It was noted that a major challenge in the plasma decontamination of granular food products was the complex surface topology of the various seed types. Roughened surfaces, cracks, crevices, and other surface features can reducing the treatment efficiency by shielding microorganisms from plasma-generated reactive species. This phenomenon has previously been noted on a variety of foods with chemical sanitizers and other non-thermal processes.[15] However, improvement of the cold plasma's inactivation efficiency was achieved by optimizing (1) the moisture level in the treatment chamber and (2) electricity voltage and frequency to increase the production of reactive species.[46] Although seed germination was decreased by the most antimicrobially effective treatment, enhanced germination was observed under milder conditions. Thus, future research will pursue an optimized balance of the impact of cold plasma treatment on pathogen survival and seed germination.

4.4.4 Enclosed, In-package Cold Plasma Systems

An interesting recent development in the field of cold plasma research is the evaluation of systems to treat commodities inside sealed packages. Currently, options for post-packaging antimicrobial interventions are limited to heating, irradiation, and some in-package chemical sanitizer generation.[47] Cold plasma generation within a package could potentially allow for treatment during shipment and again at the point-of-sale, not only as a terminal processing step in a food packaging line.

Yong *et al.*[48] evaluated a closed container with a DBD apparatus attached to the lid. This arrangement allowed the generation of cold plasma at the top of the closed chamber. When agar plates were treated within this closed container (53 mm high), *E. coli*, *S.* Typhimurium, and *L. monocytogenes* showed comparable reductions of 3.57, 6.69, and 6.53 log cfu g^{-1} with 60 s, 45 s, and 7 min treatments, respectively. The system was less effective when the pathogens were inoculated on cheese. On the food product, *E. coli*, *S.* Typhimurium, and *L. monocytogenes* counts were reduced by 2.67, 3.10, and 1.65 logs at 60 s, 45 s, and 7 min, respectively.

In a study of a closed-packaged system using minimal-gap spaced aluminum electrodes, Patil *et al.*[27] showed that a direct plasma exposure treatment of 70 kV for 60 s applied to *Bacillus atrophaeus* spores attached to a spore strip achieved a greater than 6 log reduction for several different gas mixtures: air, 90% N_2–10% O_2, and 65% O_2–30% CO_2–5% N_2. However, when an indirect treatment was used (where the strip was placed *below* the electrodes and *under* the plasma field, rather than *between* the electrodes and *inside* the plasma field), the same 60 s treatment yielded reductions as low as 2.1 log. The relative humidity of the container was seen to be a significant factor, primarily due to the generation of ozone as determined by optical absorption spectroscopy. Although this system demonstrated a notable reduction of *B. atrophaeus* spores, the experiments were conducted *in vitro*, with the spores on a suspending medium strip. It was suggested that sterilization of medical devices may be a viable application for the DBD system as designed.[27]

Another study of in-package cold plasma treatment examined bacteria suspended in artificial media (phosphate-buffered saline in a 96-well plate).[49] The treatment yielded a greater than 7 log reduction of *E. coli* following 20 s of direct or 45 s of indirect cold plasma. As has been observed elsewhere in this chapter, there is typically a differential efficacy for *in vitro* testing *versus* food commodity testing. To realize the potential of this system for food processing, further research using actual commodities will elucidate the potential applicability of this specific type of in-package cold plasma apparatus.

Packaged strawberries were treated with a 60 kV DBD system for 5 min.[50] Aerobic mesophilic bacteria, yeasts, and molds of strawberries were reduced, and declined to 2 log less than the untreated control after 24 h of post-treatment storage. There were no significant changes in strawberry respiration rate, color, or firmness. Using a different kind of DBD device, *E. coli*, *S.* Typhimurium, and *L. monocytogenes* were inoculated on cherry tomatoes and strawberries air dried and sealed inside a rigid polypropylene container.[51] Indirect exposure to the plasma discharge at a voltage of 70 kV led to a reduction of pathogen populations on tomatoes by 3.1 log (10 s), 6.3 log (60 s), and 6.7 log (120 s). This represented a near-complete sanitation process. However, longer treatment times were required for strawberries with a more complex surface. A treatment for 300 s reduced *E. coli*, *S.* Typhimurium, and *L. monocytogenes* by 3.5, 3.8 and 4.2 log cfu per sample, respectively. The background microflora was also reduced by cold plasma treatment.

Incorporation of cold plasma technology with commercial packaging is a key area of study.[33] Although data generated using research-grade packaging materials (Petri plates, stomacher bags, milfoil packs, *etc.*) are useful for the elucidation of efficacy and mechanisms, the performance of commercial packaging in these cold plasma systems is of critical importance for scale-up and commercialization. In a study using a conventional commercial polyethylene terephthalate plastic container, DBD treatment was applied to Romaine lettuce inoculated with *E. coli* O157:H7.[52] The DBD process conditions were 47.2 kV at 1.1 kHz for 5 min. The inactivation of *E. coli* O157:H7, total mesophilic aerobes, and yeasts and molds was determined during 7 days of refrigerated storage. Color, carbon dioxide generation, weight loss, and surface morphology of the DBD-treated lettuce were also evaluated. The reductions in the numbers of *E. coli* O157:H7, total mesophilic aerobes, and yeasts and molds during storage were 0.8–1.5, 0.7–1.9, and 0.9–1.7 log cfu g^{-1}, respectively. The DBD treatment had no discernible effect on leaf color, carbon dioxide generation, weight, and surface morphology of lettuce during storage. There was evidence of sublethal injury, suggesting that the cold plasma process could effectively be combined with other antimicrobial interventions.[52]

In order to scale cold plasma up to pilot scale and eventual commercialization, in-package treatment systems must be effectively applied to bulk packaged material, such as commercially bagged and containerized salads, fruits, vegetables, nuts, *etc.* It is uncertain if bulk packaged material, such as lettuce leaves, will interact equally with cold plasma discharges between leaf pieces within a plasma field. This would result in non-homogeneous DBD cold plasma processing that would cause damaging arc discharges that would over-process some spots. Therefore, there may be sensory damage in over-processed areas, while pathogens would survive in untreated areas. A study of bulk Romaine lettuce in a commercial plastic clamshell container mapped the inactivation of *E coli* O157:H7 on multiple stacks of lettuce leaves (one, three, five, or seven layers) treated in a DBD in-package treatment system.[53] A 10 min treatment at 42.6 kV reduced *E. coli* O157:H7 in the one-, three-, and five-layer configurations by 0.4–0.8 log cfu g^{-1} lettuce, with no significant impact of placement within the container and antimicrobial efficacy. In the stacking with seven layers, the top layer showed a greater degree of reduction (1.1 log cfu g^{-1} lettuce), but with shaking of the container the inactivation was uniform. The cold plasma treatment did not significantly change the surface morphology, color, respiration rate, or weight of the leaf pieces for leaves in any position within the bulk stack. Future research will address the issues related to even larger packages of bulk fruits and vegetables, and also the possibility of repeated treatments applied to packaged commodities. Optimization of this sequential and/or repeated cold plasma intervention will be a key area of inquiry in coming years.

4.5 Conclusion

Cold plasma is a flexible, effective antimicrobial process. The literature in this field is expanding rapidly, with research teams around the world focusing effort and attention on the creation and validation of new cold plasma technologies and processing protocols. As this sanitizing process does not use aqueous chemical solutions, it has great potential as a water- and chemical-free food safety intervention for fresh and fresh-cut fruits and vegetables, nuts, powders, meats, seafood, and other commodities. In addition to the use of cold plasma as a direct treatment of food commodities, on a food processing line, or as a terminal post-packaging step, the future applications of cold plasma will also likely expand into the sanitation of food contact surfaces, packaging, and equipment.

Acknowledgements

The author wishes to thank Drs D. Geveke and G. Richards for their reviews of the original text. Mention of trade names or commercial products in this chapter is solely for the purpose of providing specific information and does not imply recommendation or endorsement by the US Department of Agriculture. USDA is an equal opportunity employer.

References

1. R. M. Callejón, M. I. Rodriguez-Naranjo, C. Ubeda, R. Hornedo-Ortega, M. C. Garcia-Parrilla and A. M. Troncoso, *Foodborne Pathog. Dis.*, 2015, **12**, 32.
2. Center for Nutrition Policy and Promotion. 2010. U.S. Dept. Agriculture – Dietary Guidelines for Americans. http://www.cnpp.usda.gov/DietaryGuidelines. Accessed 3 Feb 2017.
3. J. M. Cevallos-Cevallos, G. Gu, M. D. Danyluk, N. S. Dufault and H. C. Ariena van Bruggen, *Int. J. Food Microbiol.*, 2012, **158**, 140.
4. Centers for Disease Control and Prevention, 2016. http://wwwn.cdc.gov/food-borneoutbreaks/. Accessed 3 Feb 2017.
5. S. Sivapalasingam, C. R. Friedman, L. Cohen and R. V. Tauxe, *J. Food Prot.*, 2004, **67**, 2342.
6. J. W. Buck, R. R. Walcott and L. R. Beuchat, *Plant Health Progress*, 2003. https://www.apsnet.org/publications/apsnetfeatures/Documents/2003/MicrobiologicalSafety.pdf. Accessed 6 Feb 2017.
7. Center for Science in the Public Interest, 2010. https://cspinet.org/new/201303251.html. Accessed 3 Feb 2017.
8. Centers for Disease Control and Prevention, 2017, https://www.cdc.gov/foodsafety/outbreaks/multistate-outbreaks/outbreaks-list.html. Accessed November 5, 2017.
9. L. Calvin, 2007, *USDA-ERA*. https://www.ers.usda.gov/amber-waves/2007/june/outbreak-linked-to-spinach-forces-reassessment-of-food-safety-practices/. Accessed 3 Feb 2017.

10. R. L. Scharff, *Produce Safety Project*, 2010. http://www.producesafetyproject.org/admin/assets/files/Health-Related-Foodborne-Illness-Costs-Report.pdf-1.pdf. Accessed 4 Feb 2017.
11. S. Hoffman, B. Maculloch and B. Batz, 2015, *USDA-ERS*. https://www.ers.usda.gov/publications/pub-details/?pubid=43987. Accessed 3 Feb 2017.
12. Food and Drug Administration, 2014. http://www.fda.gov/Food/GuidanceRegulation/HACCP/ucm082063.htm. Accessed 3 Feb 2017.
13. J. C. Faith and M. P. Doyle, *Curr. Opin. Biotechnol.*, 2010, **21**, 126.
14. A. O. Olaimat and R. A. Holley, *Food Microbiol.*, 2012, **32**, 1.
15. D. Ukuku, D. J. Geveke, L. Chau and B. A. Niemira, *Int. J. Food Microbiol.*, 2016, **231**, 86.
16. T. Z. Jin, M. Huang, B. A. Niemira and L. Cheng, *J Food Process. Preserv.*, 2016, **41**, e12871.
17. B. A. Niemira, *Annu. Rev. Food Sci. Technol.*, 2012, **3**, 125.
18. H. Lee, J. E. Kim, M.-S. Chung and S. C. Min, *Food Microbiol.*, 2015, **51**, 74.
19. S. Min, S. H. Roh, B. A. Niemira, J. E. Sites, G. Boyd and A. Lacombe, *Int. J. Food Microbiol.*, 2016, **237**, 114.
20. S. K. Pankaj, C. Bueno-Ferrer, N. N. Misra, V. Milosavljević, C. P. O'Donnell, P. Bourke, K. M. Keener and P. J. Cullen, *Trends Food Sci. Technol.*, 2014, **35**, 5.
21. H. I. Yong, H. J. Kim, S. Park, K. Kim, W. Choe, S. J. Yoo and C. Jo, *Food Res. Int.*, 2015, **69**, 57.
22. H. J. Kim, H. I. Yong, S. Park, K. Kim, W. Choe and C. Jo, *Food Control*, 2015, **47**, 451.
23. A. Fernández, E. Noriega and A. Thompson, *Food Microbiol.*, 2013, **33**, 24.
24. G. Isbary, T. Shimizu, Y. F. Li, W. Stolz, H. M. Thomas, G. E. Morfill and J. L. Zimmermann, *Expert Rev. Med. Devices*, 2003, **10**, 367.
25. S. Bekeschus, J. Kolata, C. Winterbourn, A. Kramer, R. Turner, K. D. Weltmann, B. Bröker and K. Masur, *Free Radical Res.*, 2014, **48**, 542.
26. M. Moisan, J. Barbeau, M. Crevier, J. Pelletier, N. Phillip and B. Saoudi, *Pure Appl. Chem.*, 2002, **74**, 349.
27. S. Patil, T. Moiseev, N. N. Misra, P. J. Cullen, J. P. Mosnier, K. M. Keener and P. Bourke, *J. Hosp. Infect.*, 2014, **88**, 162.
28. T. Moiseev, N. N. Misra, S. Patil, P. J. Cullen, P. Bourke, K. M. Keener and J. P. Mosnier, *Plasma Sources Sci. Technol.*, 2014, **23**, 065033.
29. M. Laroussi and X. Lu, *Appl. Phys. Lett.*, 2005, **87**, 113902.
30. S. G. Joshi, M. Cooper, A. Yost, M. Paff, U. K. Ercan, G. Fridman, G. Friedman, A. Fridman and A. D. Brooks, *Antimicrob. Agents Chemother.*, 2011, **55**, 1053.
31. M. A. Lieberman and A. J. Lichtenberg, *Principles of Plasma Discharges and Materials Processing*. Hoboken, NJ: Wiley-Interscience, 2005.
32. L. Han, S. Patil, D. Boehm, V. Milosavljević, P. J. Cullen and P. Bourke, *Appl. Environ. Microbiol.*, 2016, **82**, 450.

33. S. H. Han, H. J. Suh, K. B. Hong, S. Y. Kim and S. C. Min, *J. Food Sci.*, 2016, **81**, T3052.
34. J. Gay-Mimbrera, M. C. García, B. Isla-Tejera, A. Rodero-Serrano, A. V. García-Nieto and J. Ruano, *Adv. Ther.*, 2016, **33**, 894.
35. M. Selcuk, L. Oksuz and P. Basaran, *Bioresour. Technol.*, 2008, **99**, 5104.
36. M. Y. Won, S. J. Lee and S. C. Min, *Innovative Food Sci. Emerging Technol.*, 2017, **39**, 25.
37. T. McHugh and B. A. Niemira, *Food Technol.*, 2016, **72**, 68.
38. A. Lacombe, B. A. Niemira, J. B. Gurtler, J. Sites, G. Boyd, D. H. Kingsley, X. Li and H. Chen, *Food Microbiol.*, 2017, **63**, 1.
39. B. A. Niemira, G. Boyd and J. Sites, *J. Food Sci.*, 2014, **79**, M917.
40. O. M. Olanya, D. O. Ukuku, B. A. Annous, B. A. Niemira and C. Sommers, *J. Food Agric. Environ.*, 2013, **11**, 86.
41. A. Lacombe, B. A. Niemira, J. B. Gurtler, X. Fan, J. Sites, G. Boyd and H. Chen, *Food Microbiol.*, 2015, **46**, 479.
42. C. Hertwig, K. Reineke, J. Ehlbeck, D. Knorr and O. Schlüter, *Food Control*, 2015, **55**, 221.
43. N. N. Misra, K. M. Keener, P. Bourke, J. P. Mosnier and P. J. Cullen, *J. Biosci. Bioeng.*, 2014, **118**, 117.
44. S. K. Pankaj, C. Bueno-Ferrer, N. N. Misra, L. O'Neill, A. Jiménez, P. Bourke and P. J. Cullen, *Innovative Food Sci. Emerging Technol.*, 2014, **21**, 107.
45. S. Deng, R. Ruan, C. K. Mok, G. Huang, X. Lin and P. Chen, *J. Food Sci.*, 2007, **72**, M62.
46. D. Butscher, H. Van Loon, A. Waskow, P. R. von Rohr and M. Schuppler, *Int. J Food Microbiol.*, 2016, **238**, 222.
47. B. A. Niemira and H. Q. Zhang, in *The Produce Contamination Problem: Causes and Solutions,* ed. K. Matthews, G. Sapers and C. Gerba, 2nd edn, Elsevier, New York, 2014, pp. 433–450.
48. H. I. Yong, H. J. Kim, S. Park, A. U. Alahakoon, K. Kim, W. Choe and C. Jo, *Food Microbiol.*, 2015, **46**, 46.
49. D. Ziuzina, S. Patil, P. J. Cullen, K. M. Keener and P. Bourke, *J. Appl. Microbiol.*, 2012, **114**, 778.
50. N. N. Misra, S. Patil, T. Moiseev, P. Bourke, J. P. Mosnier, K. M. Keener and P. J. Cullen, *J. Food Eng.*, 2014, **125**, 131.
51. D. Ziuzina, S. Patil, P. J. Cullen, K. M. Keener and P. Bourke, *Food Microbiol.*, 2014, **42**, 109.
52. S. C. Min, S. H. Roh, G. Boyd, J. E. Sites, X. Fan, J. Uknalis and B. A. Niemira, *J. Food Prot.*, 2017, **80**, 35.
53. S. C. Min, S. H. Roh, B. A. Niemira, G. Boyd, J. E. Sites, J. Uknalis and X. Fan, *Food Microbiol.*, 2017, **65**, 1.

CHAPTER 5

Supercritical Fluid Pasteurization and Food Safety

SARA SPILIMBERGO,*[a] MICHAEL A. MATTHEWS[b] AND ALLESANDRO ZAMBON[a]

[a] Department of Materials Industrial Engineering, University of Padova, Via Marzolo 9, 35131 Padova, Italy; [b] Department of Chemical Engineering, University of South Carolina, 301 Main Street, Columbia, SC 29208, USA
*Email: sara.spilimbergo@unipd.it

5.1 Introduction

The scientific community's interest in supercritical fluids (SCFs) stems from their physicochemical properties, which are intermediate between those of liquids and gases. Supercritical fluids have a density similar to that of liquids, which assures good solvent properties, but a viscosity and a diffusivity that render them more similar to gases. These gas properties allow SCFs to have a greater permeation capability than liquids. In addition, their specific chemical structures and properties may impart additional benefits to their use as SCFs. Carbon dioxide has many potentially useful properties for food applications, and is already used in food and beverage processing as a bactericide. Other fluids (N_2, N_2O, *etc.*) in their supercritical state have also been investigated for use in food applications. A number of journal reviews have already been generated on the use of compressed CO_2 for treating a variety of foods.[1–5]

The objective of this chapter is to present the fundamental scientific principles of SCFs and to discuss their use in pasteurization and related food technologies. Much of the discussion will focus on carbon dioxide (CO_2)

Green Chemistry Series No. 53
Alternatives to Conventional Food Processing: 2nd Edition
Edited by Andrew Proctor

Published by the Royal Society of Chemistry, www.rsc.org

because of the large amount of research and development that this gas has received as an SCF. Both liquid and supercritical CO_2 have potential in food applications and so the more generic term "dense-phase" CO_2 is sometimes used. Researchers should pay careful attention to the process conditions to determine whether the fluid is actually in the liquid or the supercritical state.

Supercritical carbon dioxide ($scCO_2$) has been regarded as a "green" solvent for many years, primarily because it is non-toxic, non-flammable, and physiologically benign, and is readily available from industrial sources at a relatively low price. Relevant $scCO_2$ physical data are compiled in Table 5.1, while the main advantages of $scCO_2$ in industrial/research application processes are summarized in Table 5.2. CO_2 is attractive for numerous implemented and proposed industrial processes.[6,7] These include the synthesis of polymeric particles for the pharmaceutical industry, micronization–polymerization of active compounds, and extraction and separation processes from liquid and solid matrices. The best-known applications are the extraction of caffeine for the production of caffeine-free coffee, and particle formation using PGSS (particles from gas-saturated solution) and RESS (rapid expansion of supercritical solution) processes.[8]

Another major application of CO_2 solvent technology is textile cleaning. The low critical temperature (31 °C) and the advantageous diffusion properties of the compressed fluid are compatible with many fabrics. Textile cleaning was first patented three decades ago using both liquid CO_2[10] and supercritical CO_2[11] and has since become an accepted commercial process.[12,13] The most important advantages for this application are the solvent power (to dissolve and remove soils) and surface wetting properties of CO_2. Liquid CO_2 at −25 °C has a surface tension of approximately 9 mN m^{-1}, and at 20 °C the surface tension decreases to only 1 mN m^{-1}.[14] Pure supercritical and gaseous CO_2 have zero surface tension. In contrast, the surface tension of water is 72 mN m^{-1}.[15,16] When CO_2 at 40 °C is dissolved in water, the surface tension of the mixture decreases to 50 mN m^{-1} near the critical point of CO_2. Above the critical point, the surface tension falls dramatically below 30 mN m^{-1} and remains constant as the pressure increases, and is proportional to the solubility of CO_2 in water.[17] Whereas non-polar substances are relatively well soluble in CO_2, the solubility of polar substances and particles is limited. Surfactants and co-solvents for overcoming this shortcoming exist, and more are still being researched and developed.[18–20]

Table 5.1 Comparison of physical properties of gases, liquids, and supercritical fluids. Modified from ref. 9.

Property	Gas	SCF	Liquid
Density/g cm^{-3}	10^{-3}	0.3	1
Viscosity/Pa s	10^{-5}	10^{-4}	10^{-3}
Diffusivity/cm^2 s^{-1}	0.1	10^{-3}	5×10^{-6}
Surface tension H_2O/CO_2/mN m^{-1}	<1	<1	72/1

Table 5.2 Supercritical CO_2 characteristics and main process advantages.

$scCO_2$ characteristics	Advantages due to its peculiar characteristics
Atoxic, odorless, colorless	No contamination of materials or environment
Critical temperature very close to ambient	Possible to process thermolabile materials
Relatively low critical pressure compared with hydrostatic pressure conditions (thousands of bar) or critical pressures of other chemical compounds	Inexpensive process
High diffusivity	Short processing time
Cheap and non-flammable gas	Easily handled
Possibility to be recycled	Decrease in running costs and no environmental drawbacks
Saturation of the reaction chamber	Inhibition of oxidative reactions
Slight changes of CO_2 temperature and pressure around the critical region alter drastically its physicochemical properties	High extraction yield, high purity of the products by simple depressurization (no residues)
Versatility in many applications	Useful in different processes: extraction, fractionation, polymerization, micronization, *etc.*
Bactericidal properties	Potential use in low-temperature pasteurization process to achieve a microbiologically stable product

5.2 Supercritical Fluids and Green Technology

A few SCFs (most usually carbon dioxide but also water, ethanol, nitrous oxide, and several others) have been investigated over the last 30 years and promoted for their potential as environmentally benign and less toxic solvents relative to common organic and halogenated solvents. Many early investigations focused on the potential of SCFs to replace hazardous solvents such as chlorofluorocarbons in the chemical and material manufacturing industries. The use of SCFs in food applications, primarily for extracting specific constituents such as caffeine, also has a long history.

In more recent years, the appellation "green" has been applied to those products and technologies that are judged to be environmentally friendly or more sustainable than other alternatives. However, "green" is a relative term, and implies that a particular choice among multiple options has, on balance, the best performance in terms of environmental considerations. Supercritical carbon dioxide ($scCO_2$), for example, is now generally accepted as a green solvent.

In this section, we present a brief primer on the term "supercritical fluid" (which refers to the thermodynamic state of a substance) and on the definitions of "green chemistry" and "green engineering." The last two terms encompass a number of ideas. Simply because a fluid material is utilized in its supercritical state does not make the substance or the process green. Instead, the fluid (*e.g.* CO_2, H_2O) in its supercritical state must be employed in a process or application (*e.g.* pasteurization) that meets the requirements or definition of "greenness."

The critical point refers to the termination of the vapor–liquid equilibrium locus (vapor pressure curve). All pure, thermally stable fluids possess a unique temperature and pressure at which the familiar liquid and vapor states become indistinguishable. This unique state is called the critical point, and is characterized by a critical temperature, T_c, critical pressure, P_c, critical density, ρ_c, and other critical properties. A mental experiment involving a familiar pressure cooker may illustrate the concept. Consider a pressure cooker filled with pure water (all air having been removed) so that only pure liquid and vapor water exist inside the vessel, at the boiling temperature of the liquid. It is well known that increasing the pressure causes a corresponding increase in the equilibrium saturation (boiling) temperature. Figure 5.1 (left) illustrates the vapor pressure curve of an arbitrary substance. At a sufficiently high pressure, the density and all thermodynamic properties of the vapor and liquid become indistinguishable and the liquid–vapor meniscus disappears. The critical point of a pure fluid represents a state of mechanical instability. A "supercritical fluid" is a fluid whose temperature and pressure are above T_c and P_c.

The scientific and practical significance of processes operating in the supercritical state is that all fluid properties, including density and solubility, become very sensitive to changes in temperature and pressure. Furthermore, the viscosity of an SCF is quantitatively nearer that of a normal liquid, but the diffusion rate is nearer that of an ordinary gas. The interfacial tension between an SCF and a liquid or solid is also fairly low. Thus, generally, an SCF solvent may have a solvent strength much like that of the ordinary liquid state, but the facile transport and penetrating ability is like

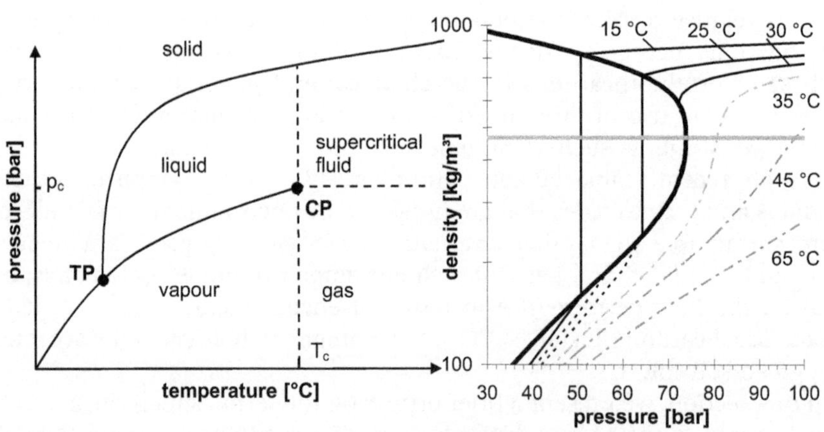

Figure 5.1 Supercritical fluids and green technology: *P–T* and density–*P* diagrams for supercritical fluids.

that of the gaseous state. The gaseous state increases the speed of extractions.

In general, SCFs give the developers of a new process additional flexibility in tailoring fluid properties to temperature and pressure. The "greenness" of the SCF depends on the inherent toxicity of the molecule. In the well-established coffee decaffeination process, for example, $scCO_2$ replaced methylene chloride (CH_2Cl_2), the appeal being the non-toxic nature of CO_2. However, the "greenness" of a solvent or of an entire process should be determined by rigorous assessment, if possible, using established principles of green chemistry and engineering. These principles have been established for the manufacture of chemicals and materials, but not formally for food processing. Still, it is instructive to consider selected principles of green chemistry and engineering as adapted to food pasteurization and processing, as outlined below.

Some relevant principles of green chemistry (adapted from ref. 21):

- Prevention: It is better to prevent the generation of waste than to treat waste generated during a process.
- Safer solvents and auxiliaries: It is better to utilize solvents and auxiliary materials that have no little or no toxicity.
- Design for energy efficiency: The energy requirements of processes should be quantified, and should be minimized to the extent possible.
- Renewable feedstocks: Where possible, renewable resources should be utilized.

Some relevant principles of green engineering (adapted from ref. 22):

- Lifecycle analysis: The entire lifecycle of a process or product, including all material and energy flows, should be considered.
- Economic, risk, and benefit analysis: All costs, including avoidance of future risks, accidents, and liability, should be considered and quantified if possible.

In light of these principles, the potential of $scCO_2$ (and other SCFs) as an agent for food pasteurization may be stated as follows:

1. CO_2 itself is a non-toxic, non-flammable substance that is a product of normal metabolism. Furthermore, it is currently used in the food industry in a variety of applications such as anaerobic packaging and carbonation.
2. CO_2 is currently captured from other industrial processes such as the manufacture of ammonia and fermentation, and is relatively inexpensive. Although CO_2 is a greenhouse gas, the amount of CO_2 used in the food industry is inconsequential compared with the amount emitted from the combustion of fossil fuels.

3. Pasteurization occurs with CO_2 under pressure and at temperatures near ambient. CO_2 is recovered by a combination of pressure reduction, refrigeration, and recompression. Thus most of the CO_2 may be recovered. The energy costs of $scCO_2$ lie in compression and refrigeration, which must be compared with the costs of heating and cooling (for thermal pasteurization) or the costs of generating electric fields or ultra-high pressure.

5.3 Current Issues in Food Pasteurization

Spoilage of food can occur during harvest, slaughter, manufacture, packaging, distribution, transport, and use by the consumer.[23] The infection chain can even start at an earlier point when feedstuff is already contaminated at harvest, as has been reported recently for vegetables such as spinach and tomatoes.

5.3.1 Food Preservation

Manufacturers of food products are currently under increasingly stringent demands concerning the production process. The entire course of manufacturing starting from raw materials to the final packing is closely monitored and inspected. This is essentially because of the growing demand for high-quality, prepackaged products, and for energy-saving and safer production processes.

Consumers and the business class today prefer fresh or mildly processed and user-friendly food products. Typically, such products are low in sugar, salt, and fat, and contain as few preservatives as possible. These demands frequently result in an excellent breeding ground for microorganisms. Microbiological stabilization of food is not concerned with the removal of all microorganisms present in the product but is intended to reduce the number of undesirable organisms below a specific critical value during the shelf-storage period of the product. However, microorganisms that can induce food alteration, degradation, and human disease (food intoxication, infection, and toxic infection) have to be dealt with differently. The most common pathogenic microorganisms are listed in Table 5.3.

The use of food preservatives and additives is viewed with increasing concern by consumers, as shown by the increasingly restrictive measures in the food regulations of industrialized countries. Apprehension about such substances is associated with the concerns for heat treatments, which are still the most widely used methods in the food industries to stabilize foodstuffs microbiologically. Thermal pasteurization (up to 80 °C) and sterilization (up to 120 °C) are successful treatments for eliminating the degenerative effects of enzymes and microorganisms, but they may also decrease food quality by causing alterations in the taste and organoleptic features of food products. The safety of certain dried foods is also of concern, and Bourdoux *et al.*[24] recently provided a comprehensive review of the safety of dried fruits

Table 5.3 Pathogenic microorganisms relevant to food spoilage.

Pathogenic microorganisms	Incubation period	Symptoms	Foods more at risk
Campylobacter jejuni Infection	3–5 days	Abdominal pains, stinking and colored diarrhea with the presence of blood, sickness, vomiting, fever	Unpasteurized raw milk, rare meat, spring water
Clostridium botulinum Intoxication	12–24 h up to 3–6 days	At first gastrointestinal symptoms, afterwards neurosis (double vision, speech difficulties, muscular weakness, paralysis)	Type A: tinned meat and vegetables Type B: products containing meat Type E: fishing products Type F: preserves containing meat and fish
Clostridium perfringens Toxic infection	8–20 h	Diarrhea, abdominal pains	White meat and red cooked meat
Listeria monocytogenes Infection	2–3 days up to 3 weeks	Fever, headache, sickness, vomiting	White and red meat, cheese, raw milk, cooked foods contaminated after cooking
Salmonella spp. Infection	24–48 h	Abdominal pains, diarrhea, vomiting, fever	Raw foods containing meat, shellfish, eggs, raw milk, cooked foods contaminated after cooking
Shigella spp. Infection	2–7 days	Abdominal pains, diarrhea with blood, fever	Adulterated foods caused by infected persons (fish products, vegetables, milk and dairy produce, ice-cream)
Staphylococcus aureus Intoxication	2–6 days	Sickness, vomiting, sweating, headache, diarrhea	Skin, milk, cream, ice-cream
Yersinia enterocolitica Toxic infection	1–7 days	Sickness, vomiting, sweating, headache, diarrhea, rheumatoid arthritis, urticaria	Raw milk, raw pig meat
Bacillus cereus Toxic infection	8–16 h	Diarrhea, abdominal pains	Meat products, soup, vegetables, puddings and sauces

and vegetables. Compressed CO2 treatment of products such as carrots has been reported, but the effect of $scCO_2$ drying on the microbial safety of such products has yet to be thoroughly evaluated. Consequently, in the future the food industry is expected to turn towards new and alternative technologies for extending shelf life while retaining all original nutritional properties of the end products.

5.3.2 Nutritional Properties

There is increasing evidence that many degenerative diseases, including cardiovascular disease, cancer, and aging, can be prevented by the consumption of fruit and vegetables. Polyphenols and vitamins are among the most discussed classes of phytochemicals that seem to be involved in the prevention of these diseases. These substances are widely distributed in the plant kingdom and are abundant in the human diet.

Phenolic compounds are a group of secondary plant metabolites and more than 8000 phenolic structures are currently known.[25] Polyphenols exhibit a wide range of biological effects, including antibacterial, anti-inflammatory, antiallergic, hepatoprotective, antithrombotic, antiviral, anticarcinogenic, and vasodilatory actions.[26] Many of these biological functions have been attributed to their free-radical scavenging and antioxidant activity. Phenolic compounds and vitamins are also responsible for major organoleptic characteristics of plant-derived foods and beverages, particularly color and taste properties.

However, polyphenols and vitamins are highly unstable and rapidly transform into a variety of reaction products when the plant cells are damaged, as occurs during food processing. The polyphenol and vitamin composition of plant-derived foods and beverages depends not only on the raw material used, but also on the extraction process and subsequent biochemical and chemical reactions. Polyphenols, for instance, are highly reactive compounds and are substrates for various enzymes, including polyphenoloxidases, peroxidases, glycosidases, and esterases. They undergo numerous enzymatic and chemical reactions during food processing and storage. Vitamins such as β-carotene are very sensitive to light and heat. These changes have great effects on food quality.

5.3.3 Innovative Techniques

A wide range of low-temperature pasteurization techniques have been developed. These methods include ionizing radiation, pulsed X-ray, pulsed light, pulsed electric fields, high-voltage discharge, magnetic fields, ultrasound, and high hydrostatic pressure (HHP) pasteurization at pressures exceeding 100 MPa. Supercritical fluid pasteurization also takes place at elevated pressure, but on the order of 30 MPa, *i.e.* not as high as in HHP pasteurization.[23,27,28] Figure 5.2 shows the range of high-pressure processes and media applied for pasteurization.

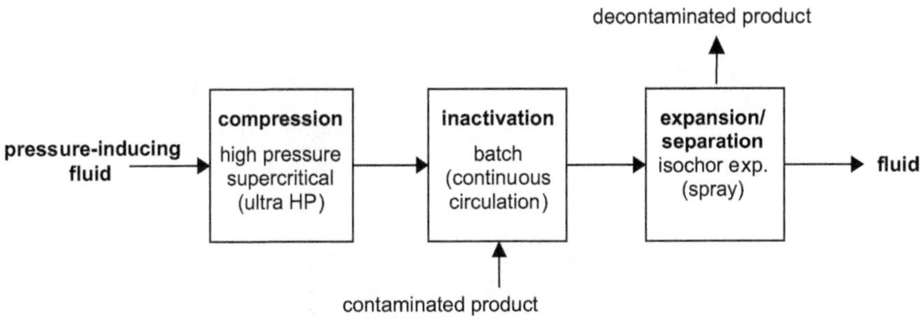

Figure 5.2 Schematic representation of processes investigated based on high pressures for pasteurization. Steps denoted in parentheses have only been investigated in research studies so far.

Pasteurization by very high hydrostatic pressure has been practiced for some time;[29–32] in HHP pasteurization, the chemical nature of the pressuring fluid does not contribute to pasteurization. In SCF pasteurization, the fluid itself is important. The use of various gases such as N_2O, Ar, and N_2 has attracted major interest. In particular, CO_2 has garnered particular interest both in early work and in recent investigations (Table 5.2) because of its unique physicochemical characteristics and microbiocidal potency.[33–35] The mechanisms of microorganism inactivation using media at high pressures and the pressure-induced damage to cell membranes, proteins, and other biogenic material have been investigated for more than half a century.[36,37] Despite all these efforts, the inactivation mechanisms remain largely unknown.[38] Two major research thrusts have evolved as a consequence: first, process research to optimize food pasteurization, *i.e.* to assure microorganism inactivation while preserving food properties, and second, basic science to elucidate the biocidal effect of the pressurized fluid media.

5.3.4 Packaging Materials

Packaging of food has played a major role ever since the emergence of convenience products. More recently, it has been stressed that the packaging material has a major impact on the preservation of food, with increasing focus on non-thermal methods.[27] Owing to the growing awareness that conventional plastic products rely on non-renewable raw materials, bioplastics have been developed for over a decade. These polymers are not only independent of fossil oil but are also usually fully biodegradable in a reasonable time span. Poly(lactic acid) (PLA) is one of the most important and industrially prevalent biopolymers.[39] The development of new packaging materials will have important implications for the pasteurization process. For example, the behavior of PLA in pressurized CO_2 has been studied extensively as it is the only biopolymer with accreditation for implant use.[40] As can be seen from Figure 5.3, amorphous PLA is strongly susceptible to

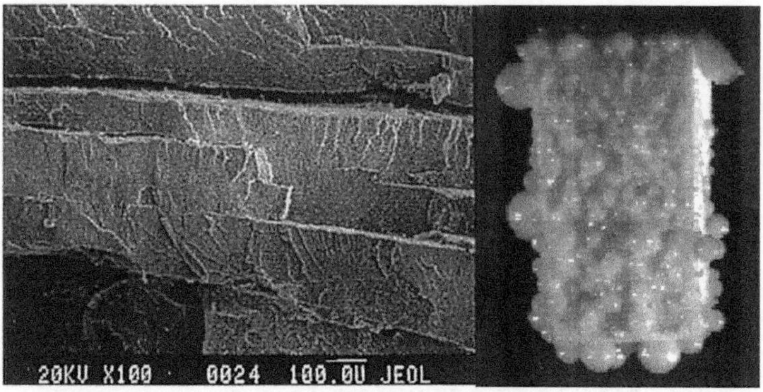

Figure 5.3 Swelling of CO_2-treated PLA films at 80 bar pressure and 35 °C for 220 min.
Reprinted with permission from K. A. Arora, A. J. Lesser and T. J. McCarthy, *Macromolecules*, 1999, **32**, 2562–2568. Copyright (1999) American Chemical Society.

foaming during $scCO_2$ treatment.[41] Polystyrene is another polymer used extensively in food packaging that is also susceptible to foaming and swelling.

Generally, polymers with a high amorphous content and glass transition temperatures below the process temperature are susceptible to foaming during CO_2 pressurization. Foaming can be controlled by adjusting both the depressurization rate and the total pressure applied to the polymer.[42–44]

5.3.5 Modified Atmosphere Packaging (MAP)

Preservation and packaging under a modified atmosphere are increasingly being used as an alternative to vacuum packaging, as they allow maintenance of the food structure, and represent an example of minimally processed food. Most manufacturers of fresh, minimally processed foods use chlorine-based washing and decontamination procedures. There is controversy about the formation of carcinogenic chlorinated compounds in water (chloramines and trihalomethanes), calling into question the use of chlorine,[45] so that alternatives to chlorine are desirable. Liquid and supercritical CO_2 can be used for pasteurization, but it is also known that CO_2 also has bacteriostatic character. Hence there may be room for exploiting CO_2 both for initial pasteurization and for subsequent bacteriostatic MAP, while avoiding the use of chlorinated compounds.

Specialty gas mixtures containing CO_2 can often be used in conjunction with low-temperature storage systems. In this way, it is possible to slow or prevent the growth and reproduction of bacteria, with no risk of the formation of carcinogenic compounds. Depending on the product type, different product mixtures (which can contain up to 80% nitrogen or oxygen) are utilized, such as for the preservation of the red color of fresh meat.[46]

5.4 Mechanisms and Biochemistry of Microbial Deactivation

It has been shown that a number of microorganisms, including bacteria, molds, viruses, and fungi can be deactivated or weakened by exposure to compressed (liquid or supercritical) CO_2. The mechanisms of deactivation are of both scientific and commercial interest. No single mechanism of action is predominant; different microorganisms apparently succumb to SCF conditions and, additionally, microorganisms at different growth phases show different inactivation rates under the same CO_2 process conditions.[47–51] In addition, multiple mechanisms may be present, and the effects of the process conditions may be confounded. Furthermore, different process conditions, including temperature, pressure, exposure time, level of agitation (or fluid shear forces), and pressure cycles, have been employed. Therefore, the design of commercial units will require careful consideration of the target microorganisms to be deactivated and the most effective process for so doing.

Several detailed reviews of the use of SCFs to deactivate microorganisms have appeared in the last decade. Zhang *et al.*[52] emphasized applications in medical sterilization and included the use of Ar, N_2, N_2O, and TFE in addition to CO_2 as the SCF, whereas Damar and Balaban[53] and Spilimbergo and Bertucco[54] focused on food processing with CO_2. Perrut[55] reviewed processes based on supercritical or liquid CO_2 for sterilization and virus inactivation.

The overall sequence of cellular events is generally as follows: (1) the SCF must dissolve into and diffuse through any external media (*e.g.* culture broth, food matrix) to the exterior of the cell; (2) the fluid must permeate across the cell exterior boundary (*e.g.* cell membrane or spore coat) to the interior of the microorganism; this permeation may cause reversible or irreversible damage to the boundary of the organism; (3) one or more intracellular events occurs that cause deactivation; the intracellular events are thought to include lowering of pH, alteration of electrolyte balance, precipitation of ions, denaturation of proteins, or extraction of water or other components that are important to cell metabolism; (4) upon lowering of pressure, the boundary of the organism may be partially or completely disrupted. Also, lowering the pressure favors the extraction of key components through the boundary to the exterior of the cell.

In the brief discussion of mechanisms that follows, we group the mechanisms under the heading of the key independent (process) variables pressure, temperature, pH, and method of mixing/agitation. The effects of these process variables on the possible microbiological mechanisms are then explained.

It is worth noting that a few recent studies have investigated the synergistic effect of CO_2 in combination with other pasteurization techniques. In particular, the use of high-power ultrasound (HPU)[56,57] and high hydrostatic pressure (HHP)[58–61] have been found to be of particular interest.

5.4.1 Pressure: Permeability, Membrane Disruption, and Extraction

High hydrostatic pressure processing uses a pressure on the order of 600 MPa. By contrast, the critical pressure of CO_2 is only 7.8 MPa and typical process conditions do not exceed 30 MPa. Therefore, the hydrostatic pressure applied to keep the fluid in the supercritical state is not, by itself, significant in the deactivation of microorganisms. Several investigators, however, have studied the rate of pressurizing or depressurizing as a means of deactivating bacteria. The earliest work by Frasier[62] employed rapid depressurization of *Escherichia coli* cells in liquid culture to physically burst the cell membrane. Several gases were used, including Ar, N_2, N_2O, and CO_2. Approximately 95–99% of the cells burst. Similar studies using N_2 were performed by Foster *et al.*[63] on six additional species, including *Staphylococcus aureus* and *Mycoplasma gallinarum.* In some species, physical disruption of cell membranes has been observed by scanning or transmission electron microscopy. Later, Spilimbergo *et al.*[32] used pressure cycling as an aid in the deactivation of bacterial spores of *Bacillus subtilis*, in a semi-continuous apparatus. Probably germination was induced by a particularly sudden change in their environmental conditions. The mechanical strength of a specific microorganism would clearly be a factor in the susceptibility of the organism to pressure-based disruption.

Increasing pressure also promotes infusion of the fluid into the microorganism, and decreasing pressure reverses this effect. The gas permeability of a given organism is key to its susceptibility to deactivation. The ability of the fluid to permeate into or out of a given organism is connected with the effects of the rate of depressurization, or of pressure cycling. For instance, Mitchell *et al.*[64] showed that the number of pressure cycles and system pressure has a significant influence on yeast inactivation, indicating that the increase in the number of pressure cycles and system pressure improve inactivation efficiency. More generally, the recent literature indicates that CO_2 and N_2O are more effective for deactivation than the chemically inert gas N_2. Since CO_2 and N_2O have similar critical conditions, it appears that both the chemical nature of the compressed fluid and the proximity to the critical point, with its enhanced diffusion and other favorable properties, play an important role in the mechanism of deactivation.

At a given temperature, the density of the SCF increases considerably with increase in pressure and, corresponding to this, the solvent strength of the fluid also increases with increase in density. Lin *et al.*[65] and Kamihira *et al.*[66] postulated that extraction of lipids has a significant affect on the deactivation of yeast cells and *E. coli* in culture media: when cells are dry, extraction of water has been cited as an essential step in cell deactivation.[34,45,66] More recently, Tamburini *et al.*[67] further explored this issue by means of flow cytometry, high-performance liquid chromatography–electrospray ionization mass spectrometry, and NMR spectroscopy, investigating the effect of CO_2 treatment on the membrane lipid profile and membrane permeability in

E. coli K12. After 15 min of $scCO_2$ treatment at 120 bar and 35 °C, the majority of bacterial cells dissipated their membrane potential (95%) and lost membrane integrity, as 81% became partially permeabilized and 18% fully permeabilized. Membrane permeabilization was associated with a 20% decrease in bacterial biovolume and with a strong (50%) reduction in phosphatidylglycerol (PG) membrane lipids, without altering the fatty acid composition and the degree of unsaturation of acyl chains. PGs are thought to play an important role in membrane stability, by reducing the motion of phosphatidylethanolamine (PE) along the membrane bilayer, therefore promoting the formation of inter-lipid hydrogen bonds.[36] The same group, examining the cell membrane integrity, demonstrated by flow cytometry (FCM) that although CO_2 treatments may assure total inactivation of different bacterial strains, a fraction of high-pressure carbon dioxide (HPCD)-treated cells maintained their metabolic activity and regrowth capacity, indicating that the treatment induces a transitory viable but not cultivable (VNBC) state.[67–69]

5.4.2 Temperature: Permeability and Extraction

It has generally been observed when using CO_2 that deactivation increases with increase in temperature. The critical point is 31.1 °C, and the literature includes experiments ranging from 0 °C (*i.e.* liquid CO_2) to 100 °C.[52] Higher temperatures may increase the fluidity and permeability of vegetative cell membranes, making them more susceptible to gas infusion. Increased permeability or fluidity will increase the rate of deactivation of the microorganism, regardless of the intracellular mechanism. Liao *et al.*[70] demonstrated that higher temperatures increased HPCD-induced outer membrane disruption and cytoplasmic membrane damage and decreased the membrane fluidity.

Another effect of temperature is related to the solvent power of the fluid. At a given pressure, the density and solvent power of CO_2 will decrease with increase in temperature, detracting from the ability of CO_2 to dissolve and extract components from the interior of cells. On the other hand, the solubility of certain species may increase with increase in temperature, overcoming the effect of decreasing CO_2 density. Hence both temperature and pressure affect the ability of the fluid to extract lipids or water, as mentioned above.

5.4.3 pH: Cell Metabolism and Protein Activity

Carbon dioxide is moderately soluble in water at elevated pressures and an equilibrium occurs between carbonate, bicarbonate, and H^+. Under practical process conditions, the pH of water with dissolved CO_2 lies between 3.1 and 3.5, so there is little ability in process design to control the pH over a wide range if CO_2 is the pasteurization fluid.[53] If the buffering ability of the cell is inadequate, large decreases in pH may inhibit cell metabolism or denature certain proteins and enzymes.[71] An interesting study[72] showed that

HPCD induced leakage loss of the proteins and DNA of *E. coli* as a function of treatment time. With regard to the protein changes, 182 proteins in the two-dimensional electrophoretic (2-DE) profile were not found in the HPCD-treated *E. coli*. Among 20 selected protein spots exhibiting significant changes in intensity, 18 were identified as 15 known proteins and two as hypothetical proteins. These proteins were involved in cell composition, energy metabolism pathways, nucleic acid metabolism, global stress regulation, and general metabolism.

Another effect of lower pH may be the precipitation of calcium or magnesium ions from intracellular solution,[73] which would interfere with the proteins that bind these ions. A recent study[74] indicated that pressurized CO_2 treatment induces the leakage of protein in *E. coli*, but the time of leakage lagged behind the time of 99% *E. coli* inactivation, so it was only the secondary phenomenon of inactivation. The inactivation of *E. coli* was related to the leakage of nucleic acids, K^+, and Mg^{2+} induced by the pressurized CO_2 treatment.

Interestingly, mineral acids producing the same pH in the culture media have not shown the same ability as CO_2 to deactivate microorganisms,[53] hence the peculiar ability of CO_2 to penetrate the cell membrane must be a key step in the deactivation sequence.[75]

On the other hand, N_2O is soluble in water but will not cause a decrease in pH, yet it has also been reported to be effective in deactivating some microorganisms.[76] More recently, Giulitti *et al.*[77,78] elucidated the mechanisms of dense gas inactivation with the aid of two novel approaches for measuring intracellular pH under high gas pressure: (i) pH-sensitive green fluorescent protein (GFP) variant of *Saccharomyces cerevisiae* as the probe and (ii) a pH-sensitive fluorophore based on fluorescein labeling of *Listeria innocua*. They demonstrated that, as a critical pressure is achieved, microbial inactivation is mainly due to pressure-induced membrane permeation. This is also stimulated by non-acidifying fluids, rather than cytoplasmic acidification.[19,20]

5.4.4 Fluid Flow and Contacting: Mass Transfer, Effect of Media, and Kinetics of Pasteurization

Although SCFs have favorable viscosity, surface tension, and diffusion that increase mass transfer rates compared with liquids, proper process equipment design can facilitate contact between the SCF and cells, decreasing the required contact time and reducing the size and cost of the equipment. Some specific equipment designs are detailed in Section 4.4.6 below. The efficiency of contacting of the fluid with the cells affects the rate of deactivation. Zhang *et al.*[52] and Damar and Balaban [53] reviewed various kinetic models for the deactivation of microorganisms in liquid culture media.

Interestingly, Ferrentino *et al.*[79] studied a model to mimic the effect of the suspending medium on the inactivation efficiency of *S. cerevisiae*. The results indicate that the equation obtained was much simpler that that obtained by correlating the microbial inactivation with pressure and temperature and, more importantly, it was independent of the suspending medium.

Jimenez *et al.*[80] showed how the deactivation kinetics of *S. aureus* embedded in a hydrogel depended on the thickness of the hydrogel layer. With no possibility of vigorously mixing the hydrogel, the deactivation kinetics decreased with increasing thickness. For any pasteurization process involving solid foods, or foods inside packaging, then the rate of diffusion of the SCF could be the rate-controlling factor. The thicker the characteristic dimension, the slower is the pasteurization. White *et al.*[81] demonstrated effective terminal sterilization of bone allografts while enclosed in terminal packaging. This is a very significant development for medical sterilization as it indicates that permeation of CO_2 through the packaging is sufficiently rapid to achieve sterilization in less than 1 h.

5.5 Applications of Supercritical Fluids for Food Preservation

Microbial and enzymatic inactivation with dense-phase CO_2 (DPCD) has been mostly applied to liquid food products, such as fruit juices, beer, wine and milk. In the decade, several studies focusing on microbial and enzymatic inactivation and the effects on food quality have been published.[82,83] They confirmed the feasibility and the effectiveness of this innovative technique towards liquid foodstuffs. The main advantage of CO_2 treatment in comparison with heat treatments consists in the low temperature applied, which has a much lower impact on the nutritional and chemical–physical properties of food.[84] In addition, compared with high hydrostatic pressure treatments, the relatively mild pressure conditions applied suggest a more easily controlled process with less expensive capital equipment. Table 5.4 summarizes the literature for numerous food products processed with liquid or supercritical CO_2; a relatively few studies involving N_2O are also noted. In several cases, an additive was used with the SCF to enhance pasteurization.

Several reviews have compiled the relevant current knowledge about the potential of $scCO_2$ as a non-thermal inactivation technology and summarized the most significant state of the art, including relevant applications and data, in both simple suspensions and complex media, for the treatment of a wide range of microorganisms in both liquid and solid substrates.[3–5,38,53,54,85–88]

Some information is available about the effects on perceived quality and nutritional properties of different liquid foods immediately after CO_2 treatment and during storage. Observations reported in the literature are scarce and conflicting and seem to depend on the food system investigated; the majority of studies have dealt with fruit juices.[53,61,89–95]

In some studies,[53,92,96–101] a sensory evaluation was conducted by panelists to understand better the consumer perspective and, of course, acceptance of CO_2-treated foods. Sensory analyses were performed on fresh control (*i.e.* untreated) and CO_2-processed juices using both difference-from-control

Table 5.4 Selected research on inactivation of microorganisms for pasteurization.

Food	Microorganism	Ref.
Liquid foods		
Fruit juices	Microbial populations	75
Orange, grapefruit, grape and apple juice	Natural flora	145, 209
Orange juice	Microbial populations	94, 96
Milk	*Listeria monocytogenes*, *E. coli*, *Brochotrix thermosphacta*, *Yersinia enterocolitica*, psychrotrophs/ *Pseudomonas fluorescens*	131, 210–214
Whole skim milk	*S. aureus*	215
Cloudy apple juice	Natural flora	189
Coconut water	Microbial populations	53, 93
Apple juice	Microbial populations	134, 189, 216–219
Watermelon juice	Natural flora	99, 220
Mandarin juice	Natural flora	102
Muscadine grape juice	Yeast and mold	100
Apple cider	*E. coli*	221
Cloudy apple juice	*E. coli*	222
Pear and apple juice	Yeast	223
Fruit juice, milk	*Enterococcus faecalis*	212
Skim and whole milk	*L. monocytogenes*	224
Peach, kiwi juice	*S. cerevisiae*, natural flora	135
Fruit juices	Natural flora	97
Carrot juice	Aerobic flora	59, 225, 226
Tomato juice	Microbial populations	227
Banana juice		228
Beer	Yeast	98
Grape juice	*S. cerevisiae/Candida stellata/ Kloeckera apiculata*	92
Litchi, lychee juice	Natural flora	229, 230
Solid foods		
Foods	*Salmonella*, *Listeria*	231
Ground beef	*E. coli*	232
Kimchi vegetable	Natural lactic acid bacteria	233
Alfalfa seed	*E. coli*	234
Grape must, tomato paste	*Pichia awry/Bacillus subtilis*	235
Shrimp, conch	Natural flora	236, 237
Oyster	Bacteria	238
Fish	Shellfish toxin	239
Ginseng	Natural flora	240
Meat	Natural flora	241
Ham, pork	Natural flora, *L. monocytogenes*	242–245
Salmon	*S. aureus*	246
Coconut	Natural flora	247, 248
Spinach	*E. coli*	249
Several meats	*E. coli*, *S. cerevisiae*	200
Carrots	Natural flora, *E. coli*	201, 226, 250
Paprika		251
Mozzarella cheese	*E. coli*, *Geobacillus stearothermophilus*, *L. monocytogenes*, spores	252
Studies using N_2O		
Apple juice	*S. cerevisiae*	101, 134
Milk	Natural flora	253

Table 5.5 Food treated with high-pressure CO_2.

Food system	Ref.
Chicken meat and shrimp	232
Whole fruit and vegetables	231
Ground beef	254
Kimchi	233, 255
Alfalfa seeds	256
Spinach leaves	249

and ranking tests. The former test was applied to determine whether a difference existed between the CO_2-treated sample and control, whereas the latter test gave information about consumers' preferences. Additionally, three studies[53,98,100] also conducted sensory analyses on heat-pasteurized samples. Sensory evaluation indicated that the CO_2-treated juices were indistinguishable from the untreated sample.[92,97,102] There were no significant differences in the flavor, aroma, and overall acceptability of these samples.[96,98,100,103] The CO_2-treated juices were ranked higher, and hence more desirable, by the sensory panel than the heat-pasteurized samples.[100,103]

Although most work has been carried out with liquid foods and juices, HPCD treatment of solid foods, particularly meat and fish, is gaining more attention. An overview of selected solid food products treated with HPCD is given in Table 5.5. Early efforts focused on seafood (shrimp, conch, oysters, salmon), and more recent work on pork and ham. Generally, a longer treatment time is needed to achieve substantial microbial inactivation as a result of the interactions between the food ingredients and intrinsic factors. Also, a longer treatment time is needed for processing solid matrices and cells (since the sample cannot be agitated during treatment) due to the limited diffusion of CO_2 into a solid.

5.5.1 Biofilms

At first consideration, biofilm formation does not play a vital role in the pasteurization of liquid foodstuffs. However, the surfaces of vessels, tubes, and other processing parts can easily support biofilm formation. If this biofilm is not removed properly, it can potentially infect the liquid product. To date, the inactivation of multilayers of microorganisms remains largely unknown; *e.g.* owing to limited diffusion of the CO_2 into the matrix of a biofilm, inactivation gradients could occur. During the past few years, research activity on biofilm formation has increased as nosocomial germs such as *Enterococcus faecalis*, *Pseudomonas aeruginosa*, and *Bacillus atrophaeus* show strong resistance against biocide agents.[104] Extracellular substances (EPS) form a complex and heterogeneous system that protects microorganisms within against biocides much better than plantonic cells; furthermore, biofilms are an optimum place for lateral gene transfer.[105]

The inactivation of biofilms using compressed carbon dioxide has only recently been addressed. It has been shown that *Bacillus mojavenis* in biofilms can be inactivated but is much more resilient than plantonic cells.[64] In addition to the *P. aeruginosa* in the biofilm, another group showed that pressurized CO_2 leads to detachment of the biofilm.[106] A hydrogen peroxide additive has also been shown to enhance sterilization in microbial biofilms.[88] Recently, the effect of HPCD on a *Staphylococcus aureus* biofilm was also studied.[107] It was found that treatment could remove the *S. aureus* biofilm to some extent but not significantly, and the effect increased with increase in pressure. The microstructure of the treated *S. aureus* biofilm was observed by scanning electron microscopy (SEM). SEM showed that the entire morphology of the biofilm was changed greatly, whereas the single cell structure was not.[107] Future research must strengthen the knowledge about the capability of CO_2 to inactivate germs in colonies and about the penetration into the EPS matrix.

5.5.2 Modeling Approaches for High-Pressure Microorganism Inactivation

For understanding the inactivation of microorganisms, mathematical models in a polynomial equation have been developed in order to fit the mechanism. However, most approaches have focused on two influencing factors, mainly concentration/time approaches. Zhang *et al.*[35] showed that the factors pressure, temperature, and type of medium have major influences on the inactivation of germs. To date, only a few studies have focused on modeling of CO_2-based inactivation, of which one considered thermodynamic aspects[32] and several considered kinetic aspects of inactivation. Whereas Karaman and Erkmen [108] concluded that first-order kinetics apply, models of higher order have also been proposed.[47–49,109–111] A completely different approach was chosen by Debs-Louka *et al.*,[112] who applied a *response surface* as an empirical description. Compared with conventional inactivation processes, the combined influence of the different variables in the SCF pasteurization process have still not been described satisfactorily and require further intensive studies.

5.5.3 Inactivation of Enzymes

Early studies on subcellular systems investigated the use of enzymes in $scCO_2$ as a non-homogeneous catalyst[33,113] and considered the prolonged stability of enzymes at high hydrostatic pressures.[114] However, the majority of recently published results concern the intended inactivation of enzymes.[87] The inactivation of enzymes plays a major role in liquid foodstuffs, as enzymes often cause undesirable changes in color or other characteristics of the product. Among these enzymes, pectinesterase (PE) causes cloud loss in some fruit juices, polyphenoloxidase (PPO) causes undesirable browning in fruits, vegetables, juices, and some seafoods, lypoxygenase (LOX) causes chlorophyll destruction and off-flavor development in frozen vegetables, and peroxidase (POD) has an important role in discoloration of foods and is used as an index of heat treatment efficacy in fruit and vegetable processing. Inactivation by CO_2 under pressure of enzymes that affect food quality has been reported by several groups.

Summaries of the recent literature, including the enzymes, the activity loss achieved, and treatment conditions, have been published.[87,115,116] Reduced activity of oxidase and esterase in apple juice was achieved by treatment at 7500 bar hydrostatic pressure[117] and amylase and protease were inactivated by using a continuous $scCO_2$ approach.[118] Kobayashi *et al.*[119] examined the inactivation of enzymes in unpasteurized sake, Zhang *et al.*[120] deactivated PPO in Pacific white shrimp, and Ortuno *et al.* [58] studied this enzyme and several others in feijoa purée. Although limited in number, studies on enzyme inactivation by CO_2 indicate its high potential, especially in fruit and vegetable juice processing where these enzymes cause quality deterioration if not inactivated. Carbon dioxide under pressure can inactivate certain enzymes at temperatures where thermal inactivation is not effective.[89] Enzyme inactivation by CO_2 could be due to many causes, such as pH lowering, conformational changes of the enzyme, and inhibitory effects of molecular CO_2 on enzyme activity. Balaban *et al.*[89] studied PE inactivation in orange juice. The pH of orange juice must be lowered to 2.4 for substantial PE inactivation, but CO_2 lowered the pH only to 3.1. Therefore, the pH lowering effect alone was not sufficient to explain enzyme inactivation. Other results[121] support this conclusion. Lipase exposure to CO_2 at ambient pressures was reported to result in lower enzyme activity due to a decrease in pH.[122] Rauh *et al.*[123] studied the effect of high hydrostatic pressure on enzymes with different sensitivities towards pressure inactivation. It has been shown that in some cases, inactivation by pressure can be higher at low temperatures.[124]

Carbon dioxide under pressure can change the isoelectric profiles and protein patterns of PPOs,[121] whereas CO_2 at atmospheric pressure does not affect these properties.[125] Chen *et al.*[121] obtained circular dichroism spectra of untreated and treated lobster, brown shrimp, and potato PPOs and showed that CO_2 under pressure caused conformational changes in the secondary structures (α-helix, β-sheet, β-turn, and random coil) of the enzymes. High pressure has also been reported to cause conformational changes in protein and enzyme molecules. Some workers even noted a marked antagonistic effect of pressure and temperature on protease in milk.[126] Hendricks *et al.*[127,128] reported that pressures around 310 MPa can cause irreversible damage to the secondary structure of proteins, but lower pressures cause no reversible changes upon depressurization. $scCO_2$ pressures are much lower than 310 MPa; therefore, the conformational changes that occur with CO_2 under pressure may not be caused by high pressure. Further research should investigate whether the conformational changes are caused by pressure alone and, if not, elucidate the other causes.

5.5.4 Processes Based on Gases Other Than CO_2

Another concern with pasteurization is the damage that may ensue to essential food ingredients such as vitamins.[117,129,130] Also, there is concern over changes in rheological, textural, and sensory properties. Specifically, the effect of pressurized media on dairy products has attracted interest.[131–133] Even though the use of pressurized CO_2 has shown good inactivation results for germs, in aqueous media CO_2 results in the formation of carbonic acid, reducing the pH to values below 4.0. For dairy products, this feature precludes the use of CO_2 for milk/cheese treatment because the pH falls below the isoelectric point of casein.

Investigations of the use of high-pressureN_2O for pasteurization purposes to avoid the pH problem have been reported.[101,134,135] The use of N_2O has been reported to be promising for microorganism inactivation. Concerning pasteurization with gases other than CO_2, and in particular N_2O, the literature is much more scant. Enomoto *et al.*[76] studied the effect of N_2O, N_2, and Ar on *S. cerevisiae* and spores of *Bacillus megaterium* in distilled water at 40–60 bar and 40 °C for 4 h and found that only N_2O assured good results in yeast inactivation. Fraser,[136] in 1951, observed a lesser reduction of *E. coli* using N_2O (compared with CO_2 after treatment for 3 min at 38 °C and 52 bar. More recently, Spilimbergo and Ciola[135] treated both peach and kiwi juice with a mixture of CO_2 and N_2O, achieving total inactivation of natural microorganisms while maintaining acceptable sensory quality.

Mun and co-workers[137–139] also examined the bactericidal effect of N_2O on different bacterial strains as a function of operating pressure, temperature, mixing intensity, and working volume ratio (defined as the ratio of sample volume to the reactor volume to be filled with supercritical fluid) in comparison with scCO_2 treatment. They found that among the reaction parameters investigated, mixing intensity appeared to be the most important operating parameter affecting the bactericidal efficiency. The release of intracellular substances in cells as a result of the scN_2O treatment was explained as one of major bactericidal actions of scN_2O. At the same time, relatively negligible changes in proteins or enzyme activities in the cells in case of the scN_2O treatment was observed compared with the scCO_2 treatment.

Nitrous oxide merits further consideration either as an alternative gas or as a mixture with dense CO_2 and perhaps other gases. Nitrous oxide is a non-toxic and relatively inexpensive, colorless, and practically odorless gas already used in the food industry as a propellant. It is known to have a biostatic effect on certain fungi.[140] No explanation of the bactericidal effects of N_2O are given in the literature. However, it is known that N_2O has a slightly narcotic effect and its solubility in water at atmospheric pressure is comparable to that of CO_2. Its inhalation causes euphoria and confusion, in addition to nausea. Its anesthetic effect is not clearly understood but is probably due to its interaction with the cerebral cell membrane, which alters the intracellular communications in the synapses. This chemical interaction, which must change the membrane fluidity, volume, and thickness and lead to a volumetric expansion of the phospholipid structure, is closely correlated with the high solubility of N_2O in lipids.

5.5.5 Subcellular Systems (Phages, Viruses, Proteins, Prions, Hazardous Macromolecular Substances)

For food pasteurization, the center of attention still lies in the inactivation of bacteria and spores and the transfer of new technology from the laboratory to pilot and industrial scales.[38,141,142] However, research on food safety has recently expanded to include non-living although biogenic materials such as enzymes (as mentioned previously[53,123,124,143–147]), and also viruses/phages[30,148–151] and, more recently, toxins.

A wide range of other biogenic threats must be taken into account when discussing pasteurization. Because many products for biotech and food industry applications are derived from large-scale bacterial fermentation, the infection of the fermenting organism by bacteriophages can become a major issue. The effect of HHP on phages was addressed in the early years of the technique, beginning with the well-described T4 coliphage[37] and later with the lactococcal and probiotic bacteria phages.[148,152] As with bacterial inactivation, it is not clear whether the inactivation is caused by increasing permeability of the enveloping protein capsid or by other mechanisms. Complex viruses, which are similar to phages, but infect not only bacteria but also mammals, have been reported to be less sensitive to pressure treatment than Gram-negative bacteria and yeasts, but more susceptible than Gram-positive bacteria.[153] It has been proven that, depending on the medium, some viruses can be inactivated at high pressures whereas some remain fully infectious.[150] This has also been reported in other studies, where adenovirus was inactivated at a pressure of 400 MPa but poliovirus did not exhibit any reduction at pressures up to 600 MPa. Baert *et al.*[149] summarized inactivation methods for viruses and listed mainly non-enveloped viruses investigated. No publication has dealt with pressures below 275 MPa for the viruses discussed in this review.

In addition to vegetative bacteria, spores, viruses, and phages, the food safety industry must also consider hazardous macromolecular substances (toxins) that are produced by many microorganisms. Toxins are usually classified as endo- or exotoxins, of which the latter are metabolites of microorganisms. Secondary fungal metabolites can be mycotoxins such as aflatoxin, which have been shown to be carcinogenic, mutagenic, and teratogenic. Even though we are aware of the dangers caused by these substances, very little work on low-temperature inactivation has been carried out. One of the first studies to deal with SCFs was the CO_2-based extraction of shellfish toxin. It was shown that solubilization of the toxin is possible, and depends on the pressure, temperature, and nature of the co-solvent.[154]

Endotoxins play a vital role in bioprocess-derived proteins in pharmaceutical production but there has been little focus on food processing.

Endotoxins are cell-wall lipopolysaccharides (LPSs) that are liberated by Gram-negative bacteria. Taken up in high concentrations into the human vascular system, they control cytokine production (*e.g.* interleukin), and can cause fever or even septic shock, and are removed by only very few sterilization processes.[155,156] Even ultrafiltration and ion-exchange approaches are often unsuccessful and as the LPSs are heat stable, they are inactivated only at temperatures above 220 °C.[157] The concentration limit for intravenal applications has been set to five endotoxin units (EU) per kilogram body weight by the European Pharmacopoeia.[158] Whereas steam sterilization is not capable of successfully removing or inactivating endotoxins,[159] Rybka *et al.*[160] and Tarafa *et al.*[161] were able to extract LPSs using $scCO_2$. In the latter study, the solvent power of the CO_2 was enhanced by adding water and a surfactant.

Prions, much like endotoxins, are highly resistant to many conventional inactivation methods and have to be handled with a combined thermal–chemical treatment.[162] Hence their inactivation cannot be assured when thermolabile products are treated. The demand for novel methods increased as the occurrence of bovine spongiform encephalopathy (BSE; mad cow disease) and variant Creutzfeldt–Jakob disease (vCJD) caused major disturbances in the food and feedstuff markets.[163] As the misfolding of the protein seems to determine how infectious prions are, publications correlating conformational changes with HHP are of interest.[164] Silva[164] discussed, on the basis of the volume cavity model, how pressure unfolding is governed by the penetration of water into the protein matrix.

5.5.6 Treatment of Solid Objects

The transfer of SCF technology from food treatment to biomedical applications was proposed by Spilimbergo *et al.*[32] Sterilization of implantable objects using gaseous, supercritical or liquid CO_2 has so far not been investigated as thoroughly as the pasteurization of liquid products. Only a few studies are known that used additives such as acetic acid,[34] triclosan, and propanol[165] for CO_2-based sterilization. However, Matthews's group has focused on implant treatment for many years and achieved promising results, particularly using H_2O_2 in trace amounts.[35,166–168] Further work, mainly concentrating the use of additives such as ethanol and hydrogen peroxide, has been summarized in a review.[168] The effectiveness of $scCO_2$ sterilization apparently correlates with the solubility of the additive in CO_2. However, applying a gas, such as ozone, as an additive overcomes this drawback and has been successfully demonstrated. The interaction of CO_2 with ozone as an additive can be seen in Figure 5.4. In such a process, production of ozone can be carried out onsite by corona discharge and, after use, the unreacted ozone can be converted to oxygen by catalytic reaction.

Further, it has been shown that ozone is effective as a sterilizing gas without CO_2 but is more efficient in a mixture with sub- and supercritical CO_2. From the literature, it is well known that glycolipids and glycoproteins,

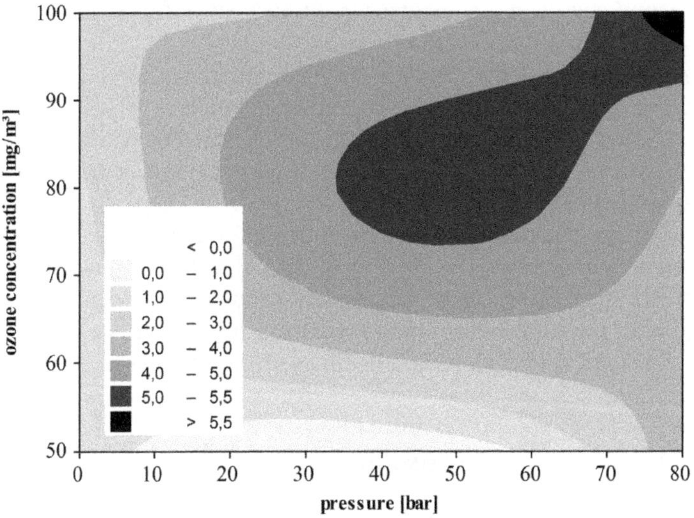

Figure 5.4 Contour diagram for inactivation [log (N_0/N)] of *B. atrophaeus* depending on ozone concentration and pressure at 35 °C for 220 min (unpublished results).

which are essential cell compounds, are oxidized by ozone.[169] Also, Christensen and Giese[170] showed that pyrimidines and purines of nucleic acids are attacked. Enzyme activity is likewise disturbed by the reaction of ozone with the sulfhydryl groups of some enzymes.[171] Therefore, is not surprising that pure ozone acts as a sterilizing agent. An interaction effect between CO_2 and ozone exists, as can be seen in Figure 5.4. This effect could be attributed to the effect of a low pH resulting from the formation of carbonic acid. However, the influence of pH is controversial in the literature.[172,173] It has not yet been clarified whether small acid-soluble spore proteins (SASPs) play a role, or whether the compressed fluid has a swelling effect of the spore coats or the pressure gradient influences the permeability of the spores. The possible role of CO_2 in additive-based sterilization must be further scrutinized.

5.5.7 Unsolved Problems to Date

Garcia-Gonzalez *et al.*[38] underlined five main problems concerning CO_2 sterilization of medical objects in their review. In addition to the widely unknown inactivation effect and concerns regarding highly resistant mutants, the following points were discussed:

1. The inactivation depends on the identity of the contaminating species. This becomes evident for inactivation of Gram-positive or -negative bacteria, fungi, and spores. It seems that spores of *B. atrophaeus* are among the most difficult to be inactivated with $scCO_2$. Hence

B. atrophaeus can be used as a bioindicator for inactivation in the $scCO_2$-sterilization process.

2. Sterilization of solid objects has not been sufficiently investigated. Only a few publications have reported successful inactivation with non-homogeneous materials such as textiles. Penetration of materials with diverse (pore) structure and thickness with CO_2 needs further research.
3. Finally, the packaging of sterilized objects has received little attention. However, it has been shown that with the use of gas/SCF as the inactivating agent, it is possible to treat prepacked objects. Commercially available packages with a paper membrane, which is easily penetrated by the fluid, and a clear HDPE film can be processed.

5.5.8 Outlook and Discussion

Food is historically understood to be the nutrition that is delivered through the human digestive system and absorbed by the bloodstream to the organs. However, medical progress now allows for a "detour" around the digestive tract to supply nutrients intravenously directly into bloodstream with KCl, sugars, and other essential substances. Intravenous delivery also allows controlled release of hormones, drugs, insulin, and other materials. Therefore, we include a discussion of new potential materials and applications of intravenous delivery systems and new challenges and sterility issues arising from this technology.

5.5.9 Materials and Composites of Future Interest

Reconstructive surgery encompasses soft tissue implants and artificial skin materials such as silicone, hydrogels, and collagen.[174,175] Polymers also play a major role in many fields of medicine. They can become part of the cardiovascular system in pacemakers and heart valves, stents (tube-like scaffolds), and also orthopedic and dental implants, very often as composites with metals such as aluminum, stainless steel, cobalt–chrome alloys or titanium.[176,177] Furthermore, in biotechnology, these materials are used for diagnostic applications, *e.g.* sensors, electrodes, and batteries. Approaches such as lab-on-a-chip technologies make use of glass, semiconductors, platinum, gold, titanium, or tantalum in extremely delicate layers,[178] for which the resistance against conventional cleaning and sterilizing methods is barely known. The range of materials should also include materials that are frequently used for membranes in hemodialysis, plasmaphoresis, oxygenation, and ultrafiltration in the medical and food industries.[179–181] It also has to be clarified whether objects with functionalized surfaces,[182] *e.g.* with protein-repelling poly(ethylene glycol) (PEG)[183] or poly(ethylene oxide),[184] or with coatings of heparin[185] and albumin,[186] can be sterilized with gas-based inactivation methods.

5.6 Commercial Aspects

5.6.1 Equipment for CO_2 Technology

Several batch, semi-continuous, and continuous systems have been developed for DPCD applications. In a batch system, CO_2 and treatment solution are stationary in a reactor during treatment. A semi-continuous system allows a continuous flow of CO_2 through the chamber, whereas a continuous system allows a flow of both CO_2 and the liquid food through the system.[116]

Early studies used batch systems.[187] A typical batch system has a CO_2 gas cylinder, a pressure regulator, a pressure vessel, a water-bath or heater, and a CO_2 release valve (Figure 5.5).

The sample is placed in the pressure vessel and the temperature is set to the desired value, then CO_2 is introduced into the vessel until the sample is saturated at the desired pressure and temperature. The sample is left in the vessel for a period of time and then the CO_2 outlet valve is opened to release the gas. Some systems contain an agitator to decrease the time needed to saturate the sample with CO_2.

In 2005, Spilimbergo and Mantoan[188] developed a multi-batch apparatus, specifically constructed for the determination of the rate of microorganism inactivation in foodstuffs by DPCD. Figure 5.6 sketches the equipment set-up. The multi-batch apparatus consists of 10 identical reactors connected in parallel, so that each experimental run provides a set of 10 experimental data obtained under identical process conditions but with different treatment times. Each reactor is connected to an on–off valve that can be used to de-pressurize it independently of the others. The 10 reactors are submerged in a single thermostated water-bath so that the temperature is uniform in all of

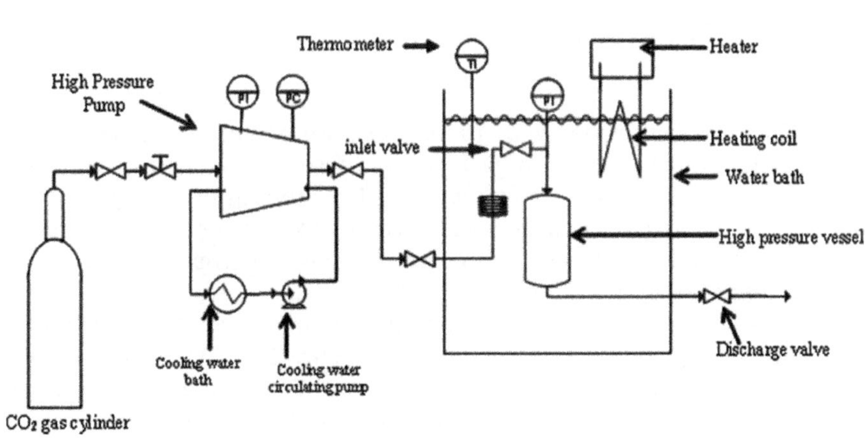

Figure 5.5 A typical batch dense-phase CO_2 system.
Reproduced from F. Dehghani, N. Annabi, M. Titus, P. Valtchev and A. Tumilar, *Biotechnology and Bioengineering*[240] John Wiley and Sons, © 2008 Wiley Periodicals, Inc.

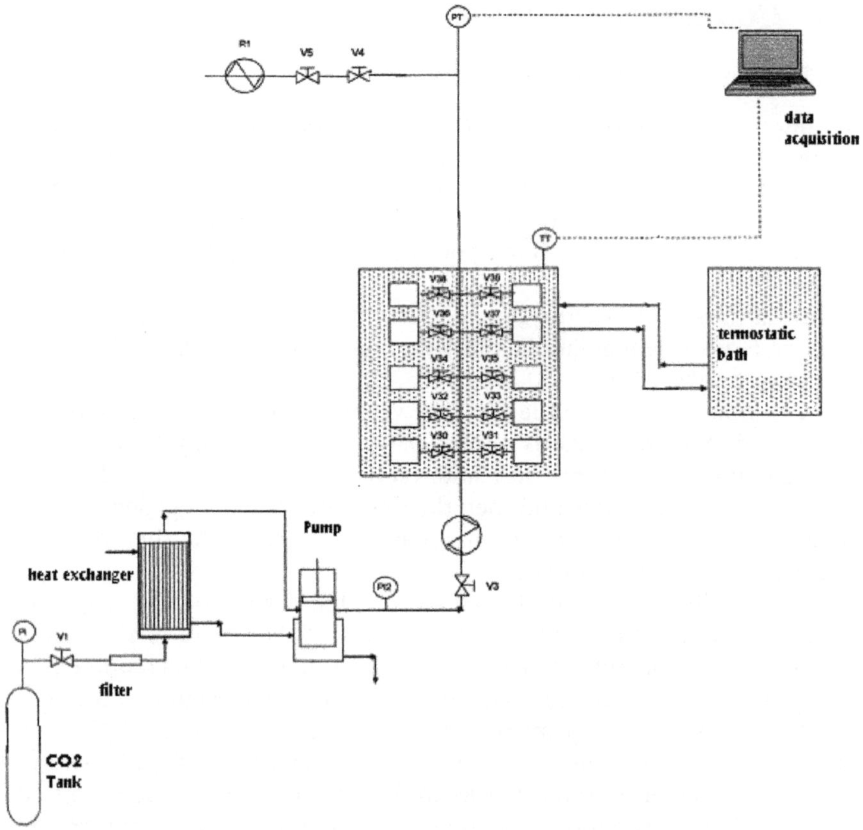

Figure 5.6 Sketch of the multi-batch system.
Reprinted with permission from S. Spilimbergo and D. Mantoan, *International Journal of Food Engineering*, 2006, **2**, 6.[189] © 2011 Walter de Gruyter GmbH & Co. KG, Berlin/Boston.

them. The bath is provided with 10 magnetic stirrers, each one serving one reactor where a magnetic bar is located. These stirrers permit sufficient mixing to ensure fast mass transport of CO_2 from the gas to the liquid phase. The CO_2 reservoir is heated by a water circuit so that the selected process temperature could be precisely controlled without thermal overheating of the samples during pressurization.

A good example of this was presented by Shimoda *et al.*,[190] who proposed the continuous-plug-flow system shown in Figure 5.7. The liquid CO_2 and the sample are simultaneously pumped into the CO_2 dissolving vessel (215 mL) with preset flow rates. Liquid CO_2 is heated to the gaseous or supercritical state while passing through an evaporator, and is then dispersed into the liquid by a stainless-steel mesh filter (10 μm pore size) attached to the bottom of the dissolving vessel. The microbubbles of pressurized CO_2 migrate upwards while being dissolved in the suspension.

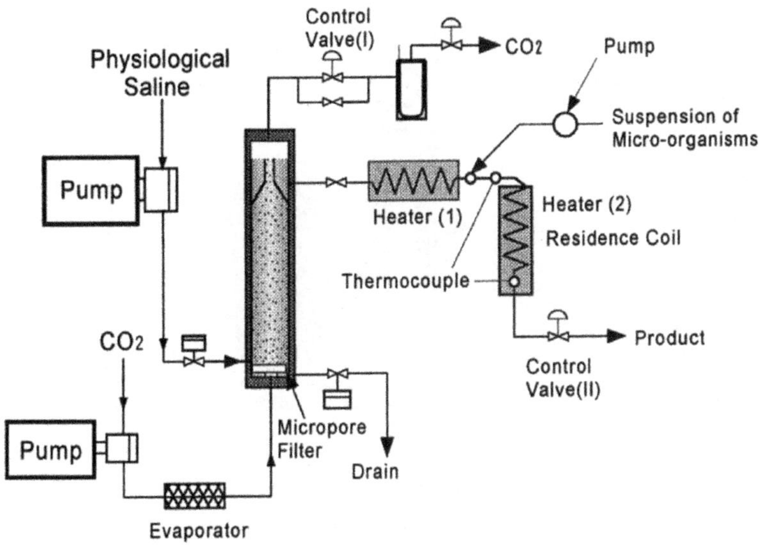

Figure 5.7 Schematic of continuous-flow system for microbial inactivation with microbubbles of pressurized CO_2.
Reproduced from M. Shimoda *et al.*, The influence of dissolved CO_2 concentration on the death kinetics of *Saccharomyces cerevisiae*, *Journal of Applied Microbiology*[190] John Wiley and Sons, © 2001 The Society for Applied Microbiology.

The saturated solution is heated to the experimental temperature by a heater (1) and soon after a suspension of microorganisms is introduced into the CO_2-saturated flow of saline. A second heater (2) keeps the temperature constant while a residence coil assures a sufficient residence time for the inactivation. After the sample has flowed through the residence coils, it is withdrawn *via* a pressure control valve (II), keeping the pressure at 1 bar.

Another interesting apparatus was conceived by PoroCrit (Berkeley, CA, USA),[191] shown schematically in Figure 5.8. The principle is similar to that described above, In the first part of the continuous apparatus, a system pressure of 75–150 bar is used to reach a dissolved CO_2 concentration of 5–6%. The peculiar feature of the system is the use of a microporous polypropylene membrane contactor to carbonate the stream rapidly to saturation. This system appears to be particularly efficient as the cartridge is very effective for carbonation, owing to the large surface area created by 15 fibers of 1.8 mm diameter, 390 mm length and 83 cm^2 surface area. It can completely saturate a water or orange juice samples at any preset flow rate. Dense CO_2 is continuously recirculated without depressurization. In this way, the fastest killing rate of a large range of bacteria and yeasts could be obtained at a particular pressure and temperature that never exceeds 45 °C.[192]

In 1999, Praxair (Burr Ridge, IL, USA) developed a continuous-flow DPCD system using the concept shown in Figure 5.9. CO_2 and the product are pumped through the system and mixed before passing through the

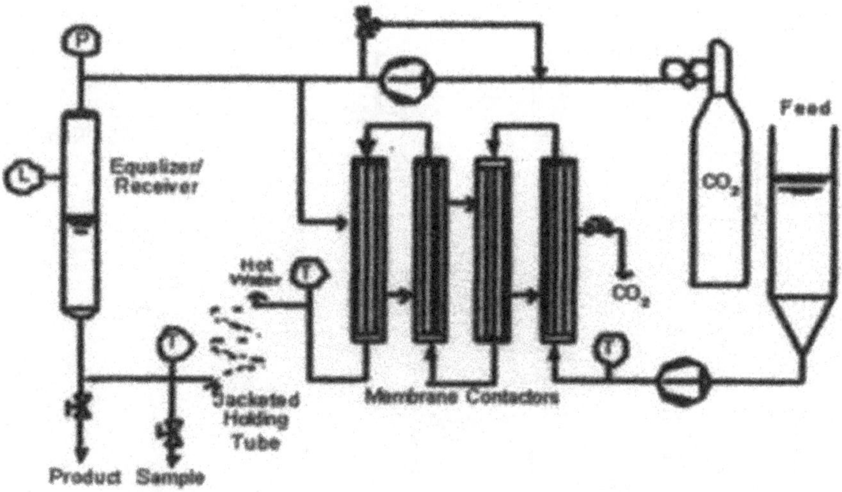

Figure 5.8 A continuous CO_2 membrane contactor system.
Reproduced with permission from M. Sims and E. Estigarribia, 6th International Symposium on Supercritical Fluids, Versailles, France, 2003.[193]

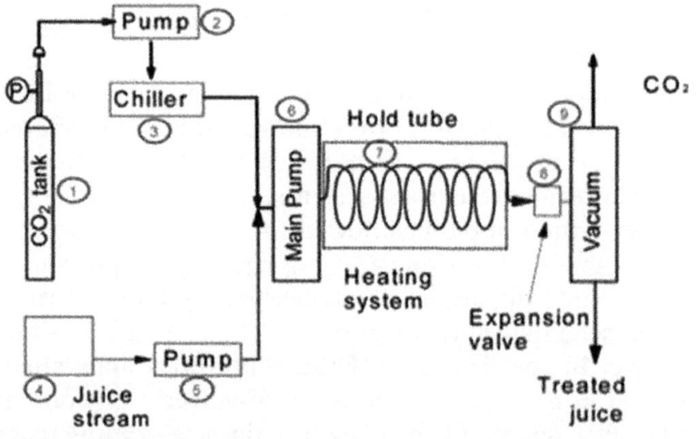

Figure 5.9 A continuous-flow dense-phase system.
Reprinted with permission from ref. 100. Copyright (2006) American Chemical Society.

high-pressure pump, which increases the pressure to the process level. The product temperature is controlled in holding coils. The residence time is adjusted by setting the flow rate of the product through the coils. At the end of the process, an expansion valve is used to release CO_2 from the mixture. It is possible to pull out the residual CO_2 in the food in a vacuum tank. This system has been shown to be very effective in killing pathogens and spoilage bacteria in a short time.[53,98,99,194,195]

Another example of a continuous pilot-scale plant with a gas–liquid CO_2 porous metal contactor was proposed by Yuk *et al.* and applied to apple cider.[196,197]

Combined systems with techniques such as pulsed electric field[198] and ultrasound[199] have been developed. In one example, ultrasound was combined to $scCO_2$ by Ortuno *et al.*,[199] demonstrating that a 95% shorter time was needed to inactivate *E. coli* in Luria–Bertani broth and orange and apple juice. The ultrasound was integrated into the process by incorporating an ultrasonic transducer with the vessel. Additional evidence of the effect of combined technologies was provided by Benedito *et al.*[200] and Ferrentino and Spilimbergo,[201] who proposed applications with HPU and $scCO_2$ for microbial and enzyme inactivation on solid and liquid products. In this case, an ultrasound system was embedded in the plant: the system consisted of transducer and a sonotrode. The application of HPU enhanced the inactivation mechanisms of microorganisms, generating a vigorous agitation that facilitated the solubilization of CO_2 and the mass transfer process. Hence, using combined techniques, reasonable industrial processing times and mild process conditions could be used that could result into cost reductions and minimize food nutritional and organoleptic changes. HPU was also integrated into a continuous plant.[57]

5.6.2 Patents

A considerable number of patents have been granted for processes that employ the CO_2 pasteurization/sterilization concept[202] for microbial inactivation; most of these concern liquid foodstuffs. Patents with a US, an EP or a WO analogue are listed in Table 5.6. Additional patents can be found elsewhere.[202] For patents filed or granted only in Japan, the reader is referred to Spilimbergo and Bertucco.[54]

5.6.3 Commercialization

Carbon dioxide pasteurization is operative and essentially ready for deployment at the commercial scale for a few liquid foods. Most of the commercialization efforts so far have been performed by Praxair (Burr Ridge, IL, USA). Praxair developed a continuous system that utilizes CO_2 processing as a non-thermal alternative to thermal pasteurization[84] based on technology licensed from the University of Florida.[203] This system has been commercialized under the Trade Mark "Better Than Fresh (BTF)." To date, Praxair has constructed four mobile BTF units for processing about 1.5 $L\,min^{-1}$ for demonstration purposes. In addition, a commercial-scale unit of 150 $L\,min^{-1}$ has also been constructed[84] and tested at an orange juice processing plant in Florida.

Pilot-scale equipment was manufactured by Mitsubishi Kakoki (Tokyo, Japan) on behalf of and according to the patents owned by Shimadzu (Kyoto, Japan)[204,205] for the continuous treatment of liquid foods. This equipment

Table 5.6 Relevant patents in the field of HPCD pasteurization/sterilization.

Patent no.	Remarks	Ref.
US 3483005	Irradiation method with pressurized CO_2 for sterilizing meat	257
US 4263253	Process for rendering solids sterile in which the solid is dissolved in a supercritical gas and then transported through a sterile filter to retain bacteria and germs	258
US 5393547 WO 90/02799	Method for inactivating enzymes in foods	203
EP 0488012 B1	Method for germ reduction in semi-prepared products containing flour	259
US 5380826	Method and apparatus for supercritical disruption of microbial cells and extraction of intracellular components	260
US 5520943 US 5667835 US 5704276 US 5869123 EP 0786513 B1 US 6616849 B1 US 6821481 B1	Different methods and apparatus to inactivate enzymes and microorganisms in liquids, or to deodorize liquid materials (foods or medicines) using microbubbles of CO_2 in batch or continuous implementation	205, 261–264
EP 0674845 B1	Method for deactivating enzymes or microorganisms in food comprising heat treatment of the enzymes or microorganisms in an aqueous medium under CO_2 or N_2O pressure, and in the presence of a reducing agent selected from the active sulfured compounds	265
US 6723365 B2	Continuous method and apparatus for reducing microbial and/or enzymatic activity in liquids	203
US 6149864	Batch method for sterilizing materials for medical applications, particularly polymers for drug delivery and implantation	266
US 2002/0122860	Continuous method and apparatus for reducing microbial and/or enzymatic activity in liquids	267
US 5996155	Process for cleaning, disinfecting, and sterilizing materials using the combination of dense-phase CO_2, UV radiation, and sterilants	15
US 2004/0120852 A1	Method and apparatus for sterilizing medical instruments	268
US 6331272 B1	Continuous method and apparatus (membrane contactor) for preserving liquids	191
US 2002/0127317 A1	Process to inhibit or reduce bacterial growth in a liquid by adding CO_2 to the liquid, and thermally inactivating the bacteria in the liquid	269

Table 5.6 (*Continued*)

Patent no.	Remarks	Ref.
WO 2004/045316 A1	Continuous method and apparatus for enzyme deactivation, microbial reduction, and removing CO_2 dissolved in the liquid to minimize the loss of scent of components	270
WO 2004/000434 A1	Continuous method for efficiently separating the dissolved CO_2 gas from the treated liquid	271
WO 03/101210 A1	Method for pressurized decontamination of meat using carbonic acid	46
WO 2004/039180 A1	Batch method for inactivating biological contaminants in liquid high-fat and/or protein food or feed products	272
US 2004/0234661 A1	Continuous method and apparatus for reducing microbial and/or enzymatic activity in a liquid beer or wine product	273
US 2005/0084581 A1	Method and apparatus (not requiring any large-sized pressure tanks) to inactivate enzymes and microorganisms in liquids	274
US 7108832	Method and apparatus for achieving 6D (6 decades) microbial reduction in (bio)materials using chemical-containing pressurized CO_2	275
US 2008171116	Pasteurization of prepackaged food using $scCO_2$	276
US 20100203206	Method for treating food products and food product treatment apparatus	277
US20120118175	Treatment to reduce microorganisms with carbon dioxide by multiple pressure oscillations	278
WO2013001117 A1	Method of inactivating microorganisms by combining ultrasonic and SCF treatments	279
US20120288614 A1	Method of treating solid and semi-solid foods to reduce microorganisms and enzymes in the food	280

consisted of a 5.8 L treating vessel through which CO_2 and liquid foodstuff were simultaneously pumped at maximal flow rates of 3.0 and 20 $kg\,h^{-1}$, respectively.

On the basis of their own patent,[191] PoroCrit also commercialized equipment (a membrane contactor consisting of several hollow-fiber membrane modules) for the continuous HPCD pasteurization of liquid foods (mainly beverages, juices, milk, and wine).

Further commercialization will occur when the market drivers, primarily food sensory properties and safety, create sufficient demand. The technological challenges for the SCF pasteurization of liquid foods do not appear to be great, given the amount of research and development carried out up to this point. The development of commercial processes for solid foods or packaged foods is not nearly so well developed.

5.6.4 Economic Aspects

Of course, an essential consideration for the adoption of an scCO_2 pasteurization process is its economic feasibility. However, there are very few papers dealing with economic aspects of CO_2 processes (extraction, reaction enzymatic catalysis) in the literature,[9,206] and, to our knowledge, no cost estimates for CO_2 pasteurization have been published. Detailed economic analyses will depend on the particular food item and market opportunity, which would be considered confidential business information. Nevertheless, economic analysis is, with the study of the effect of stabilization treatments on perceived quality, of the utmost importance because it is the key factor for consumer product acceptance. An example of an economic analysis is a preliminary analysis of CO_2 technology in the medical sterilization area that was published by Kaiser *et al.*[207] This analysis compared a hypothetical CO_2 sterilizer with steam autoclaves, ethylene oxide sterilization, and hydrogen plasma sterilization in a clinical (hospital) setting.

To determine economic feasibility, first an overview should be made of the total costs of the process lines (*e.g.* investment and running costs) and a comparison with the current thermal process lines in use. Spilimbergo's group has attempted a preliminary economic analysis for apple juice pasteurization. A rough estimation indicates that the production cost of apple juice pasteurized with a CO_2 continuous plant such as that patented by Praxair[208] are higher than the cost for juice pasteurized thermally with a traditional tube-in-tube heat exchanger. The calculation is based on the same feed throughput per year. On the other hand, an increase of a few cents per liter for the juice product compared with the current price in the market could guarantee both a product of better quality and amortization of the additional costs (confidential data, unpublished). Additional studies are clearly needed to develop the basis for economic evaluation, particularly concerning energy consumption and possible savings.

5.7 Conclusion

A number of studies have established some of the basic science behind SCF pasteurization, particularly for CO_2 as the SCF. Several applications to liquid foods have been demonstrated at the pilot-plant or near-commercial scale. Related work has shown that disinfection or even sterilization of solid objects is possible; these results have potential application in the cleaning and disinfection of food preparation equipment.

With further comparative analysis with existing methods, SCFs may prove to be the basis for green (or "greener") pasteurization technology. scCO_2 pasteurization is a low-temperature process that allows a dramatic reduction or complete elimination of persistent, toxic chemicals or irradiation. Furthermore, the relatively low pressure of an SCF process is comparable to approaches using high hydrostatic pressure and it is favorable due to energy and safety concerns. High-pressure technology for supercritical treatment is commercially available,[7] and adoption of the technology in other areas such as textile cleaning and extraction processes provides confidence that there are no insurmountable barriers for application in liquid food pasteurization.

Currently, several industrial sectors are considering the application of CO_2 technology in the processing of foods and biodegradable pharmaceutical and cosmetic materials. However, none of the interested sectors are currently using $scCO_2$ in their primary their pasteurization process. Since DPCD prototype equipment is already widely available, the extensive use of CO_2 in a non-thermal pasteurization treatment process is probably a matter of just a few years away. In view of the realization of this goal, we believe that next future research developments in CO_2 treatment should now move to explore new issues, such as the following:

1. The retention of vitamins and modifications of cell enzymes and volatile compound loss (sensorial analysis in different substrates).
2. The precise inactivation mechanism in order to optimize the process (operating conditions) and minimization of costs at the industrial scale.
3. The effect of CO_2 processing on food (different and new), pharmaceutical, and cosmetic components.[54]

References

1. M. Sihvonen, E. Järvenpää, V. Hietaniemi and R. Huopalahti, *Trends Food Sci. Technol.*, 1999, **10**, 217–222.
2. F. J. Senorans, E. Ibanez and A. Cifuentes, *Crit. Rev. Food Sci. Nutr.*, 2003, **43**, 507–526.
3. G. Ferrentino and S. Spilimbergo, *Trends Food Sci. Technol.*, 2011, **22**, 427–441.
4. G. Ferrentino, S. Spilimbergo and A. Bertucco, *Funct. Food Ingredients Nutraceuticals*, 2015, **13**, 427–454.
5. A. Rawson, B. K. Tiwari, N. Brunton, C. Brennan, P. J. Cullen and C. P. O'Donnell, *Food Rev. Int.*, 2012, **28**, 253–276.
6. A. B. Jarzebski and J. J. Malinowski, *Process Biochem.*, 1995, **30**, 343–352.
7. E. J. Beckman, *J. Supercrit. Fluids*, 2004, **28**, 121–191.
8. A. König, in *Kristallisation in der industriellen Praxis*, ed. G. Hofmann, Wiley VCH, 2005, pp. 237–262.
9. P. G. Jessop, W. Leitner, *Chemical Synthesis Using Supercritical Fluids*, Wiley VCH, Weinheim, 1999.
10. R. L. Maffei, *U. S. A. Pat.*, 4,012,194, 1977.
11. E. Schollmeyer, D. Knittel, H.-J. Buschmann and R. Kosfeld, *Verfahren zum Reinigen bzw*, 1990, DE3904514 A1.
12. General-Electric, *Focus Surfactants*, 2002, **2002**, 5.

13. U. HPC, *Focus Surfactants*, 2002, **2002**, 5.
14. C. R. Weast, M. J. Astle, ed., *Handbook of Chemistry and Physics*, CRC Press, Boca Raton, 1981.
15. Y. Sun and B. Y. Shekunov, *J. Supercritical Fluids*, 2003, **27**, 73–83.
16. J. W. King and L. L. Williams, *Curr. Opin. Solid State Mat. Sci.*, 2003, **7**, 413–424.
17. Y. Sutjiadi-Sia, P. Jaeger and R. Eggers, *J. Supercritical Fluids*, 2008, **46**, 272–279.
18. M. J. E. van Roosmalen, G. F. Woerlee and G. J. Witkamp, *J. Supercritical Fluids*, 2004, **30**, 97–109.
19. M. J. E. van Roosmalen, M. van Diggelen, G. F. Woerlee and G. J. Witkamp, *J. Supercritical Fluids*, 2003, **27**, 97–108.
20. J. Galy, K. Sawada, B. Fournel, P. Lacroix-Desmazes, S. Lagerge and M. Persin, *J. Supercritical Fluids*, 2007, **42**, 69–79.
21. P. T. Anastas and J. C. Warner, 2000.
22. A. E. Marteel, J. A. Davies, W. W. Olson and M. A. Abraham, *Annu. Rev. Environ. Resour.*, 2003, **28**, 401–428.
23. G. W. Gould, *Int. J. Food Microbiol.*, 1996, **33**, 51–64.
24. S. Bourdoux, D. Li, A. Rajkovic, F. Devlieghere and M. Uyttendaele, *Compr. Rev. Food Sci. Food Saf.*, 2016, **00**, 1–11.
25. L. Bravo, *Nutr. Rev.*, 1998, **56**, 317–333.
26. E. K. Middleton and C. Theoharides, *Pharmacol. Rev.*, 2000, **52**, 673–751.
27. C. Morris, A. L. Brody and L. Wicker, *Packag. Technol. Sci.*, 2007, **20**, 275–286.
28. J. Raso and G. V. Barbosa-Canovas, *Crit. Rev. Food Sci. Nutr.*, 2003, **43**, 265–285.
29. P. Butz and H. Ludwig, *Physica B + C*, 1986, **139–140**, 875–877.
30. T. Nakagami, T. Shigehis, T. Ohmori, S. Taji, A. Hase, T. Kimura and K. Yamanishi, *J. Virol. Methods*, 1992, **38**, 255–261.
31. F. Moerman, B. Mertens, L. Demey and A. Huyghebaert, *Meat Sci.*, 2001, **59**, 115–125.
32. S. Spilimbergo, N. Elvassore and A. Bertucco, *J. Supercritical Fluids*, 2002, **22**, 55–63.
33. Y. Taniguchi and S. Makimoto, *J. Mol. Catal.*, 1988, **47**, 323–334.
34. M. Kamihira, M. Taniguchi and T. Kobayashi, *Agric. Biol. Chem.*, 1987, **51**, 407–412.
35. J. Zhang, T. A. Davis, M. A. Matthews, M. J. Drews, M. LaBerge and Y. H. An, *J. Supercritical Fluids*, 2006, **38**, 354–372.
36. D. Fraser, *Nature*, 1951, **167**.
37. L. Solomon, P. Zeegen and F. A. Eiserling, *Biochim. Biophys. Acta, Biophys. Incl. Photosynth.*, 1966, **112**, 102–109.
38. L. Garcia-Gonzalez, A. H. Geeraerd, S. Spilimbergo, K. Elst, L. Van Ginneken, J. Debevere, J. F. Van Impe and F. Devlieghere, *Int. J. Food Microbiol.*, 2007, **117**, 1–28.
39. E. T. H. Vink, K. R. Rábago, D. A. Glassner and P. R. Gruber, *Polym. Degrad. Stab.*, 2003, **80**, 403–419.

40. H. M. Powell, O. Ayodeji, T. L. Summerfield, D. M. Powell, D. A. Kniss, D. L. Tomasko and J. J. Lannutti, *Biomaterials*, 2007, **28**, 5562–5569.
41. I. Tsivintzelis, A. G. Angelopoulou and C. Panayiotou, *Polymer*, 2007, **48**, 5928–5939.
42. L. J. M. Jacobs, M. F. Kemmere and J. T. F. Keurentjes, *Green Chem.*, 2008, **10**, 731–738.
43. L. J. M. Jacobs, S. A. M. Hurkens, M. F. Kemmere and J. T. F. Keurentjes, *Macromol. Mater. Eng.*, 2008, **293**, 298–302.
44. L. J. M. Jacobs, M. F. Kemmere, J. T. F. Keurentjes, C. A. Mantelis and T. Meyer, *AIChE J.*, 2007, **53**, 2651–2658.
45. C. I. Wei, M. O. Balaban, S. Y. Fernando and A. J. Peplow, *J. Food Prot.*, 1991, **4**, 189–193.
46. World International Patent Organization Patent, *WO Pat.* 03/101210 A1, 2002.
47. J. M. Silva, A. A. Rigo, I. A. Dalmolin, I. Debien, R. L. Cansian, J. V. Oliveira and M. A. Mazutti, *Food Control*, 2013, **29**, 76–81.
48. M. Peleg and C. M. Penchina, *Crit. Rev. Food Sci. Nutr.*, 2000, **40**, 159–172.
49. C. Ortuno, M. Balaban and J. Benedito, *J. Supercrit. Fluids*, 2014, **90**, 18–26.
50. C. Ort, M. Teresa Martinez-Pastor, A. Mulet and J. Benedito, *J. Supercritical Fluids*, 2012, **63**, 8–15.
51. L. Garcia-Gonzalez, A. H. Geeraerd, J. Mast, Y. Briers, K. Elst, L. Van Ginneken, J. F. Van Impe and F. Devlieghere, *Food Microbiol.*, 2010, **27**, 541–549.
52. J. Zhang, T. A. Davis, M. A. Matthews, M. J. Drews, M. LaBerge and Y. H. H. An, *J. Supercrit. Fluids*, 2006, **38**, 354–372.
53. S. Damar and M. O. Balaban, *J. Food Sci.*, 2006, **71**, R1–R11.
54. S. Spilimbergo and A. Bertucco, *Biotechnol. Bioeng.*, 2003, **84**, 627–638.
55. M. Perrut, *J. Supercritical Fluids*, 2011, **66**, 359–371.
56. C. Ortuno, M. T. Martinez-Pastor, A. Mulet and J. Benedito, *Food Res. Int.*, 2013, **51**, 474–481.
57. I. Paniagua-Martinez, A. Mulet, M. A. Garcia-Alvarado and J. Benedito, *J. Food Eng.*, 2016, **181**, 42–49.
58. C. Ortuno, T. Duong, M. Balaban and J. Benedito, *J. Supercrit. Fluids*, 2013, **82**, 56–62.
59. S. J. Park, J. I. Lee and J. Park, *J. Food Sci.*, 2002, **67**, 1827–1834.
60. L. Wang, J. A. Pan, H. M. Xie, Y. Yang and C. M. Lin, *Int. J. Food Microbiol.*, 2010, **144**, 118–125.
61. S. J. Park, H. W. Park and J. Park, *J. Food Sci.*, 2003, **68**, 976–981.
62. D. Fraser, *Nature*, 1951, **167**, 33.
63. J. W. Foster, R. M. Cowan and T. A. Maag, *J. Bacteriol.*, 1962, **83**, 330.
64. A. C. Mitchell, A. J. Phillips, M. A. Hamilton, R. Gerlach, W. K. Hollis, J. P. Kaszuba and A. B. Cunningham, *J. Supercrit. Fluids*, 2008, **47**, 318–325.
65. H. M. Lin, Z. Y. Yang and L. F. Chen, *Biotechnol. Prog.*, 1992, **8**, 458–461.

66. M. Kamihira, M. Taniguchi and T. Kobayashi, *Agric. Biol. Chem.*, 1987, **51**, 407–412.
67. S. Tamburini, A. Ballarini, G. Ferrentino, A. Moro, P. Foladori, S. Spilimbergo and O. Jousson, *Microbiology*, 2013, **159**, 1056–1066.
68. S. Spilimbergo, P. Foladori, D. Mantoan, G. Ziglio and G. Della Mea, *Process Biochem.*, **45**, 647–654.
69. S. Tamburini, P. Foladori, G. Ferrentino, S. Spilimbergo and O. Jousson, *J. Appl. Microbiol.*, 2014, **117**, 440–450.
70. H. M. Liao, F. S. Zhang, X. J. Liao, X. S. Hu, Y. Chen and L. Deng, *Int. J. Food Microbiol.*, 2010, **144**, 169–176.
71. S. I. Hong and Y. R. Pyun, *Int. J. Food Microbiol.*, 2001, **63**, 19–28.
72. H. M. Liao, F. S. Zhang, X. S. Hu and X. J. Liao, *Microbiology-Sgm*, 2011, **157**, 709–720.
73. H. M. Lin, Z. Y. Yang and L. F. Chen, *Chem. Eng. J. Biochem. Eng. J.*, 1993, **52**, B29–B34.
74. C. Yao, X. Li, W. Bi and C. Jiang, *J. Basic Microbiol.*, 2014, **54**, 858–865.
75. G. J. Haas, H. E. Prescott, E. Dudley, R. Dik, C. Hintlian and L. Keane, *J. Food Safety*, 1989, **9**, 253–265.
76. A. Enomoto, K. Nakamura, K. Nagai, T. Hashimoto and M. Hakoda, *Biosci. Biotechnol. Biochem.*, 1997, **61**, 1133–1137.
77. S. Giulitti, C. Cinquemani, A. Quaranta and S. Spilimbergo, *J. Supercritical Fluids*, 2011, **58**, 385–390.
78. S. Giulitti, C. Cinquemani and S. Spilimbergo, *Biotechnol. Bioeng.*, 2011, **108**, 1211–1214.
79. G. Ferrentino, M. O. Balaban, G. Ferrari and M. Poletto, *J. Supercrit. Fluids*, 2010, **52**, 151–160.
80. A. Jimenez, J. Zhang and M. A. Matthews, *Biotechnol. Bioeng.*, 2008, **101**, 1344–1352.
81. A. White, D. Burns and T. W. Christensen, *J. Biotechnol.*, 2006, **123**, 504–515.
82. G. Gunes, L. K. Blum and J. H. Hotchkiss, *J. Food Prot.*, 2006, **69**, 12–16.
83. X. F. Liu, B. Q. Zhang and T. J. Li, *Chin. J. Chem. Eng.*, 2005, **13**, 140–143.
84. K. A. Connery, P. Shah, L. Coleman and B. Hunek.
85. L. Rao, X. F. Bi, F. Zhao, J. H. Wu, X. S. Hu and X. J. Liao, *Crit. Rev. Food Sci. Nutr.*, 2016, **56**, 1808–1825.
86. W. F. Hu, L. Y. Zhou, Z. Z. Xu, Y. Zhang and X. J. Liao, *Crit. Rev. Food Sci. Nutr.*, 2013, **53**, 145–161.
87. Z. Wimmer and M. Zarevucka, *Int. J. Mol. Sci.*, 2010, **11**, 233–253.
88. A. Checinska, I. A. Fruth, T. L. Green, R. L. Crawford and A. J. Paszczynski, *J. Microbiol. Methods*, 2011, **87**, 70–75.
89. M. O. Balaban, A. G. Arreola, S. Pekyardimci, J. S. Chen and M. R. Marshall, *Abstr. Pap. Am. Chem. Soc.*, 1991, **202**, 28–AGFD.
90. S. Lim, Y. Vagiz and M. O. Balaban, *Food Sci. Biotechnol.*, 2006, **15**, 13–18.
91. J. M. Boff, T. T. Truoung, D. B. Min and T. H. Shellhammer, *J. Food Sci.*, 2003, **68**, 1179–1184.

92. G. Gunes, L. K. Blum and J. H. Hotchkiss, *J. Sci. Food Agric.*, 2005, **85**, 2362–2368.
93. M. Cappelletti, G. Ferrentino, I. Endrizzi, E. Aprea, E. Betta, M. L. Corollaro, M. Charles, F. Gasperi and S. Spilimbergo, *J. Food Eng.*, 2015, **145**, 73–81.
94. S. Fabroni, M. Amenta, N. Timpanaro and P. Rapisarda, *Innovative Food Sci. Emerging Technol.*, 2010, **11**, 477–484.
95. L. Zhou, X. Bi, Z. Xu, Y. Yang and X. Liao, *Crit. Rev. Food Sci. Nutr.*, 2016.
96. A. G. Arreola, M. O. Balaban, C. I. Wei, A. Peplow, M. Marshall and J. Cornell, *J. Food Qual.*, 1991, **14**, 275–284.
97. M. O. Balaban, D. Kincal, S. Hill, M. R. Marshall, R. Wildasin, *IFT Annual Meeting Book of Abstracts*, 2001 June 23–27.
98. G. Folkes, PhD thesis, University of Florida, USA, 2004.
99. M. Lecky and M. O. Balaban, *IFT Annual Meeting Book of Abstracts*, 2005 July 15–20.
100. D. Del Pozo-Insfran, M. O. Balaban and S. T. Talcott, *J. Agric. Food Chem.*, 2006, **54**, 5468–5473.
101. F. Gasperi, E. Aprea, F. Biasioli, S. Carlin, I. Endrizzi, G. Pirretti and S. Spilimbergo, *Food Chem.*, 2009, **115**, 129–136.
102. M. Lecky and M. O. Balaban, *IFT Annual Meeting Book of Abstracts*, 2005 July 15–20.
103. S. Damar and M. O. Balaban, *IFT Annual Meeting Book of Abstracts*, 2005 July 15–20, 54F-2.
104. D. Campoccia, L. Montanaro and C. R. Arciola, *Biomaterials*, 2006, **27**, 2331–2339.
105. H.-C. Flemming, T. R. Neu and D. J. Wozniak, *J. Bacteriol.*, 2007, **189**, 7945–7947.
106. S. M. Mun, J. S. Jeong, J. E. Kim, Y. W. Lee and J. Y. Yoon, *Biofouling*, 2009, **25**, 473–479.
107. S. Zhang, Y. Hu, Y. Zou, J. Chen, X. J. Liao and B. Han, *Afr. J. Microbiol. Res.*, 2013, **7**, 736–744.
108. H. Karaman and O. Erkmen, *Food Microbiology*, 2001, **18**, 11–16.
109. C. Cinquemani, C. Boyle, E. Bach and E. Schollmeyer, *J. Supercrit. Fluids*, 2007, **42**, 1–6.
110. C. Ortuno, A. Quiles and J. Benedito, *FRIN*, 2014, **62**, 955–964.
111. H. Liao, X. Kong, Z. Zhang, X. Liao and X. Hu, *Food Microbiol.*, 2010, **27**, 94–100.
112. N. L. E. Debs-Louka, G. Abraham, V. Chabot and K. Allaf, *Appl. Environ. Microbiol.*, 1999, **65**, 626–631.
113. H. Knez, *J. Supercrit. Fluids*, 2002, **23**, 29–42.
114. V. V. Mozhaev, *Biotechnol. Bioeng.*, 1996, **52**, 320–331.
115. S. Spilimbergo, PhD thesis, Universitá degli Studi di Padova, 2002.
116. G. Ferrentino, PhD thesis, Universita degli studi die Salerno, 2008.
117. V. P. Valdramidis, W. D. Graham, A. Beattie, M. Linton, A. McKay, A. M. Fearon and M. F. Patterson, *Innovative Food Sci. Emerging Technol.*, 2009, **10**, 396–404.

118. T. Yoshimura, M. Furutera, M. Shimoda, H. Ishikawa, M. Miyake, K. Matsumoto, Y. Osajima and I. Hayakawa, *J. Food Sci.*, 2002, **67**, 3227–3231.
119. F. Kobayashi, H. Ikeura, S. Odake and Y. Hayata, *Innovative Food Sci. Emerging Technol.*, 2013, **18**, 108–114.
120. L. Zhang, S. C. Liu, H. W. Ji, C. H. Zhang, C. J. Deng, W. H. Cao, W. J. Mao and J. L. Gao, *Innovative Food Sci. Emerging Technol.*, 2011, **12**, 635–641.
121. J. S. Chen, M. O. Balaban, C. I. Wei, M. R. Marshall and W. Y. Hsu, *J. Agric. Food Chem.*, 1992, **40**, 2345–2349.
122. S. Fadiloglu and O. Erkmen, *J. Food Eng.*, 2001, **52**, 331–335.
123. C. Rauh, A. Baars and A. Delgado, *J. Food Eng.*, 2009, **91**, 154–163.
124. G. I. Katsaros, P. Katapodis and P. S. Taoukis, *J. Food Eng.*, 2009, **91**, 42–48.
125. P. C. Chen, Y. P. Chen and D. S. H. Wong, *Fluid Phase Equilib.*, 1993, **83**, 175–182.
126. C. Bilbao-Sáinz, F. L. Younce, B. Rasco and S. Clark, *Innovative Food Sci. Emerging Technol.*, 2009, **10**, 314–320.
127. M. Taniguchi, H. Suzuki, M. Sato and T. Kobayashi, *Agric. Biol. Chem.*, 1987, **51**, 3425–3426.
128. M. Hendrickx, L. Ludikhuyze, I. Van den Broeck and C. Weemaes, *Trends Food Sci. Technol.*, 1998, **9**, 197–203.
129. T. F. Calix, G. Ferrentino and M. O. Balaban, *J. Food Sci.*, 2008, **73**, E439–E445.
130. M. K. Bull, K. Zerdin, E. Howe, D. Goicoechea, P. Paramanandhan, R. Stockman, J. Sellahewa, E. A. Szabo, R. L. Johnson and C. M. Stewart, *Innovative Food Sci. Emerging Technol.*, 2004, **5**, 135–149.
131. M. M. Calvo and E. Balcones, *Milchwissenschaft-Milk Sci. Int.*, 2001, **56**, 366–369.
132. D. A. Tisi, Cornell University, 2004.
133. B. Juan, A. J. Trujillo, V. Guamis, M. Buffa and V. Ferragut, *Int. Dairy J.*, 2007, **17**, 248–254.
134. S. Spilimbergo, D. Mantoan and A. Dalser, *J. Supercrit. Fluids*, 2007, **40**, 485–489.
135. S. Spilimbergo and L. Ciola, *Int. J. Food Sci. Technol.*, 2010, **45**, 1619–1625.
136. D. Fraser, *Nature*, 1951, **167**, 33–34.
137. S. Mun, J.-S. Hahn, Y.-W. Lee and J. Yoon, *Int. J. Food Microbiol.*, 2010, **144**, 372–378.
138. S. Mun, J. Kim, S.-J. Ahn, Y.-W. Lee and J. Yoon, *Int. J. Food Microbiol.*, 2012, **153**, 15–20.
139. S. Mun, Y.-W. Lee and J. Yoon, *J. Ind. Eng. Chem.*, 2011, **18**, 504–508.
140. A. A. H. Qadir, *Postharvest Biol. Technol.*, 2001, **22**, 279–283.
141. T. Parton, A. Bertucco, N. Elvassore and L. Grimolizzi, *J. Food Eng.*, 2007, **79**, 1410–1417.
142. B. Vinneras, A. Bjorklund and H. Jonsson, *Bioresour. Technol.*, 2003, **88**, 47–54.

143. L. Zhou, Y. Zhang, X. Hu, X. Liao and J. He, *Food Chem.*, 2009, **115**, 449–455.
144. I. Aguiló-Aguayo, G. Oms-Oliu, R. Soliva-Fortuny and O. Martín-Belloso, *Food Chem.*, 2009, **116**, 59–65.
145. M. Tsuji, Y. Sato and Y. Komiyama, *J. Jpn. Soc. Food Sci. Technol.-Nippon Shokuhin Kagaku Kogaku Kaishi*, 2005, **52**, 528–531.
146. F. Gui, J. Wu, F. Chen, X. Liao, X. Hu, Z. Zhang and Z. Wang, *Food Chem.*, 2007, **100**, 1678–1685.
147. F. Gui, F. Chen, J. Wu, Z. Wang, X. Liao and X. Hu, *Food Chem.*, 2006, **97**, 480–489.
148. M. L. Capra, F. Patrignani, A. D. L. Quiberoni, J. A. Reinheimer, R. Lanciotti and M. E. Guerzoni, *Int. Dairy J.*, 2009, **19**, 336–341.
149. L. Baert, J. Debevere and M. Uyttendaele, *Int. J. Food Microbiol.*, 2009, **131**, 83–94.
150. D. H. Kingsley, H. Chen and D. G. Hoover, *Virus Res.*, 2004, **102**, 221–224.
151. N. Wilkinson, A. S. Kurdziel, S. Langton, E. Needs and N. Cook, *Innovative Food Sci. Emerging Technol.*, 2001, **2**, 95–98.
152. O. Moroni, J. Jean, J. Autret and I. Fliss, *Int. Dairy J.*, 2002, **12**, 907–913.
153. J. C. Cheftel, *Food Sci. Technol. Int.*, 1995, **1**, 75–90.
154. J. C. Gonzalez, M. R. Vieytes, J. M. Vieites and L. M. Botana, *J. Am. Oil Chem. Soc.*, 2001, **78**, 77–81.
155. F. Brétagnol, H. Rauscher, M. Hasiwa, O. Kylián, G. Ceccone, L. Hazell, A. J. Paul, O. Lefranc and F. Rossi, *Acta Biomater.*, 2008, **4**, 1745–1751.
156. M. E. P. Goad, D. L. Goad, M. H. Wanda, G. R. Colin and A. W. Matthew, in *Handbook of Toxicologic Pathology*, 2nd edn, Academic Press, San Diego, 2002, pp. 459–477.
157. K. H. Wallhäuser, *Praxis der Sterilisation, Desinfektion – Konservierung*, Georg Thieme Verlag, Stuttgart, New York, 5th edn, 1995.
158. D. Petsch and F. B. Anspach, *J. Biotechnol.*, 2000, **76**, 97–119.
159. H. K. Kayser, K. A. Bienz, J. Eckert, R. M. Zinkernagel, *Medizinische Mikrobiologie*, Thieme Verlag, Stuttgart, 1998.
160. J. Rybka, P. Grycko, J. D. C. Francisco, A. Gamian and E. S. Dey, *J. Supercrit. Fluids*, 2008, **45**, 51–56.
161. P. J. Tarafa, E. Williams, S. Panvelker, J. Z. Zhang and M. A. Matthews, *J. Supercrit. Fluids*, 2011, **55**, 1052–1058.
162. D. M. Taylor, *J. Hosp. Infect.*, 1999, **43**, S69–S76.
163. G. Fichet, *Lancet*, 2004.
164. J. L. Silva, *Biochim. Biophys. Acta*, 2002, 250–265.
165. A. Schmidt, K. Beermann, E. Bach and E. Schollmeyer, *J. Cleaner Prod.*, 2007, **42**, 881–885.
166. A. Jiménez, G. L. Thompson, M. A. Matthews, T. A. Davis, K. Crocker, J. S. Lyons and A. Trapotsis, *J. Supercrit. Fluids*, 2007, **42**, 366–372.
167. J. D. Hemmer, M. J. Drews, M. LaBerge and M. A. Matthews, *Journal*, 2007, **80B**, 511–518.

168. J. Zhang, S. Burrows, C. Gleason, M. A. Matthews, M. J. Drews, M. LaBerge and Y. H. An, *J. Microbiol. Methods*, 2006, **66**, 479–485.
169. D. B. M. Scott and E.C. Lesher, *J. Bacteriol.*, 1963, **85**, 567–576.
170. E. Christensen and A. C. Giese, *Arch. Biochem. Biophys.*, 1954, **51**, 208–216.
171. B. D. Goldstein and E. M. McDonagh, *Environ. Res.*, 1975, **9**, 179–186.
172. P. M. Foegeding, *Food Microbiol.*, 1985, **2**, 123–134.
173. S. Farooq, R. S. Engelbrecht and E. S. K. Chian, *Water Res.*, 1977, **11**, 737–741.
174. W. Pompe, H. Worch, M. Epple, W. Friess, M. Gelinsky, P. Greil, U. Hempel, D. Scharnweber and K. Schulte, *Mat. Sci. Eng. A*, 2003, **362**, 40–60.
175. L. Thorrez, J. Shansky, L. Wang, L. Fast, T. VandenDriessche, M. Chuah, D. Mooney and H. Vandenburgh, *Biomaterials*, 2008, **29**, 75–84.
176. G. Riepe, C. Heintz, E. Kaiser, N. Chakfé, M. Morlock, M. Delling and H. Imig, *Eur. J. Vasc. Endovasc. Surg.*, 2002, **24**, 117–122.
177. Y. H. An, F. I. Alvi, Q. Kang, M. Laberge, M. J. Drews, J. Zhang, M. A. Matthews and C. R. Arciola, *Int. J. Artif. Organs*, 2005, **28**, 1126–1137.
178. M. Zhang, in *Biomaterials and Tissue Engineering*, ed. D. Shi, Springer, Berlin, 2004, pp. 83–137.
179. N. Krasteva, U. Harms, W. Albrecht, B. Seifert, M. Hopp, G. Altankov and T. Groth, *Biomaterials*, 2002, **23**, 2467–2478.
180. H.-G. Hicke, I. Lehmann, G. Malsch, M. Ulbricht and M. Becker, *J. Membr. Sci.*, 2002, **198**, 187–196.
181. M. Ulbricht, *Polymer*, 2006, **47**, 2217–2262.
182. J. M. Goddard and J. H. Hotchkiss, *Prog. Polym. Sci.*, 2007, **32**, 698–725.
183. M. A. Cole, N. H. Voelcker, H. Thissen and H. J. Griesser, *Biomaterials*, 2009, **30**, 1827–1850.
184. H. Chen, L. Yuan, W. Song, Z. Wu and D. Li, *Prog. Polym. Sci.*, 2008, **33**, 1059–1087.
185. H. P. Wendel and G. Ziemer, *Eur. J. Cardio-Thorac. Surg.*, 1999, **16**, 342–350.
186. B. Ríhová, *Adv. Drug Delivery Rev.*, 1996, **21**, 157–176.
187. S. I. Hong and Y. R. Pyun, *J. Food Sci.*, 1999, **64**, 728–733.
188. S. Spilimbergo and D. Mantoan, *Biotechnol. Prog.*, 2005, **21**, 1461–1465.
189. S. Spilimbergo and D. Mantoan, *Int. J. Food Eng.*, 2006, **2**, Article 6.
190. M. Shimoda, J. Cocunubo-Castellanos, H. Kago, M. Miyake, Y. Osajima and I. Hayakawa, *J. Appl. Microbiol.*, 2001, **91**, 306–311.
191. United States Patent, *U. S. Pat.* 6,331,272 B1, 2001.
192. M. Sims and E. Estigarribia, Venice, Italy, 2002.
193. M. Sims and E. Estigarribia, *6th International Symposium on Supercritical Fluids*, Versailles, France, 2003.
194. D. Kincal, W. S. Hill, M. O. Balaban, K. M. Portier, C. I. Wei and M. R. Marshall, *J. Food Sci.*, 2005, **70**, M249–M254.

195. Y. Yagiz, S. L. Lim, M. O. Balaban, New Orleans, La. Chicago, 2005.
196. H.-G. Yuk and D. J. Geveke, *Food Microbiol.*, 2011, **28**, 377–383.
197. H.-G. Yuk, D. J. Geveke and H. Q. Zhang, *Int. J. Food Microbiol.*, **138**, 91–99.
198. G. Pataro, G. Ferrentino, C. Ricciardi and G. Ferrari, *J. Supercrit. Fluids*, 2010, **54**, 120–128.
199. C. Ortuno, M. T. Martinez-Pastor, A. Mulet and J. Benedito, *Innovative Food Sci. Emerging Technol.*, 2012, **15**, 31–37.
200. J. Benedito, C. Ortuno, R. I. Castillo-Zamudio and A. Mulet, *Phys. Procedia*, 2015, **70**, 824–827.
201. G. Ferrentino and S. Spilimbergo, *J. Supercrit. Fluids*, 2015, **105**, 170–178.
202. A. M. Sikin and S. S. Rizvi, *Recent Pat. Food, Nutr. Agric.*, 2011, **3**, 212–225.
203. United States Patent Office Patent, *U. S. Pat.* 5,393,547, 1988.
204. United States Patent Office Patent, *U. S. Pat.* 5,520,943, 1993.
205. United States Patent Patent, *U. S. Pat.* 5,520,943, 1996.
206. G. Brunner, *J. Food Eng.*, 2005, **67**, 21–33.
207. H. Kaiser, L. Warner and M. A. Matthews, *Med. Device Diagn. Ind.*, 2001, 140.
208. United States Patent Office Patent, *U.S. Pat.*, 6,723,365, 2004.
209. T. Parton, E. Rocchi, S. Spilimbergo, N. Elvassore and A. Bertucco, Advances in High Pressure Bioscience and Biotechnology II, *Proceedings of the International Conference on High Pressure Bioscience and Biotechnology, 2nd, Dortmund, Germany, Sept. 16–***19**, 2002, 2003, 419–424.
210. H. M. Lin, N. J. Cao and L. F. Chen, *J. Food Sci.*, 1994, **59**, 657–659.
211. O. Erkmen, *Int. J. Food Microbiol.*, 2001, **65**, 131–135.
212. O. Erkmen, *J. Sci. Food Agric.*, 2000, **80**, 465–470.
213. O. Erkmen, *Food Sci. Technol. Int.*, 2001, **7**, 245–250.
214. B. G. Werner and J. H. Hotchkiss, *J. Dairy Sci.*, 2006, **89**, 872–881.
215. O. Erkmen and S. Karatas, *J. Food Eng.*, 1997, **33**, 257–262.
216. H. M. Liao, L. Y. Zhang, X. S. Hu and X. J. Liao, *Int. J. Food Microbiol.*, 2010, **137**, 81–87.
217. S. Niu, Z. H. Xu, Y. D. Fang, L. Y. Zhang, Y. J. Yang, X. J. Liao and X. S. Hu, *Innovative Food Sci. Emerging Technol.*, 2010, **11**, 91–97.
218. C. D. Porto, D. Decorti and F. Tubaro, *Int. J. Food Sci. Technol.*, 2010, **45**, 1821–1827.
219. Z. H. Xu, L. Y. Zhang, Y. T. Wang, X. F. Bi, R. Buckow and X. J. Liao, *J. Food Eng.*, 2011, **104**, 577–584.
220. Y. Liu, X. S. Hu, X. Y. Zhao and H. L. Song, *Innovative Food Sci. Emerging Technol.*, 2012, **13**, 112–119.
221. G. Gunes, L. K. Blum and J. H. Hotchkiss, *J. Food Prot.*, 2006, **69**, 12–16.
222. H. Liao, X. Hu, X. Liao, F. Chen and J. Wu, *Int. J. Food Microbiol.*, 2007, **118**, 126–131.
223. T. Parton, N. Elvassore, A. Bertucco and G. Bertoloni, *J. Supercrit. Fluids*, 2007, **40**, 490–496.

224. O. Erkmen, *Food Microbiol.*, 2000, **17**, 589–596.
225. H. M. Liao, X. Z. Kong, Z. Y. Zhang, X. J. Liao and X. S. Hu, *Food Microbiol.*, 2010, **27**, 94–100.
226. X. F. Bi, Y. T. Wang, F. Zhao, Y. Zhang, L. Rao, X. J. Liao, X. S. Hu and Z. J. Sun, *Food Control*, 2014, **41**, 139–146.
227. T. Parton, C. Toniolo, N. Elvassore, A. Bertucco, Versailles, France, 2003.
228. Y. S. Yu, G. S. Xiao, J. J. Wu, Y. J. Xu, D. B. Tang, Y. L. Chen, J. Wen and M. Q. Fu, *Innovative Food Sci. Emerging Technol.*, 2013, **18**, 95–100.
229. M. M. Guo, J. J. Wu, Y. J. Xu, G. S. Xiao, M. W. Zhang and Y. L. Chen, *Eur. Food Res. Technol.*, 2011, **232**, 803–811.
230. H. Li, L. Zhao, J. H. Wu, Y. Zhang and X. J. Liao, *Food Microbiol.*, 2012, **30**, 139–145.
231. C. I. Wei, M. O. Balaban, S. Y. Fernando and A. J. Peplow, *J. Food Prot.*, 1991, **54**, 189–193.
232. U. Sirisee, F. Hsieh and H. E. Huff, *J. Food Process. Preserv.*, 1998, **22**, 387–403.
233. S. I. Hong, W. S. Park and Y. R. Pyun, Food Sci. Technol.-*Lebensm.-Wiss. Technol.*, 1997, **30**, 681–685.
234. A. M. Mazzoni, R. R. Sharma, A. Demirci and G. R. Ziegler, *J. Food Saf.*, 2001, **21**, 215–223.
235. T. Parton, A. Bertucco and G. Bertoloni, *Ital. J. Food Sci.*, 2007, **19**, 425–437.
236. M. Chen, X. Sui, X. Ma, X. Feng and Y. Han, *J. Sci. Food Agric.*, 2015, **95**, 1016–1023.
237. H. W. Ji, L. Zhang, S. C. Liu, X. J. Qu, C. H. Zhang and J. L. Gao, *Int. J. Food Microbiol.*, 2012, **156**, 44–49.
238. D. A. F. Meujo, D. Kevin, J. J. Bowling, J. Liu and M. T. Hamann, *Planta Med.*, 2008, **74**, 342.
239. J. C. Gonzalez, O. I. Fontal, M. R. Vieytes, J. M. Vieytes and L. M. Botana, *J. Agric. Food Chem.*, 2002, **50**, 400–405.
240. F. Dehghani, N. Annabi, M. Titus, P. Valtchev and A. Tumilar, *Biotechnol. Bioeng.*, 2009, **102**, 569–576.
241. Y. M. Choi, Y. C. Ryu, S. H. Lee, G. W. Go, H. G. Shin, K. H. Kim, M. S. Rhee and B. C. Kim, *LWT – Food Sci. Technol.*, 2008, **41**, 317–322.
242. G. Ferrentino, S. Balzan and S. Spilimbergo, *Int. J. Food Microbiol.*, 2013, **161**, 189–196.
243. G. Ferrentino, S. Balzan and S. Spilimbergo, *Food Bioprocess Technol.*, 2013, **6**, 1164–1174.
244. S. Spilimbergo, C. Martina and F. Giovanna, 2014.
245. B. A. E. Yun Young, K. I. M. Nam Hee, K. I. M. Kyoung Heon, K. I. M. Byoung Chul and R. H. E. E. Min Suk, *J. Food Saf.*, 2011, **31**, 48–53.
246. M. Cuppini, J. Zeni, J. Barbosa, E. Franceschi, G. Toniazzo and R. L. Cansian, *Food Sci. Technol.*, 2016, **36**, 8–11.
247. G. Ferrentino, S. Balzan, A. Dorigato, A. Pegoretti and S. Spilimbergo, *J. Food Sci.*, 2012, **77**, E137–E143.

248. G. Ferrentino, A. Belscak-Cvitanovic, D. Komes and S. Spilimbergo, *J. Chem.*, 2013, **2013**, 703057.
249. Q. X. Zhong, D. G. Black, P. M. Davidson and D. A. Golden, *J. Food Prot.*, 2008, **71**, 1015–1017.
250. X. F. Bi, J. H. Wu, Y. Zhang, Z. H. Xu and X. J. Liao, *Innovative Food Sci. Emerging Technol.*, 2011, **12**, 298–304.
251. J. Casas, J. Tello, F. Gatto and L. Calvo, *J. Supercrit. Fluids*, 2016, **116**, 57–61.
252. A. M. Sikin, M. Walkling-Ribeiro and S. S. H. Rizvi, *Food Control*, 2016, **70**, 174–182.
253. S. Spilimbergo, *J. Food Eng.*, 2011, **105**, 193–195.
254. G. J. Haas, H. E. Prescott, E. Dudley, R. Dik, C. Hintlian and L. Keane, *J. Food Saf.*, 1989, **9**, 253–265.
255. S.-I. Hong and W.-S. Park, *Biosci., Biotechnol. Biochem.*, 1999, **63**, 1119–1121.
256. A. M. Mazzoni, R.R. Sharma, A. Demirci and G.R. Ziegler, *J. Food Saf.*, **21**, 215–233.
257. United States Patent, *U. S. Pat.* 3,483,005, 1966.
258. United States Patent, *U. S. Pat.* 4,263,253, 1981.
259. German Patent, *EP* 0 488 012 B1, 1990.
260. United States Patent, *U. S. Pat.* 5,380,826, 1992.
261. United States Patent, *U. S. Pat.* 5,667,835, 1997.
262. United States Patent, *U. S. Pat.* 5,704,276, 1998.
263. United States Patent, *U. S. Pat.* 5,869,123, 1999.
264. United States Patent, *U. S. Pat.* 6,616,849 B1, 2003.
265. European Patent, *Eur. Pat.* 0674845 B1, 1994.
266. International Patent, *WO Pat.* 99/66960, 1999.
267. International Patent, *WO Pat.* 02/03816 A1, 2002.
268. United States Patent, *U. S. Pat.* 6,610,251 B1, 2003.
269. United States Patent Application, *U. S. Pat. Appl.* 2002/0127317 A1, 2001.
270. World Intellectual Property Organization, *WO Pat.* 2004/045316 A1, 2002.
271. World International Patent Organization, *WO Pat.* 2004/000434 A1, 2002.
272. World International Patent Organization, *WO Pat.* 2004/039180 A1, 2002.
273. United States Patent Office, *U. S. Pat. Appl.* 2004/0234661 A1, 2003.
274. United States Patent Office, *U. S. Pat. Appl.* 2005/0084581 A1, 2003.
275. United States Patent Office, *U. S. Pat. Appl.* 7,108, 832 B2, 2005.
276. United States Patent Office, *U. S. Pat. Appl.* 20080171116 A1, 2008.
277. United States Patent Office, *U. S. Pat. Appl.* 20100203206, 2010.
278. United States Patent Office, *U. S. Pat. Appl.* 20120118175, 2012.
279. Spain Patent, *Spanish Pat. Appl. ES* 2011–31099, 2013.
280. United States Patent, *U. S. Pat. Appl.* 201113103334, 2012.

CHAPTER 6

Developments in the Processing of Foods and Natural Products Using Pressurized Fluids

KEERTHI SRINIVAS*[a] AND JERRY W. KING[b]

[a] Bioproducts, Sciences and Engineering Laboratory, Washington State University, Tri-Cities, 2710 Crimson Way, Richland, WA 99354, USA;
[b] Critical Fluids Symposia and University of Arkansas, 1965 E. Spinel Link #7, Fayetteville, AR 72701, USA
*Email: keerthi.srinivas@tricity.wsu.edu

6.1 Introduction

The use of supercritical fluids (SCFs), in particular supercritical carbon dioxide ($scCO_2$), has been developed for the past four decades into a useful, environmentally benign processing methodology, particularly with respect to its application to food and agricultural products. Since its initial application to decaffeinate coffee and provide hop extracts for the beer industry,[1] the use of supercritical fluid extraction (SFE) employing $scCO_2$ and occasionally propane has grown considerably. In the mid- to late-1960s, the term "green" processing had not yet been coined in terms of its current definition, and it was not until later (in the 1990s) that the above processing option became "green",[2] and was considered under the umbrella of "environmentally sustainable"[3] technologies.

Several excellent, fairly current reviews involving the SFE of foods and natural products[4–8] have been published in the past and the goal of this chapter is limited to addressing advances and applications of sub- and

Green Chemistry Series No. 53
Alternatives to Conventional Food Processing: 2nd Edition
Edited by Andrew Proctor

Published by the Royal Society of Chemistry, www.rsc.org

supercritical fluid extraction techniques in various parts of food and natural product processing. In discussing the advances in critical fluid processing, it should be noted that almost 150 SFE processing plants exist worldwide today on various processing scales that employ SFE.[9] The focus of this chapter is plainly on "what are the newer developments that show the most promise in extending the range of applications of these versatile compressed fluids" in the food processing sector. In this chapter, four areas are discussed: sub- and supercritical fluids, coupled unit processing using critical fluids, the use of sub- and supercritical fluids for extractions and reactions, and the potential use of SCFs in food biotechnology and natural products processing. However, before delving into each of these topics, we first describe the difference between sub- and supercritical fluids, followed by the current state of the commercial development of SFE with respect to plants and products.

6.2 Supercritical *Versus* Subcritical Fluids

Unlike SCFs, subcritical fluids are, by definition, compressed fluids below their critical temperatures, yet kept in the liquid state and used above their boiling points through the application of sufficient pressure. Subcritical fluids are also referred to as "hot" liquid solvents or pressurized liquid solvents, and when employed in the field of analytical chemistry using a technique known as pressurized solvent extraction (PSE) or accelerated solvent extraction (ASE). To define better the subcritical region for these compressed fluids, it is instructive to examine a phase diagram, such as that for water, shown in Figure 6.1. Here the well-known SCF region for water is shown in right upper corner of the two-dimensional *P–T* phase diagram, a region defined by water's critical pressure ($P_c = 218$ atm) and temperature ($T_c = 374$ °C), as it is for other fluids. In Figure 6.1, the line connecting the

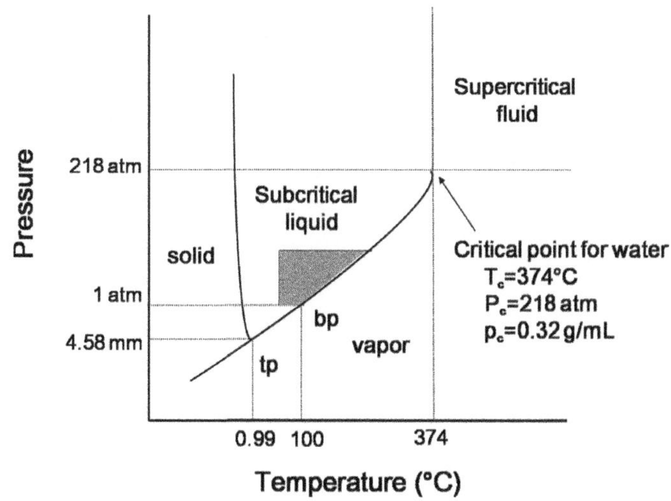

Figure 6.1 Phase diagram for water as a function of temperature and pressure.

critical point for water and its boiling point is known as the V–L curve, which establishes the boundary between the liquid and vapor states. Hence operating at a specific temperature will require a corresponding applied pressure, if the subcritical water is to remain in the liquid state. This type of phase diagram is common for all fluids, *e.g.* carbon dioxide's phase diagram has a format identical with Figure 6.1 except that the critical pressure (72 atm) and critical temperature (31 °C) are much lower and so, of course, would the proportion of the axis or V–L curve. Such data have been

(a)

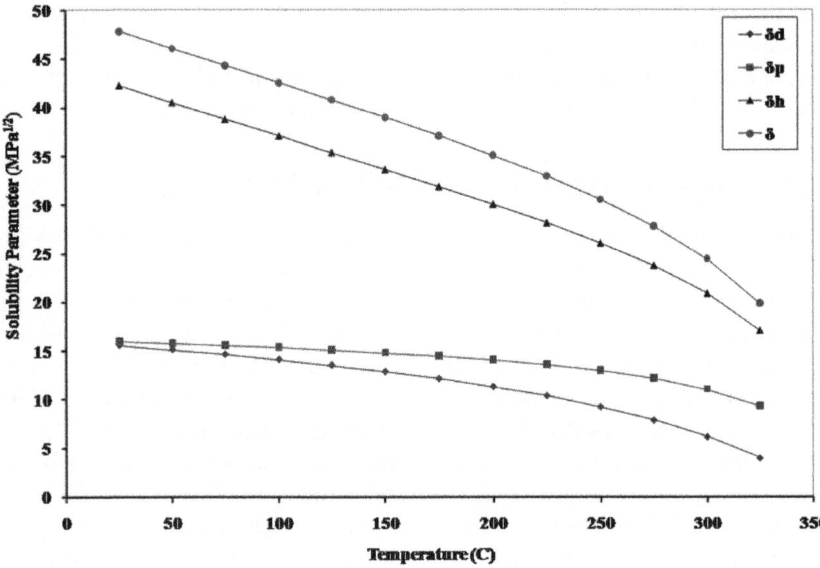

(b)

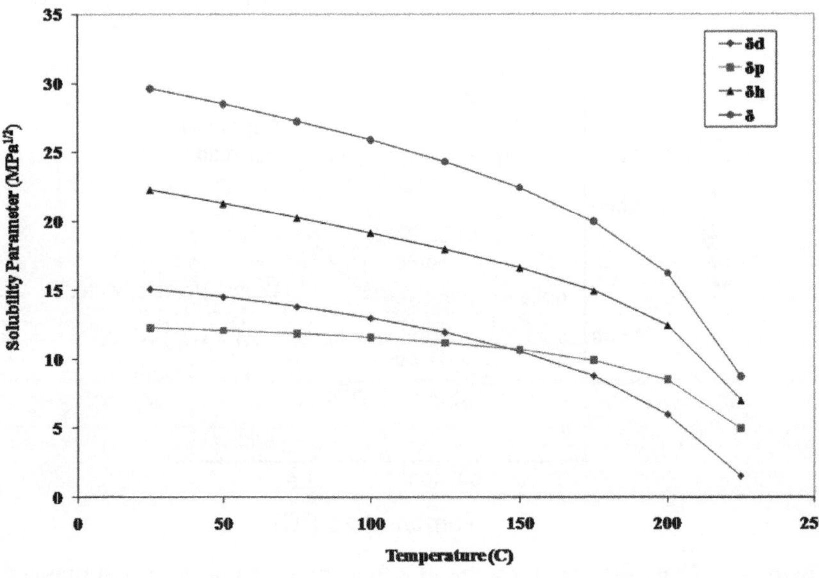

(c)

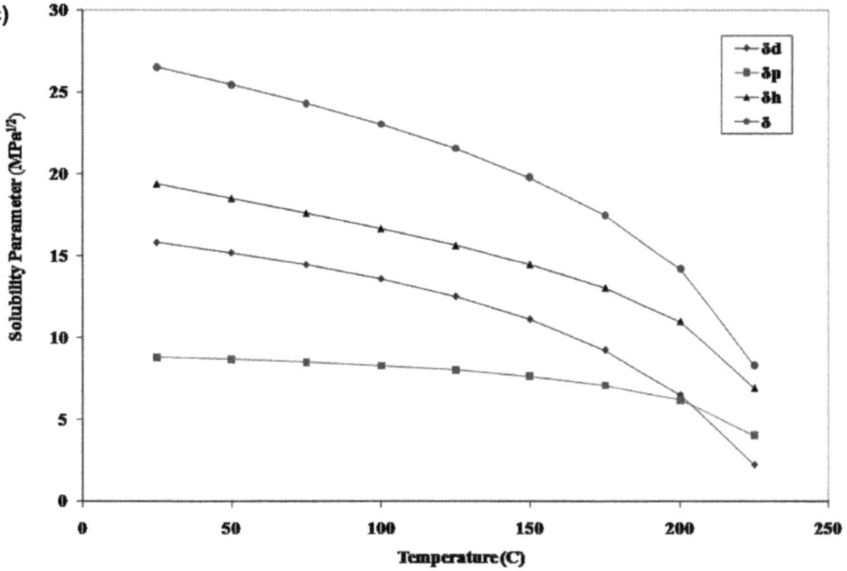

(d)

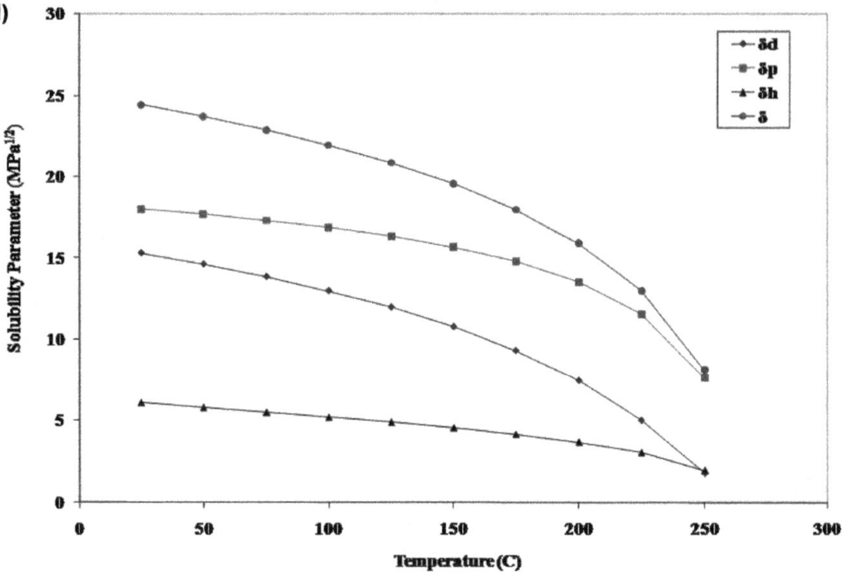

Figure 6.2 Total and three-dimensional solubility parameter variation with temperature for four subcritical fluids: (a) water, (b) methanol, (c) ethanol, and (d) acetonitrile.

extensively measured through phase equilibria studies under different temperature and pressure conditions.

Depending on the use of the subcritical fluid, operations such as extractions or reactions can be conducted anywhere along the V–L curve, and the choice of conditions may be partially dictated by the thermal stability of the solutes under consideration and the desired end products. It should be noted, however, that the solvent power of subcritical fluids is highly dependent on the temperature whereas pressure has a negligible effect on the resultant solvent properties. For the subcritical water extraction of many food and natural products, the subcritical region would lie below 200 °C, and hence would require specific minimal pressures (as determined by the V–L curve) to remain in the liquid state. For example, when conducting an extraction with subcritical water at 120 °C, a pressure exceeding 2 atm or more is required to maintain water in its liquid state. Therefore, unlike extractions with $scCO_2$, fairly modest pressures are usually employed when using subcritical fluid extraction.

As indicated above, unlike SCFs, where the fluid's solvent strength is defined by both the applied temperature and pressure (and hence the fluid's density), subcritical liquids such as water exhibit a monotonic dependence of solvent power over an extended temperature range, and only at temperatures approaching the T_c does the fluid's solvent power change drastically. Figure 6.2a–d illustrate the variations in total and three-dimensional solubility parameters[10] for four subcritical fluids, water, methanol, ethanol, and acetonitrile, as a function of temperature. We prefer the solubility parameter, which is the square root of the cohesive energy density of a compound, as a more accurate representation of the solvent power of a compound rather than the often-cited variation of the dielectric constant with temperature,[11] and have demonstrated that it can be used to optimize solubility and enhance extraction yields when using subcritical fluid media.[12] According to the three-dimensional solubility parameter concept,[10] the total solubility parameter (δ_T) of a solute/solvent is a function of the solubility parameter due to dispersion (δ_d), polar (δ_p), and hydrogen bonding (δ_h) forces present in the solute/solvent. In Figure 6.2a, it can be seen that a change in the total solubility parameter of water with temperature is largely a function of the decrease in hydrogen bonding propensity, since δ_d and δ_p vary much less with increase in temperature. Similar trends are also recorded for subcritical methanol, ethanol, and acetonitrile (Figure 6.2b–d); however, owing to their lower T_cs, both the total and component solubility parameters deviate from a linear dependence with temperature at higher temperatures, and interestingly their δ_d and δ_p variations with temperature intersect at a given temperature. Hence, by manipulating the subcritical fluid's temperature, one can change the contributions of various intermolecular forces that govern solute solubility in such subcritical fluids.

Subcritical water is of particular interest as a processing agent for food or natural product extractions, for several reasons. First, water is a complementary "green" processing agent with CO_2, as shown in Figure 6.3, and can be utilized as an environmentally benign, sustainable processing medium. Second, subcritical water can solubilize polar compounds and, as commented previously, have its solvent power adjusted by varying the temperature. Third, by varying the extraction or reaction temperature,

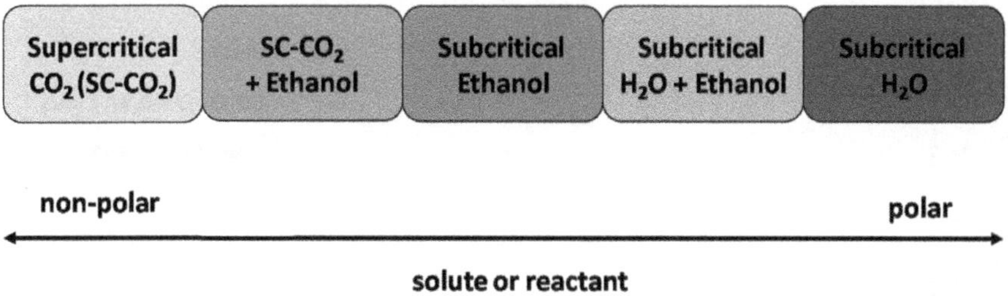

Figure 6.3 "Green" high-pressure solvents or reaction media.

subcritical water can emulate, to a certain extent, the solvent characteristics of aprotic solvents and water–ethanol mixtures. In Figure 6.3, it should be noted that "subcritical ethanol" is placed as a "natural" bridging solvent between compressed CO_2 and subcritical water. This is because ethanol, unlike methanol and hexane, which have significant applications in food processing, is a "generally recognized as safe" (GRAS)-approved solvent and a sustainable solvent medium that can be produced *via* existing and optimized fermentative processes. In short, all of the solvents and their possible combinations as shown in Figure 6.3 are compatible with human food consumption.

Subcritical water, depending on the chosen temperature, exhibits solubility parameters between 33 and 47 $MPa^{\frac{1}{2}}$, values that match those calculated or determined for a number of polar solutes contained in food and natural products.[13] On the basis of matching the solubility parameters between the target solute and the subcritical fluid, *i.e.* "like dissolves like," subcritical water is an excellent solvent medium for generic classes of solutes such as carbohydrates, proteins, glycosides, and flavonoids. However, other subcritical fluids may be better, depending on the chemical characteristics of the target solute. Figure 6.4 shows the relationship between the temperature-dependent solubility parameter for subcritical water or ethanol and the value for betulin, a naturally occurring antioxidant found in birch bark. One can see that the solubility parameter trends for subcritical water *versus* betulin are not commensurate with a high degree of solubilization of the target solute, whereas the solubility parameters for betulin and subcritical ethanol cross at 90 °C, almost identical with the temperature where the optimal amount of betulin is extracted using subcritical ethanol.[14] This is based on the well-defined solubility parameter concept,[10] where degree of solubilization of a solute in solvent is higher if the solubility parameter of the solute is closer to that of the solvent under particular temperature and pressure conditions.

When utilizing subcritical fluids, an increase in the extraction temperature increases both the solute solubility and the diffusion coefficient (a measure of enhanced mass transfer) exponentially, with the largest increases occurring after the boiling point of the solvent has been exceeded. Figure 6.5a and b illustrate the large increase in solute solubility in

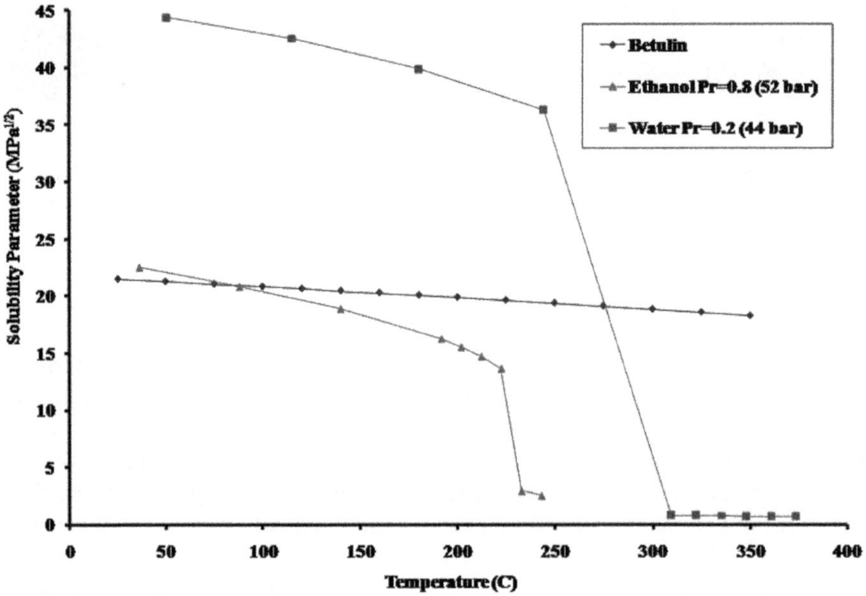

Figure 6.4 Solubility parameter *versus* temperature for the betulin–ethanol and betulin–water systems.

subcritical water with increase in temperature for both silybinin, a silymarin compound found in milk thistle,[15] and quercetin, a well-known flavonoid found in many foods that exhibits antioxidant properties.[16] Note that these solubility trends are independent of the solute concentration units or temperature scale and are fairly universal from one solute type to another. Figure 6.6a and b show the trend for the solute diffusion coefficients (D_{12}) in subcritical water for both common amino acids and sugars that are found in foods and botanical matter. Note that in Figure 6.6a, the absolute D_{12} values of amino acids increase exponentially as the temperature of subcritical water is increased from 25 to 175 °C. If the same data are plotted relative to their D_{12} values at 25 °C, there is over a 45-fold increase in D_{12}. Such a trend is illustrated in Figure 6.6b for the sugars, which exhibit a 35-fold increase in D_{12} as the water temperature is raised to 175 °C. The increases in solute solubilities and D_{12} values with increase in temperature suggest that very large and fast solute extraction fluxes can be realized by using subcritical fluids for extraction.

An example of the "fast flux" principle in action for subcritical water is shown in Figure 6.7a and b using subcritical water for the extraction of B vitamins from spent cider yeast.[17] Here both the vitamin and protein content in the aqueous extract increase as a function of extraction temperature above the boiling point of water. Note that in Figure 6.7a, pressure has a negligible effect on the vitamin content extracted from the cider yeast, which in commensurate with our earlier comments about the effect of pressure on the

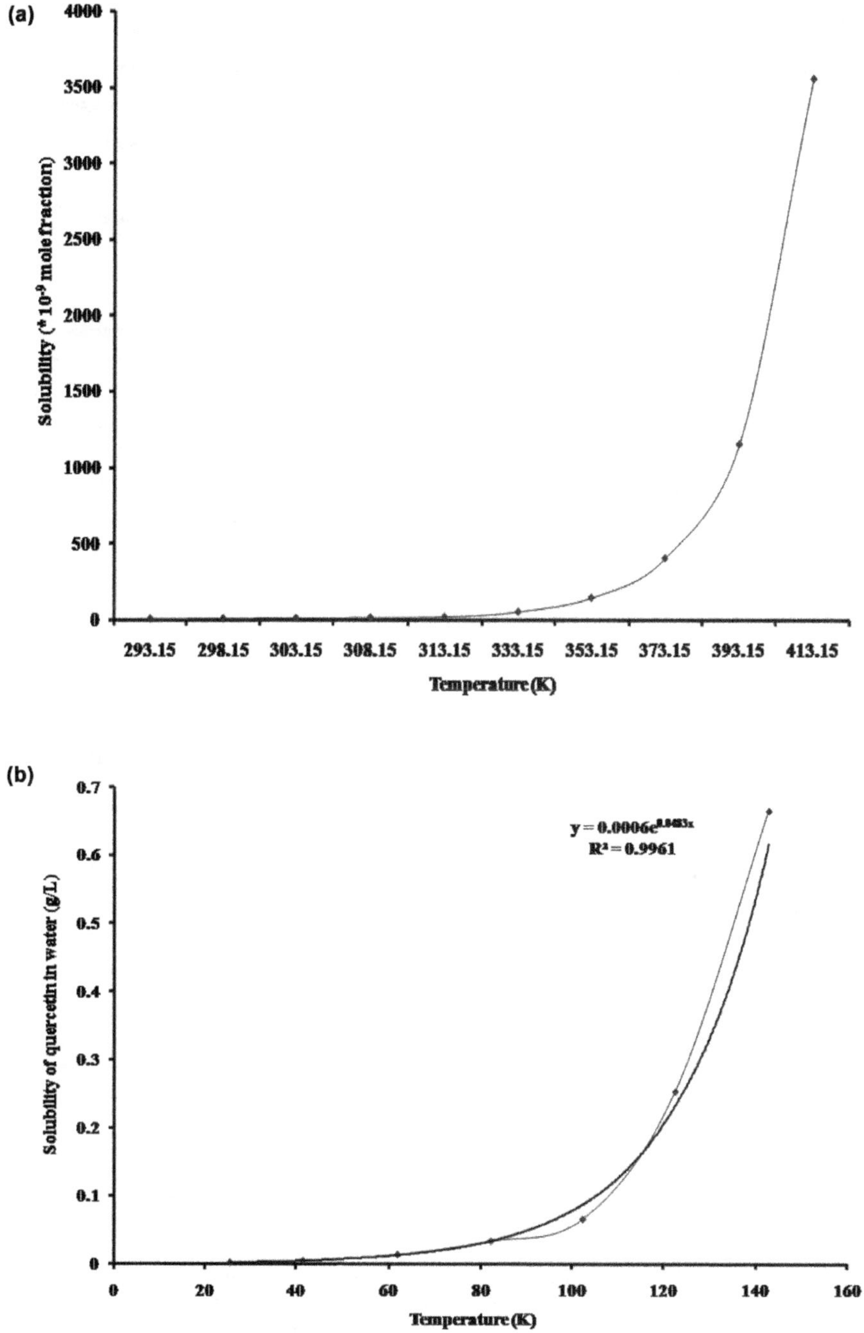

Figure 6.5 Solute solubility as a function of temperature in subcritical water for (a) silybinin and (b) quercetin.

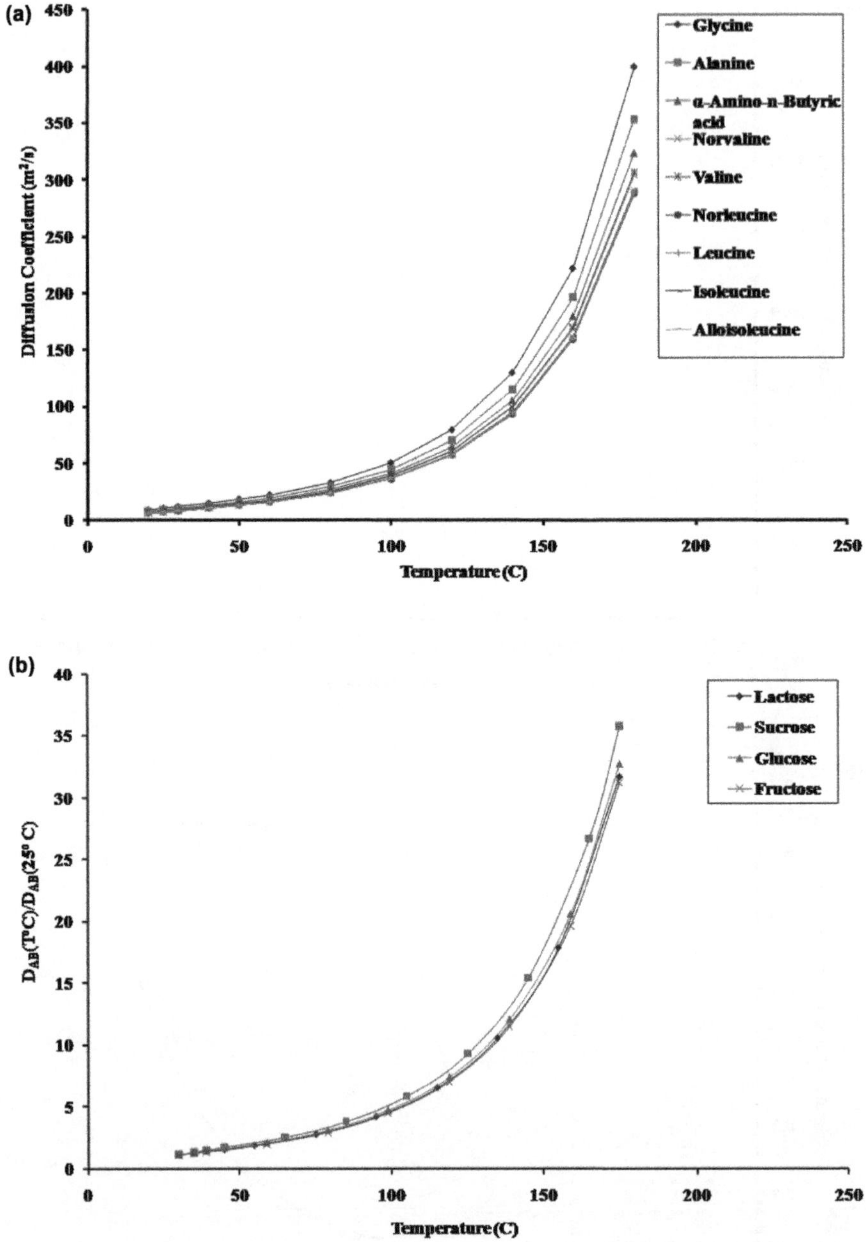

Figure 6.6 (a) Amino acid diffusion coefficients in water as a function of temperature: both experimental data and extrapolated values. (b) Relative change in the diffusion coefficient of sugars in water as a function of temperature.

solvent power of subcritical fluids. In Figure 6.7b, the proximate protein content as measured by nitrogen content is reduced as the extraction temperature of the subcritical water is increased, suggesting that hydrolysis may

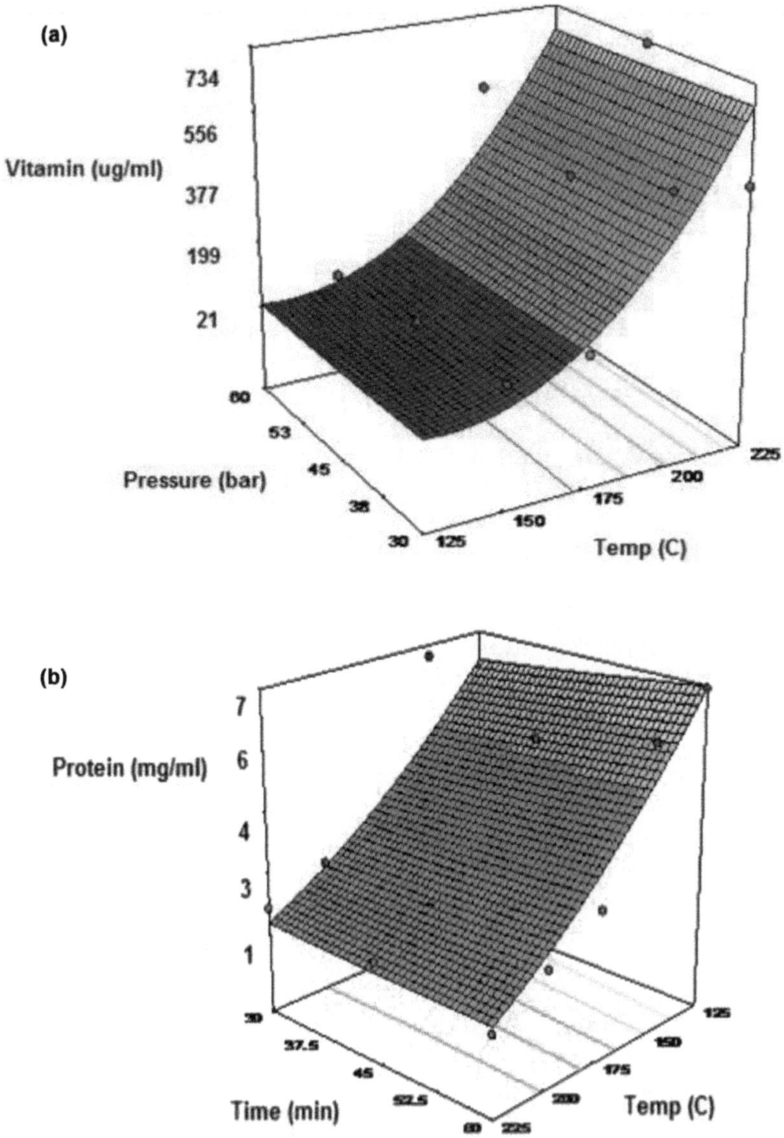

Figure 6.7 (a) Subcritical water treatment of spend cider yeast – effect of pressure and temperature on vitamin recovery (treatment time = 45 min). (b) Subcritical water treatment of spend cider yeast-soluble protein fraction in extract as a function of temperature.

be playing a role in the release of the vitamin moieties from the cider yeast matrix. This is an excellent example of the simultaneous extraction–reaction mode in which the hydrolytic degradation of the sample matrix facilitates recovery of the targeted vitamin compounds. Additional applications of subcritical water and similar fluids have been summarized in several reviews.[18–21]

6.3 Current Status of Supercritical Fluid Processing with CO_2

The use of $scCO_2$ extraction for the processing of foodstuffs has several well-known advantages:

- the production of solvent-free extracts
- non-flammability of $scCO_2$ as solvent
- processing of temperature-sensitive materials
- aroma retention and fortification *via* removal of CO_2
- antimicrobial properties of resultant products
- low energy associated with solvent ($scCO_2$) recovery
- selective separation of solutes from starting substrates.

With respect to the last statement, $scCO_2$ is not a "miracle" solvent and has a relative low dissolution capacity for most solutes relative to liquid solvents, except for substrates containing lipid moieties, where extracts obtained at high extraction pressures and temperatures are similar to those obtained *via* conventional liquid solvent extraction. However, $scCO_2$ extraction can be made more versatile by adjustment of the density of the fluid by manipulation of the applied pressure and temperature, making it more selective for targeted solutes (*e.g.* caffeine from coffee beans) or classes of solutes responsible for a desired property in food compounding (*e.g.* coffee flavor or oil). The "solvent power" of $scCO_2$ is enhanced by the application of more polar co-solvents and/or conditioning the starting substrate, by the addition of moisture, which acts as an "*in situ*" co-solvent effect responsible for greater selectivity during the extraction (*i.e.* $scCO_2$ extraction of moist coffee beans is selective towards caffeine).

Using $scCO_2$ extraction as a fractionating method permits the isolation of food extracts having different properties for different intended applications. For example, "designer" extracts of commodity and specialty spices can be obtained, frequently by employing CO_2 extraction pressures in the region of 100 bar to obtain the essential oil for aroma purposes, while elevating the CO_2 pressure to $\sim$275 bar results in an oleoresin-based extract for spice application. Similarly in the processing of hops, a "flavor" extract can be obtained for use in alcoholic beverages while the hop oil is diverted to other uses. All hop extract produced today intended for the beverage flavoring market is obtained *via* $scCO_2$ processing. Not always is the intended product the extract, *e.g.* decaffeination of coffee and tea produces beans and leaves used as a commodity food product. However the extract, caffeine, which is solvent-free, finds use as a pharmaceutical product.

$scCO_2$ extracts are numerous throughout the world in applications aside from coffee decaffeination and the production of hop extract. Table 6.1 lists some of the generic applications for $scCO_2$ extraction in industry. Most of the extracts listed in Table 6.1 are obtained by the straightforward cyclic $scCO_2$ extraction process described by the author and others in numerous

Table 6.1 Current industries using supercritical carbon dioxide for the production of high-value products with multiple applications.

Type of extract	Company	Website
Saw palmetto, chia, and perilla extract	Valensa, USA	www.valensa.com
Corn oil	MOR Supercritical, USA	www.reyntek.com
Natural extracts	Feyecon, The Netherlands	www.feyecon.com
Oils and phytochemicals	Flavex, Germany	www.flavex.com
Oil and phytochemicals	Proderna Biotech, India	www.proderna.com
Botanical and cannabis extracts	Eden Labs, USA	www.edenlabs.com
Oleoresins, botanical extracts, natural antioxidants, and oils	Kancor, India	www.kancor.com
Flavors, spices, botanical extracts, and oils	UMax, South Korea	iumax.koreasme.com
Soybean-based oils	Europa Crown, UK	www.europacrown.com
Pigments (capsicum, marigold)	Mohri Oil Mill Co., Japan	www.mohri-seiyu.jp
Oleoresins and phytochemicals from rice	Five Kings Cereal Industry Co., Taiwan	www.5-king.com
Phytochemical extracts	NATECO$_2$, Germany	www.nateco2.de
Spices, essential oils, and botanical extracts	Synthite, India	www.synthite.com
Aromatics, flavors, and nutraceuticals	Hitex, France	www.hitex-co2.com
Food supplements, flavors, fragrances, and nutraceuticals	Separex, France	www.separex.fr
Nutraceuticals	Phasex, USA	www.phasex4scf.com
Essential oils, oleoresins, and botanical extracts	Co$_2$ Extracts, India	www.co2extract.com
Algae-based extracts	NatureXtracts, Portugal	joomla.naturextracts.com
Botanical extracts	Honsea Sunshine Biotech, China	www.honseabio.com

reviews,[22–24] or by using $scCO_2$ extraction conditions employing stepwise extraction (fractionation) using different pressures and temperatures.

The use of $scCO_2$ can be manipulated in various ways for the processing and/or treatment of foods, which is covered in later sections in this chapter. Sesame oil, a high-price specialty oil, is used throughout Asia in many food cooking applications, and the demand for it has resulted in the construction of two industrial-scale plants in South Korea.[25] These plants, however, serve slightly different purposes. One facility, constructed in 2005 by Natex, produces about 2000 tons per year of sesame oil by employing two 3000 L batch extractors. The properties of the extracted sesame oil, aside from a higher extraction yield, are its brighter color, richer flavor, higher content of tocopherols, sesamol, and sesamin in the extract, and lower phosphorus content, as found in many $scCO_2$-extracted vegetable oils.[25] Other prophylactic properties are also conferred by extraction (treatment) with $scCO_2$, such as obtaining a pasteurized product that exhibits a longer shelf life. The other South Korean CO_2-based plant also processes sesame oil after it has been roasted, to obtain a "roasted" flavor extract for use as a food additive. Both of these plant-scale processes illustrate the versatility of SFE using $scCO_2$ on one food substrate to produce two different products.

In summarizing the current state of the art with respect to food processing using $scCO_2$, it is not only utilized for the classical purposes mentioned above, but is also driven by the perception of the process and resulting products being more "natural" and hence preserving or improving upon the original properties as they occur in the originally agricultural or natural product. The ability of CO_2 extraction to obtain different product qualities from the same starting material, as illustrated by the sesame oil example cited above, are attributes of the SFE process, and account for its success to date. We will now explore how CO_2-based extraction can be combined in various processing formats and be coupled to advantage with other compressed fluids to yield additional products using the same and new processing equipment.

6.4 Subcritical Fluids for Processing of Food and Natural Products

The use of subcritical water in place of polar or aprotic organic solvents for the extraction of natural antioxidants that can be concentrated and compounded into foodstuffs has been a driving force in the field. As noted by Pereira and Meireles[26] for this reason, we shall focus on this area and describe several studies that typify the use of water along with ethanol for these purposes. A number of these studies made use of pressurized solvent extractors developed initially for analytical purposes, and popularly referred to in the literature as ASE or PSE. The use of this instrumentation allows for a rapid assessment of the extraction potential and optimization by using pressurized fluids above their boiling point under pressure, *e.g.*, subcritical water[27] or ethanol.[28] However, there are differences and limitations in extrapolating PSE- or ASE-derived results for scaled-up batch or continuous extractions,[18] namely that the excessive pressures used in the ASE and PSE protocols are not always needed and are a waste of energy due to compression costs upon scaling the process up. In addition, the typical ASE protocol involves the use of several quasi-repetitive batch extractions followed by several "purges" of the extraction cell with solvent – a process that would be difficult to emulate on an engineering scale. In addition, the use of a dispersant, which is frequently mixed with the sample matrix, *e.g.* sea sand, would not be amenable in a scaled-up extraction process. Nonetheless, as a combinatorial-based screening tool on the laboratory scale, both ASE and PSE have proven their worth in the field.

Antioxidant extractions have been reported on rosemary,[28] for the recovery of polyphenolic compounds from juicing and winery by-products,[29] of phenolics from components of flax,[30] anthocyanins in berries[31] and red grape skins,[32] and catechins in tea leaves and grape seeds.[33,34] An excellent overall and recent review of efforts aimed at antioxidant extraction and fractionation using subcritical water and/or pressurized ethanol–water mixtures was provided by Hassas-Roudsari *et al.*[35] with a particular emphasis on antioxidants extracted from canola meal. Here, as in other studies, an applicable temperature range is frequently between 80 and 160 °C, a condition predictable using the Hansen solubility parameter/sphere approach reported by the present authors.[36] Aside from using high-purity water, it is essential that the water be deoxygenated for use in subcritical water extraction in order to reduce degradation of the antioxidant moieties during extraction. As shown by Duan *et al.*,[37] who applied subcritical water or ethanol mixtures for the removal of silymarin antioxidants from milk thistle, thermal degradation with increasing temperature must be taken into account when using subcritical fluid extraction, although ethanol has a prophylactic effect in retarding degradation of the targeted solutes. The impact of the presence of ethanol is shown in Figure 6.8, where to differing extents all of the silymarin compounds suffer decreasing half-lives in hot water with increasing water content of the extraction solvent.

However, degradation of bioactive compounds during hot subcritical liquid extraction can be minimized by limiting the residence time of the matrix and the extracted antioxidant during the extraction process. As shown by Duan and co-workers,[37,38] there is a finite amount of time before degradation of the extracted compounds sets in, as indicated in Figure 6.9. Here simulation of the extraction of silybinin B with water at 120 °C shows that removal of the target solute can be readily accomplished by adding an increasing amount of water to the sample to effect close to 99% extraction of the silybinin B. However, once the silybinin B is in the hot water, there is a finite amount of time before it starts to degrade (~30 min in the case under consideration), so it is important to optimize or reduce the extracted solute's residence time in the extraction device. This dualistic phenomenon has also been observed for the extraction of mannitol from olive leaves[39] as a function of temperature, which suggests that since the solubility of solutes in subcritical water increases with increase in temperature, accelerating the water flow, *i.e.* reducing the residence time in the extraction vessel, would result in higher yields of the target solute. The effect of temperature is significant when extracting thermally labile natural products such as antioxidants from food compounds. Studies by Monrad *et al.*[40,41] on using subcritical water–ethanol solvents for the extraction of anthocyanins and procyanidins from grape pomace indicated maximum yields between 80 and 120 °C using an ASE system. These studies indicated that pressurized water had a negative impact on anthocyanin and procyanidin yield, primarily due to thermal degradation. Significant improvements in anthocyanin and procyanidin yields were achieved using a semi-continuous hot–cold extraction technique,[42] where water was heated to subcritical temperatures and allowed to contact grape pomace placed under room temperature conditions. Time-based analysis indicated that at 140 °C and a water flow rate of 15 mL min^{-1}, there was considerable thermal degradation of the anthocyanins present in grape pomace after the eighth minute, with

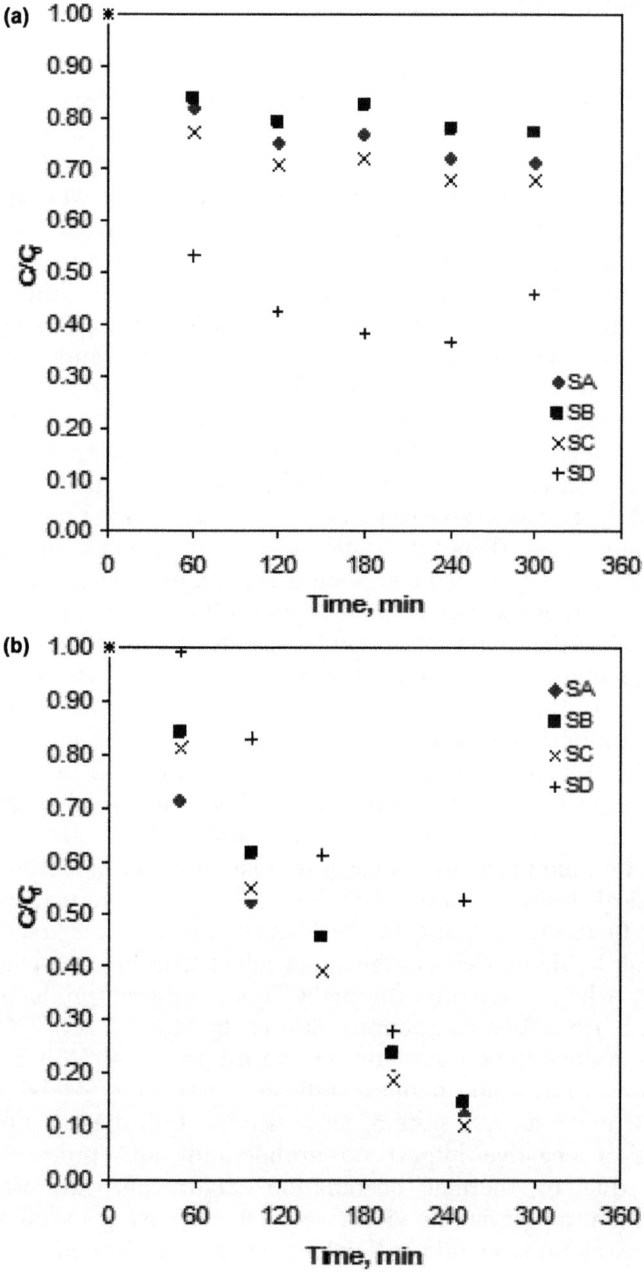

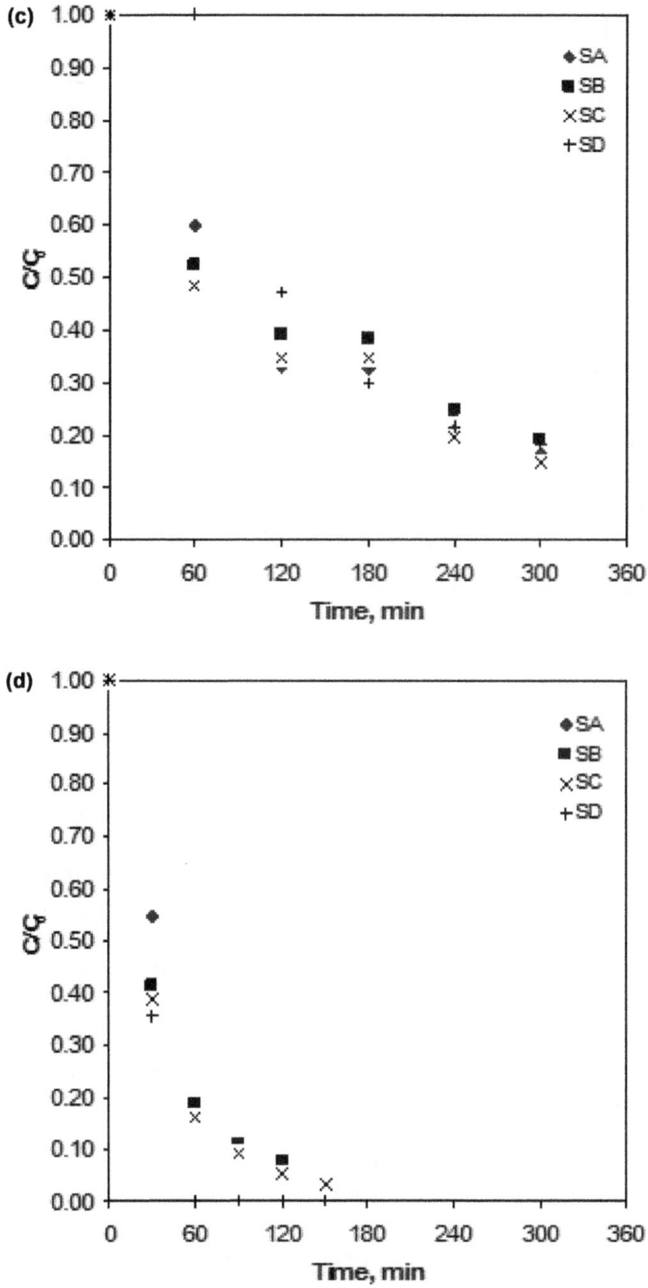

Figure 6.8 Degradation of silymarin compounds with time (C/C_0) as a function of percentage of ethanol in the extraction solvent [(◆, SA) silybinin A; (■, SB) silybinin B; (×, SC) silychristin; (+, SD) silydianin] at pH 5.10 and 140 °C: (a) 95%, (b) 50%, (c) 25%, and (d) 5% v/v ethanol.

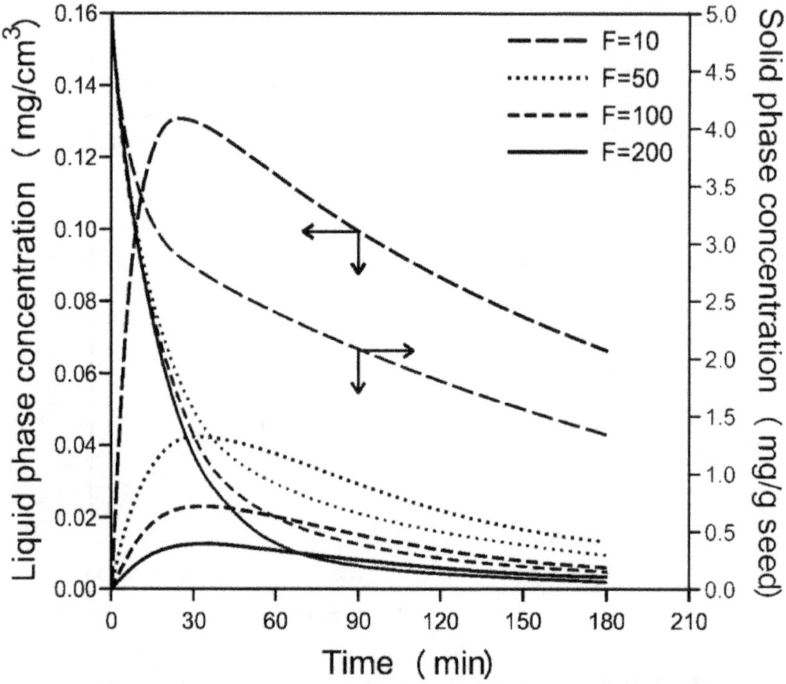

Figure 6.9 Simulation of the extraction of silybinin B with water at 120 °C for various solvent-to-solids ratios.

almost 85% of the total anthocyanins extracted from the grape pomace in the first 8 min.[42]

The presence of co-extractants during subcritical water extraction is fairly common and usually increases with increase in temperature. Kim and Mazza[30] also found increased protein and carbohydrate levels in subcritical water extracts from flax seed as the extraction temperature was increased from 100 to 160 °C. This suggests the possibility of reactions between the protein and carbohydrate moieties, *i.e.* the Maillard reaction, which often produces increased color and flavor intensity in extracts. With respect to the latter, it has been observed by Clifford[43] that flavor precursors could be generated *via* the subcritical water extraction of hops.

The fractionation of naturally derived extracts using subcritical water is also possible primarily through a change in the extraction temperature of water. Ibanez's group in Spain[29,44] demonstrated enrichment effects for polyphenolic compounds that are dependent on the extraction temperature, hence class selectivities based on solvation become possible. In their classic study on rosemary, they showed segregation of the less polar components found in rosemary by using higher extraction temperatures. Similarly, the same group demonstrated that polyphenolic oligomer fractionation is possible by adjusting the temperature of the subcritical water, while our group reported that by using subcritical water–ethanol gradients, there is an

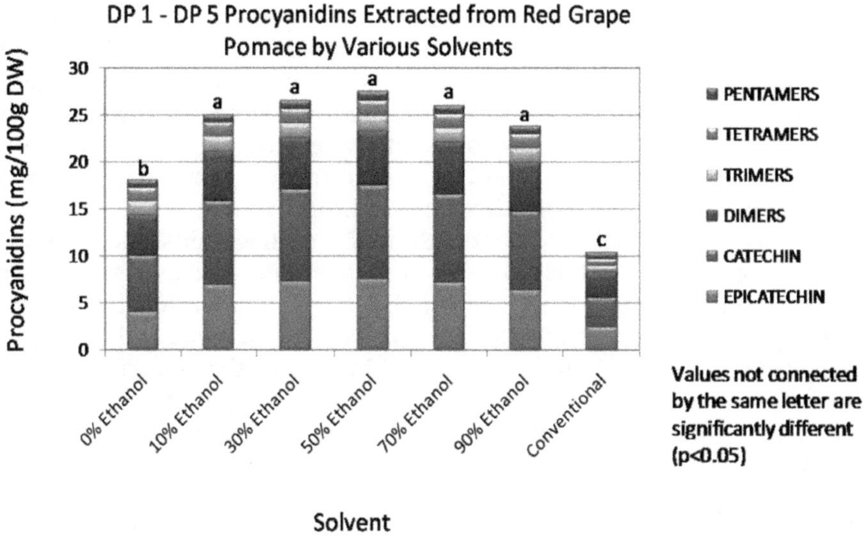

Figure 6.10 Procyanidin composition of grape pomace extract for ethanol–water extracts *versus* conventional solvent extraction. The order of the boxes in each column from top to bottom follows the entries in the key on the right.

increase in the extraction of procyanidins from anthocyanins from red grape pomace. For example, a 50 : 50 ethanol–water solvent extraction mixture could be optimized to extract the bioactive monomers and dimers preferentially from waste grape pomace.[40] This effect can be seen in Figure 6.10, where epicatechin and catechin and procyanidin dimers were extracted much better with 50 : 50 ethanol–water mixture relative to conventional solvent extraction at 25 °C.

The fractionation principle using subcritical water has also been extended into the essential oil field, particularly through the pioneering studies of Basile and Clifford[45] and others.[46] In this case, subcritical water shows a preference in extracting oxygenated terpenoid compounds from the essential oils *versus* the hydrocarbon constituents, thereby improving the essential oil flavor. Others have applied subcritical water processing to coriander,[47] cumin,[48] and oregano,[49] with similar results. The use of subcritical water extraction invites comparisons with steam distillation and competitive methodology,[45–49] where it has been noted that extraction yields are somewhat lower when using hot pressurized water, although subcritical water extraction is quicker and yields a higher quality of extract. This results in the subcritical water extraction being less expensive, particularly in terms of energy costs, as noted by Clifford and Basile.[50]

The addition of buffering or chemical agents to subcritical water has been reported in a few cases. Log K_w for water varies with temperature, but adjustment of the pH *via* this mechanism requires fairly high temperatures, which may not be amenable for extraction of thermally labile solutes. Kim and Mazza[30] studied primarily the effect of pH change on the mass transfer of carbohydrates and phenolics from flax shives and concluded that the enhanced extraction of phenolic moieties was due to hydrolysis of the lignin in the flax shives. Srinivas *et al.*[51] found that when using acetic and tartaric acids to modify water–ethanol mixtures, greater amounts of acylated anthocyanins were extracted from grape pomace under subcritical conditions. This study indicated that 80 vol.% ethanol in water was the optimal solvent for extracting anthocyanins and flavonols from grape pomace, with an increase in anthocyanin yield as high as 158% under optimized conditions (compared with the conventional extraction technique) when modifying the solvent with acetic acid. Other studies using different acidic and alkaline additives in the subcritical water hydrolysis of whey protein isolate indicated a maximum amino acid yield using 0.83 M sodium bicarbonate at 264 °C with a hydrolysis time of 29 min.[52] These studies indicated that a higher degree of hydrolysis was obtained under alkaline conditions, but under acidic conditions a greater degree of hydrolysis was obtained using lactic acid as additive with subcritical water compared with acetic acid.

6.5 Multi-fluid and Unit Operation Processing Options

The concept of embracing discrete processing steps using more than one critical fluid and also utilizing different types of unit operations that use environmentally benign agents such as $scCO_2$ and subcritical water has been discussed[23–25,50,53–57] over the past decade, and is being slowly implemented in research investigations and pilot plant–processing plant construction. The concept of using multi-fluid processing consists in staging and sometimes combining one or more fluids to achieve an end goal; for example, a sample matrix could be treated sequentially with $scCO_2$, followed by $scCO_2$ combined with a co-solvent, and ultimately with subcritical water.[21] The concept of multiple fluid processing is based on the tunable solvent characteristics of different super- and subcritical fluids by controlling the experimental conditions so as to achieve maximum extraction of different value-added components from the same food or natural product matrix. The order of sequential fluid application can be varied depending on the anticipated composition of extractable or reactive solutes present in the matrix being processed and the severity of the conditions and their effect on the matrix during the sequential process.

As indicated in Section 6.3, the use of SFE with $scCO_2$ is well known in the food processing industry, but the past decade and a half has seen increasing application of fractionation techniques such as column-based enrichment of target solutes, or similarly the application of supercritical fluid chromatography on a preparative or plant scale.[58] The present authors also noted that by integrating one of a number of diverse methods for producing fine particles onto the end of the SFE process, that direct powder or encapsulated ingredients can be realized embracing the "green" critical fluid platform.[54]

An example of the versatility of a multiple critical fluid processing platform has been demonstrated with $ScCO_2$ extraction of oil from rice bran.[59] Extracted rice bran oil can be further fractionated using a packed column in either a co-current or counter-current mode to yield fractions enriched in nutraceutical ingredients.[60] Treatment of rice with compressed CO_2 in a vessel to reduce spoilage and improve properties is currently industrially practiced in the Far East, and includes the removal of pesticide and metal residues from the rice matrix. Finally, rice bran oil can also be extracted with subcritical water, provided that the exposure time is short, to fractionate the lipophilic components.

The studies of Goto's group in Japan[61] are an excellent example of integrated multi-fluid processing. As shown in Figure 6.11a, $scCO_2$, depending on its density, can be used to extract non-polar solutes and, when combined with a co-solvent, more polar moieties such as flavonoids and glycosides from a sample matrix. This selective extraction can then be followed by applying hot water extraction to remove more polar solutes, including macromolecular species such as proteins or saccharides from the matrix. This concept has been applied in the format illustrated in Figure 6.11b, where either CO_2 or H_2O can be dispersed in hot water or $scCO_2$, respectively. When these two compressed fluids are applied in this way to a sample matrix as shown in Figure 6.11b, a fractionating effect is achieved, which combines the co-solvent effect of water with the pH effect of CO_2 for selectively extracting targeted solutes from a natural product.

The application of sequential extraction with $scCO_2$ followed by subcritical water to a target matrix has been demonstrated by several researchers. Mannila *et al.*[62] applied slightly subcritical CO_2 (30 °C, 80 atm) for removal of the bioactive component, hyperforin, from St John's wort. This was followed by subcritical water treatment at pH 7.0 using a weak bicarbonate buffer to isolate the more polar component, hypercin, in good yield from the above matrix. Often $scCO_2$ is applied prior to subcritical water extraction to remove a non-polar oil or lipophilic ingredient from a plant matrix. D'Antuono and Favati[63] applied this sequential fluid procedure to grape seeds and marc, by delipidating the seeds to recover the oil with $scCO_2$ at 80 °C and 60 MPa, in preparation for subcritical water extraction of the defatted seeds to recover polyphenolic solutes. Application of subcritical water between 145 and 180 °C and over a pH range of 4.5–7.8 was optimal for the recovery of the polyphenolic solutes, such as gallic acid and procyanidin B1 and B2. A similar approach was taken by del Valle *et al.*[64] for the $scCO_2$ and subcritical water fractionation of the essential oils from boldo, a South American plant, followed by subcritical water extraction (SWE) of boldine, a phytoactive ingredient in the plant matrix.

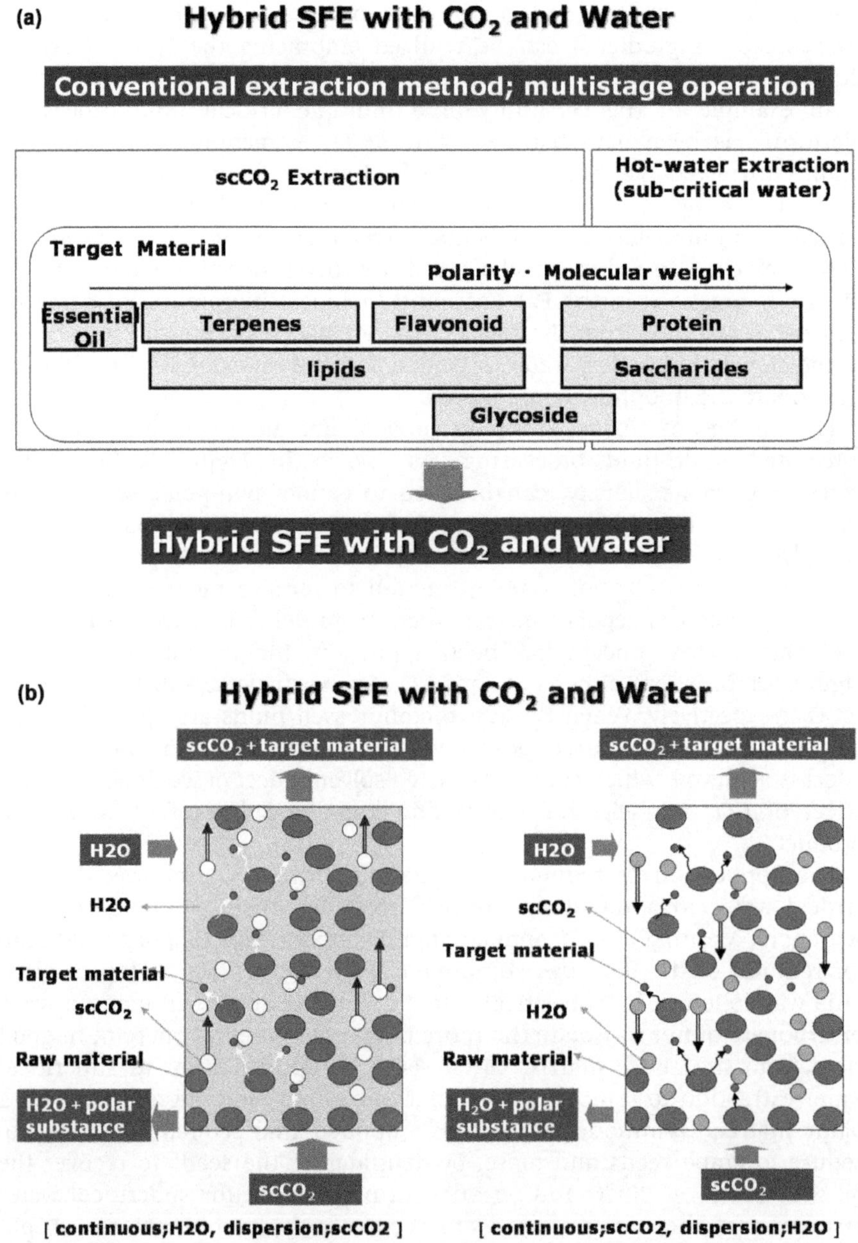

Figure 6.11 Hybrid SFE using CO_2 and water: (a) selective extraction; (b) continuous-dispersion approach.

A slightly more sophisticated study was carried out by Smith *et al.*[65] on separating the various constituents of cashew nut and cashew nut shell liquid (CNSL) with $scCO_2$ and subcritical water. There are several noteworthy features of this "green" processing of cashew,[65] namely the use of view cells to observe the extraction and reaction phenomena and also compression–decompression cycling of $scCO_2$ during the SFE of cashew nut shells. Using the latter phenomena, they were able to achieve greater fluxes of CNSL as a function of the mass of CO_2 used due to rupture and swelling of the nut shell matrix, thereby facilitating higher recoveries of CNSL (94% recovery). The alkenyl phenolic compounds in CNSL, such as anacardic acid, exhibit antitumor and antimicrobial activities. After extraction of alkenyl phenolic compounds from the shell; the residual matter can be treated with high-temperature subcritical water for liquefaction/gasification fuel purposes.

The use of column-based fractionations employing $scCO_2$ and a co- or counter-current contacting phase has been applied to a number of problems, and an extensive review on this subject was provided by Brunner.[66] This fractionation approach is dependent on having a liquid-state substrate at the temperature of fractionation in contact with the supercritical, or in some cases liquid, carbon dioxide fluid. The column is usually packed and has a temperature gradient superimposed along it to improve the fractionation of the liquid components. Applications have ranged from the mundane, such as the removal of hexane from a vegetable oil,[67] to the fractionation of biologically active components in vegetable oil hexane extracts, such as olive oil.[68] Both the extract and raffinate from the column fractionation can be used to isolate targeted solutes. Removal of fatty acids from lipid mixtures to yield vegetable oils enhanced in tocopherol[69] or phytosterol[70] content from wheat germ and rice bran oils, respectively, has been achieved using column-based methods.

An excellent example of tandem processing was provided by Brunner's group for the rectification of the minor constituents in cold-pressed palm oil.[71] As shown in Figure 6.12, the cold-pressed palm oil is transesterified with a suitable alcohol not only to improve the volatility of the minor lipid components, but also to enhance their fractionation inside the column. A number of $scCO_2$–methyl or ethyl ester binary systems are well characterized with respect to phase equilibria and thermodynamic solubility so that the fractionating column can be optimized to yield high separation factors between the esterified and non-esterified solutes. Thus, as shown in Figure 6.12, the tocopherol and/or carotenoid concentration in cold-pressed palm oil can be enriched by $scCO_2$ column fractionation from 500–1000 ppm in the base oil to 20–50 wt% in the extract phase. Further enrichment of the resultant fractions is possible by multi-stage SFE in which selective enrichment is afforded by changing the density of the $scCO_2$ in a multi-vessel or -settler array, or by integrating adsorbents into the SFE step.[72] Enrichments to 70–100% of the component tocopherols or carotenoids are possible by introducing this step. Refinement to even purer components can be accomplished by preparative supercritical chromatography, such as has been demonstrated by our group for phospholipids from the lecithin fraction derived from crude soybean oil.[73]

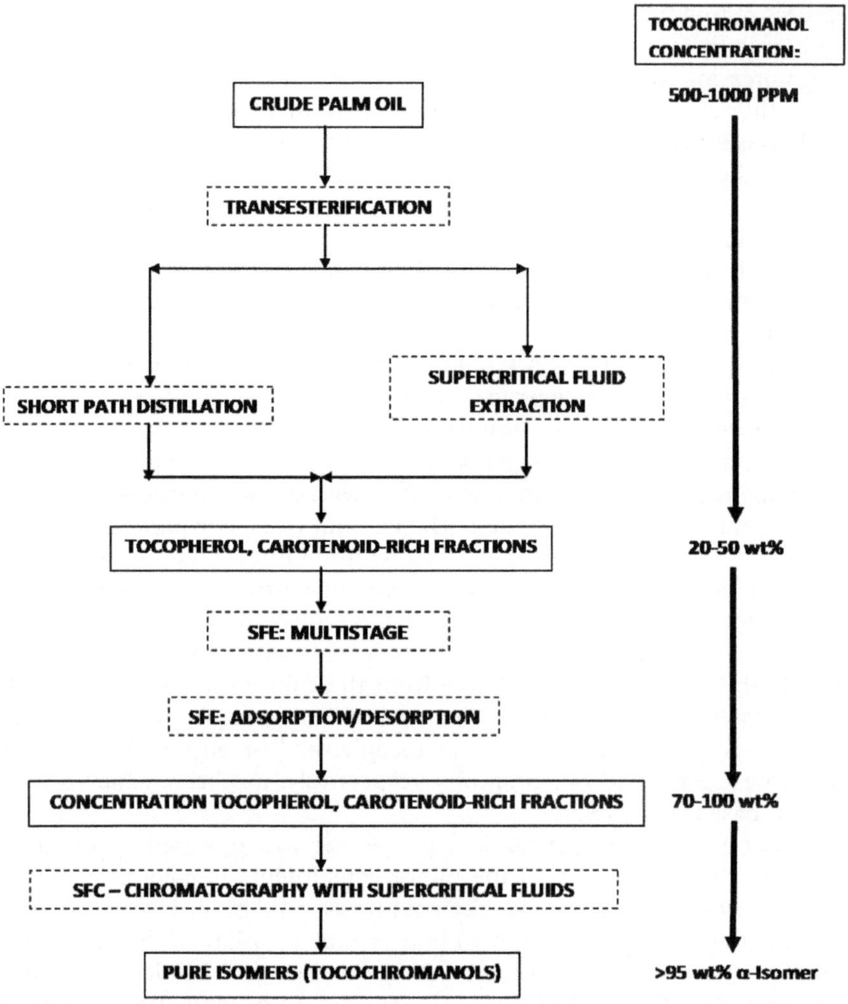

Figure 6.12 Fractionation of minor palm oil constituents by transesterification reaction followed by multi-stage supercritical fluid-based processing.

Reaction chemistry in sub- and supercritical fluid media has been integrated with SFE and SWE systems in addition to the above-noted scheme for enrichment of low levels of minor constituents from vegetable oil or side-stream products, such as deodorizer distillate.[72] Of the reaction-type possibilities,[20] probably enzymatic and hydrogenation reactions and reactions conducted in subcritical water are the most relevant to green food processing. Reactions of analytical significance conducted in critical fluids also find a place in food analysis and also serving as templates for scale-up, *e.g.* esterification reactions.[73] Despite their inherently high cost and sensitivity to environmental factors, enzymatic reactions show considerable promise for the production of higher cost, specialty chemicals used in the food and flavor industry, particularly chiral compounds. Most of the schemes for integrating reaction chemistry in critical fluids for the production of food products have been conducted with bench-scale apparatus, and have not moved into the production stage to date. However, several examples are provided below to illustrate the promise that this type of tandem process holds.

Baig *et al.*[74] reported a study on critical fluids as a process environment for adding value and functionality to sunflower oil, *i.e.* a model system for biorefining. Developed as part of a European Technology Platform for Sustainable Chemistry, sunflower oil has been subjected to continuous-flow hydrolysis using subcritical water at temperatures of 250–390 °C and pressures of 100–200 bar to produce fatty acids. This reaction step was then followed by using a supported enzyme, Lipozyme TL IM, to convert the fatty acids to ethyl esters utilizing a continuous-flow reactor with $scCO_2$. The latter reaction has also been used to prepare the ethyl esters of fish oil for ω-3 and -6 enrichment using a column-based $scCO_2$ method, the resultant products having enjoying commercial success in the nutraceuticals market. Vegetable oils can also be converted *via* a supercritical fluid reaction (SFR) by using $scCO_2$ for the SFE of the target seed oil followed by enzymatic glycerolysis or transesterification, depending on the point at which the reaction was concluded.[75,76] Again, these end products are amenable to supercritical fluid fractionation (SFF) techniques, such as column fractionation based on the molecular weight and vapor pressure differences in the solutes contained in the reaction mixtures.[77]

A particularly interesting study involving the subcritical water extraction of onion waste to recover the natural antioxidant quercetin was reported by Turner *et al.*[78] Subcritical water was considered a preferred extraction solvent since the solubility of quercetin in $scCO_2$ is fairly low. A thermostable β-glucosidase was employed to hydrolyze the quercetin glycosides recovered from the onion waste *via* SWE because of their high temperature stability at 90 °C. Therefore, the resultant process is environmentally sustainable since it only uses water as a solvent and enzymes as catalysts.

Membrane-based fractionation is another alternative for integration with critical fluids, although it has not been investigated extensively for food processing. Membranes have been integrated with recycling $scCO_2$ during SFE processes as a means to separate targeted solutes from a compressed CO_2 stream without resorting to reducing the pressure to precipitate the solutes.[79] Sims and Estigarribia developed the PoroCrit process,[80,81] largely based on liquid CO_2 processing using concentric membranes to enrich solutes in aqueous streams by partitioning them into liquid CO_2. Later, this approach was also used as a method to sterilize orange juice (see Section 6.6). Other food applications include a process patented by Sarrade *et al.*[82] for fractionating triglycerides dissolved in $scCO_2$ or propane, a particular example being fish oil fatty acids noted previously. Other lipid-based supercritical membrane (SCM) processes were reported by Sarmento *et al.*[83] for the enrichment of polyphenols extracted from cocoa seeds *via* $scCO_2$

extraction, both with and without an ethanol entrainer at 8–15 MPa and 40 °C. Using commercially available membranes on a pilot-plant scale, they obtained polyphenol yields of over 40%. Ethanol-modified scCO_2 was also used by Beltran *et al.*[84] to separate antioxidants from grape extracts by coupling SFE with SCM.

Multiple fluid processing through the coupling of a critical fluid-based extraction process directly with fine particle production began in the mid-1990s, where many studies employing over 15 separate methods were applied to produce micrometer or sub-micrometer particles with a narrow size distribution. Initially utilized in the pharmaceutical industry,[85] eventually it was applied in the food and particularly the nutraceutical fields. Several reviews on this technology focusing on food applications, such as those by Rodriguez-Meizoso and Plaza,[86] Cocero *et al.*,[87] Weidner[88] and Jung and Perrut,[89] covering both the published and patent literature. The high-pressure concentrated powder form (CPF) pioneered by Weidner has found applications in the food industry, including powered soaps, chocolate, and grape extracts. For example, cocoa butter and chocolate have been micronized on finely powdered carriers for use in food compounding applications[88] and anthocyanins in grape marc[89] have been treated with water or 50 : 50 ethanol–water mixture and precipitated onto a silica carrier using the CPF process. This process seems compatible with some of the aqueous ethanolic extraction studies discussed in Section 6.4. Although still in the experimental stage, an SWE step has been combined with an online generation process to produce fine particles in a process called WEPO (water extraction and particle formation online). Aside from using subcritical water between 150 and 200 °C for the SWE stage, the aqueous extract is mixed with scCO_2 in a spray chamber to produce aerosols with particle sizes from as small as 4 μm to as large as 140 μm of rosemary or oregano extract, which is a suitable particle size range for use in developing functional foods.[90] In another study, WEPO was used in optimizing the extraction of high-value quercetin extracts from onions, with optimized conditions of 120 °C, a CO_2 flow rate of 10 mL min^{-1} and nitrogen pressures of 80 and 12.5 bar, to achieve a fine and constant spray pattern resulting in the formation of well-defined spherical particles smaller than 4 μm.[91] These studies indicated that whereas the antioxidant capacity and concentration of quercetin derivatives from onions were comparable between WEPO and tandem pressurized fluid extraction–freeze-drying, the particle sizes obtained through freeze-drying were found to be significantly larger and distributed unevenly. Another type of micronization system for particle formation from natural product extracts such as *p*-coumaric acid is the automatic rapid injection or ARISE system.[92] The ARISE system, using scCO_2 as antisolvent with acetone, at 298 K and 9.5 MPa resulted in particles with sizes up to 23 times smaller than with the current particle formation technology while having product recoveries as high as 96%. The ARISE process works, in principle, similarly to a combination of GAS (gas antisolvent) and PGSS (particles from gas-saturated solutions) where a bioactive compound is dissolved in an organic solvent and then mixed with $scCO_2$ as an antisolvent or solvent, respectively, and allowed to particulate through an injection nozzle through instantaneous depressurization. The PGSS process has been scaled up to produce as much as 1500 kg of particles of product per hour with a projected cost of powders ranging between \$0.25 kg^{-1} and \$1.33 kg^{-1} depending on the annual production time and CO_2-to-powder ratio.[93] Further descriptions and benefits of various food and bioactive compound powders produced through critical fluid particle technologies have been summarized by King.[94]

6.6 Multi-phase Fluids for Sustainable and "Green" Food Processing

The concept of multi-phase fluids for food processing is unlike the multi- or tandem fluid processing discussed in Section 6.5, in that multi-phase fluids involve having a gas or gaseous-like fluid dissolved in a condensed liquid state, predominantly sub- and/or supercritical fluid simultaneously, rather than sequentially as described in Section 6.5. Hence, if one looks at the phase diagram for water in Figure 6.1, dissolving carbon dioxide in a subcritical liquid is often referred to as conferring "enhanced fluidity" on the subcritical liquid and, depending on molecular interactions, the dissolved gas may yield what is referred to in the literature as a "gas expanded" liquid.[95] For food processing, the use of CO_2 dissolved in water provides a "green" processing medium, and CO_2 imbibed in certain organic solvents can fulfill a similar role.

First, we will discuss gas-expandable liquids (GELs), which are expanded in volume upon dissolution of a gas or SCF. The attractiveness of this concept is similar to that for the use of an SCF, *i.e.* solute solubility is pressure dependent and separation of the two phases can be achieved by depressurization. Some of the advantages of using GELs are similar to those of SCFs, such as enhanced transport properties in solvents and higher gas diffusivity, but the optimal experimental conditions can be milder than that required for sub- and supercritical fluids. It is also possible to experience an equilibrium in which the introduction of a gas or fluid such as CO_2 can introduce multiple immiscible phases whose formation is dependent on the type of gas or liquid chosen and the applied pressure. This was first observed in the 1950s by Elgin and Weinstock[96] and has recently been resurrected as the OATS process, which is particularly applicable to separating reaction products and catalysts, such as enzymes, *via* phase immiscibility.[97]

There are three types of GELs:[98] a Type 1 liquid exhibits negligible expansion with the introduction of a critical fluid such as $scCO_2$, whereas a Type 2 liquid expands significantly with the introduction of the gaseous component, and a Type 3 liquid usually exhibits intermediate swelling phenomena, examples being ionic liquids and the swelling of petroleum oils in $scCO_2$. Hence the principal effect of adding CO_2 to water is to acidify the medium, which will be discussed shortly. It should be noted that one can have a ternary system consisting of gas–organic liquid–water, a medium that can be acidified while also exhibiting some degree of swelling. Although the use of the traditional pH scale cannot strictly be applied to organic liquids, arbitrary standards have been developed; for example, the pH of a carbon dioxide–methanol–water system can reach a pH as low as 4.2. Thus reactions that are catalyzed by hydronium ions can be affected in the gas-expanded mixtures mentioned.

As with SCFs, GELs have adjustable solvent properties, and the introduction of the gaseous components also improves mass transfer and mixing of components relative to what would be experienced in the neat liquid. In terms of dielectric constant, these GELs exhibit a range of permittivity between those of the neat liquid and gaseous component as a function of the mole fraction of gas, *e.g.* from the dielectric constant of ethanol to $\varepsilon \approx 10$ at a CO_2 mole fraction of 0.5.[99] Such a range of values encompass a range of moderate to non-polar liquid polarities. Li and Maroncelli[100] commented that by using GELs, over 95% of the solubility parameters of common organic liquids can be equaled. For use in the processing of foods or production of ingredients, $scCO_2$–alcohol systems[101] are of particular interest because of the reaction possibilities noted above and the fact that CO_2 can combine with *n*-alkanols to form alkylcarbonic acid catalysts, analogous to the formation of carbonic acid in aqueous solution. Although other gaseous components can be used, CO_2 is preferred for many well-cited reasons, included its prophylactic effect and fire-suppressing properties when used with organic liquids at higher temperatures.

Despite the noted benefits,[102,103] and the citations given in the following, GELs have not been applied extensively in food processing operations, which suggests a fertile area for more research and development. Strictly, even a molten lipid with copious amounts of imbibed CO_2 is a gas-expanded fluid. This concept has been studied in terms of regulating the melting point of cocoa butter with compressed carbon dioxide.[104] Separations of some relevance to food materials processing have made use of the switchable solvent principle,[105] in which the polarity of one of two binary phases can be changed relative to the other by injection of a gaseous fluid component into the system. This principle has been applied to the extraction of soybean oil from soya flakes using ethanol and/or an ionic liquid in conjunction with $scCO_2$. Injection of gaseous CO_2 can also be used to separate triglycerides from hexane, suggesting that CO_2 is an antisolvent that could be used to phase separate the two components after traditional solvent extraction of a vegetable oil. A similar principle has also be used fractionate mixtures of citric and oxalic acids[106] and for the purification of carotenoid isomers.[107]

Gas-expanded fluid mixtures as noted above provide opportunities to conduct reactions of potential importance in the food industry. Simple esterifications[108] along with transesterifications primarily devoted to making methyl esters[109,110] have been reported. With regard to the latter, the practicality of producing biodiesel *via* such a method might be questioned; however, ethyl esters of lutein or fish oils are other possibilities that could be explored. Essential oil components have also been reacted using carbonic acid catalysis, for example, the cyclization of citronellal[111] in a water–CO_2 system, which is pH dependent, and the esterification of β-pinene.[112] Such reactions may be applicable to the production of food flavors. Eckert *et al.*[113] described a process designed to win fine chemicals from lignin using GELs. Although the original intent of their studies was more within the scope of biomass utilization or a biorefinery operation, the use of gas-expanded alcoholic solvents produced phenolic compounds, such as vanillin, which have application as food flavoring agents.

It was noted above that Type 1 GELs show negligible volumetric expansion with gas absorption and that the primary effect of CO_2 absorption is to decrease the pH of an aqueous solution due to the carbonic acid equilibria in solution. Hunter and Savage[114] mathematically formalized the relationship between pH and the various equilibria constants and CO_2 gas pressure, P_{CO_2}, as

$$[H^+] = (K_W + K_{a_1}[CO_2(aq)])^{1/2} = \left(K_W + \frac{K_{a_1}}{K_H} P_{CO_2}\right)^{1/2} \tag{6.1}$$

where K_w is the dissociation constant for water, K_H the overall dissociation constant for the carbonic acid equilibrium, and K_{a_1} the first dissociation constant for carbonic acid. To use eqn (6.1) to calculate pH requires CO_2 pressure data over extended pressure and temperature ranges. We have summarized relevant CO_2 gas solubility data in water from four reliable sources[25] from 0 to 150 °C between 0.1 and 60 MPa. For carbonated water solutions, as the CO_2 pressure is increased, the solution pH decreases, whereas an increase in the aqueous solution temperature at a given pressure usually increases the pH. Graphical plots of these trends have been generated at relative low pressures and temperatures by Chuang and Johannsen[115] while pH trends with CO_2 pressure at quite higher temperatures are provided by van Walsum and Shi.[116] We have replotted some of these data in Figure 6.13, showing the "linear" dependence of pH on temperature from 100 to 260 °C from 1 to 600 bar because of its relevance and application in subcritical water extraction of anthocyanins.[117] Note that in Figure 6.13, $pH < 3$ is predicted at temperatures below 150 °C. It should be noted from the limited available data that in sugar solutions[118] or grapefruit juice,[119] less carbon dioxide is absorbed as the concentration of sugars is increased in the aqueous solution matrix.

The formation of carbonic acid in solution upon dissolution of CO_2 provides a pressure-dependent method of solution pH control, applicable to the separation of compounds by protonation of a solute or, in the case of biopolymers, according to their isoelectric point characteristics. By using carbonic acid in certain cases, minerals acids such H_2SO_4 and HCl can be avoided, creating a more environmentally benign extraction or reaction medium. An example of carbonated water extraction selectivity is provided from our own research in facilitating the removal of pH-sensitive solutes

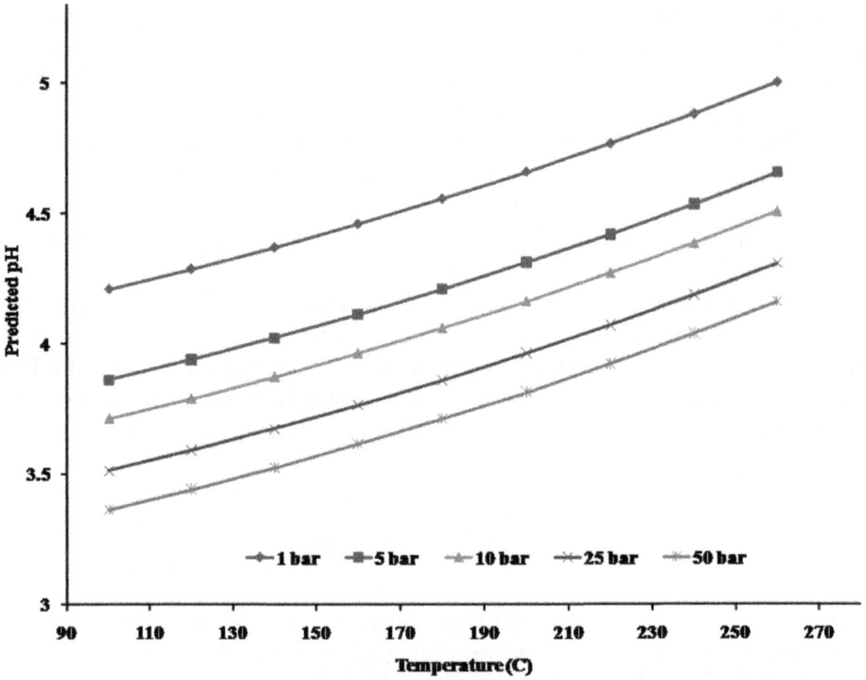

Figure 6.13 H_2O–CO_2 system: predicted pH *versus* pressure and temperature.

such as anthocyanins from grape pomace in their flavylium cation form at acidic pH, as shown in Figure 6.14. Using cyanidin-3-*O*-glucoside as a model anthocyanin, it can be seen in Figure 6.14 that several pH-sensitive forms of the compound exist, ranging from the red flavylium cation form to the blue quinoidal base form. Batch extraction of freeze-dried red grape pomace with subcritical water at 120 °C using 150 bar increments from 0 to 600 bar CO_2 pressure produced an increasingly intense purple extract with increasing CO_2 pressure, consistent with the estimated pH range of 2.8–3.2 from Figure 6.13, and the formation of intermediate species between the flavylium cation and quinoidal base forms of the constituent anthocyanin forms.

The extraction results using carbonated subcritical water are shown in Figure 6.15, where the percentage total anthocyanins recovered as determined by the summation of the constituent high-performance liquid chromatography (HPLC) peaks is plotted against extraction time at different CO_2 pressures. It is apparent that a higher degree of water carbonation aids in the recovery of anthocyanins whereas increasing the extraction time decreases the amount of available anthocyanins. From Figure 6.15, it appears that about 55% of the available anthocyanins can be extracted in 7.5 min, and probably more, if the heat-up time of the cell to 120 °C could be decreased even further. This is consistent with an independent study[35] which showed that model anthocyanin compounds degrade within minutes at these temperatures, unless a very fast extraction can be achieved. It is interesting that

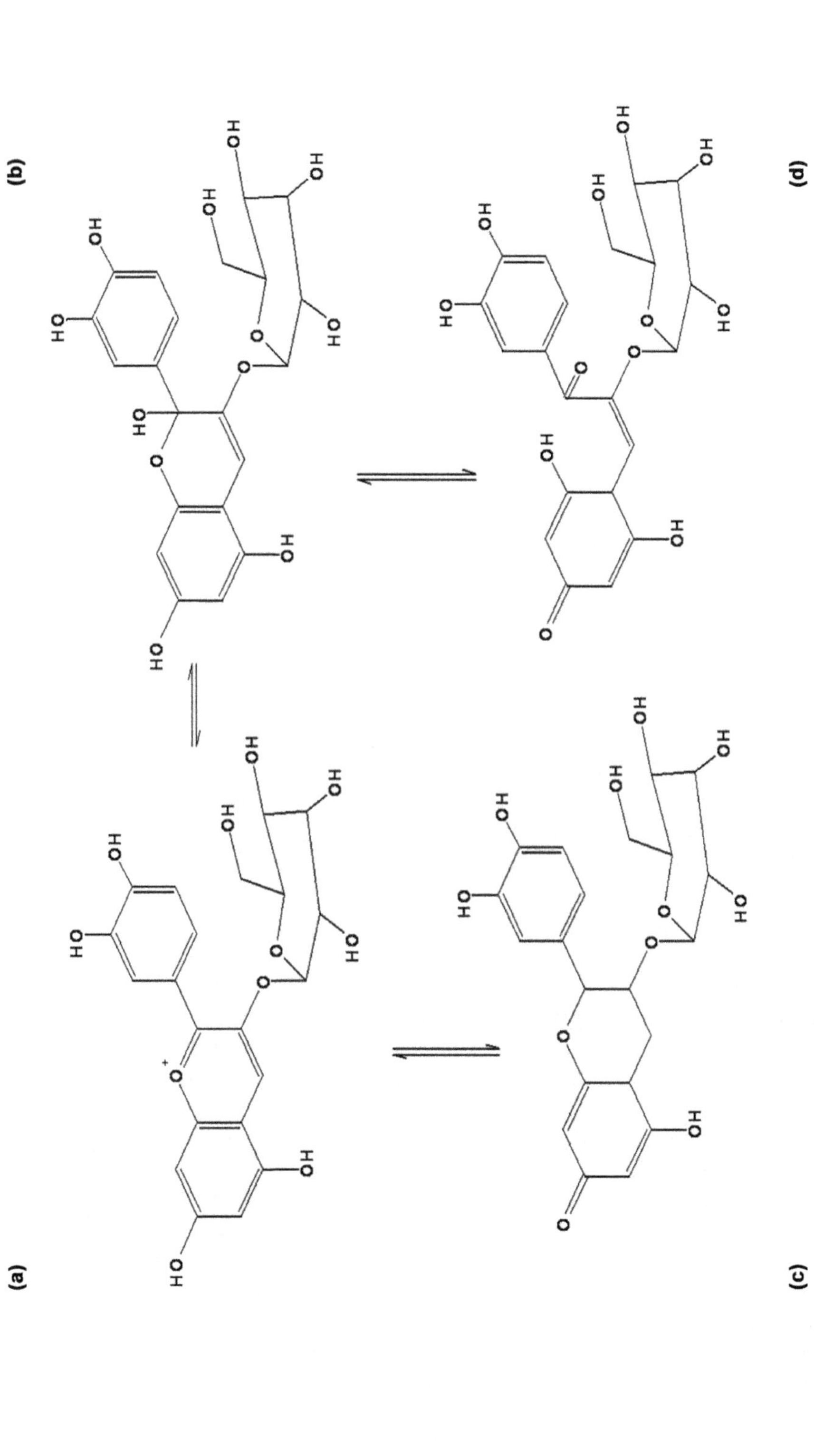

Figure 6.14 Equilibria for forms of cyanidin-3-*O*-glucoside: (a) flavylium cation form; (b) carbinol pseudobase form; (c) quinoidal base form; (d) chalcone pseudobase form.

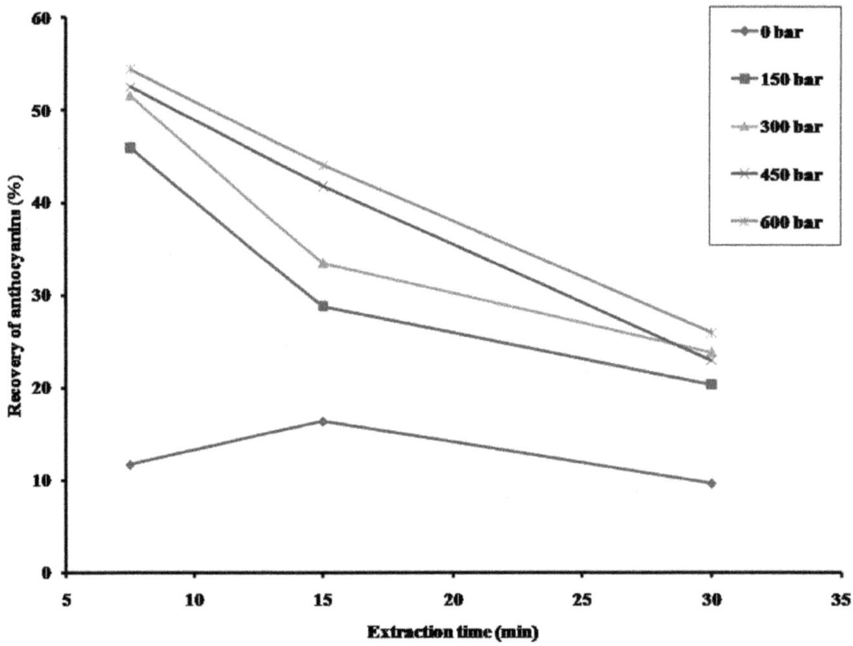

Figure 6.15 Recovery of anthocyanins as a function of extraction time and CO_2 pressure in carbonated water at 120 °C.

the water at high levels of carbonation may obviate the use of bisulfate ion or SO_2, which are often used as preserving agents in the processing of grape-derived products despite their asthma-inducing properties when consumed.

Other pH-dependent food or ingredient processing methods can benefit from the carbonic acid equilibria previously described. Soybean protein precipitation was studied by Thiering *et al.*[120] using CO_2 as a "volatile electrolyte." Historically, protein fractions have been obtained *via* selective precipitation based on their isoelectric points (which are dependent on their amino acid composition) using mineral acids accompanied by centrifugation of the resultant precipitates. An alternative method using CO_2 at 50 bar pressure permits the fractionation of proteins amenable to precipitation in the pH range 4.0–8.5. Using defatted soybean protein as a model matrix, Hofland *et al.*[121] optimized soy protein precipitates by studying the precipitation curves *versus* pH at various pressures of CO_2 in aqueous solution, achieving the best yields at an isoelectric point equivalent to a pH of 4.8. These experiments were run up to soy protein concentrations exceeding 50 $g\,L^{-1}$ in a relative simply batch reactor device.

Another aspect of the carbonation studies of Thiering *et al.*[122] was the production of a specific particle size range and morphology when using high-pressure CO_2. By accurately controlling the CO_2 pressure, the precipitates for typical constituent soybean proteins such as glycinin and β-conglycinin can be achieved at 95 and 80% concentrated fractions, respectively, and precipitate yields of 28 and 21%, respectively.[122] Scanning electron microscopy (SEM) studies on the resultant soy protein precipitates showed that carbonated water-induced precipitants have a distinctly different morphology to those produced using sulfuric acid precipitation.

Similar pH pressure-induced precipitations have been reported in connection with casein[123–126] and cheese whey[127] processing. In support of these studies, the solubility of CO_2 in milk at high pressures was measured by Tomasula and Boswell[128] between 25 and 50 °C and at pressures up to 6.9 MPa. The recorded solubility of CO_2 in milk was found to increase with increase in pressure and to decrease with increase in temperature. Complete precipitation of casein from milk was achieved at 38 and 50 °C using pressures $>$5.5 MPa, a condition in which the solubility of CO_2 in milk equaled that observed in neat water. Although beyond the scope of this chapter, Tomasula and co-workers also studied the properties of films made from CO_2-precipitated casein[129] and provided details on a continuous process[130] for producing the above precipitates, including an economic analysis.[131]

6.7 Brief Introduction to High-pressure Pasteurization in Food Processing

Aside from food protein precipitation, high pressure has been used to pasteurize a number of food matrices. The antimicrobial and antibacterial properties of $scCO_2$ used in processing of foods have been studied extensively and, owing to the enormity of the literature involved,[132,133] are covered in a separate chapter in this book. However, a brief summary to pasteurization of foods using $scCO_2$ will be presented here to serve as an introduction to the concept. A now rather old but still very interesting review on the effect of CO_2 on the growth and metabolism of microorganisms was provided by Dixon and Kell,[134] discussing their effects on common foods and their processing. The attractiveness of using high-pressure CO_2 in reducing microbial counts in foodstuffs is that it represents a cold pasteurization technique relative to thermal retorting, thereby allowing retention of the original properties of the food matrix with respect to flavor, color, *etc.* A critical updated review was provided by Garcia-Gonzalez *et al.*[135] Relative to high-pressure food processing, the requisite pressures of CO_2 are much lower (20 MPa *versus* 300–600 MPa), and $scCO_2$ is unique relative to other high-pressure fluids, such as nitrogen and nitrous oxide, in reducing microbial counts due to the carbonic acid equilibria cited previously. Carbon dioxide in its supercritical state is consistently capable of lowering microbial counts *versus* subcritical CO_2, *i.e.* liquid CO_2, on the same type of food matrix, the trend being ascribed to the superior mass transport properties that allow it to penetrate the morphological regions of the food matrix.

The versatility of microbial deactivation with $scCO_2$ is facilitated by using moist foodstuffs facilitating the formation of carbonic acid, and the rate – particularly of CO_2 depressurization – which, if abrupt, can rupture cell walls. These benefits suggest that the previously cited subcritical water processing at slightly above 100 °C could be a very effective method for sterilizing an aqueous extract *in situ*, an effect that has not yet been tested. If a high rate of microbial deactivation were to occur during subcritical water extraction, it could probably be due to a combination of thermal deactivation and carbonic acid lowering of the microbial count. Some support for the dual benefit of carbonated subcritical water extraction can be seen from the results of Del Pozo-Insfran *et al.*,[136] who compared dense CO_2-treated grape juice with the same juice treated by heat pasteurization, in terms of shelf life, anthocyanin content, total phenolics, and antioxidant capacity. Dense CO_2 treatment consistently showed higher values *versus* thermally treated juice in all of the above properties, which allowed an equivalent microbial kill count without sacrificing phytochemical content.

The thermodynamic basis of aqueous scCO_2 treatment of fruit and vegetable products is based once again on the solubility of CO_2 in water as a function of CO_2 pressure, temperature, and CO_2 partial pressure above the liquid being treated. Additional high-pressure CO_2 solubility data in neat water and juices[119] have been reported by Balaban's group[137] based on a stirred autoclave solubility device, and measurements made of CO_2 sorption into neat water, real orange and apple juice, and model solutions simulating the fruit juices. In general, the CO_2 solubility was suppressed slightly in the real and simulated juice matrices. Computer simulation of estimating CO_2 solubilities in the above liquids was performed using a combination of the electrolyte NRTL model employing the Redlich–Kwong equation of state. It should be noted that these models were applied for CO_2 pressures below 20 MPa at 40 °C. Slightly higher CO_2 pressures were applied for reed grapefruit juice pasteurization, resulting in 5 log reductions in yeast and retention of many of the proximate parameters and physicochemical properties relative to the untreated juice. Of particular interest in the processing of fruit juices is combining high-pressure food processing with CO_2 deactivation of pectin methylesterase[138] and polyphenoloxidase, *Lactobacillus platarum*, and *Escherichia coli.*[139] The inclusion of 12–15 psig of CO_2 to juice solutions containing the above moieties led to reductions in enzyme activity and *L. platarum* count relative to the neat juice; however no effect was recorded for *E. coli*. A detailed description of the food substrates and microbes of concern for CO_2-treated substances has been compiled in greater detail[140] along with the patent literature on the current state of the art in high-pressure fluid pasteurization.[141]

6.8 Extraction *Versus* Reaction Using Pressurized Fluids

scCO_2 extraction has been advocated as a non-invasive, environmentally benign process and has been predominantly used for extractions. The prophylactic effect of an "inert" atmosphere of CO_2 has also often been cited as one of the benefits when conducting SFE, particularly when extracting oxygen-sensitive solutes. Although both liquid and supercritical CO_2 are commonly thought of as non-reactive media, CO_2 has been used as both a reaction medium[142] and a reagent,[143] and in the presence of water is a potent catalytic medium, particularly with respect to conducting hydrolytic-based reaction chemistry. Unlike CO_2, sub- and supercritical water has been predominantly used as a reaction medium owing to the high temperatures involved, with supercritical water having significant destructive properties. This has been exploited in its use as a reaction medium for both organic[144] and inorganic[145] synthesis, conversion of biomass to functional monomers,[146] and the total destruction of chemical wastes or toxicants to CO_2, H_2O, *etc.*[147] Reactions in subcritical water were summarized by Liotta *et al.*[148] and there is a substantial literature on the subject. The role of pressurized water can be diverse, as noted by the authors, and several possibilities are listed as follows:

- to alter an extract composition selectively
- to depolymerize a polymeric moiety
- to depolymerize biopolymers to functional monomers
- to degrade depolymerized products to seed mixtures for fermentation
- to facilitate polymerizations.

As indicated previously, CO_2 in water can be tuned for different purposes based on the temperature and pressure conditions used, and different applications are discussed in the following subsections.

6.8.1 Extraction of Organic Acids and Lipids Using Supercritical Carbon Dioxide

The extraction of non-polar solutes from natural matrices using supercritical CO_2 has been extensively studied and discussed. The application of SFE using $scCO_2$ for the extraction of commodity and specialty vegetable-based oils has been practiced for over 40 years, with significant research and development occurring in the early to mid-1980s.[149] The solubility of triglyceride-based oils is pressure and temperature dependent,[150] with maximum oil solubility occurring in the range 70–80 MPa and 80–100 °C. The book by King and List,[151] although somewhat outdated, provides a comprehensive overview of oil and lipid extraction with $scCO_2$ through 1996, and the relevant extraction approaches and equipment. Engineering process economic and design analysis has indicated that batch or batch semi-continuous $scCO_2$ extraction of commodity seed oils is too expensive, although the previous quoted maximum solubility conditions are optimal with respect to lowering oil production costs. The situation is not quite the same for specialized oils having a higher market value, and indeed of the over 150 critical fluid-based processing plants worldwide, many process specialty oils on a routine basis, as noted in Section 6.3, using semi-continuous batch extraction. The pressure range of batch extractors has also generally increased from the 300–350 bar range to a 550–700 bar operational limit.

Despite the above issues and partial successes, a continuous $scCO_2$ process in which the material to be extracted is fed continuously into an extraction vessel has been elusive. Eggers's group in Germany[152] was perhaps the first to suggest a feasible way of processing oilseeds continuously using SFE. In this continuous method, the seeds are introduced into a hopper, which feeds them into an auger barrel where the seed mass is compressed mechanically into a plug. The formation of a feed plug is important, because it becomes a barrier against the escape of high-pressure CO_2 that is being introduced into the high-pressure extraction vessel as the seed mass is slowly introduced into the extraction vessel containing internal baffling to induce mixing and contact between the oilseed and $scCO_2$. Upon extraction and dissolution of the seed oil in the $scCO_2$, the oil-laden extract dissolved in $scCO_2$ is exhausted from the extraction vessel into the CO_2 recycle–heat exchanger loop for reuse. The deoiled seed flakes exit from the bottom of the vertically held extraction vessel *via* a reverse auger arrangement in which total or partial depressurization of the CO_2 from the seed mass can be achieved. The continuous-feed pilot plant at the Technische Universität Hamburg-Harburg was only partially successful in operation, and no further scale-up was done owing to leakage in the system and unstable seed plugs in the auger conveying devices, which incidentally are also called screw cage presses.

The concept of utilizing a modified expeller in conjunction with supercritical or liquid CO_2 injection can be ascribed to Foidl,[153] who developed the process at a Latin American research center. The physicochemical basis of the process is still not well understood, but involves the addition of liquefied CO_2 to the seeds or meal inside an expeller barrel to aid in the oil extraction process. The hydraulic compression on the seed meal creates considerable pressure and heat on the seed matrix, resulting in the conversion of the added CO_2 to its supercritical state. The hot compressed CO_2 partially solvates the seed oil, akin to what occurs when SFE is performed with $scCO_2$, but more importantly dilutes or expands the expressed oil, enhancing its removal from the seed or meal bed. In this regard, it might be considered that CO_2 is acting as a displacement fluid, very similar to its role in enhanced oil recovery projects.[154] This CO_2-assisted expression process has been commercialized by Crown Iron Works in Minneapolis, MN, as the HIPLEX process, and expellers modified for CO_2 addition presses have been further developed by Harburg-Freudenberger,[155] initially with a unit capable of a capacity of 25 tons per day.

Photographs of a HIPLEX pilot unit (a Harburg-Freudenberger Model EP08 press) are shown in Figure 6.16. In the current configuration, liquid CO_2 is fed into the barrel of the expeller as shown in Figure 6.16a. As remarked earlier, the CO_2 acts as diluent; pressurized CO_2 dissolves in the oil and reduces the viscosity to one-tenth, thus improving oil separation at

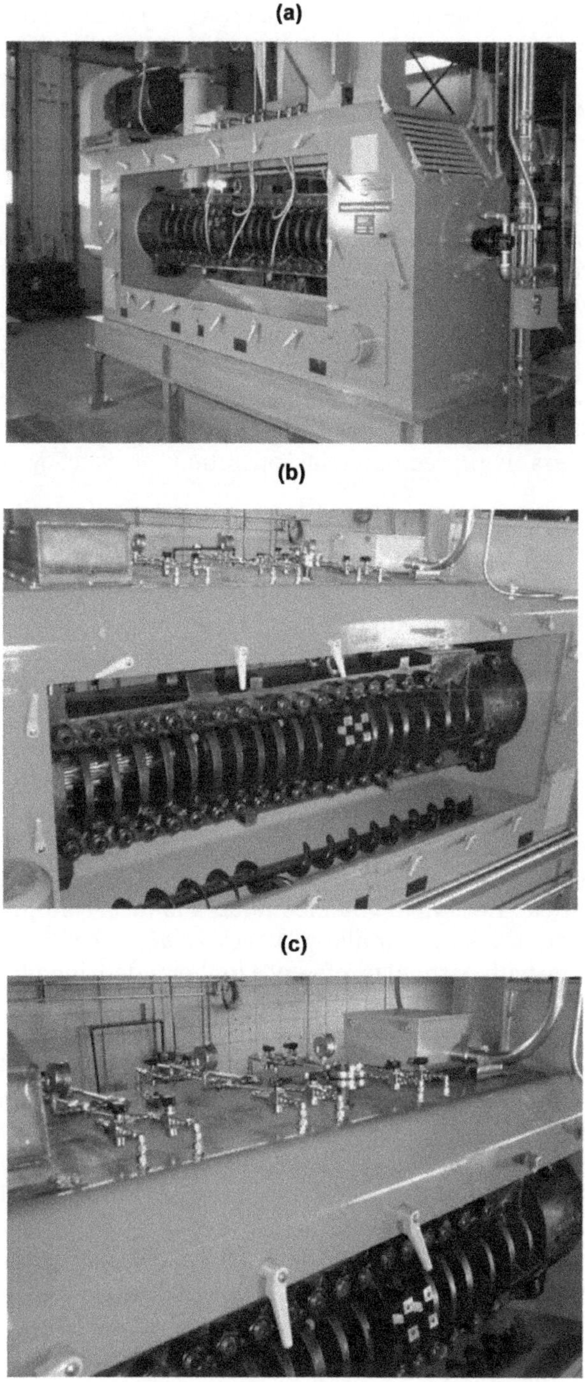

Figure 6.16 Photographs of HIPLEX continuous extraction system.

lower temperatures. The valving associated with the CO_2 delivery system is shown on top of the expeller in Figure 6.16b and c; the oil to-CO_2 ratio is typically 3 : 1. The operation consists in monitoring typical screw cage parameters once the press has stabilized, and collection of samples of the cake and oil after injection of CO_2 into the press. Termination of the CO_2 injection results in a noticeable difference in the cake within minutes of turning off the CO_2 (Figure 6.16c). For both soybean and canola, there is a significant improvement in oil recovery when using CO_2 injection *versus* pressing without CO_2 injection, and also in oil yield. This translates, for example, when processing 200 tons per day of 19% oil-laden soybeans into 38 tons per day oil yield, or an 8–10% improved oil recovery – an additional 3.0–3.8 tons per day of soybean oil. Similarly, for canola, for 200 tons per day press processing 42% oil-laden canola, there is a 2–3% improved oil recovery, resulting in an additional 1.68–2.52 tons per day of canola oil. The HIPLEX process is in commercial operation at SafeSoy Technologies, Ellsworth, IA.

In more recent years, elegant research on the gas-assisted mechanical expression (GAME) has been reported from The Netherlands.[156] A number of publications have appeared on GAME, including supporting-phase equilibria studies, such as the solubility of $scCO_2$ in cocoa butter[157] and the effect of CO_2 dissolution on the physical properties of molten cocoa.[158] Considerable detail on the GAME research program can be found in the PhD thesis of Venter.[159] Essentially, Venter's thesis is concerned with the effect of increasing CO_2 solubility in cocoa butter on the expression yields of cocoa. The study made use of a modified hydraulic press to study cocoa butter expression yields with and without CO_2 injection. With CO_2 injection, the optimal cocoa butter yield was 87.1% at 100 °C and 10 MPa CO_2 pressure, and a mechanical pressure of 50 MPa on the platen. This yield may be contrasted with a 71.8% yield obtained under identical conditions but in the absence of CO_2 pressure. Similarly, Voges *et al.*[160] reviewed gas-assisted oilseed pressing with particular reference to the underlying physicochemical principles and rapeseed oil extraction.

Extension of the GAME approach as applied to oilseeds such as sesame, linseed, rape seed, palm kernel, and jatropha were reported by Willems *et al.*[161] As they noted, the combined advantages of CO_2 sorption under pressure in various triglyceride-based oils and the reduction in oil viscosity achieved using GAME reduce the amount of CO_2 required *versus* conventional SFE and also the required CO_2 pressure (10 MPa). This latter feature results in energy savings primarily in the need to recompress and recycle more CO_2 relative to a conventional SFE process operating between 40 and 70 MPa. This study has put on a firmer footing our understanding of the basis of gas-assisted expressions, which is lacking in the patent literature, and also provides data for modeling to optimize GAME processing configurations.[162]

As shown in Figure 6.17, the concept embodying elements of SFE coupled with an extruder referred to as C-REX can be facilitated by the

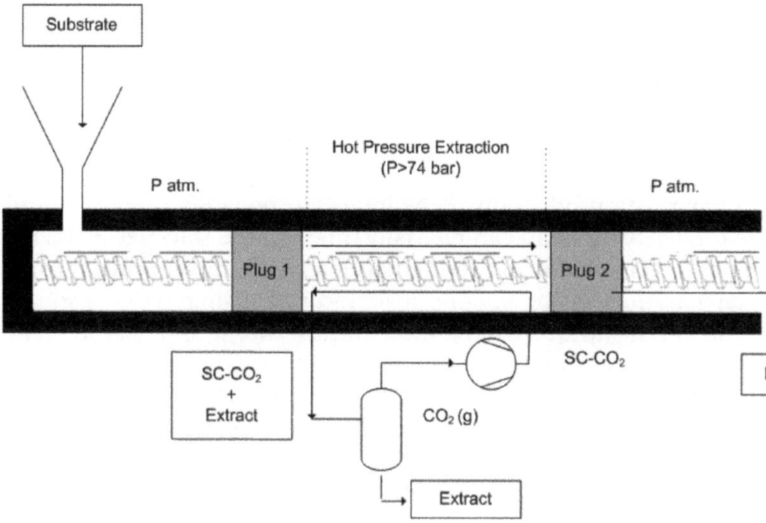

Figure 6.17 Schematic diagram of a continuous supercritical extrusion–extraction process.

formation of two oilseed plugs at both the beginning and the end of the extruder. $scCO_2$ is introduced into the barrel of the extruder in a recycle loop so that the compressed CO_2 is countercurrently contacted with the oilseed substrate as it passes down the extruder barrel. In this regard, it is very similar to the process developed by Eggers in the late 1980s.[152] The SFE recycle system is attached to a modified Clextral BC45 twin-screw extruder capable of being operated at 250 bar pressure. The material to be processed is facilitated by screws designed to create the plugs mentioned above. $scCO_2$ is injected just upstream of the second plug (Plug 2) using a membrane pump capable of delivering up to 500 $kg\,h^{-1}$ of CO_2. Up to 10 $kg\,h^{-1}$ of material can be processed; processed materials included caraway seed at 80 bar and hops at 120 bar. Although extruders have shown promise in the processing of finished food products,[163] their attendant expense and lower throughputs make them less attractive than the expeller-based processes described above for large-scale processing of commodity food products such as oilseeds.

Unlike lipids and fatty acids, lower carboxylic acids such as acetic, citric, and lactic acids are comparatively more polar in nature but have significant applications in the food industry. For example, acetic acid has been used as a food additive in the form of vinegar, citric acid is a primary constituent in a number of fruits and is used predominantly as a food preservative, and lactic and propionic acids are used in baked goods and cheese. Recently, significant efforts have been made to produce these low molecular weight carboxylic acids using anaerobic fermentation processes but they face significant disadvantages through increased downstream processing costs of separation and purification of the final product. In fact, one study indicated that downstream processing costs in such processes can be as high as 30–40% of the total operating costs.[164] Several reviews have been published on innovations in the production and recovery of these carboxylic acids.[165–169] In many of these industrial processes, the key requirements for a process with good product recovery include, but are not limited to, high product purity (>99.5%), high product recovery (>90%), low energy and solvent requirements, and low waste production.[169] As indicated previously, $scCO_2$ fulfills most of the requirements and can be easily tuned to the selective extraction of the carboxylic acids from the fermentation broths. Optimization of the experimental parameters for organic acid recovery using $scCO_2$ can also indicate if the SFE technique will be more or less energy and cost intensive than other recovery techniques.

One of the present authors was involved in designing and optimizing a semi-continuous high-pressure carbon dioxide extraction system for obtaining acetic acid from fermentation broth, where it was found that a maximum recovery of 93% was obtained under optimized conditions of 2 g min^{-1} CO_2 solvent flow rate, 14.8 MPa, 45 °C, and a residence time of 5 h.[170] Although the optimized conditions for such a high recovery were reasonable when compared with $scCO_2$ extraction of lipids and fatty acids from food matrices, kinetic studies indicated that the $scCO_2$ extraction of acetic acid from an aqueous solution was extremely concentration dependent, as shown in Figure 6.18. These studies indicated that as the concentration of the acetic acid in the aqueous solution or fermentation broth decreases (with time), the solvent utilization efficiency measured as milligrams of acid extracted per gram of CO_2 decreases non-linearly. This trend is significant since it can help determine the efficiency of using $scCO_2$ to extract volatile fatty acids (VFAs) such as acetic and propionic acids based on the initial concentration of the VFAs in the fermentation broth. Similar trends were also seen by Henczka and Djas,[171] where the recovery was four times higher at higher initial acetic acid concentrations. It is also possible, however, to increase the solvent utilization efficiency through multiple passes of the $scCO_2$ through the fermentation broth, since although the concentration of VFAs extracted by CO_2 decreases with the VFA concentration actually present in the bath, there is still the ability of CO_2 to dissolve more VFA since solubility is mainly a temperature- and pressure-dependent variable.

Another means of increasing the carboxylic acid recovery from fermentation broth is through the use of a diluent that forms a matrix with the organic acids, thereby, increasing its solubility in $scCO_2$. Several studies have been carried out using tri-*n*-octylamine as diluent for the extraction of acetic acid,[172] propionic acid,[173] and citric acid,[174] and tributyl phosphate as diluent for the extraction of citric acid[175] and lactic acid.[176] These "solute-philic" diluents are primarily added to alter the polarity of CO_2 and also form a Lewis acid–base complex with the hydrophilic carboxylic acids (especially citric and lactic acids), which has a higher solubility in CO_2. Studies have indicated that using a diluent such as tributyl phosphate

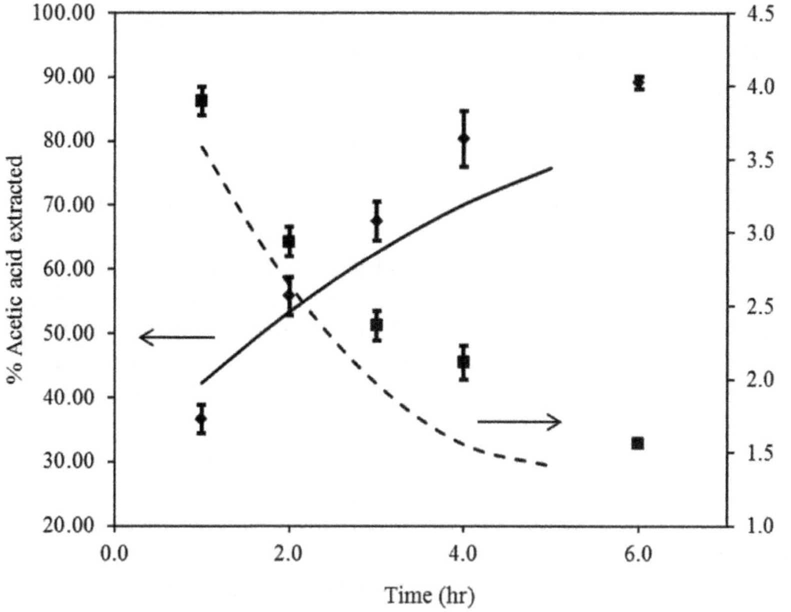

Figure 6.18 Comparison of dynamic confirmation experimental results with the response surface model fit to the optimization data at 1500 psi (10.3 MPa), 35 °C, 85.8 g L^{-1} initial acetic acid concentration, and 2.03 g min^{-1} solvent flow rate. Lines indicate statistical model and points indicate experimental results for percentage of acetic acid extracted (—, model; ◆, data), and solvent utilization efficiency (—, model, ■, data). Reprinted from *Journal of Supercritical Fluids*, **95**, B. G. Garrett *et al.*, Design and optimization of a semicontinuous high pressure carbon dioxide extraction system for acetic acid, 243–251. Copyright (2014), with permission from Elsevier.

can increase the solubility of lactic acid and citric acid in $scCO_2$ by at least 7–10-fold.[175] It is important, however, that the selected diluent does not react with any other component of the fermentation broth and can be separated from the carboxylic acids upon depressurization. Usually, organic solvents such as decanol and dodecanol have been used to facilitate the deconstruction of the acid–diluent matrix upon depressurization of the $scCO_2$ phase.[176] The advantage of the technique is that the organic phase can then be separated and dissolved in $scCO_2$ and reused for further extraction of organic acids from the fermentation broths. Although the recent literature has not revealed a convincing technique for the low-cost and low energy-intensive extraction of carboxylic acids from the fermentation broth using $scCO_2$, when compared with traditional salt precipitation techniques, these studies have indicated that there is great potential to have a high-purity organic acid product with low waste production using SFE, and further studies are under way to optimize the technology before comparing techno-economics with the current state of the art.

6.8.2 Subcritical Water Extraction/Reaction to Produce Thermally Labile Substances from Natural Product Matrices

A number of polyphenolic and flavonoid moieties can be produced from biopolymers by hydrolysis in water. As remarked previously, gas-expanded solvents such as alcohols can be used to treat lignin for the production of vanillin.[113] Carbonic maceration is a technique employed in the wine production industry, involving pumping CO_2 into a sealed container containing grape clusters, which stimulates at an intercellular level the production of a number of fermentation-derived products, including flavor compounds such as ethyl cinnamate and ethyl and methyl vanillate. This in essence involves a mixture of CO_2 in water that is used commercially to produce fruity-type wines.

Depending on the thermal stability of a solute at a particular temperature and its susceptibility to follow a certain reaction pathway (such as hydrolysis in aqueous media), a competition will exist between extraction of the solute and its conversion through reaction into another compound. This is a very general occurrence – often the kinetics associated with the reaction are very slow or barely detectable – but on the other hand some solutes start to undergo a rapid molecular transformation at the onset of the start of the extraction. Modeling of this competitive situation was performed by Duan[177] for solutes undergoing SWE in which the change in solute solid-phase concentration, C_{Si}, with time was given in differential form by

$$\frac{\mathrm{d}C_{Si}}{\mathrm{d}t} = -\, k_{pi}a_{p}(m_iC_{Si} - C_{ti}) \tag{6.2}$$

where k_{pi} = overall mass transfer of the solute being extracted from the solid phase, a_p = specific surface area of solid, m = thermodynamic distribution coefficient, and C_{ti} = concentration of the solute in the solid phase at time t. The mass balance differential equation for the solute concentration in the liquid phase as a function of time, $\mathrm{d}C_{Li}/\mathrm{d}t$, is

$$\frac{\mathrm{d}C_{Li}}{\mathrm{d}t} = k_{pi}a_{p}\left(\frac{\rho_L}{\rho_S}\right)\frac{1}{F}(m_iC_{Si} - C_{Li}) - K_{Di}C_{Li} \tag{6.3}$$

where ρ_L = density of the liquid phase, ρ = density of the solid phase, F = mass ratio of water to solids, and K_{Di} = degradation rate constant for the ith component. The change in C_{Si} as a function of time should decrease with time in the manner depicted in Figure 6.19, theoretically approaching zero with exhaustive extraction of the solute. For a non-reactive or non-degrading solute, the concentration of solute in the liquid phase should asymptote towards a constant value with respect to the time of extraction – again, as shown in Figure 6.19. When a solute reacts or is degraded during the conditions of the extraction, eqn (6.3) predicts through the $K_{Di}C_{Li}$ term the loss of the solute with time due to its reacting to form another compound.

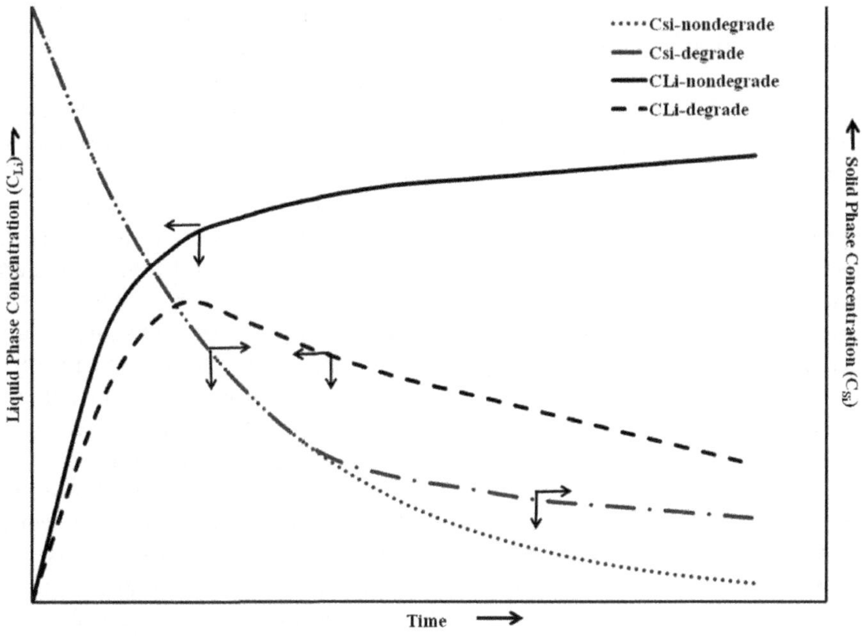

Figure 6.19 Concentration of the solute in the solid phase *versus* liquid phase for extraction accompanied by reaction of the solute.

The shape of the dC_{Li}/dt curve in this case will show a maximum in C_{Li} and then the solute concentration in the liquid phase will continue to be reduced with respect to time. C_{Si} under such conditions will appear to show incomplete extraction whereas in reality it can never approach zero since a portion of it is being converted to another product.

These extraction–reaction curves will vary with solute type depending on its susceptibility to undergoing a reaction. Hence the maxima in the dC_{Li}/dt curve, as remarked previously, may vary from several seconds to 1 h (Section 6.4), suggesting that most of the extraction should be completed before dC_{Li}/dt. The extent of solute degradation can often be reduced by using a higher water-to-solids ratio, but this will result in a more diluted extract. In many cases, solute degradation can be modeled using first-order reaction kinetics. As noted in Section 6.4, a solute's propensity to undergo degradation can be solvent dependent.

An excellent overview on the generation of antioxidants from lignocellulosic materials problem has been published previously.[178] Numerous water-based treatment and reactive methods have often been used, yielding a similar array of mixed hydrolysis products, including sugar oligomer mixtures, low molecular weight phenolic compounds, fatty acids, and peptides or amino acids from naturally occurring proteins. Within the context of food processing, Maillard reaction products as flavor precursors are well known, and this reaction can be facilitated both in SWE and in $scCO_2$ medium.

Eucalyptus grandis treatment studies with subcritical water[179] for the production of antioxidant compounds are of interest owing to the widespread propagation of the tree throughout the world. As was noted previously, the antioxidant capacity of these hot water extracts–reaction mixtures is partly dependent on the antioxidant or bioassay employed. The main phenolic compound identified in the hot water extracts obtained between 160 and 300 °C was pyrogallol obtained from the degradation of tannic acid, which exhibited a higher antioxidant capacity than butylated hydroxytoluene (BHT). Rodriquez-Meizoso *et al.*[180] applied SWE to isolate nutraceuticals from oregano, measuring the antioxidant activity of the extracts taken at various temperatures up to 200 °C. Maximum antioxidant activity was recorded at 200 °C, suggesting that degradation at this high temperature might be producing oligomeric compounds with antioxidant capacity. In general, the higher the temperature at which the SWE was conducted, the more likely there would be an increase in the antioxidant capacity of the resultant extracts. Buranov and Mazza[181] treated flax shives and wheat or corn bran using alkaline hydrolysis and pressurized solvents and were successful in isolating ferulic acid and vanillin. A decomposition mechanism was proposed for the production of vanillin and vanillic acid in pressurized water–ethanol from a lignin phenolic–carbohydrate complex at 180 °C and 750–1000 psi applied pressure. Owing to the increase in the solubility of ferulic acid in ethanol, a purification scheme was tested using ethanol concentrations above 30% v/v for isolating pure ferulic acid.

A unique application of reaction chemistry in subcritical water was reported by Yamamoto *et al.*[182] for producing polyphenols from livestock or animal excrement. Two interesting conclusions came from this study. First, the production of a particular polyphenolic compound was dependent on the extraction–reaction temperature. For example, the maximum vanillin yield occurred at 250 °C, whereas lower temperatures yielded more protocatechuic acid (200 °C) and vanillic acid (150 °C). Second, the polyphenolic content of excrement was species dependent: herbivorous animals produced polyphenolics for SWE, which was not the case with carnivores.

Salak *et al.*[183] characterized extracts from the subcritical water treatment of rice bran residual biomass over the temperature range 100–360 °C using 10 min treatment periods. The presence of nine polyphenolic compounds was monitored by HPLC and, as reported by Watchararuji *et al.*,[184] each polyphenolic compound that was produced had its own specific temperature that was optimal for its production. There was a direct correlation between the amount of total phenolic compounds as measured by HPLC and the antioxidant activity of the extract. Studies on the subcritical water treatment of rice bran have been reviewed by Pourali *et al.*[185] owing to the importance of utilizing the residual biomass, which represents up to 30% of the rice bran production in many Asian countries.

6.8.3 Subcritical Water Hydrolysis to Deconstruct Biopolymers Such as Proteins and Carbohydrates

Additional extraction–reaction studies have been performed on crude agricultural materials, ill-defined substrates, and recyclable materials having mixed functionalities, such as carbohydrates and proteins. Subcritical water hydrolysis of rice bran or soybean meal if performed in the temperature range 200–220 °C for 10–30 min can yield proteinaceous compounds, amino acids, and sugars.[184] Agricultural materials containing large amounts of cellulose or hemicellulose, when hydrolyzed with subcritical water,[186] produce useful oligomeric saccharide mixtures plus glucose and xylose – sweeteners well known in the food industry. However, upon further heating for extended periods and depending on the applied pressure and temperature, both fragmentation and dehydration products from the reaction of monomeric sugars can occur to produce a number of fragmentation products, such as some of the fermentation-derived acids, common acids, or dehydration products such as furfural and 5-hydroxymethylfurfural, whose formation often causes off-flavor notes to appear in certain food products. Lignin is of particular interest, since a variety of phenolic compounds are available when it is reacted in an appropriate medium from a diverse array of feedstocks, such as corncobs, *Eucalyptus*, barley husk, grape pomace, and rice husks. The lower overall solubility parameter for lignin36 suggests that alcohols would be an appropriate medium for extracting–reacting this polyphenolic-based polymer.

Adachi's group[187,188] focused studies on the hot water extraction of wheat and rice bran using an array of extraction–reaction temperatures between 50 and 260 °C, and an excellent review of the production of functional compounds from agricultural by-products was published by Wiboonsirikul and Adachi.[189] These studies featured relatively short exposure (extraction) times of 5 min, maximum protein and carbohydrate yields being obtained at 200 °C, although the maximum free radical-absorbing activity as judged from the DPPH (2,2-diphenyl-1-picrylhydrazyl) assay occurred with extracts taken at 240–250 °C. They tabulated results on SWE for 24 different agricultural products according to extraction temperature, time, pressure, and the analytical methods employed to characterize the extracts.

The production of mixtures of amino acids from protein substrates, including biomass, has been investigated by several research groups. Yoshida's group in Japan[190–193] studied the simultaneous extraction–reaction of organic and amino acids from substrates such as fish meal. Two prominent organic acids could be obtained from subcritical water hydrolysis: lactic acid at approximately 237 °C and pyroglutamic acid at 300 °C. Amino acids such as cystine, alanine, glycine, and leucine were produced in the temperature range 240–350 °C, optimally at 270 °C with a 5 min exposure time. Further study of individual amino acid yields *versus* reaction time supported the following purification scheme: liquefy fish meal at 200 °C for the production and recovery of lactic acid and histidine, followed by production of the above four other amino acids at 270 °C with a reaction time of 6 min, followed by a longer holding time of 30 min for maximum pyroglutamic acid yield. Residual oil and fat in the fish meal can also be hydrolyzed at 300 °C to give potentially useful fatty acids, such as eicosapentanoic acid.[193]

A similar theme was also pursued by Brunner's group in Germany for model proteins and waste proteinaceous materials. Rogalinski *et al.*[194] studied amino acid production from bovine serum albumin (BSA) using very short reaction times (30 s) at 310 °C and 250 bar. Further optimization of the reaction conditions indicated that the highest amino acid yield from BSA hydrolysis was achieved at 290 °C and a reaction time of 65 s. The addition of CO_2 to the subcritical water also led to an increase in the amino acid yield due to acceleration of the acid-catalyzed hydrolysis reaction. By saturating the subcritical water 100% with CO_2 at 250 °C and 250 bar for a 5 min reaction time, the yield of amino acids from BSA could be increased from 30 to 150 mg g^{-1} BSA. These conditions were put to good use in the hydrolysis of poultry feathers containing keratin to produce a mixture of amino acids at extended reaction times consisting predominately of alanine and glycine and also significant amounts of tyrosine, valine, and leucine. Cheng *et al.*[195] applied subcritical water hydrolysis to fish and chicken wastes and also hair and feathers for amino acid production over the temperature range 180–320 °C. It was found that the amino acid composition of the hydrolyzate could be influenced by the substitution of nitrogen for carbon dioxide during the hydrolysis. Hydrolysis of spent brewer's yeast waste (see Section 6.4) has also been attempted to produce a hydrolyzate not for food use, but as a nutrient source for yeast growth.[196]

To summarize this section, we refer to Figure 6.20, in which several scenarios for reacting recyclable biomass are shown with respect to the end products obtained. Since such waste steams will probably contain the three major food composition groups, oils/fats, proteins, and carbohydrates, a mixed array of potential products can result. Extraction–reaction using

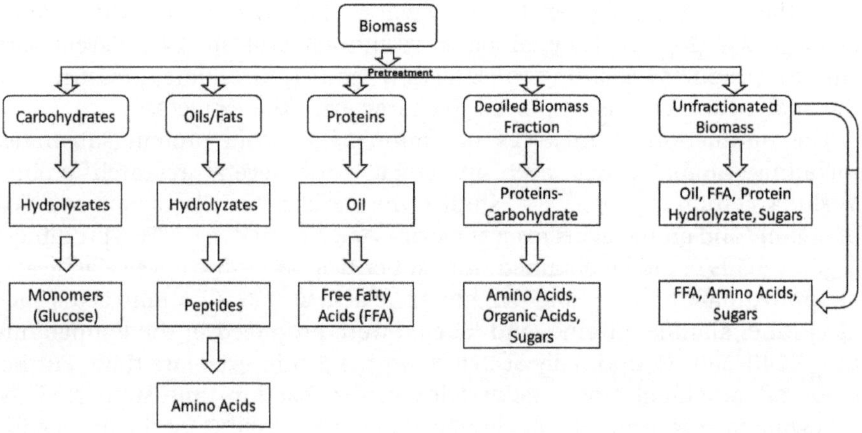

Figure 6.20 Subcritical water hydrolysis options on neat and fractionated biomass.

subcritical water can be terminated at an intermediate stage. For example, hydrolysis of protein-containing matter can be stopped at the peptide stage to produce useful food products, and this has been demonstrated on whey protein isolate.[197] Intermediate carbohydrate compositions already exist in the food industry, primarily as corn syrups, and these can also be obtained *via* subcritical water hydrolysis. Partially fractionated biomass, such as delipidated fractions, will undergo extensive hydrolysis to produce primarily mixtures of amino acids, organic acids, and sugars. Obviously, these product mixtures represent a challenging separation problem if higher purity or discrete compounds are desired for use in foods, but enrichment of particular components could be accomplished by membrane-based separation techniques or, as mentioned earlier, by selective temperature-dependent hydrolysis. Hypothetical processing scenarios have been advanced for separating individual chemical groups and discrete solutes[192] and a large subcritical water processing facility focused on converting recyclable materials was constructed at an eco-town in Japan. Subcritical water processing can be accomplished using similar equipment and plant processing facilities as now used for $scCO_2$ extraction, and our group is working on a continuous processing method employing subcritical water to extract value-added compounds from food processing waste streams.

6.9 Conclusions

The status of SFE using $scCO_2$ is confirmed by the respectable number of processing plants that now employ this environmentally benign technology. Extraction or fractionation of more polar food-related solutes using polar co-solvents along with $scCO_2$ is also well researched, but faces some practical limitations that can be overcome by the use of polar SCFs. Among these, subcritical water has attracted considerable attention over the past several years both as a temperature-dependent solvent medium and as an extractive–reactive fluid depending on the desired end product and temperature stability of the target solutes or biomass matrix. Subcritical water and some subcritical alcohols can be coupled advantageously for sequential multi-fluid processing of agriculturally derived substrates. This approach, combined with the coupling of unit processing operations such as extraction–reaction–fractionation, opens up numerous options for processing food ingredients in a "green" manner.

Multi-phase fluids in which a gas is combined with a liquid phase to produce a combination of sub- and supercritical fluids can be used in all of the above unit operations. Carbonic acid is an attractive "natural" catalyst that can be used in place of mineral acids not only for performing hydrolytic chemistries, but also other syntheses such as hydrogenation of fats and oils.[198] Coupling CO_2 with subcritical water allows the food engineer to exploit the reactive properties of hot compressed water to yield an array of food ingredients and nutraceuticals.[199] Continuous processing using expellers has not only been optimized and established for the SFE of oils from oilseeds, but has also been extended to the subcritical fluid extraction of antioxidants from food waste[200] due to the increased mass transport while the raw material passes through the expeller.

References

1. G. Brunner, *Gas Extraction*, Springer, New York, 1987.
2. J. M. DeSimone and W. Tumas, *Green Chemistry Using Liquid and Supercritical Carbon Dioxide*, Oxford University Press, New York, 2003.
3. C. A. Eckert, D. Bush, J. S. Brown and C. L. Liotta, *Ind. Eng. Chem. Res.*, 2000, **39**, 4615.
4. J. W. King, *C. R. Chim.*, 2004, **7**, 647.
5. N. T. Dunford, J. W. King and G. R. List, in *Extraction Optimization in Food Engineering*, ed. C. Tsai and G. Liadakis, Marcel Dekker, New York, 2003, pp. 57–93.
6. J. W. King, in *Lipid Biotechnology*, ed. T. M. Kuo and H. W. Gardner, Marcel Dekker, New York, 2002, pp. 663–687.
7. J. L. Martinez, *Supercritical Fluid Extraction of Nutraceuticals and Bioactive Compounds*, CRC Press, Boca Raton, FL, 2008.
8. M. Mukhopadhyay, *Natural Extracts Using Supercritical Carbon Dioxide*, CRC Press, Boca Raton, FL, 2000.
9. *Drug Delivery and Supercritical Technology*, ed. J. P. York, U. B. Kompella and B. V. Shekunov, Marcel Dekker, New York, 2004.
10. C. M. Hansen, *Hansen Solubility Parameters – A Users Handbook*, CRC Press, Boca Raton, FL, 2007.
11. A. J. M. Lagadec, D. J. Miller, A. V. Lilke and S. B. Hawthorne, *Environ. Sci. Technol.*, 2000, **34**, 1542.
12. K. Srinivas, J. W. King, J. Monrad, L. Howard and C. M. Hansen, *J. Food Sci.*, 2009, **74**, E342.
13. J. W. King, L. R. Howard, K. Srinivas, Z. Y. Ju, J. Monrad and L. Rice, *Proceedings of the 5th International Symposium on Supercritical Fluids (Super Green 2007)*, Seoul, South Korea, November 28–December 1, 2007, Proceedings KL04.
14. M. Co, P. Koskela, P. Eklund-Åkergren, K. Srinivas, J. W. King, P. J. R. Sjöberg and C. Turrner, *Green Chem.*, 2009, **11**, 668.
15. S. N. Wallace, D. J. Carrier and E. C. Clausen, *Appl. Biochem. Biotechnol.*, 2003, **108**, 1599.
16. J. H. de Vrie, P. L. Karin Janssen, P. C. Hollman, A. W. van Staveren and M. B. Katan, *Cancer Lett.*, 1997, **114**, 141.
17. A. Bahari, I. Tonnies, M. N. Biag, G. A. Leeke, R. C. D. Santos, C. Ztezl, J. W. King, E. Galley, R. Heathcote and S. Bowra, *Proceedings of the 11th European Meeting on Supercritical Fluids*, Barcelona, Spain, May 4–7, 2008, P_NP_31.
18. J. W. King, in *Modern Extraction Techniques: Food and Agricultural Samples*, ed. C. Turner, American Chemical Society, Washington, DC, 2006, p. 79.

19. E. Ong, J. Cheong and D. Goh, *J. Chromatogr. A*, 2006, **112**, 92.
20. M. Herraro, A. Cifuentes and E. Ibanez, *Food Chem.*, 2006, **98**, 136.
21. J. W. King and K. Srinivas, *J. Supercrit. Fluids*, 2009, **47**, 508.
22. E. Reverchon and I. De Marco, *J. Supercrit. Fluids*, 2006, **38**, 146.
23. F. Temelli and O. N. Ciftci, *J. Supercrit. Fluids*, 2015, **96**, 77–85.
24. R. Smith, H. Inomata and C. Peters, in *Supercritical Fluid Science and Technology Series*, ed. E. Kiran, vol. 4, Elsevier, Oxford, UK, 2013.
25. K.-P. Yoo and R. Fukuzato, *Proceedings of the 8th International Symposium of Supercritical Fluids*, Kyoto, Japan, November 5–8, 2006, Abstract PL-3.
26. C. G. Perreira and M. A. A. Meireles, *Food Bioprocess Technol.*, 2009, DOI: 10.1007/s11947-009-0263-2.
27. C. Bergeron, S. Gafner, E. Clausen and D. J. Carrier, *J. Agric. Food Chem.*, 2005, **53**, 3076.
28. E. Ibanez, A. Kubatova, F. J. Senorans, S. Cavero, G. Reglero and S. B. Hawthorne, *J. Agric. Food Chem.*, 2003, **51**, 375.
29. M. Garcia-Marino, J. C. Rivas-Gonzalo, E. Ibanez and C. Garcia-Moreno, *Anal. Chim. Acta*, 2006, **563**, 44.
30. J.-W. Kim and G. Mazza, *J. Agric. Food Chem.*, 2006, **54**, 7575.
31. J. W. King, R. D. Grabiel and J. D. Wightman. *Proceedings of the 6th International Symposium on Supercritical Fluids – Tome 1*, Versailles, France, April 28–30, 2003, p. 409.
32. Z. Y. Ju and L. R. Howard, *J. Food Sci.*, 2005, **70**, S270.
33. Z. Pineiro, M. Palma and C. G. Barroso, *J. Chromatogr. A*, 2004, **1026**, 19.
34. M. Palma and L. T. Taylor, *J. Chromatogr. A*, 1999, **849**, 117.
35. M. Hassas-Roudsari, P. R. Chang, R. B. Pegg and R. T. Tyler, *Food Chem.*, 2009, **114**, 717.
36. J. W. King, K. Srinivas, J. M. del Valle and J. C. de la Funete, *Proceedings of the 8th International Symposium on Supercritical Fluids* (ISSF 2006), Kyoto, Japan, November 5–8, 2006. Proceedings No. OC-2-17.
37. L. Duan, D. J. Carrier and E. C. Clausen, *Appl. Biochem. Biotechnol.*, 2004, **114**, 559.
38. L. Duan, S. N. Wallace, A. Engelberth, J. K. Lovelady, E. C. Clausen, J. W. King and D. J. Carrier, *Appl. Biochem. Biotechnol.*, 2009, **158**, 362.
39. S. M. Ghoreishi, R. G. Shahrestani and S. H. Ghaziaskar, *World Acad. Sci. Eng. Technol.*, 2008, **43**, 114.
40. J. K. Monrad, L. R. Howard, J. W. King, K. Srinivas and A. Mauromoustakos, *J. Agric. Food Chem.*, 2010, **58**, 2862.
41. J. K. Monrad, L. R. Howard, J. W. King, K. Srinivas and A. Mauromoustakos, *J. Agric. Food Chem.*, 2010, **58**, 4014.
42. J. K. Monrad, K. Srinivas, L. R. Howard and J. W. King, *J. Agric. Food Chem.*, 2012, **60**, 5571.
43. A. A. Clifford, Personal communication, March 2000.
44. M. Garcia-Marino, J. Rivas-Gonzalo, E. Ibanez and C. Garcia-Moreno, *Anal. Chim. Acta*, 2006, **563**, 44.

45. A. Basile and A. A. Clifford, *Proceedings of the 5th Meeting on Supercritical Fluids – Tome 2*, Nice, France, March 23–25, 1998, 635–640.
46. M. Z. Ozel, F. Gogus and A. C. Lewis, *Food Chem.*, 2003, **82**, 381.
47. M. H. Eikani, F. Golmohammad, M. Mirza and S. Rowshanzamir, *J. Food Proc. Eng.*, 2007, **30**, 255.
48. M. H. Eikani, F. Golmohammad and S. Rowshanzamir, *J. Food Eng.*, 2007, **80**, 735.
49. R. S. Ayala and M. D. L. de Castro, *Food Chem.*, 2001, **75**, 109.
50. A. A. Clifford and A. Basile, *Proceedings of the 5th Meeting on Supercritical Fluids – Tome 2*, Nice, France, March 23–25, 1998, 723–728.
51. K. Srinivas, J. W. King, J. K. Monrad, L. R. Howard and D. Zhang, *Ital. J. Food Sci.*, 2011, **23**, 90.
52. A. D. Espinoza and R. O. Morawicki, *J. Agric. Food Chem.*, 2012, **60**, 5250.
53. J. W. King, in *Supercritical Carbon Dioxide: Separations and Processes*, ed. A. S. Gopalan, C. M. Wai and H. K. Jacobs, American Chemical Society, Washington, DC, 2003, p. 104.
54. J. W. King, in *Healthful Lipids*, ed. C. Akoh and O. M. Lai, AOCS Press, Champaign, IL, 2005, pp. 173–200.
55. J. W. King, D. Zhang, K. Srinivas, Y.-J. Wang and L. R. Howard, *Proceedings of the 9th International Symposium on Supercritical Fluids (ISSF 2009)*, Arcachon, France, May 18–20, 2009, Oral #C01.
56. S. R. M. Moreschi, A. J. Petenate and M. A. A. Meireles, *J. Agric. Food Chem.*, 2004, **52**, 1753.
57. C. Schacht, C. Zetzl and G. Brunner, *J. Supercrit. Fluids*, 2008, **46**, 299.
58. P. Ramírez, T. Fornari, F. J. Señoráns, E. Ibañez and G. Reglero, *J. Supercrit. Fluids*, 2005, **35**, 128.
59. M. S. Kuk and M. K. Dowd, *J. Am. Oil Chem. Soc.*, 1998, **75**, 623.
60. J. W. King and N. T. Dunford, *Sep. Sci. Technol.*, 2002, **37**, 451.
61. M. Yamagata, M. Goto and R. Fukuzato, *Method for Extracting Target Substance and Apparatus for Extracting Target Substance*, *WO Pat.* 026446, March 6, 2008.
62. M. Mannila, H. Kim, C. Isaacson and C. M. Wai, *Green Chem.*, 2002, **4**, 331.
63. I. D'Antuono and F. Favati, *Proceedings of the 14th Workshop on the Developments in the Italian PhD Research on Food Science Technology and Biotechnology*, University of Sassari, Oristano, September 16–18, 2009.
64. J. M. del Valle, T. Rogalinski, C. Zetzl and G. Brunner, *Food Res. Int.*, 2005, **38**, 203.
65. R. L. Smith Jr., R. M. Malaluan, W. B. Sentianto, H. Inomata and K. Arai, *Proceedings of the 9th Asian Pacific Confederation of Chemical Engineering, Green Processing and Sustainability*, September 29–October 3, 2002.
66. G. Brunner, *J. Food Eng.*, 2005, **67**, 21.

67. F. J. Eller, S. L. Taylor and D. E. Palmquist, *J. Agric. Food Chem.*, 2007, **55**, 2779.
68. J. Tabera, A. Guinda, A. Ruiz-Rodriguez, F. J. Senorans, E. Ibanez, T. Albi and G. Reglero, *J. Agric. Food Chem.*, 2004, **52**, 4774.
69. P. Zacchi, J. Daghero, P. Jaeger and R. Eggers, *Braz. J. Chem. Eng.*, 2006, **23**, 105.
70. N. T. Dunford and J. W. King, *J. Food Sci.*, 2000, **65**, 1395.
71. A. Birtigh, M. Johannsen, G. Brunner and N. Nair, *J. Supercrit. Fluids*, 1995, **8**(1), 46.
72. M.-H. Chuang and G. Brunner, *J. Supercrit. Fluids*, 2006, **37**(2), 151.
73. S. L. Taylor, J. W. King, L. Montanari, P. Fantozzi and M. A. Blanco, *Ital. J. Food Sci.*, 2000, **12**(1), 65.
74. M. N. Baig, R. Alenezi, G. A. Leeke, R. C. D. Santos, C. Zetzl, J. W. King, D. Pioch and S. Bowra, *Proceedings of the 11th Meeting on Supercritical Fluids*, Barcelona, Spain, May 4–7, 2008.
75. K. Rezaei and F. Temelli, *J. Supercrit. Fluids*, 2001, **19**, 263.
76. C. Turner, L. C. Whitehand, T. Nguyen and T. McKeon, *J. Agric. Food Chem.*, 2004, **52**, 26.
77. A. Weber, E.-H. Lee, S.-K. Shin and B.-S. Chun, *Chem. Eng. Technol.*, 2007, **30**(6), 732.
78. C. Turner, P. Turner, G. Jacobson, K. Almgren, M. Waldeback, P. Sjorberg, E. N. Karlsson and K. E. Markides, *Green Chem.*, 2006, **8**, 949.
79. S. Sarrade, C. Guizard and G. M. Rios, *Desalination*, 2002, **144**, 137.
80. M. Sims, *Membr. Technol.*, 1998, **97**, 11.
81. M. Sims and E. Estigarribia, *Proceedings of the 6th International Symposium on Supercritical Fluids*, April 28–30, 2003, 1457.
82. S. Sarrade, M. Carles, C. Perre and P. Vignet, *Process and Installation for the Separation of Heavy and Light Compounds by Extraction using a Supercritical Fluid and Nanofiltration*, *U. S. Pat.* 5,961,835, October 5, 1999.
83. L. A. V. Sarmento, R. A. F. Machado, J. C. C. Petrus, T. R. Tamanini and A. Bolzan, *J. Supercrit. Fluids*, 2008, **45**, 64.
84. S. Beltran, M. T. Sanz, B. Santamaria, R. Murga and G. Salazar, *Electron. J. Environ., Agric. Food Chem.*, 2008, **7**(8), 3270.
85. R. Docherty, T. Kougoulos and K. Horspool, American Pharmaceutical Review, October 2009, 34.
86. I. Rodriguez-Meizoso and M. Plaza, in *High Pressure Fluid Technology for Green Food Processing*, ed. T. Fornari and R. P. Stateva, Springer, Switzerland, 2015.
87. M. J. Cocero, A. Martin, F. Mattea and S. Varona, *J. Supercrit. Fluids*, 2009, **47**(3), 546.
88. E. Weidner, *J. Supercrit. Fluids*, 2009, **47**(3), 556.
89. J. W. King, in *Drug Delivery and Supercritical Technology*, ed. P. York, U. B. Kompella and B. V. Shekunov, Marcel Dekker, New York, 2004, p. 579.
90. T. Vatai, M. Skerget, Z. Knez, S. Kareth, M. Wehowski and E. Weidner, *J. Supercrit. Fluids*, 2008, **45**, 32.

91. J. M. Andersson, S. Lindahl, C. Turner and I. Rodriguez-Meizoso, *Food Chem.*, 2012, **134**, 1724.
92. N. R. Foster, F. Kurniawansyah, A. Tandya, C. Delgado and R. Mammucari, *Chem. Eng. J.*, 2017, **308**, 535.
93. S. B. Iversen, T. Larsen, V. Luthje, K. Felsvang and E. Lack, *Method and Apparatus for Producing Micro-Emulsions*, *U. S. Pat.* 20090062407 A1, March 5, 2009.
94. J. W. King, *Annu. Rev. Food Sci. Technol.*, 2014, **5**, 215.
95. K. W. Hutchenson, A. M. Scurto and B. Subramaniam ed. *Gas-Expanded Liquids and Near Critical Media*, American Chemical Society, Washington, DC, 2009.
96. J. C. Elgin and J. J. Weinstock, *J. Chem. Eng. Data*, 1959, **4**, 3.
97. C. A. Eckert, C. L. Liotta, D. Bush, J. S. Brown and J. P. Halett, *J. Phys. Chem. B*, 2004, **108**, 18108.
98. P. G. Jessop and B. Subramaniam, *Chem. Rev.*, 2007, **107**, 2666.
99. J. P. Hallett, C. A. Eckert and C. L. Liotta, in *Gas-Expanded Liquids and Near Critical Media*, ed. K. W. Hutchenson, A. M. Scurto, and B. Subramaniam, American Chemical Society, Washington, DC, 2009, p. 131.
100. H. Li and M. Maroncelli, *J. Phys. Chem. B*, 2006, **110**, 21189.
101. T. Fornari, E. J. Hernandez and G. Reglero, *J. Supercrit. Fluids*, 2009, **51**, 115.
102. G. R. Aiken and M. Poliakoff, *Green Chem.*, 2009, **11**, 1083.
103. A. P. Abbott, E. G. Hope, R. Mistry and A. M. Stuart, *Green Chem.*, 2009, **11**, 1536.
104. H. Hammam and B. Sivik, *J. Supercrit. Fluids*, 1993, **6**, 223.
105. L. Phan, H. Brown, J. White, A. Hodgson and P. G. Jessop, *Green Chem.*, 2009, **11**, 53.
106. A. Shishikura, K. Kanamori, H. Takahashi and H. Kinbara, *J. Agric. Food Chem.*, 1994, **42**, 1993.
107. C. J. Chang, A. D. Randolph and N. E. Craft, *Biotechnol. Prog.*, 1991, **7**, 275.
108. L. A. Blanchard and J. F. Brennecke, *Green Chem.*, 2001, **3**, 17.
109. V. T. Wyatt and M. J. Haas, *J. Am. Oil Chem. Soc.*, 2009, **86**, 1009.
110. D. M. Ginosar, R. V. Fox and L. M. Petkovic, *Production of Biodiesel Using Expanded Gas Solvents*, *U. S. Pat.* 7,514,575 B2, April 7, 2009.
111. H. Cheng, X. Meng, R. Liu, Y. Hao, Y. Yu, S. Cai and F. Zhao, *Green Chem.*, 2009, **11**, 1227.
112. T. S. Chamblee, R. R. Weikel, S. A. Nolen, C. L. Liotta and C. A. Eckert, *Green Chem.*, 2004, **6**, 382.
113. C. Eckert, C. Liotta, A. Ragauskas, J. Hallett, C. Kitchens, E. Hill and L. Draucker, *Green Chem.*, 2007, **9**, 545.
114. S. E. Hunter and P. E. Savage, *AIChE J.*, 2008, **54**, 516.
115. M. Chuang and M. Johannsen, *Proceedings of the 9th International Symposium on Supercritical Fluids* (ISSF 2009), Arcachon, France, May 18–20, 2009, Poster #P004, p. 152.

116. G. P. van Walsum and H. Shi, *Bioresour. Technol.*, 2004, **93**, 217.
117. J. W. King and R. D. Grabiel, *Isolation of Polyphenolic Compounds from Fruits or Vegetables Utilizing Subcritical Water Extraction. U. S. Pat.* 7,208,181, April 24, 2007.
118. G. V. Una, F. C. Romero, G. P. Goncalves and J. P. Lorenzo, *J. Chem. Eng. Data*, 1994, **39**, 639.
119. G. Ferrentini, M. L. Plaza, M. Ramirez-Rodrigues, G. Ferrari and M. O. Balaban, *J. Food Sci.*, 2009, **74**, E333.
120. R. Thiering, G. Hofland, N. Foster, G.-J. Witkamp and L. van de Wielen, *Biotechnol. Prog.*, 2001, **17**, 513.
121. G. W. Hofland, A. de Rijke, R. Thiering, L. A. M. van der Wielen and G.-J. Witkamp, *J. Chromatogr. B.*, 2000, **743**, 357.
122. R. Thiering, G. Hofland, N. Foster, G.-J. Witkamp and L. van de Wielen, *Biotechnol. Bioeng.*, 2001, **73**, 1.
123. P. J. Jordan, K. Lay, N. Ngan and G. F. Rodley, *N. Z. J. Dairy Sci. Technol.*, 1987, **22**, 247.
124. C. V. Santos, P. M. Tomasula and M. J. Kurantz, *J. Food Sci.*, 1999, **64**, 400.
125. G. W. Hofland, M. van Es, L. A. M. van der Wielen and G.-J. Witkamp, *Ind. Eng. Chem. Res.*, 1999, **38**, 4919.
126. P. M. Tomasula, J. C. Craig, R. T. Boswell, R. D. Cook, M. J. Kurantz and M. Maxwell, *J. Dairy Sci.*, 1995, **78**, 506.
127. P. M. Tomasula, N. Parris, R. T. Boswell and R. O. Moten, *J. Food Proc. Preserv.*, 1998, **22**, 463.
128. P. M. Tomasula and R. T. Boswell, *J. Supercrit. Fluids*, 1999, **16**, 21.
129. P. M. Tomasula, N. Parris, W. Yee and D. R. Coffin, *J. Agric. Food Chem.*, 1998, **46**, 4470.
130. P. M. Tomasula, J. C. Craig and R. T. Boswell, *J. Food Eng.*, 1997, **33**, 405.
131. P. M. Tomasula, J. C. Criag and A. J. McAloon, *J. Dairy Sci.*, 1998, **81**, 3331.
132. *Ultra High Pressure Treatments of Foods*, ed. M. E. Hendrix and D. Knorr, Kluwer Academic, New York, 2001.
133. T. Norton and D.-W. Sun, *Food Bioprocess Technol.*, 2008, **1**, 2.
134. N. M. Dixon and D. B. Kell, *J. Appl. Bacteriol.*, 1989, **67**, 109.
135. L. Garcia-Gonzalez, A. H. Geeraerd, S. Spilimbergo, K. Elst, L. Van Ginnekin, J. Debevere, J. F. Van Impe and F. Devlieghere, *Int. J. Food Microbiol.*, 2007, **117**, 1.
136. D. Del Pozo-Insfran, M. O. Balaban and S. T. Talcott, *J. Agric. Food Chem.*, 2006, **54**, 5468.
137. T. F. Calix, G. Ferrentino and M. O. Balaban, *J. Food Sci.*, 2008, **73**, E439.
138. T. T. Truong, J. M. Boff, D. B. Min and T. H. Shellhammer, *J. Food Sci.*, 2002, **67**, 3058.
139. H. Corwin and T. H. Shellhammer, *J. Food Sci.*, 2002, **67**, 697.
140. S. Spilimbergo and A. Bertucco, *Biotechnol. Bioeng.*, 2003, **84**, 627.
141. M. O. Balaban and T. Duong, *Agri. Agri. Sci. Proc.*, 2014, **2**, 2.
142. E. J. Beckman, *J. Supercrit. Fluids*, 2004, **28**, 121.

143. C. M. Rayner, R. S. Oakes, T. Sakakura and H. Yasuda, in *Green Reaction Media in Organic Synthesis*, ed. K. Mikami, Blackwell Publishing, Oxford, UK, 2005, p. 125.
144. C. M. Rayner, *Org. Process Res. Dev.*, 2007, **11**, 121.
145. P. E. Savage, S. Gopalan, T. I. Mizan, C. J. Martino and E. E. Brock, *AIChE J.*, 1995, **41**, 1723.
146. M. Sasaki, Z. Fang, Y. Fukushima, T. Adschiri and K. Arai, *Ind. Eng. Chem. Res.*, 2000, **39**, 2883.
147. J. Abeln, M. Kluth, G. Petrich and H. Schmieder, *High Pressure Res.*, 2001, **20**, 537.
148. C. L. Liotta, J. P. Hallett, P. Pollet and C. A. Eckert, in *Organic Reactions in Water: Principles, Strategies and Applications*, ed. U. M. Lindstrom, Blackwell Publishing, Oxford, UK, 2007, p. 256.
149. E. Stahl, K.-W. Quirin and D. Gerard, *Dense Gases for Extraction and Refining*, Springer-Verlag, Heidelberg, 1988.
150. *Supercritical Fluid Technology in Oil and Lipid Chemistry*, ed. J. W. King and G. R. List, AOCS Press, Champaign, IL, 1996.
151. T. L. Mounts, *J. Am. Oil Chem. Soc.*, 1983, **60**, 406A.
152. R. Eggers, in *Supercritical Fluid Technology in Oil and Lipid Chemistry*, ed. J. W. King and G. R. List, AOCS Press, Champaign, IL, USA, 1996, p. 35.
153. N. Foidl, *Device and Process for the Production of Oils or Other Extractable Substances*, *U. S. Pat.* 5,939,571, August 17, 1999.
154. R. Sandrea and R. F. Nielsen, *Dynamics of Petroleum Reservoirs Under Gas Injection*, Gulf Publishing Company, Houston, TX, 1974.
155. B. MacKinnon and H. Boeck, Abstracts of the 98th AOCS Annual Meeting, Quebec City, Canada, May 13–16, 2007, p. 15.
156. M. J. Venter, P. Willems, N. J. M. Knipers and A. B. de Haan, *J. Supercrit. Fluids*, 2006, **37**, 350.
157. J. Stoldt and G. Brunner, *Fluid Phase Equilib.*, 1998, **146**, 269.
158. M. J. Venter, P. Willems, S. Kareth, E. Weidner, N. J. M. Kuipers and A. B. de Hann, *J.* Supercrit. Fluids, 2007, **41**, 195.
159. M. J. Venter, Gas Assisted Mechanical Expression of Cocoa Nibs, Ph.D. Thesis, University of Twente, June 28, 2006.
160. S. Voges, R. Eggers and A. Pietsch, *Sep. Purif. Technol.*, 2008, **63**, 1.
161. P. Willems, N. J. M. Kuipers and A. B. de Haan, *J. Supercrit. Fluids*, **45**, 298.
162. P. Willems, N. J. M. Kuipers and A. B. de Haan, *J. Food Eng.*, 2009, **90**, 238.
163. S. S. H. Rizvi, S. J. Mulvaney and A. S. Sokhey, *Trends Food Sci. Technol.*, 1995, **6**, 232.
164. A. J. J. Straathof, in *Comprehensive Biotechnology*, ed. M. Moo-Young, Elsevier, 2011.
165. R. Datta and M. Henry, *J. Chem. Technol. Biotechnol.*, 2006, **81**, 1119.
166. K. L. Wasewar, A. A. Yawalkar, J. A. Moulijn and V. G. Pangarkar, *Ind. Eng. Chem. Res.*, 2004, **43**, 5969.

167. G. S. Dhillon, S. K. Brar, M. Verma and R. D. Tyagi, *Food Bioprocess Technol.*, 2011, **4**, 505.
168. J. Eastoe, S. Gold, S. Rogers, P. Wyatt, D. C. Steytler, A. Gurgel, R. K. Heenan, X. Fan, E. J. Beckman and R. M. Enick, *Angew. Chem.*, 2006, **118**, 3757.
169. C. S. Lopez-Garzon and A. J. J. Straathof, *Biotechnol. Adv.*, 2014, **32**, 873.
170. B. G. Garrett, K. Srinivas and B. K. Ahring, *J. Supercrit. Fluids*, 2014, **95**, 243.
171. M. Henczka and M. Djas, *J. Supercrit. Fluids*, 2016, **110**, 154.
172. M. Djas and M. Henczka, *Proceedings of the 14th European Conference on Mixing*, Warsaw, 10–13 Sept 2012.
173. A. Keshav, K. L. Wasewar and S. Chand, *Sep. Purif. Technol.*, 2008, **63**, 179.
174. M. Djas and M. Henczka, *J. Supercrit. Fluids*, 2016, **117**, 59.
175. J. S. Wang, J. Herman, Y. Enokida and C. M. Wai, *J. Supercrit. Fluids*, 2005, **36**, 98.
176. A. F. Morales, J. Albet, G. Kyuchoukov, G. Malmary and J. Molinier, *J. Chem. Eng. Data*, 2003, **48**, 874.
177. L. Duan, *Extraction of Silymarins from Milk Thistile, Silybum Marianum, Using Hot Water as Solvent*, Ph.D. Thesis, University of Arkansas, May 2005.
178. P. T. V. Rosa, J. C. Parajo, H. Dominguez, A. Moure, B. Diaz-Reinoso, R. L. Smith, Jr., M. Toyomizu, L. J. Florusse, C. J. Peters, M. Goto, S. Lucas and M. A. A. Meireles, in *Extracting Bioactive Compounds for Food Products: Theory and Applications*, ed. M. A. A. Meireles, CRC Press, Boca Raton, FL, 2008, p. 269.
179. A. Kulkarni, S. Suzuki and H. Etoh, *J. Wood Sci.*, 2008, **54**, 153.
180. I. Rodriguez-Meizoso, F. R. Marin, M. Herrero, F. J. Senorans, G. Reglero, A. Cifuentas and E. Ibanez, *J. Pharm. Biomed. Anal.*, 2006, **41**, 1560.
181. A. U. Buranov and G. Mazza, *Food Chem.*, 2009, **115**, 1542.
182. M. Yamamoto, Y. Futamura, K. Fujioka and K. Yamamoto, *Int. J. Chem. Eng.*, 2008, DOI: 10.1155/2008/603957.
183. F. Salak, O. Pourali and H. Yoshida, Proceedings of the World Congress on Engineering and Computer Science WCECS, San Francisco, USA, October 20–22, 2009.
184. K. Watchararuji, M. Goto, M. Sasaki and A. Shotipruk, *Bioresour. Technol.*, 2008, **99**, 6207.
185. O. Pourali, F. S. Asghari and H. Yoshida, *Food Chem.*, 2009, **115**, 1.
186. G. Brunner, *J. Supercrit. Fluids*, 2009, **47**, 373.
187. J. Wiboonsirikul, Y. Kimura, M. Kadota, H. Morita, T. Tsuno and S. Adachi, *J. Agric. Food Chem.*, 2007, **55**, 8759.
188. M. Kataoka, J. Wiboonsirikul, Y. Kimura and S. Adachi, *Food Sci. Technol. Res.*, 2008, **14**, 553.
189. J. Wiboonsirikul and S. Adachi, *Food Sci. Technol. Res.*, 2008, **14**, 319.
190. O. Tavakoli and H. Yoshida, *Green Chem.*, 2006, **8**, 100.

191. W. Abdelmoez, T. Nakahasi and H. Yoshida, *Ind. Eng. Chem. Res.*, 2007, **46**, 5286.
192. O. Tavakoli and H. Yoshida, *Ind. Eng. Chem. Res.*, 2006, **45**, 5675.
193. H. Yoshida, M. Terashima and Y. Takahashi, *Biotechnol. Prog.*, 1999, **15**, 1090.
194. T. Rogalinski, S. Herrmann and G. Brunner, *J. Supercrit. Fluids*, 2005, **36**, 49.
195. H. Cheng, X. Zhu, C. Zhu, J. Qian, N. Zhu, L. Zhai and J. Chen, *Bioresour. Technol.*, 2008, **99**, 3337.
196. W. Lamoolphak, M. Goto, M. Sasaki, M. Suphantharika, C. Muangnapoh, C. Prommuag and A. Shotipruk, *J. Hazard. Mat.*, 2006, **B137**, 1643.
197. A. Espinoza and R. O. Morawicki, *Proceedings of Institute of Food Technologists (IFT) Annual Meeting and Expo*, New Orleans, LA, June 28 – July 1, 2008, Poster# 095–12.
198. E. Weidner, C. Brake and D. Richter, in *Supercritical Fluids as Solvents and Reaction Media*, ed. G. Brunner, Elsevier, London, 2004, p. 269.
199. J. W. King, K. Srinivas, Y.-W. Lu, O. Guevara, D. Zhang and Y.-J. Wang, *J. Supercrit. Fluids*, 2012, **66**, 221.
200. J. K. Monrad, M. Suarez, M. J. Motilva, J. W. King, K. Srinivas and L. R. Howard, *Food Res. Int.*, 2014, **65**, 77.

CHAPTER 7

High Hydrostatic Pressure Food Processing: Potential and Limitations

STEPHANIE JUNG*[a] AND CAROLE TONELLO-SAMSON[b]

[a] Department of Food Science and Nutrition, California Polytechnic State University, 1 Grand Avenue, San Luis Obispo, CA 93407, USA; [b] Hiperbaric, Calle Condado de Trevino 6, Pol. Industrial Villalonquejar, 09001 Burgos, Spain
*Email: stjung@calpoly.edu

7.1 Introduction

7.1.1 Rationale for the Interest in High-pressure Processing

Interest in non-thermal alternative food processing technologies, including but not limited to gamma-irradiation, high-pressure processing (HPP), high-pressure homogenization, pulsed electric field, ohmic treatment, cold plasma, ozone, and pulsed electric light methods, has increased steadily over the past few years, because their uses address both consumers' requests and food industry needs. These technologies have attracted food industry interest primarily because they increase food safety and potentially increase shelf life without chemical preservatives and/or thermal treatment that often impair product quality. The commercial use of these technologies around the world varies geographically, and some of the limitations in their implementation include high investment costs, the need for a better understanding of process parameters, and challenging regulatory requirements.[1]

Green Chemistry Series No. 53
Alternatives to Conventional Food Processing: 2nd Edition
Edited by Andrew Proctor

Published by the Royal Society of Chemistry, www.rsc.org

HPP has the advantage of being well perceived by consumers relative to other novel food processing technologies,[2–4] which contributes greatly to the success of the commercialization of pressurized products. HPP was identified as one of the top technologies having emerged or having commercial applications in a recent survey comparing it with 10 other novel and emerging food processing technologies, mainly in North America and Europe. The market seems to be gaining increasing interest in South America, Asia-Pacific, and Africa.[1]

During HPP treatment, the pressure is applied rapidly, uniformly, and independently of the shape and according to Pascal's principle. The size of the products results in fewer challenges when the production is scaled up. This is an advantage of this technology compared with thermal treatment. HPP could (i) address consumers' requests for a product with high nutritional value, without chemical preservatives and with a fresh taste, and (ii) be used for the development of products with novel or unique attributes owing to its unique effect on food constituents. As a result, this technology has found some niche markets, such as a guacamole treatment, in natural ready-to-eat meat production, and smoothies and juices processing, and is often referred to as a technology than could deliver functional foods.

7.1.2 Brief Description of Processing Steps and Concept of Adiabatic Heating

Industrial vessels can reach pressures up to 600 MPa, *i.e.* ~87 000 psi, whereas in laboratory-scale processing pressures of up to 1000 MPa, *i.e.* ~145 000 psi, could be reached. There are two categories of systems.[5] The first system is used to treat products directly, *i.e.* without packaging, and therefore does not use a transmission fluid. This type of system can only be used for products that can be pumped into the vessel. From a food safety perspective, the main disadvantage of these systems is the risk of post-contamination after treatment and before packaging, for which an aseptic environment is necessary. This is similar to what is required with an ultra-high-temperature (UHT) treatment. From a processing equipment perspective, there may be difficulties in ensuring reliable hygienic design under pressure.

The second category is systems for prepackaged foods and can be used for broader applications and are preferred by the food industry. The product is prepackaged in its final container before treatment, which eliminates potential post-treatment microbial contamination. The choice of the packaging material is important; it must be flexible to withstand pressure, maintain its physical integrity, and have excellent heat sealability.[6] Many studies have investigated the impact of treatment conditions on the mechanical, barrier, and mass transfer properties of a wide range of packaging materials.[7–9]

The prepackaged food is introduced into the vessel and pressure is transmitted to the product through a pressurization fluid, which is usually water. The properties of the pressurization fluid and the processing conditions applied to transmit the pressure are crucial parameters that affect the overall effect of the treatment. Application of pressure results in compression of foods due to a decrease in the average intermolecular distance between molecules, which generates heat called "adiabatic heat." This adiabatic heating is mainly a function of initial temperature, compressibility of the material, *i.e.* its composition, target pressure, and rate of heat loss to the surroundings.[10] The importance of these parameters may differ depending on the processing conditions and food products.[11] The increase in temperature due to adiabatic heating can be low, as is the case with water, ~3 °C per 100 MPa at 20 °C, or as high as 8.5 °C per 100 MPa for milk fat treated at 20 °C.[12] Owing to the difference in adiabatic heating between the pressurization fluid, the product that is being pressurized, and the wall of the steel vessel, a temperature gradient develops during treatment. This temperature gradient results in heat transfer during the holding time, which must be known to assure adequate microbial inactivation.[13] The temperature gradient and temperature increase of the product are commonly considered to be negligible when using large processing equipment and the product and water are at refrigerated temperature.

The quasi-adiabatic temperature increase upon compression (δ_s, °C per 100 MPa) can be determined experimentally by applying the following equation:[10]

$$\delta_s = \frac{T_f - T_i}{\Delta P} \times 100 \tag{7.1}$$

where T_f is the temperature of the sample at the target pressure, T_i the temperature of the food at the beginning of the process, and ΔP the pressure difference between the targeted pressure and atmospheric pressure. Once δ_s has been determined for a specific set of conditions, the equation can be used to predict T_f.

The increase in temperature upon compression has been maximized for treatments leading to shelf-stable pressurized products.[12] Although usually it is not a desired feature of the HPP treatment as it may be responsible for undesired physicochemical reactions that reduce nutritional and/or food quality attributes, heat can be combined with pressure in a process denoted pressure-assisted thermal sterilization (PATS), pressure-assisted thermal processing (PATP), high-pressure thermal sterilization (HPTS), or high-pressure, high-temperature (HPHT) processing. During PATS, adiabatic heating is used to increase the temperature of the preheated food rapidly and uniformly to between 90 and 120 °C. This temperature is reached in a few minutes, whereas it takes much longer to achieve a similar temperature with a conventional retort system, *e.g.* during canning.

Pressure food pasteurization is usually performed by applying pressures of 500–600 MPa for 1–10 min at an initial temperature of 0–42 °C, whereas sterilization requires a treatment of 700–1000 MPa for 2–5 min, or a multi-pulse treatment, at an initial temperature of 70–90 °C.[13] For PATS treatment, a preheating step and cooling step of the food are required. The preheating step has been identified as a critical stage of the process as it affects the target temperature during pressurization, which is key in assuring food safety.[14] The difference in the temperature profile between canning and PATS is more marked than that between UHT and HPP PATS. PATS was accepted at the beginning of 2009 by the US Food and Drug Administration (FDA)[15] as a new sterilization process for the production of mashed potatoes (a low-acid food). Inactivation of spores of different species and strains in buffer and food products using PATS has been demonstrated.[16–18] Although quality attributes and nutritional data are needed on PATS products, the available studies showed better quality of PATS products compared with canned products. When quality attributes of various low-acid foods (*e.g.* chicken breast, salmon fillet, egg omelet, fried potato wedges, and green beans) treated by PATS were compared with retorted products, PATS foods had better sensory attributes than retorted products, with a fresher, less processed flavor in all the pressurized foodstuffs.[13] Optimization of the PATS processing parameters and formulation for the production of scrambled egg patties with optimum quality attributes has also been achieved.[14,19–21] The successful application of PATS to a specific food product also depends on the effects of the treatment on enzyme activity and has been summarized for the enzymes related to the quality of fruits and vegetables by Ludikhuyze *et al.*[22]

The remainder of this chapter focuses on the use of pressure treatment with temperature lower than or at ambient temperature, and will not cover the impact of PATS on food.

7.1.3 Is HPP a Green (Environmentally Friendly) Technology?

Several factors contribute to HPP often being described as a green technology. During HPP of prepackaged foods on an industrial scale, the transmission fluid is water, and the water is recycled. The pressure is usually generated with the use of high-pressure intensifiers that use electricity to produce the pressure. It takes a few minutes to reach the desired pressure depending on the characteristics of the high-pressure pump and the target pressure level. Once the pressure has been reached, no additional energy is required to maintain the pressure. This is in contrast to a thermal treatment, where energy needs to be constantly provided for the duration of the treatment. Once the food has been treated for the desired time or holding time, *i.e.* time under the desired pressure, the pressure release takes place in a range of times from a few seconds to a few minutes without additional energy being required. This depends on the pressure-release device installed on the HPP machine. This is another advantage relative to thermal treatments, which often involves additional energy to decrease the product temperature after the required processing time, to avoid a major decrease in the quality attributes of the final product due to over-exposure to high temperature. Overall, HPP is often referred to as a green, environmentally friendly technology because of the lower energy requirement compared with thermal treatments, the recycling of the pressurization fluid, and the lack of waste production (waste-free process).[23,24] There are, however, some conflicting data in the literature regarding the lower energy requirement associated with HPP. The energy requirement of HPP has been compared with four other alternative food processing methods: membrane filtration (MF), pulsed electric field (PEF), ultraviolet radiation (UV), and conventional high-temperature short-time (HTST) processing.[25] Based on energy consumption, it was concluded that, for the parameters considered in that study, MF and UV methods consume less energy than HTSF, PEF, and HPP methods.

A significant decrease in, or absence of, chemical preservatives in the formulation of pressurized food could also justify this technology being considered as a "green" technology.

7.2 HPP as an Efficient Tool for Food Microbial Safety and Shelf Life Extension

7.2.1 Food Safety

Inactivation of food-borne pathogens by HPP has been extensively studied under a broad range of conditions. Reviews covering the effects of pressure on bacteria and fungal cells, bacterial spores, parasites, and viruses are available,[26–28] while others focus on one aspect of the technology. For example, Buzrul[29] summarized the impact of multi-pulsed high pressure on microbial inactivation whereas Baptista *et al.*[30] and Gänzle and Liu[31] focused on the inactivation under pressure of *Staphylococcus aureus* and *Escherichia coli*, respectively.

Bacterial inactivation with pressure involves membrane disruption, repair of the damaged cells, damage to the nuclear material, and enzymatic deactivation.[32] Numerous studies have shown that pressure inactivation of microorganisms is related to the processing conditions, mainly pressure level, temperature, and dwell time, and a combination of microorganism factors including bacterial species and strain, growth phase (exponential *versus* stationary), and growth temperature.[26,33–39] At the matrix level, food composition and environmental conditions such as pH, water activity, and salt and sugar content are the key parameters that will affect the microbial inactivation efficiency.[26,38,40,41] Fat content had no impact on the immediate reduction in *E. coli* levels in milk and *Listeria innocua* and *Salmonella* levels in meat.[18,42] It has been reported in many studies that Gram-positive bacteria (such as *Listeria monocytogenes*, lactic acid bacteria, and *Micrococcus*) are generally more resistant to pressure than Gram-negative bacteria (such as *Salmonella* spp. and *E. coli*). Stationary-phase cells are more baroresistant than exponential growth-phase cells. Damage to the cell membrane after pressure seems to be the major structural change impacting microbial inactivation. A more robust cell envelope of the Gram-positive bacteria that contains a high percentage of peptidoglycan and teichoic acids (40–49%), *versus* only 10–20% of peptidoglycan and no teichoic acids in Gram-negative cells, could explain their different barotolerances.[36] However, all of the parameters stated above cannot predict the effect of pressure on microbial inactivation in foods and contradictory data can be found in the literature. For example, *L. monocytogenes* (Gram-positive bacteria) when treated at 250 MPa in a fish slurry was more resistant to pressure than *E. coli* (Gram-negative bacteria) (*D* values of 40.1 *versus* 20.7 min). However, at 400 MPa, *E. coli* was more resistant than *L. monocytogenes* (*D* values of 3.19 *versus* 1.49 min).[43]

Variability in the baroresistance of various microorganisms in a food matrix relative to a synthetic medium has been observed in many cases. The extent of the difference is dependent on the food matrix.[18,38,40] Therefore, to avoid underestimation of the effect of HPP on microbial inactivation, the behavior of the pathogen in real food systems must be investigated to identify appropriate processing conditions.[44] Microorganisms are usually more barosensitive in buffer solutions as nutrients play a baroprotective role. Georget *et al.*[28] recently reviewed the impact of pressure on microbial inactivation of food with non-uniform or low water activity (a_w). The review demonstrated the importance of taking into account the matrix composition. A few examples illustrating the importance of matrix composition on microbial inactivation are described below. *E. coli* treated at 300 MPa for 5 min had a logarithmic cycle reduction of 1.0, 1.6, and 3.4 in milk (more nutrients), peptone water, and buffer solution, respectively. Casein and lactose contributed to the baroprotective effect of milk.[19] After treatment at 375 MPa, the decimal reduction times (*D* values) of *L. monocytogenes* in phosphate-buffered saline (PBS), UHT milk, and raw chicken mince were 8.9, 17.2, and 5.0 min, respectively.[38]

However, because of the complexity of food products and the potential interactions between components during processing, microorganism studies in buffer or synthetic media are important starting points in identifying and understanding the precise mechanisms responsible for the microorganism's inactivation during treatment. Precise mechanisms resulting in a loss of microbial activity have not yet been elucidated, despite an extensive literature reporting the efficacy of HPP on the inactivation of microorganisms.[35] The use of analytical methods with conventional cultivation techniques should contribute to a better understanding of the changes that occur during treatment. For example, flow cytometric measurements were effective in comparing the effects of pressure and thermal treatment on *Lactobacillus rhamnosus* ATCC 53103 cellular activities, or integrity and population heterogeneities.[45]

Mathematical models are the basis of predictive microbiology and have been developed as powerful tools to predict the effects of treatment on microbial inactivation.[46,47] The difficulties in establishing these models include potential recovery of the microorganisms during storage, owing to their ability to recover from their pressure-induced injuries. The recovery of sublethally injured cells during HPP has been reported in many studies and is a major food safety concern.[48] Microorganisms are often determined using only selective media to estimate the effect of high-pressure treatment, but injured cells may not grow on selective media and, therefore, there is a risk of mistakenly considering a food to be free of microorganisms.[48,49] This increases the risk of overestimating the efficacy of pressure treatment. The extent of recovery of sublethality-injured cells depends on many factors, including food composition, processing conditions, and storage conditions, which justify the need to determine sublethal recovery for a specific food product. In a study on pressure-treated milk during storage, Bozoglu *et al.*[50] identified two types of injury, I1 and I2, that were attributed to structural damage and/or cell membrane injury and metabolic injury, respectively. Injury of *L. monocytogenes* inoculated in whole UHT milk occurred after a 400 MPa treatment for 1–1800 s at room temperature, regardless of the temperature at which the microorganisms were grown (4–43 °C) or growth phase (mid- or late-exponential and mid-stationary).[34]

Inactivation of food-borne pathogens (except spores) and spoilage microorganisms could be achieved with pressure alone, the higher pressure level applied usually resulting in increased inactivation. Although elevated pressure is beneficial for food safety and shelf life extension, it increases energy costs and may have detrimental effects on the food quality and nutritional attributes. Hurdle treatment such as mild heat treatment and addition of chemical or natural substances could be a viable alternative to the application of elevated pressure. Such combinations of treatment have been increasingly reported over the past few years. Indeed, a decrease in pressure level to promote microbial inactivation could be beneficial for food quality and also from an economical perspective, resulting in lower operation costs due to time and energy reductions. Some examples of hurdle technologies are summarized below. For example, when nisin (500 IU mL^{-1}), a natural antimicrobial agent, was added to milk prior to treatment at 500 MPa and 20 °C for 5 min, an 8.5 log reduction of *L. innocua* and an 8.3 log reduction of *E. coli* were observed *versus* 3.8 and 6.5 log reductions, respectively, when pressure was applied alone.[51] Similarly, a mixture of two bacteriocins (pediocin AcH and nisin A) added prior to HPP (345 MPa, 25 °C, 10 min) to four pathogens (*S. aureus*, *L. monocytogenes*, *E. coli*, and *Salmonella typhimurium*) increased the viability loss of the pathogens and was partially correlated with their bactericidal effect on injured cells.[37] When nisin and potassium lactate were added to the formulation of a cooked ham, treatment at 600 MPa and 31 °C for 6 min and refrigeration at $\leq$6 °C were an effective system to control *L. monocytogenes* and *S. aureus* during 3 months of storage.[52] Combinations of HPP with essential oils and their components have also been studied for their potential synergistic effects.[53] A 5 log reduction of *E. coli* 0157:H7 in apple juice was observed when 200 $\mu L\, L^{-1}$ of (+)-limonene addition was combined with a 300 MPa treatment of 20 min. Similarly, carbon dioxide addition to poultry sausage packaging reduced inactivation of several microorganisms after treatment at 350 MPa for

10 min at room temperature. It was hypothesized that the cellular damage due to pressure facilitated the penetration of CO_2, which impacted their metabolism and growth.[54] Processing of carrot juice and apple juice by an ultrasonic treatment (25 kHz and 70% amplitude at 20 °C for 60 min) followed by an HPP treatment (450 MPa for 10 min) resulted in no residual total plate counts and yeast and mold counts, which were not obtained when the treatments were applied alone.[55]

Hurdle technologies, including the combination of HPP and irradiation, addition of nisin or lysozyme, and a combination of heat, have also been studied for inactivation of spores.[56–60] Pressure alone, up to 1500 MPa, does not inactive spores and spore baroresistance is a challenge that needs to be overcome to produce shelf-stable pressurized food products (*i.e.* products that can be stored at ambient temperature). The mechanisms involved during pressure inactivation of spores and the effects of processing parameters and potential hurdle approaches were reviewed by Black *et al.*[61]

Although HPP is applied to many food products that are commercialized, there are no standardized treatment conditions that have been established by regulatory authorities as is the case for thermal treatment.[27] The establishment of process criteria for categories of food and pathogens by authorities could contribute to the increasing use of HPP. However, it should not replace validation studies of specific food products to establish the efficiency of the treatment but could be a contributor for food companies to consider the use of the technology.

7.2.2 Shelf Life

One of the major advantages of HPP is its potential to extend the shelf life of food products without altering their fresh sensory and nutritional attributes. Many studies have been performed on a wide variety of food products to assess the potential of the processing parameters such as pressure level, dwell time, and processing temperature and environmental conditions such as pH, water activity, and constituent composition and concentration on the extension of product shelf life. The majority of these studies showed that each food product is unique and studies are needed to find processing conditions to produce a safe food product with extended shelf life and optimum quality attributes. HPP usually doubles or triples the shelf life of ready-to-eat (RTE) meat products or meals. The shelf life extension is generally more than tripled in the case of vegetable products (juices, fruit purées, avocado products).

HPP is one of the most attractive preservation technologies available to the meat industry to preserve product quality and safety.[62–64] The importance of the composition of meat products has been illustrated by comparing the effects of a treatment at 600 MPa for 6 min at 16 °C (initial temperature) on aerobic total count, psychrotrophs, and lactic acid bacteria in sliced vacuum-packed cooked ham, sliced vacuum-packed dry-cured ham, and sliced vacuum-packed marinated beef loin, followed by up to 120 days of storage at 4 °C.[65,66] Whereas low water activity of dry-cured ham decreased the efficiency of the treatment, its composition seemed to inhibit the recovery of the cells during storage. When RTE meats (low-fat pastrami, Strasburg beef, export sausage, Cajun beef) were treated at 600 MPa and 20 °C for 3 min, aerobic and anaerobic mesophiles, lactic acid bacteria, yeasts, and molds were undetectable or detected at low levels throughout storage (98 days at 4 °C).[67] Treatment at 600 MPa for 10 min increased the shelf life of Spanish blood sausage by 15 days, which was equivalent to the shelf life extension obtained by adding 3% of potassium and sodium lactate.[68]

Most of the studies on fruits, fresh-cut fruits, fruit purées, and fruit juices have been focused on the effect of HPP on the quality attributes of these products or reported the impact of treatment after inoculation of food safety-related microorganisms.

The treatment of avocado (avocado purée and guacamole) is one of the success stories of food applications using HPP.[69] Thermal treatment of these avocado-based products significantly impairs their quality attributes, which reduces consumer acceptability. The alternative option of adding antibrowning or antioxidant agents did not satisfy consumers' desire for a preservative-free product. HPP maintained the quality attributes of avocado products after treatment and efficiently decreased polyphenol-oxidase activity enzyme, which is responsible for the browning of avocado products. It also significantly increased the shelf life of these highly perishable products, usually by 5 days with refrigerated storage. After treatment at 345, 517, and 689 MPa for 10, 20, and 30 min at either 5, 15, or 25 °C, the standard plate count and yeast and mold counts of avocado purée were below the detection limit for 35–110 days, independent of the storage temperature.[70] Similarly, the shelf life of pressurized guacamole was extended from 5 to up to 30 days.[71] The effect of processing freshly cut pineapple under conditions of 200, 270, and 340 MPa pressure at 4, 21, and 38 °C for 5, 15, 40, and 60 min on the immediate reduction of total plate counts and total yeast and mold counts were compared by Aleman *et al.*[72] The decimal reductions varied from less than one to almost three. Similarly, a 2 log colony-forming units (cfu) mL^{-1} reduction was obtained after treatment of mangos at 600 MPa and 3 °C for 1 min. After 9 weeks of refrigerated storage, pressurized mango had a lower bacterial count of 13 cfu mL^{-1} that the control on the day of treatment, which was 725 cfu mL^{-1}.[73] Cashew apple juice treated at 400 MPa for 3 min had no aerobic mesophilic, yeast or filamentous fungi detected after 8 weeks of refrigerated storage, while an untreated control sample reached about 6 log cfu mL^{-1} after 3 weeks of storage.[74] Total counts and yeast, mold, and coliform counts of an apple–broccoli juice treated at 500 MPa and 15 °C for 10 min and stored for up to 30 days were comparable to the values obtained with the same juice submitted to heat pasteurization (80 °C for 20 min).[75] Increasing dwell time significantly increased the effect of a 400 MPa treatment on the viable aerobic mesophilic population of tofu, which was reduced by 0.31, 1.56, and

2.38 log units after 5, 30, and 45 min of treatment, respectively.[76] A shorter dwell time of 1 or 5 min had no significant effect on the total bacterial count of soymilk. However a treatment at 600 MPa and 25 °C increased the soymilk shelf life by up to 14 days compared with untreated soymilk.[77] Similarly, an increased dwell time from 1 to up to 4 min of wheat dough treated at pressures from 100 to 250 MPa did not further improve the effect of treatment on total aerobic mesophilic bacteria, molds, and yeasts.[78] The shelf life of pressurized (400 MPa and 25 °C for 30 min) whole and skim milk stored at 7 °C was increased from 30 to 45 days, compared with 15 days for the untreated samples.[79]

The potential of HPP as a means to decrease the dependence on additives to stabilize dressing-type products was reported by Waite *et al.*[80] With a treatment at 600 MPa for 5 min at 8 °C (initial temperature), followed by 8 weeks of storage at 26 °C, the pressurized sample contained about 2 log cfu g^{-1} (mainly lactic acid bacteria), compared with 7 log cfu g^{-1} for the untreated samples, while no significant difference was observed between the pressurized sample and a control after up to 16 weeks at 4 °C. Inoculated spoilage microorganism *Pediococcus acidilactici* was suggested as a microorganism to study in order to optimize the processing conditions of ranch dressings because it was found to be the most baroresistant in the study.

HPP treatment of shellfish has been very successful and offers several benefits, including an increased meat yield of products such as lobster, oysters, and clams and easier shucking of products such as oysters and clams, which greatly improves labor costs. In addition to these advantages, improved appearance and sensory attributes of the products can be obtained in addition to the potential benefit of microbial inactivation.[81,82] Murchie *et al.*[83] reviewed the effect of HPP on microbiological and quality aspects of shellfish. Release of the adductor muscle of oysters, which determines the ease of shucking, was dependent on pressure level and dwell times: 88 and 100% detachment were obtained after treatment at 242 MPa for 2 min and 311 MPa for 0 min (no holding time), respectively.[81] The microbial shelf life of pressurized fish and shellfish is usually increased with HPP. For oysters, HPP reduced the total aerobic counts on the day of treatment. Some variability was observed in the lag phase and recovery of spoilage microorganisms during storage.[81,82,84] The total bacterial count during storage of pressurized oysters varied from no change to an increase. This increase led to pressurized samples having higher bacterial counts than raw oysters.[81,82,84]

A few studies have looked at HPP as a way to store food under hyperbaric conditions to increase its shelf life – in other words, as a replacement for the traditional refrigerated storage. The concept is food storage at temperatures close to ambient temperature and at a much lower pressure than that required for microbial inactivation for increased food safety and shelf life, *i.e.* 600 MPa. The pressure, once achieved, involves minimum additional energy needing to be maintained. This represents a major energy saving compared with other preservation methods, such as freezing or refrigeration. It was shown that with storage for up to 60 h at 100 MPa and at room temperature, growth of microorganisms in watermelon juice can be avoided.[85] Comparison of pressures of 25, 50, 100, and 200 MPa applied to strawberry juice at 20 °C for up to 15 days emphasized that treatments at 25 and 50 MPa were not sufficient to provide microbial stability after decompression.[86] Both duration of storage and pressure applied determined whether the treatment could provide microbial inhibition and/or microbial reduction, and to what extent microorganisms are injured. Level of injury of microorganisms under pressure is important as it impacts their recovery once back at ambient pressure.

7.3 Nutritional and Toxicological Aspects of Pressurized Foods

It is generally considered that HPP at moderate temperatures retains most of the nutritional value of food, which is one of the main advantages of using high-pressure treatment rather than pasteurization or sterilization.[87,88] Many studies have reported the impact of processing conditions during HPP on plant food materials such as fruits and vegetables, cereals, legumes, and nuts that contain significant amounts of health-related compounds. Studies includes investigation of the fate of vitamins and carotenoids, polyphenols, and antioxidants during processing. Koutchma *et al.*[89] published an extensive review on the effects of HPP on quality and health-related constituents of fresh juice products. They concluded that vitamin C has been the most studied vitamin with an average residual content of 92%, thus displaying high barostability. Total phenols showed similar stability, whereas anthocyanins had a lower barostability, with an average residual content of 86%. Average values were obtained based on a review of 50 peer-reviewed articles. The fate of vitamins has been determined in both model systems and in foods. Vitamin stability varies depending on the food matrix and the vitamin. Vitamin C content after high-pressure treatment was maintained in model systems and in foods such as strawberry purée[90] and germinated cowpea[91] and no significant losses of vitamins B_1 and B_6 were observed after high-pressure treatment (up to 600 MPa for 30 min in model systems).[92] Effects of HPP on many water- and fat-soluble vitamins can be found in a review by Oey *et al.*,[93] which also reported on the impact of the treatment on antioxidant properties.

Little is known about the effect of high pressure on most other biologically active compounds. Nevertheless, Jung *et al.*[94] demonstrated that high-pressure treatment did not affect the isoflavone content of soymilk, but the isoflavone profile can be altered due to treatment.

Although the effect of high-pressure treatment on the content of health-related compounds is of interest, human studies are necessary to study the bioavailability of various compounds in pressurized foods. However, human studies are expensive and complex, and only a few have been performed. *In vitro* bioavailability studies offer an alternative to human studies to understand better the effect of various processing parameters on bioavailability or bioaccessibility of biologically active compounds in pressurized food products. Some bioavailability studies include the impact of HPP on *Prosopis chilensis* seed, split peas, whole white beans, brown rice, and Granny Smith apples. The fate of minerals in pressurized foods has also been investigated. Minerals are not chemically changed by the treatment, but their bioavailability or bioaccessibility is affected.[93,95] The changes to the matrix that occur during treatment and the new interactions between the minerals and the other food constituents are key parameters affecting the bioavailability of mineral-rich foods. A treatment at 300 MPa increased the bioaccessibility of manganese, phosphorus, and magnesium of uncooked pregerminated brown rice. A higher pressure (500 MPa), however, decreased their bioaccessibility. An increased release of the minerals in the matrix was observed at both pressures, but at 500 MPa the gelatinization of starch resulted in the encapsulation of the minerals.[96] The study of Briones-Labarca *et al.*[97] on the processing of Granny Smith apples showed that although the free calcium content was improved after treatment at 500 MPa for 2, 4, or 8 min (from 30 mg per 100 mg for the control to up to 122 mg per 100 mg), its bioaccessibility/availability decreased from 14% to about 1%. This is a good illustration of increasing levels of a health-related compound not being a useful indicator of bioavailability or a basis for health claims.

HPP had a significant impact on the profile of resistant and digestible starch of *P. chilensis* seed, and significantly decreased antinutritional factors such as trypsin inhibitors and flatulence-causing oligosaccharides in peas and beans.[98,99]

Formation of food processing contaminants (FPCs) such as acrylamide and furan occurs during conventional heat processing and has been under recent scrutiny.[100] Temperatures favoring the formation of FPCs can be reached during PATS and therefore PATS can result in the formation of these compounds.[95] A retarding effect of high pressure on the overall Maillard reaction was observed in asparagine–glucose model systems, with acrylamide concentrations generated during a conventional thermal treatment of 6500 ppb being reduced to about 1700 ppb during a 60 min treatment at 115 °C and 600 MPa.[101] Similarly in food systems (tuna in sunflower oil, sardines in olive oil, tuna in brine), a decrease in furan formation was observed for various temperature–time combinations for a 600 MPa treatment compared with a conventional treatment for 28 min at 115 °C.[102] When similar comparative experiments were performed at a pilot scale (55 L pressure vessel), it was confirmed that reduction of furan with pressure/temperature treatments can be reduced by 41–98% compared with what was obtained by retorting.[103]

In addition to reducing food processing contaminants, high-pressure thermal sterilization could also inactive spores and maintain food quality, offering the food industry a valuable alternative to thermal sterilization.[95,102,103]

Functional foods containing prebiotics such as inulin or fructo-oligosaccharides (FOS) have been of increasing interest to health-conscious consumers. Thermal treatment favors the hydrolysis of these compounds, therefore limiting the use of this processing technology in providing microbial safety and shelf stability of these products. HPP was explored as an approach that could deliver prebiotic foods. Apple purées showed a 60% increase in hydrolysis of inulin for thermally processed purées compared with HPP-treated purées,[104] and the FOS content was maintained in pressurized (450 MPa, 5 min) cranberry juices.[105]

Another important application that HPP offers in terms of improving "health attributes" of food is the possibility of decreasing the salt content. Excessive consumption of sodium chloride is associated with the development of several chronic diseases such as cardiovascular diseases, hypertension, and neurological diseases.[106] Dairy products, *e.g.* cheese, and meat products such as patties, sausage, bacon, and ham are products that contribute large amounts of salt in our diet. Studies on the reduction of salt in cheese products are scarce,[106] and those pertaining to meat products are discussed in Section 7.4.1.1.

7.4 Quality Attributes of Pressurized Food Products

7.4.1 Textural and Rheological Properties

A wide range of texture modifications can occur during pressure treatment, which depend mainly on the microscopic and macroscopic modifications of the food constituents. The nature of textural changes depends on the processing conditions and the properties of the starting materials, such as the concentration and/or state of the proteins, *e.g.* raw or cooked products. Minimizing fruit and vegetable texture changes is desirable. However, significant texture changes are needed in meat patties and fish-based gel processing. The impact of HPP on the textural and rheological properties can be summarized by food category.

7.4.1.1 Meat

Bouton *et al.*[107] carried out extensive work in the 1970s and 1980s on the potential of this technology to increase beef muscle tenderness. If the treatment is applied to meat in pre-rigor conditions, *i.e.* a few hours after the animal's death, and at a mild temperature, an increase in tenderness was reported. This increased tenderness of pressurized meat might have some economic benefit; however, the logistical challenges of the short time window for meat processing are an important problem. Some companies have explored treating the entire carcass and/or an large part of the carcass of various animals, but to our knowledge, no food industry has yet implemented this approach. The difference in size of the carcass and the smaller diameter of the processing vessel are a technical difficulty limiting implementation.

Once the meat has reached the post-rigor state, however, the use of HPP at low temperature or room temperature often decreased meat tenderness. The hardness of post-rigor beef increased from about 4000 to 7000 g after treatment at 400 MPa and 20 °C. The texture was not further affected by an increase in pressure to 800 MPa.[108] A similar trend was observed in chicken breast muscle.[109] An increased texture of beef muscle after treatment at 520 MPa was maintained during refrigerated aging. After 17 days of storage, pressurized biceps femoris muscles had a shear value of about 92 compared with 66 for the untreated sample.[110] However, an increase in tenderness of bovine muscle treated at up to 500 MPa and 8 °C for 5 min was reported by Ichinoseki *et al.*[111]

Textural changes of pressurized meat are the result of the effect of the treatment on the myofibrillar proteins[112] and enzymatic systems, mainly cathepsins and calpains, which are involved in developing meat tenderness.[110,113,114] The tenderizing effect on connective tissue is still being debated, as results are contradictory.[111,115] A clear understanding of how the variations in these enzymatic systems correlate with structural changes of myofibrillar and collagen components and how they affect the tenderness of pressurized meat is still needed.

HPP can be a powerful tool for increasing post-rigor meat tenderness when combined with a mild thermal treatment. A significant increase in meat tenderness was observed when beef and chicken muscle were pressurized from 400 to 800 MPa, at an initial temperature of 70 °C.[108,109]

The effect of high-pressure treatment on restructured meat depends on its formulation. Consumers' awareness of the detrimental health effect of a high-sodium diet has increased the demand for low-sodium products. In addition, consumers are seeking low-fat, phosphate-free products. However, salt and phosphate are key ingredients of processed meat products that contribute to textural attributes such as cohesiveness and hardness. HPP might be useful in restoring some of these properties that are lost with low-salt and low-phosphate meat products.[111,116,117] Sikes *et al.* demonstrated improved properties, including increased hardness and reduced cooking loss, of a 1% NaCl reduced-salt beef sausage batter pressurized at 200 MPa and 10 °C for 2 min.[118] Low pressure was also beneficial in reducing cooking loss and improving the rheological properties of 1–2% NaCl pork sausage.[119] A combination of pressure and heat (400 MPa at 70 °C for 30 min) decreased the weight loss and hardness of 1.5% NaCl and 1% NaCl + 0.3% tripolyphosphate pork batters.[120] The hardness and cutting force of chicken meat batters with egg protein (low fat, low sodium, phosphate free) treated at 700 MPa and 40 °C for 30 min were similar to those obtained with thermal treatment (75 °C for 30 min). However, these can be improved and increased with addition of transglutaminase before HPP treatment.[121] However, HPP at 600 MPa and 10 °C for 6 min had limited benefits on the sensory attributes of transglutaminase-restructured dry-cured hams with reduced salt content.[122] Increasing the pressure from 100 to 200 MPa at 4 °C for 30 min for the treatment of carrageenan-added restructured NaCl reduced-salt pork showed an improvement with increased binding strength. Furthermore, addition of glucono-δ-lactone (GDL) allowed the amount of carrageenan required to obtain palatable binding properties of restructured meat to be decreased.[123,124] Addition of carrot fiber or potato starch hydrocolloids with a 600 MPa treatment produced 1.2% reduced-salt pork sausages with water binding capacity, color, and texture similar to those of a 1.8% salt pork sausage.[125] O'Flynn *et al.*[126] took a different approach by investigating the use of pressurized pork meat processed at 150 MPa for 5 min and used the meat in the formulation of reduced-salt and phosphate breakfast sausages. They applied HPP to the meat as an ingredient, instead of applying it to the final product. Organoleptic, functional, and quality properties of the sausages were affected by the addition of pressurized meat and level of added salt (0.1, 1.0, 1.5, 2.0, or 2.5%). The effect of HPP at various stages of cooked ham production was compared using ham with reduced sodium chloride and potassium chloride replacement. Treatment at 100, 300, and 600 MPa applied to the raw meat or after its injection was detrimental to water retention of the cooked ham. However, a treatment at 100 MPa after tumbling had a positive impact on the quality attributes.[127] With the replacement of some sodium chloride and potassium chloride (0.2%), and a 100 MPa HPP treatment, 45% salt reduction was achieved without impairing drip loss, color, or cooking loss.[127] HPP has tremendous potential in processing ham at various production stages, with the use of a combination of alternative ingredients, to decrease significantly the salt concentration of salt-rich food products. There was an increase in saltiness perception with HPP that was related to the degree and nature of proteins, preprocess steps, quantity of fat, and formulation.[128] Gaining a better understanding of this increase in salt perception could help in the formulation of these food products.

7.4.1.2 Fish

Modulation of HPP process parameters applied alone or in combination with thermal treatment can produce fish fillets and restructured fish products with a wide variety of textural properties and improved microbial safety. Treatment of cod fillet at 400 MPa for 20 min increased hardness by 64% compared with its raw control, and by fivefold compared with cooked muscle, whereas pressure treatment before cooking did not modify the hardness of the product relative to the cooked meat.[129] It was concluded that heat-labile hydrogen bonding of proteins was induced by pressure. In contrast, thermal treatments favored the formation of a protein structure stabilized by disulfide and hydrophobic interactions. Lower pressure (<400 MPa) applied for 5 min, on the other hand, slightly decreased the hardness of sea bass fillet, whereas pressures up to 500 MPa had no significant impact. After 14 days of storage, the sea bass fillet treated at 500 MPa for 5 min had textural properties, such as chewiness and gumminess, that were similar to those of the control on the day of treatment. However, these parameters decreased significantly for the control during storage.[130] The use of three 5 min 200–300 MPa pressure pulses to treat salmon, tuna, and desalted bacalao (dried salt cod) carpaccios did not confer major advantages over the same treatment applied for 15 min.[131] Both treatments increased the shear strength of salmon and tuna carpaccios, but decreased that of desalted bacalao. Use of HPP for pretreatment of barramundi (*Lates calcarifer*) muscles at 150, 200, 250 and 300 MPa for 3 min at an initial temperature of 5 °C before freezing to −18 °C for up to 18 weeks showed that pressures below <200 MPa can be applied to increase hardness and delay lipid oxidation during frozen storage without conferring a cooked appearance on the pressurized fish. Similarly, HPP was beneficial for the production of mild smoked rainbow trout fillets (up to 600 MPa, 5 min, room temperature) with only slight sensory differences observed by a sensory panel, relative to the controls.[132]

Many studies have observed distinct mechanisms of protein aggregation for pressure- and heat-induced fish gels, leading to products with unique properties. After 20 days of refrigerated storage, the hardness and adhesiveness of cod fish sausages pressurized at 350 MPa at 7 °C for 15 min were significantly greater than those of the control on the day of treatment. Addition of chitosan promoted hardness and adhesiveness.[133] Minced albacore tuna muscle paste with improved hardness was obtained after treatment at 275 and 310 MPa for 2, 4, and 6 min. This was maintained during refrigerated and frozen storage for up to 22 and 93 days, respectively.[134] Heat-induced arrowtooth flounder gels prepared from fish paste pretreated at 400 and 600 MPa (for 1 and 5 min) were significantly harder and less fragile and had higher values of springiness, cohesiveness, and chewiness than heat-induced gels from unpressurized paste.[135] Tilapia gels with a wide variety of texture attributes were obtained by varying the combination of pressure (200 MPa for 60 min) and heat treatment (50 °C for 60 min). The potential of applying pressure and pressure combined with thermal treatment for the production of surimi gels has also been reported.[136,137] Reducing the salt content in surimi-type gels can be achieved without compromising their color or texture during storage by adding 0.1 g per 100 g of amino acid to a 300 MPa treatment. The development of off-flavor in the treated samples in some conditions can, however, limit their shelf life to 14 days.

7.4.1.3 Fruits and Vegetables

The food industry is facing a major challenge of addressing the growing consumer demand for safe fresh fruit and vegetable products prepared without chemical preservatives. Extending the shelf life of these products without the use of chemicals or decreasing product and nutritional quality and nutritional attributes has driven interest in non-traditional processing. HPP can be applied as a total replacement of conventional preservation methods or as a pretreatment method. Texture changes of fruits and vegetables involve enzymatic and non-enzymatic degradation of cell-wall polymers, particularly pectin. Texture changes of pressurized products result from a combination of both physical and enzymatic changes induced by the treatment. The following parameters impact fruit and vegetable texture and should be investigated when HPP is considered: product flexibility to withstand compression; cell membrane permeability changes; and starch gelatinization and cell-wall polymer changes.[138]

Pectin methylesterase (PME) and polygalacturonase (PG) are the two major pectin-degrading enzymes involved in textural changes. The effect of pressure treatment on PME and PG after treatment of cherry tomato at 600 MPa for 20 min, ranged from barotolerance, with no effect of pressure, to almost total inactivation.[139] Non-enzymatic changes that occur in pressurized fruits and vegetables include tissue deformations/disruption and cell permeability modifications, which have been observed in many fruits and vegetables.[93,138,140]

Perishable green photosynthetically active products such as lettuce might benefit from the use of HPP treatment to improve their food safety and increase their shelf life. However, a pressure level greater than 150 MPa irreversibly damaged the leaf tissue of lamb's lettuce, as indicated by the decline in photosynthetic efficiency obtained with chlorophyll fluorescence image analysis.[141] This result seems to indicate that pressure alone might not be a viable option for increasing the food safety of perishable green products, as limited microbial inactivation would certainly occur at this pressure level.

Treatment of pears at temperatures from 20 to 100 °C during treatment at 600 MPa for 1, 3, or 5 min illustrated the versatility of HPP.[142] Fresh-like pear products for immediate consumption could be obtained at 20 and 40 °C, while processing at 100 °C could be applied to produce pears for food service or as a food ingredient with a soft texture. These quality changes in treated pears correlated with residual PME, polyphenoloxidase, and peroxidase activities. The importance of the right selection of a fruit variety was demonstrated in the treatment of peaches by HPP under conditions of 500, 600, and 700 MPa for 1 and 5 min at room temperature. Treatment at 600 MPa did not cause a detrimental change in texture in two peach varieties. However, the variety having the lower initial polyphenoloxidase activity was recommended in order to be able to obtain total inactivation of the enzyme by the treatment, with less risk of browning of the pressurized pears.[143] Strawberries are very sensitive fruits with a short shelf life, and could therefore benefit from a processing technology to extend shelf life without quality loss. A treatment at 400 MPa for 5 min had less detrimental impact on the texture of strawberries compared with a thermal treatment at 75 °C for 20 min. The hardnesses of the pressure-processed and thermally processed strawberries were ~6 N and ~2 N, respectively. However, there was a significant decrease during subsequent storage, with 4 °C being more detrimental than 25 °C.[144]

7.4.1.4 Plant and Egg Products

Plant proteins, especially soy proteins, are increasingly used by the food industry to replace more expensive proteins from animal sources, and there has been a steady interest in further extending the use of soy proteins by using HPP treatment to modify their functional properties. The proteins β-conglycinin and glycinin are the two major soy proteins. Soy protein isolate (SPI) is at least 90% protein and a mixture of these two proteins is used extensively by the food industry for their gelling properties in restructured meat products. If SPI proteins are pretreated with HPP, their heat gelation is significantly modified, with a significant decrease in their G' value, a measure of gel stiffness (rigidity).[145] High-pressure pretreatment at 200 and 400 MPa promoted the formation of weaker heat-induced gels independent of the nature of the proteins. HPP favored the interactions between the proteins and calcium, leading to stronger SPI gels during heating, which improved product quality.

When soy protein gels were produced by placing 20% soy proteins or SPI dispersions under pressure, the hardness of the pressure-induced gels was lower than that of gels obtained with thermal treatment. Furthermore, an increase in pressure from 300 to 700 MPa did not greatly affect the gel texture.[146] Tofu is conventionally prepared from soymilk and has a lower protein content of about 5%, which is less than in the dispersions usually used for the formation of soy protein gel. Adding a coagulant, such as GDL or calcium sulphate, to soymilk before the pressure treatment led to the formation of a tofu directly due to the pressure.[147–150] The properties of these tofu gels differed from those of heat-induced gels, creating products with unique characteristics. The hydrolysis rate of GDL in water and in soymilk was increased under pressure and the coagulation time necessary to obtain tofu under pressure was significantly reduced relative to that in the conventional thermal method (unpublished personal data).

It was found that 30–60% w/w soy glycinin, when treated at 600 MPa, displayed extensive unfolding, whereas it maintained most of its native conformation when treated at the same pressure but at a solid level of 70 and 80% w/w. More information is needed to understand better the structural changes of this protein. It is a key ingredient in the development of nutritious food products such as smoothies and protein bars.[151]

Partial unfolding of egg albumen and yolk proteins induced by HPP produced more elastic gels than heat-induced gels. The yolk coagulated at a lower pressure of 400 MPa than the albumen, which coagulated at 600 MPa. The pressure threshold was dependent on the egg characteristics, temperature, and salt and sucrose contents. However, research on HPP of eggs is rather limited, but there have been egg processing studies under high pressure, high temperature conditions.[152]

7.4.1.5 Dairy Products

Commercially available pressurized dairy products are less developed relative to the meat and juice sectors. However, there is extensive literature on the effect of HPP on dairy proteins and products such as milk, yogurt, cheese, and whey.[153–156] The interest in HPP dairy products was primarily in improving product shelf life and microbial safety. Nevertheless, the impact of treatment of milk protein has also been extensively reported. Pure β-lactoglobulin and α-lactalbumin are baro-sensitive and resistant, respectively.[157] Modifications of dairy protein structures and interactions create opportunities to develop innovative products.

The potential of HPP has been studied as a pretreatment, a processing step, or a post-treatment intervention. Several studies have shown the benefits of pretreatment of milk with HPP prior to yogurt or cheese making.[158–160] Increased firmness of ewe milk yogurt was observed after treatment at 350 or 500 MPa and 10, 25, or 55 °C, whereas decreased firmness was obtained at lower pressure levels compared with yogurt prepared from heat-treated skim milk.[161] Increased firmness occurred with increase in pressure level for acid-set gels prepared from pressurized milk, which was accompanied by an increased syneresis resistance at pressures greater than 200 MPa.[162] Alternatively, although the treatment of milk fortified with whey proteins at 600 MPa for 15 min increased the rheological attributes G' and G'' of the corresponding yogurt, there was decreased firmness and cohesiveness of the gels compared with a yogurt prepared from heat-treated milk.[163] Pretreatment of milk at 586 MPa (three 1 min cycles) for Cheddar cheese production could shorten the coagulation time of rennet from 20 to 12 min.[157] However, the texture of this cheese was rated lower than that of traditionally prepared cheese. Both the pressure level and temperature applied to the milk affected the rheological properties of Cheddar cheese, with treatment at 483 MPa and 10 °C leading to a more solid-like cheese.[159] Goat cheese prepared from 500 MPa-treated milk was firmer on the first day of storage at 14 °C than cheese prepared from pasteurized milk, but with no significant difference after 30–60 days of ripening.[164]

Product modifications induced by HPP can also be used to produce directly a cheese with unique properties, without further processing. For example, a cheese curd could be obtained directly under pressure from skim milk when previously acidified with GDL. Aggregates structures formed from GDL-added skim milk under pressure ranged from floccules after treatment at 100 MPa for 40 min to coarse floccules after treatment at 200 MPa for 10 min and fine-stranded coagulum after treatment at 200 MPa for 20 min.[165] However, higher pressure levels, which would promote protein denaturation, were not investigated.

Whey protein isolate (WPI) and whey protein concentrate (WPC) gels with unique properties can be produced directly by pressure. The minimum pressure level required is usually above 300 MPa. The properties of the gel produced by pressure processing are affected by the nature of the protein, protein concentration, pH, presence of salt, pressure, temperature, holding time, and rate of depressurization, and are usually softer than thermally induced gels.[166–172] For example, to produce a gel with 10% WPI, a pressure of 600 MPa was required, whereas to obtain a gel with >18% WPC, a pressure of 400 MPa was needed.[172] When a pressure of 800 MPa was applied to 20% WPI and WPC solutions, the resulting WPI gel had a porous network structure like that of a honeycomb, whereas the WPC gel had an irregular granular network like that of coral.[172] The rate of depressurization can also affect the texture of pressure-induced gels, depending on the type of protein pressurized. By varying the pressure release rate from 20 to 600 MPa min^{-1}, Fertsch *et al.* observed a significant difference in the firmness of pressure-induced casein gels whereas no change was observed for the whey protein gel.[170] These changes were attributed to the difference in the chemical bonding involved in the gels, namely mainly non-covalent bonding in casein gels and covalent disulfide bonds in whey protein gels.

HPP has also been directly applied to dairy products to improve processing steps such as brining and ripening of cheese making and the product functionality.[155] While the treatment's potential to accelerate the brining step seems limited,[156,173,174] conflicting evidence has been reported regarding the effect of HPP on accelerating cheese ripening.[155] The difference could be attributed to variations in the processing condition and cheese properties such as age at time of treatment. Rynne *et al.*[175] demonstrated that treatment of a 1-day-old full-fat cheddar cheese at 400 MPa at room temperature for 10 min has limited value in accelerating cheese ripening. Nevertheless, aminopeptidase activity was reported to increase by a factor of three after a 20 min treatment at 200 MPa and 20 °C. This could lead to a faster ripening of white cheese, although no data were provided to verify this effect.[176] Cheddar cheese treated at 400 MPa for 10 min maintained its firmness, whereas its fracture stress and fracture strain were increased, which was attributed to the impact of the pH increase induced by the treatment.[175] Softening of immature mozzarella cheese and half-fat Cheddar cheese was obtained after treatment at 200 MPa for 60 and 120 min, respectively. Therefore, the treatment was beneficial in shortening the maturation time and in compensating for the fat reduction of the immature mozzarella and half-fat Cheddar cheese, respectively.[177,178] La Serena cheese treated at 300 or 400 MPa and 10 °C for 10 min after 2 days of ripening had a firmer texture after 60 days of ripening than the corresponding control. This was not a desired attribute for this cheese variety, as shown by the lower sensory panel score.[179] Treatment after 50 days of ripening did not affect the cheese, determined after 60 days of ripening, but did improve microbial safety. Ripening time also affected the rheological and textural properties of Gouda cheese treated at 400 MPa and 14 °C for 1 h.[180] The treatment decreased the hardness and increased gumminess, as shown by the texture profile analysis of the pressurized cheese on the day of pressure treatment, but there was no significant difference after 42 days of ripening.

HPP treatment increased the viscosity of fruit yogurt at an initial temperature of 10 °C and then treated at 550 MPa for 10 min.[181] Increased viscosity was also reported for yogurt treated at pressures from 250 to 400 MPa at 20 °C for 15 min. Viscosity was maintained after storage at 4 °C for 20 days.[182] Treatment at 400 MPa, however, impaired the sensory characteristics of the yogurt, which had no flavor and underwent syneresis.

To conclude, HPP can offer opportunities not only in terms of manufacturing structured dairy products with unique sensorial properties,[183] but also in terms of increased food safety.[184] However, treatment of dairy products by HPP is still in its infancy despite the tremendous opportunities that this technology provides (see Section 7.6, Commercial Applications of HPP). In the past, the technical limitations of the technology were one of the major problems, but with the tremendous advances in recent years, the dairy food sector and associated niche market could soon become one of the fastest growing markets using HPP.

7.4.2 Functional Properties

The functionality of whey, soy, and egg proteins as food emulsifiers and foaming agents and in enhancing rheological properties is industrially important. The property changes that occur in pressure-treated proteins differ from thermally-induced changes because of the different effects of HPP on protein structure and interactions between food components.

The functionality of whey protein isolate, concentrate, and skim milk can be modulated with HPP. These changes are due to the effect of the treatment on the molecular structures and interactions of key milk and whey proteins.[185–189] Improvements in WPC and WPI foaming stability were obtained with HPP, and increased from 2.24 min for the WPI control to 11.35 min after a 10 min treatment at 600 MPa.[185] The treatment had a different effect on their foaming properties. After a 600 MPa treatment, foaming was improved and reduced for WPI and WPC, respectively. When hen egg yolk low-density lipoproteins (LDLs) were pressurized at 200, 400, and 600 MPa and 20 °C for 10 min, the solubility of the dispersions was maintained, whereas the viscosity at a low shear stress of emulsions made from these dispersions, when treated at pH 8.0, was dramatically reduced after treatment at 600 MPa.[190] The foaming properties of egg white proteins treated at pressures from 400 to 700 MPa at 10–60 °C for 20 min were dependent on the pH during treatment. It ranged from an improved foaming ability, when treatment resulted in a high level of protein unfolding with residual protein solubility, to increased foam stability and density for a high level of protein unfolding accompanied by limited residual protein solubility.[191] The effect of pressures ranging from 100 to 400 MPa for 30 min on the rheological properties of whole liquid egg, albumen, and yolk was reported by Ahmed *et al.*[192]

The stability of an emulsion prepared from pressurized soymilk (500 and 600 MPa at 25 °C for 10 min) was significantly increased relative to that of an emulsion prepared from untreated soymilk. The viscosity of the treated soymilk can be maintained at a level similar to that of the control with pH adjustment.[193] Pressure treatment also favored interactions between soymilk components that decreased the sedimentation of particles, therefore increasing the physical stability of pressurized soymilk. This increase was maintained for up to 28 days of refrigerated storage.[77] The emulsifying properties of glycinin, β-conglycinin, SPI, and lupin meals were affected by pressure treatment, pH, and dispersion concentration, which affected the extent of changes due to the treatment.[194–196]

7.4.3 Color

The effect of HPP on the color of food products varies from none to extensive, depending on the product being treated and the processing conditions. One of the major problems with meat treatment to improve food safety and shelf life is the detrimental effect of pressure on raw meat color, which is an important criterion for retail purchasers. Whereas pressure processing can affect the appearance of meat, the color change is minimal or non-existent if the meat is thermally cooked prior to pressure treatment.[112]

The red color of meat is due to the presence of the heme protein myoglobin. The underlying molecular processes of myoglobin changes in pressurized pork meat were studied by resonance Raman spectroscopy. This is a particularly suitable technique as it can selectively probe the vibrational spectrum of the myoglobin heme, while the remainder of the protein and meat matrix does not contribute to the spectrum.[197] It was reported that HPP caused the formation of metmyoglobin from oxymyoglobin and that the treatment of meat with a low oxymyoglobin-to-deoxymyoglobin ratio may have advantages, including the fact that pressure treatment of deoxymyoglobin did not change the oxidation state in intact meat tissue.

Color changes due to pressure processing have also been reported for fish and shellfish.[83] The total color difference (ΔE) of oysters increased from 2.25 to 4.59 and 5.89 after treatment at 260 MPa (3 min), 500 MPa (15 min), and 800 MPa (5 min), respectively.[82] These changes were primarily due to an increase and decrease in the sample's lightness and redness value (*i.e.* increasing green color), respectively. Increases in L^* values (lightness), which decreased the transparency of the sample, often giving it a cooked appearance, were also reported for pressurized minced tuna muscle, sea bass fillets, and cold-smoked salmon.[198] These changes in color were usually maintained during refrigerated storage. Increases in L^* values were often related to protein aggregation. The extent of color changes was related to the fish species. For example, after treatment at 300 MPa and 7 °C for 15 min, the ΔE value of desalted bacalao carpaccio was 4.70 compared with 17.96 and 20.96 for carpaccios of salmon and tuna, respectively,[131] the L^* values being the parameter most affected. Heat-induced gels prepared from pressurized treatments at 400 or 600 MPa for 1 or 5 min of arrowtooth flounder displayed color changes relative to untreated samples. This was a function of the cooking conditions, suggesting that the color difference of these fish gels due to pressure were maintained after thermal treatment.[135]

Changes in the L^* values of pressurized milk were reported, with treatment usually decreasing L^*. This decrease in pressurized milk lightness was attributed to the disintegration of the casein micelles into fragments of smaller particle size, which increased the milk translucency.[154,155] Skim milk fortified with whey proteins treated at 85 °C for 20 min had an L^* value of 84.73 compared with 73.63 after treatment at 600 MPa for 15 min.[163] When thermal treatment of milk at 72 °C for 15 s prior to cheese formation was replaced with a pressure treatment at 500 MPa and 20 °C for 15 min, the resulting cheeses had different L^*, a^*, and b^* values. The ΔE value calculated based on raw milk was higher for cheese prepared from pasteurized milk than pressure-treated milk, namely 2.02 and 0.97, respectively, after 30 days of ripening.[164] The surface color of pressurized fresh goats' milk cheese processed at 500 MPa and 10 and 25 °C for 5, 15, and 30 min resulted in low ΔE values, varying between 0.47 and 1.75, b^* yellowness values being the parameter affected the most.[199] An increase in yellowness, from 27 for the untreated sample to 35, was also observed after treatment of immature mozzarella cheese at 200 MPa for 60 min.[177] Yellower and greener 1-day-old full-fat Cheddar cheese as a result of a treatment at 400 MPa for 10 min was observed by Rynne *et al.*[175]

Cano and de Ancos reported that HPP, unlike conventional thermal sterilization treatment, maintains the flavor and color of pressurized fruit jams.[200] Similar observations were made for strawberry and blackberry purées pressurized up to 600 MPa at 20 °C for 15 min. The color retention was greater in pressure-treated samples than in thermally processed samples, this being more important for the strawberry purée.[90] The impact of pressure on the color of strawberry jam pressurized at 400 MPa and 25 °C for 5 min varied with pectin concentration (0.1–10% w/w) and was related to the effect of treatment on anthocyanin degradation.[201] The color of mushrooms after treatment at pressures as high as 950 MPa for 5 min was similar to that of blanched mushrooms, if they were evacuated before treatment, *i.e.* air was replaced with water with or without citric acid and sodium sulphite. Treatment applied without prior evacuation resulted in dark-brown mushrooms due to browning enzyme and substrate release and subsequent oxidation.[202] An increase in pressure from 400 to 600 MPa had minor effects on the color parameters of carrot and tomato purée.[90]

7.4.4 Flavor

Flavor changes of pressurized food products are undesirable. The fruit and juice flavor profile is often similar to that of a fresh product. However, in some cases flavor is modified, leading to beneficial or detrimental effects. In some cases, increased lipid oxidation can result in off-flavor.[203–205] The acceleration of meat lipid oxidation was dependent on the type of meat, storage conditions, and pressure level, the critical pressure level being between 300 and 600 MPa. For example, treatment at 500 MPa for 5 or 10 min was identified as the critical pressure level for lipid oxidation for the treatment of chicken breast muscle.[206] As meat is often industrially treated at 600 MPa for microbial inactivation, lipid oxidation needs to be considered. Strategies to delay lipid oxidation have been identified and include the use of radical scavengers such as butylated hydroxyanisole (BHA) and butylated hydroxytoluene (BHT), and also citrate and ethylenediaminetetraacetic acid (EDTA), which are metal chelators. However, the use of these chemical additives is counter to the green image often associated with HPP. Therefore, the potential of natural antioxidants from herbs and spice has been investigated. Rosemary extract, which contains compounds having metal-chelating properties, had a protective effect against pressure-induced lipid oxidation (600 MPa for 10 min at an initial temperature of 10 °C) in cooked chicken (95 °C for 1 h).[207] Garlic, on the other hand, seems to have a pro-oxidant effect when added to processed chicken meat.[208]

The generation of volatile compounds in milk when subjected to pressures of 482, 586, and 620 MPa at 25 and 60 °C for 1, 3, and 5 min was compared with those obtained with thermal treatments at 25, 60, and 80 °C for 1, 3, and 5 min.[209] Low-temperature HPP caused minimal changes to the volatile composition of milk, but the generation of volatile compounds in pressure-heated milk differed from that in samples that had only been heated. The flavor binding and flavor release of the four flavor compounds benzaldehyde, heptanone, octanone, and nonanone in whey protein concentrate treated at 600 MPa and 50 °C for holding times of 0, 10, or 30 min revealed the importance of the processing conditions and nature of the compounds on modifications of these parameters.[210]

The volatile components of guava juice treated at 600 MPa and 25 °C for 15 min were similar to those of the fresh guava juice, suggesting that pressurized treatment preserved the original juice flavor.[211] During storage at 4 °C, the volatile flavor components of the pressurized samples were very similar to those of the untreated samples for up to 30 days. For longer storage at 4 °C, or storage at 25 °C, the volatile compounds were affected by residual enzyme activity. When subjected to treatment at pressures up to 500 MPa and 20 °C for 20 min, strawberry aromatic volatile compounds were maintained, but an increase in pressure to 800 MPa led to the appearance of new compounds.[212] Zabetakis *et al.* observed the importance of the pressure level and storage temperature at 4, 20, and 30 °C for the modifications induced by the treatment by following the effect of treatment at 200–800 MPa on three acid and three furanone strawberry flavor compounds.[213]

7.4.5 Allergenicity/Antigenicity

Many proteins, although contributing to the nutritional and functional qualities of food products, can cause allergies. According to the US Food and Drug Administration (FDA), the eight most common allergenic foods are milk, eggs, fish, crustacean shellfish, tree nuts, peanuts, wheat, and soybeans. About 150 food allergen-related deaths are reported annually in the USA.[48] Understanding the impact of food processing on allergen structure and allergenic potential is vital, but investigations are still in their infancy.[214] Increased protease digestibility of proteins pressurized prior to or during enzymatic treatment has been reported in soybean proteins, whey proteins, β-lactoglobulin, and ovalbumin.[215–220] The activity of proteolytic enzymes can be enhanced by treatment because of the effect of HPP on protein unfolding and structure. The reduced antigenicity of pressurized whey protein hydrolyzates that could be relevant for the development of hypoallergenic infant formulae was reported by Penas *et al.*[221] HPP also produced short segments of IgE-binding regions of β-lactoglobulin, which might result in a reduced allergenicity of the protein.[217] Chicon *et al.* observed an increase in hydrolysis of β-lactoglobulin by pepsin after treatment at 400 MPa of whey protein isolate, but this increase was lost during refrigerated storage.[222] In addition, the peptic digests of the HPP samples did not lead to a decrease in reactivity against IgE from human sera of allergic patients. Red kidney bean protein isolate processed at up to 600 MPa and 25 °C for 20 min displayed a significantly modified protein structure. However, the structural changes did not improve *in vitro* trypsin digestibility, which was attributed to protein aggregation.[223] Soaking rice in water at 300 MPa increased the release of allergenic proteins. When protease was added to the treatment, allergenic proteins almost completely disappeared from the rice.[224] The allergenicity of the main allergen in peach, Pru p 3, was determined not only *in vitro* but also *in vivo* in a study by Lavilla *et al.*[225] The results demonstrated that the skin response with pressurized peach extract was stronger for more than half of the 17 patients tested, showing that HPP increased the protein allergenicity.

7.5 Pressure-assisted Extraction of Food Components

Recently, there has been increased interest in HPP as a tool to extract biologically active compounds from foods. This technique is sometimes referred to as high-pressure extraction (HPE). Compound modifications could be caused by increased permeability of pressurized cells due to mass transfer changes, enhanced solubility of ingredients, and changes in the structure of the food matrix.[226] Parameters that have been most studied to improve the extraction yield include the nature of the extraction solvent, the pressure level, the dwell time, and the solvent-to-material ratio.[226–229]

Caffeine extraction from green tea leaves in 50% ethanol was increased 14.8-fold after treatment at 500 MPa for 1 min, leading to a yield similar to that obtained with heat reflux extraction and ultrasonic extraction, both of which required much longer processing times.[229] Similarly, the extraction yields of flavonoids from propolis, a resinous material produced by bees,[228] and lychee fruit pericarp,[230] salidroside from *Rhodiola sachalinensis*, and ginsenosides were higher than with more conventional energy-consuming methods.[231–233] Prasad and co-workers also reported the benefits of using HPE to increase the extraction yield of bioactive compounds from longan fruit pericarp, which had higher antioxidant and anticancer activities than the extract obtained with conventional extraction.[234,235] For grape by-products, in addition to enhancing the total phenolic content twofold and the antioxidant capacity of the extract threefold, treatment at 600 MPa and 70 °C for 1 h, in 50% ethanol, led to the selective extraction of the acylated anthocyanin monoglucosides.[236] Treatment of onions at 100–400 MPa and

5 °C for 5 min increased the extractability of quercetin-4′-glucoside, total quercetin, and quercetin-3,4′-diglucoside by 33, 26, and 17%, respectively.[237] The extraction of total carotenoids (extracted with tetrahydrofuran) from tomato purée was unchanged after treatment at pressures up to 600 MPa and 20 °C for 60 min, but readily extractable carotenoids (extracted with petroleum ether) were reduced by about 80%. Extraction with heat-treated tomato purée was also significantly reduced with this solvent.[238] High-pressure treatment of soybeans at 100–700 MPa and 25 °C for 10 min prior to water extraction did not improve isoflavone recovery.[94] Similarly, extraction of gelatin from fish skins was not increased by HPP, but the processing time was decreased and gelatin with unique properties was obtained.[239] Aqueous oil and protein extraction was significantly reduced after treatment of soybean flakes at 500 MPa and 25 °C initial temperature for 15 min, owing to the aggregation and precipitation of the proteins.[240]

Although HPP is not currently been used at the industrial level to extract food compounds, the increasing interest of the food industry in more sustainable and more environmentally friendly processes might open up new opportunities for HPP as a tool to enhance the extractability of compounds without the use of solvents and add value to agricultural and food waste.

7.6 Commercial Applications of HPP

The first commercialization of HPP products, a variety of fruit desserts, occurred in 1990 in Japan, but the number of HPP facilities started to increase significantly after 2000.[241] Initially, many aspects of the process were still unknown, the equipment was often at a prototype stage and did not always meet US federal food standards, and productivity was low. Only a few innovative companies that benefited from HPP (safety, shelf life, freshness), invested in the technology. Over the years, the design, reliability, and technical parameters of HPP industrial equipment improved, and the interest of the food industry in the technology increased. The technology addresses the consumer's interest in fresh, refrigerated, minimally processed food, such as "cold pressed" juices and RTE products.[242] Because these parameters occurred simultaneously (improved equipment, industry's interest, consumers' requests), the number of HPP machines in production has steadily increased to reach around 50 per year in the last 3 years (Figure 7.1). Most of these industrial HPP machines are manufactured by the two largest HPP equipment vendors, Hiperbaric (Spain) and Avure (USA). Table 7.1 summarizes some companies manufacturing industrial-scale HPP equipment along with those manufacturing laboratory scale HPP vessels. A list of the main academic centers using HPP and some video links explaining HPP technology are also provided.

From the day the first HPP equipment was installed to the end of 2016 more than 400 HPP industrial-size machines, from 35 to 525 L, have been installed. About 25 machines installed before the millennium have been dismantled, leaving 385 industrial HPP machines in use around the world,

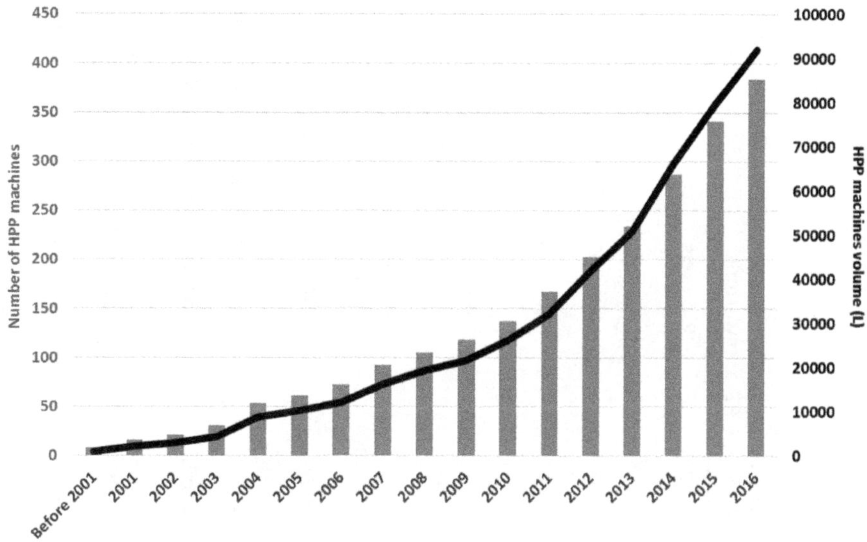

Figure 7.1 Accumulated number of HPP industrial machines in production (column) and their corresponding accumulated volume (black line).

representing a total volume of more than 90 000 L (Figure 7.1). Around 25 of these machines are used for R&D (in technical centers, universities, or industrial pilot plants), so around 360 HPP machines, owned by around 200 companies worldwide, are in use for the production of commercial HPP food products (Figure 7.2).

The 200 companies owning HPP machines vary from large international groups to very small enterprises with less than 10 employees (Table 7.2). More than 300 additional companies in the world use HPP technology without owning HPP equipment; they use an HPP tolling company, which provides access to HPP equipment, or HPP co-processing or tolling companies offer access to commercial HPP machines to food processors on a "pay as you go" basis (Table 7.2). Customers of these companies pay a fee per cycle or, for large quantities of products, a price per unit or per quantity (kg or lb) can be negotiated. The tolling facility only processes products for other companies whereas co-processing companies have their own pressurized products on the market. They have one or several machines with their own production and have their own brands, and they pressurize products for companies usually located nearby, to amortize the equipment's cost more quickly.

The Japanese were pioneers in commercializing pressurized food products in the 1990s, followed by France and Spain. Now the USA, Canada, and Mexico have about 50% of the worldwide HPP equipment (about 200 machines) and feature the most advanced commercialize pressurized food processing equipment. The USA has about 35% of the total number of industrial HPP machines, which are mainly used for processing juices and beverages, RTE meat products, and dips. About 9% of the HPP industrial

Table 7.1 Companies manufacturing high-pressure equipment for the food and biosciences sectors, examples of laboratories and technical centers providing high-pressure technology services, and links to videos of HPP applications.

Company/institution	Website	Country	Products
HPP manufacturers			
Avure Technologies	http://www.avure-hpp-foods.com/	USA	Industrial scale
Bao Tou Kefa	http://www.btkf.com/Item/list.asp?id=62	China	Laboratory and industrial scale
Kobelco	http://www.kobelco.co.jp/	Japan	Laboratory and industrial scale
Hiperbaric	http://www.hiperbaric.com/	Spain	Industrial-scale equipment
Resato International	http://www.resato.com/	The Netherlands	Laboratory-scale equipment
Sanshui Yinhe	http://en.sxssyh.com/	China	Laboratory and industrial scale
Stansted Fluid Power (SFP).	https://www.highpressurefoodprocessor.com/	UK	Laboratory-scale equipment
UHDE High Pressure Technologies	http://www.uhde-hpt.com/	Germany	Laboratory and industrial scale
Unipress	http://www.unipress.waw.pl/	Poland	Laboratory-scale equipment
Academic			
Agri-Food and Biosciences Institute (AFBI)	https://www.afbini.gov.uk/articles/high-pressure-processing	UK	Product testing and evaluation
Alberta Food Processing Center	http://www1.agric.gov.ab.ca/$department/deptdocs.nsf/all/fpdc12917	Canada	Product testing and evaluation
Anfaco	http://www.anfaco.es/	Spain	Product testing and evaluation
AZTI (Marine and Food Technological Centre)	http://www.azti.es/	Spain	Product testing and evaluation
Canadian Centre for Fisheries Innovation (CCFI)	http://www.ccfi.ca/	Canada	Product testing and evaluation
CDBQ	http://www.cdbq.net/	Canada	Product testing and evaluation

Centa IRTA	http://www.irta.cat/en-us/RIT/Centres/pages/CENTA.aspx	Spain	Product testing and evaluation
Cornell University	https://foodscience.cals.cornell.edu/about-us/facilities/geneva-facilities/hpp-validation-lab	USA	Validations studies
DIL (German Institute of Food Technologies)	http://www.dil-ev.de	Germany	Product testing and evaluation
Food Science Australia	https://www.csiro.au/	Australia	Product testing and evaluation
Institute for Food Safety and Health (IFSH)	https://www.ifsh.iit.edu/	USA	Product testing and validation studies
INTAEX	http://cicytex.juntaex.es/es/centros/intaex	Spain	Product testing and evaluation
Itacyl	http://www.itacyl.es	Spain	Product testing and evaluation
NARO-NFRI	http://www.naro.affrc.go.jp/english/nfri/organization/06food_engineering_division/08/index.html	Japan	Product testing and evaluation
Ohio State University	https://fst.osu.edu/research/facilities#Pressure	USA	Product testing and evaluation
Oniris	http://www.oniris-nantes.fr/	France	Product testing and evaluation
TEI Larrissa	https://www.teilar.gr/index_en.php	Greece	Product testing and evaluation
TU Berlin	http://www.tu-berlin.de	Germany	Product testing and evaluation
University of Nebraska	http://fpc.unl.edu/pilot_plants	USA	Product testing and evaluation
University of Aveiro	http://www.ua.pt/ptaltapressao/	Portugal	Product testing and evaluation
University of Burgundy-Welience	http://www.welience.com/fr/agroalimentaire-bioindustriel/plateformes/hall-technologie-alimentaire	France	Product testing and evaluation
Virginia Tech	http://www.fst.vt.edu/facilities/high-pressure-processing.html	USA	Product testing and evaluation
VUPP (Food Research Institute Prague)	https://www.vupp.cz/	Czech Republic	Product testing and evaluation
Videos on HPP industrial processing available online			
Hiperbaric users	https://www.youtube.com/user/HiperbaricHPP		
Avure users	https://www.youtube.com/user/AvureTV		

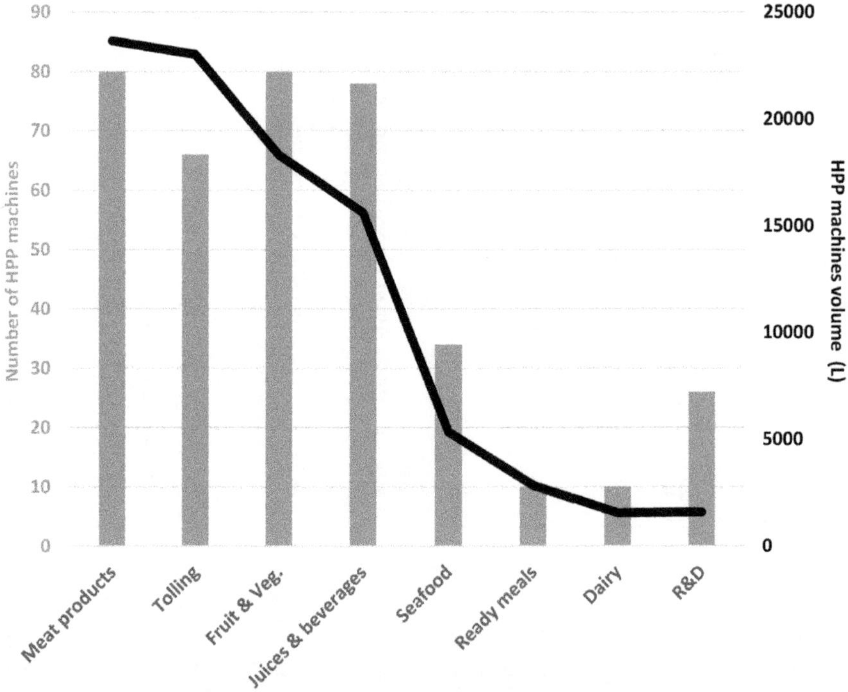

Figure 7.2 Distribution of HPP industrial size machines (columns) and associated volume (black line) *versus* food sectors or activities for 2016.

equipment is located in Mexico, for the production of pressurized avocado (guacamole). Mexico is the leading country for the manufacture of pressurized guacamole, which is mostly produced for export all over the world. About 6% of the HPP industrial equipment is in Canada, where the main use of the technology is for RTE meat products, mainly because HPP is recommended by the United States Department of Agriculture (USDA) as an efficient technology for the control of *L. monocytogenes* in RTE meat and poultry products.[243] The technology helps Canadian companies maintain strict control of food microbial pathogens in order to export to their principal export market, the USA. Europe has about 22% of the HPP machines, namely about 85 units, which are mainly used in Mediterranean countries such as Spain, Italy, France, Portugal, Greece, Cyprus, and Lebanon, but also in The Netherlands, UK, Ireland, Germany, Belgium, Switzerland, Czech Republic, Hungary, Rumania, Estonia, and Finland. After initial pioneering efforts in France and Spain in the 1990s, HPP has had difficulties in gaining acceptance in Europe, mainly because of a novel food regulation, EC 258/97, published in January 1997. This regulation described HPP as an example of a "novel process" that could modify the characteristics of processed products. This means that this regulation requires each user of a novel technology to provide evidence that food application was microbiologically and toxicologically safe and that the process had no detrimental effect on the

Table 7.2 Summary of some food companies using HPP technology and applications.

Food sector	Company	Website	Country	Products
Dairy	Rodilla	http://www.rodilla.com/	Spain	Cheese and mayonnaise based fillings for sandwiches
	Made by Cow	www.madebycow.com.au/	Australia	Raw cow milk
	Villa de Patos	http://villadepatos.com/	Mexico	Raw cow milk
Juices and beverages	Beskyd	http://www.beskyd.cz/vyrobky/44-zeleninove-stavy-refit.html	Czech Republic	Fruit and vegetable juices
	Bolthouse	http://www.bolthouse.com/	USA	Fruit and vegetable juices
	Blue Print	http://blueprint.com/	USA	Vegetable juices and cleansers
	ColdPress	http://www.cold-press.com/	UK	Apple juices, smoothies, and sparkling fruit water
	Cruzoe	http://www.cruzoe.fr	France	Fruit and vegetable juices
	Daily Greens	https://drinkdailygreens.com/	USA	Fruit and vegetable juices
	Evolution Fresh	http://www.evolutionfresh.com/	USA	Fruit and vegetable juices and smoothies, lemonades
	FreshBev	https://www.drinkripe.com/	USA	Juices
	Frubaça	http://www.copa.pt/	Portugal	Fruit and vegetable juices and fruit purées
	Fruitapeel	http://fruitapeel.co.uk/	UK	Juices and smoothies
	FruityLine	http://fruity-line.nl/	The Netherlands	Juices and smoothies
	Go Fruselva	http://www.fruselva.com/	Spain	Fruit and vegetable juices and fruit purées
	Hoogesteger	http://www.hoogesteger.nl/	The Netherlands	Fruit and vegetable juices and nut milks
	Inside Out	http://www.insideoutng.com.au/	Australia	Non-dairy milks
	Juste Pressé	http://justepresse.fr/	France	Fruit and vegetable juices
	Malk	http://malkorganics.com/	USA	Nut milks and coffee
	Invo	https://www.invococonutwater.com/	USA	Coconut water
	PDM – B'Fresh	http://www.b-fresh.co.uk/	UK	Juices and smoothies
	Preshafood	http://preshafood.com.au/	Australia	Fruit purées and juices

Table 7.2 (*Continued*)

Food sector	Company	Website	Country	Products
	Savsé	https://savse.com/	UK	Fruit and vegetable smoothies, baby foods
	Suja	https://www.sujajuice.com/	USA	Juices, waters, and drinking vinegars
	Secret Squirrel	http://www.squirrelbrew.com/	USA	Cold-brew coffee
	Temple Turmeric	http://templeturmeric.com/	USA	Juices and drinking vinegars
	Tio Gazpacho	https://www.tiogazpacho.com/	USA	Cold soups
	Tea Riot	http://teariot.com	USA	Teas with juices
	Ugo Juices (Kofola)	http://ugo.cz/	Czech Republic	Vegetable and fruit juices
	Ulti (Granini)	http://www.ulti.fr/	France	Juices and smoothies
	WTRMLN	https://wtrmlnwtr.com/	USA	Watermelon juice
Fruits and vegetables	Aspasia	https://www.aspasia-france.com/	France	Vegetable dips
	Born Pure	https://www.bornpure.com.au	Australia	Baby foods
	Calavo Growers	http://www.calavo.com/	USA	Guacamole
	Cedarlane	http://www.cedarlanefoods.com/	USA	Hummus, dips, and wet salads
	Echigo Seika	http://www.echigoseika.co.jp	Japan	RTE rice and cereals
	Fresherized Foods	https://www.eatwholly.com/	USA	Guacamole
	Garden Fresh Salsa	http://www.gardenfreshsalsa.com/	USA	Hummus, dips, and salsas
	Good Foods	http://goodfoods.com/	USA	Guacamole, hummus, and salads
	Hope	https://hopefoods.com/	USA	Hummus
	Once Upon a Farm	https://uponafarm.com/	USA	Baby foods
	Pure Spoon	https://purespoon.com	USA	Baby foods
	Sandridge	http://www.sandridge.com/	USA	Wet salads, soups, RTE meals
	Simplot Foods	http://www.simplotfoods.com	USA	Avocado products
	SimplyFresco	http://www.simplyfresco.com	USA	Pasta sauces and salsas
	Verfruco	http://www.verfruco.com/	Mexico	Avocado products
Meat	Applegate Farms	http://applegate.com/	USA	Organic RTE meat products
	Abraham (Bell Foods Group)	https://www.anainc.us/	Germany	Smoked and cured hams

	Campofrio	http://www.campofrio.es/	Spain	Dry-cured meats
	Cargill	http://fressure.com	USA	Raw ground beef patties
	Creta Farms	http://www.cretafarm.gr/	Greece	Sliced deli meats
	Espuña	http://www.espuna.es/es/innovacion/alta-presion.aspx	Spain	Sliced ham and tapas
	Ferrarini	http://www.ferrarini.com/	Italy	Dry-cured meat products
	Foster Farms	http://www.fosterfarms.com/	USA	Sliced poultry products
	Fresh Pet	https://freshpet.com/	USA	Pet food
	GoodHeart	http://goodheart.com/	USA	RTE meat products
	Hormel	https://www.makethenaturalchoice.com/	USA	Sliced deli meats, RTE meats
	Infantis	http://www.freshpress.gr/	Greece	Sliced deli meats
	Jamcal	http://www.jamcalalimentacion.es/	Spain	Dry-cured meats
	Jennie-O	http://www.jennieofoodservice.com	USA	RTE turkey products
	Maple Lodge	http://www.maplelodgefarms.com	Canada	Sliced deli meats, RTE meats
	Moira Mac's	http://www.moiramacs.com.au/	Australia	RTE poultry meat products
	MRM	http://www.mrm.es/	Spain	RTE meat meals and sliced meats
	Nature Variety's	http://www.naturesvariety.com/	USA	Pet food
	Organic Prairie	http://www.organicprairie.coop/	USA	Sliced deli meats
	Oscar Meyer (Kraft)	http://www.oscarmayer.com/en/natural	USA	Oscar Mayer sliced deli meats
	Perdue Farms	http://www.perdue.com/	USA	RTE poultry products
	Primal Pet Foods	http://www.primalpetfoods.com/	USA	Pet food
	Sadler's	http://www.sadlerssmokehouse.com/	USA	Smoked meats
	Santa Maria	http://www.santamariafoods.ca/	Canada	Sliced cured and cooked meats
	Sofina	http://www.sofinafoods.com/	Canada	RTE meat products
	Stella and Chewys	https://www.stellaandchewys.com/	USA	Pet food
	Wellshire	http://www.wellshirefarms.com/	USA	Sliced meats
	West Liberty Foods	http://www.wlfoods.com/	USA	RTE meat products
	Zwanenberg	http://www.zwanenberg.nl/en/about-us/innovation/	The Netherlands	Raw beef products
Seafood	Cinq Degrés Ouest	http://www.5do.fr/	France	Raw lobster and clams
	Clearwater Seafoods	http://www.clearwater.ca/	Canada	Frozen raw lobster
	Delpierre		France	Fresh fish fillets

Table 7.2 (*Continued*)

Food sector	Company	Website	Country	Products
		http://www.delpierre.com/nos-produits/rayons/poissons-natures.php		
	Ghezzi	http://www.ghezziitalia.com	Italia	Ready-to-cook salted cod and stockfish
	Mitsunori	http://www.mitunori.co.jp/	Japan	Clams, whelks
	Motivatit Seafoods	http://www.motivatit.com/	USA	Oysters
	Goose Point	http://www.goosepoint.com/	USA	Oysters
	Prestige Oyster	http://www.prestigeoysters.com/home	USA	Oysters
	Riverside Lobster	http://riversidelobster.ca/	Canada	Raw lobster
	Shediac Lobster	http://www.shediaclobster.ca/	Canada	Raw lobster
	Shucks Maine Lobster	http://www.shucksmaine.com/	USA	Raw lobster
	Westmorland Fisheries	http://www.westmorlandfisheries.ca/	Canada	Raw lobster
Tool processing	American Pasteurization Company	http://www.americanpasteurizationcompany.com	USA	HPP
	Ameriqual	http://www.ameriqual.com/hpp/	USA	HPP
	APA Processing	http://apaprocessing.com/	Spain	HPP and juice bottling
	Deli 24	http://www.deli24.co.uk/	UK	HPP
	Eddy Packing	http://www.eddypacking.com	USA	HPP and meat product co-packing
	Fresh Advantage	http://www.fresherpressure.com/	USA	HPP
	Green Plant	http://greenplantmiami.com/	USA	HPP
	Highland Fresh Foods	https://www.highlandfreshfoods.com/	Guatemala	HPP and avocado product co-packing

HPP Atlantique	http://www.hppatlantique.fr/	France	HPP
HPP Food Services	http://www.hppfs.com	USA	HPP
HPP Italia	https://www.hppitalia.com/	Italy	HPP
HPP Los Angeles	http://hpplosangeles.com/	USA	HPP
HPP South Africa	http://sahpp.co.za/	South Africa	HPP
HPP Tolling	http://www.hpptolling.com/	Ireland	HPP
Idro	http://idro.es/	Spain	HPP
Kee Fresh	http://www.kfs-foodtech.com/	Taiwan	HPP
Lineage	http://hpp.lineagelogistics.com/	USA	HPP
Longfresh	http://www.longfresh.com.au/	Australia	HPP
Maryland Packaging	http://marylandpackaging.com/	USA	HPP and juice manufacturing and bottling
Natural Food Works	http://nfworks.com/	USA	HPP and juice bottling
Natur + l XTD	http://www.naturelxtd.com	Canada	HPP
Nutrifresh	http://www.nutrifreshservices.com/	USA	HPP and juice manufacturing and bottling
Pascal Processing	http://www.pascalprocessing.com/	The Netherlands	HPP
Press Foods	https://www.pressfoods.mx/	Mexico	HPP and juice manufacturing and bottling
Safe Pac	http://www.safepac.biz	USA	HPP and meat product co-packing
Stay Fresh Foods	http://www.stayfreshfoods.com/	USA	HPP
SterilParma	https://www.youtube.com/watch?v=oCD8vhh2GMg	Italy	HPP
Texas Food Solutions	http://texasfoodsolutions.com/	USA	HPP
True Fresh	https://truefreshhpp.com/	USA	HPP
Universal	http://universalpasteurization.com/	USA	HPP

product's nutritional quality. Unfortunately, this regulation did not provide guidance as to what evidence was necessary to meet its requirements.[244] European countries provided different answers to food companies that were trying to understand better the legal requirements for commercializing a pressurized food product. Some countries considered that the technology had no significant impact on the food characteristics or the nutritional value or toxicity of the food and required minimal justifications. Other countries were not able to provide any recommendations on how to justify its use. Many European companies interested in HPP were discouraged by the lack of regulatory direction and the unpredictable outcome of an apparently tedious and expensive legal process. France required the French Food Safety Authorities to review the evidence provided by a food company. However, after some French companies went through the required process, France published a public notice in 2010 stating the benign nature of food processed at pressures up to 600 MPa for 5 min at chilled or ambient temperature.[245] This notice also established that there was no risk of packaging components migrating into food or any increased risk of food lipid oxidation during processing. Consequently, French food producers were no longer required to submit a request before commercially processing food by HPP. This decision had consequences for the rest of Europe. The European Community (EC) Treaty stipulates free circulation for Community goods throughout all countries of the EC. Therefore, since HPP food was authorized for commercialization in France, which is an EC country, all EC countries were then allowed to use HPP. Table 7.3 summarizes the public notices of Food Safety Authorities on pressure-processed foods in France, the UK, and Canada.

Health Canada, the department of the government of Canada with responsibility for national public health, announced in 2016 that commercial launching of pressurized food no longer required prenotification. Health Canada stated that:

> HPP is no longer a novel process... However, the manufacturer or importer of HPP-treated food products is still responsible for ensuring that these products are in compliance with all applicable regulatory requirements... For all packaging materials associated with foods treated with HPP, it is the responsibility of the manufacturer to ensure that such materials are appropriate for their HPP treatment and do not pose a health risk to the consumer.

Aside from North America and Europe, the remaining HPP industrial equipment for food production is in Asia (60 machines, *i.e.* 16%), China (about 25 machines), Korea (12 machines) and Japan (10 machines). Taiwan, Thailand, Vietnam, Singapore, Malaysia, India, and the United Arab Emirates also have a few machines. Australia and New Zealand together have about 20 machines. South America and Central America account for 15 HPP installations located in Chile, Brazil, Peru, Guatemala, and Dominican Republic. Africa is the continent with the fewest HPP machines in use, only three, all located in South Africa.

Table 7.3 Food Safety Authority notices (Novel Foods files and Letters of Non-Objection) on pressure-processed foods.

Country	Year	Company	Product	Ref.
Canada	2004	Leahy Orchards	Apple sauce and apple sauce–fruit blends	278
	2006	Santa Maria Foods	RTE meats and poultry	279
	2006	Maple Leaf	RTE meat-containing entrées, meat-containing salads, and meat products	280
	2010	Gridpath Solutions	RTE meat and poultry products	281
	2012	Gridpath Solutions	Avocado pulp, guacamole, and tomato-based salsas	282
	2013	Gridpath Solutions/Cargill	Raw ground beef	283
	2013	Drinkme Beverage	Fruit- and vegetable-based smoothies	284
	2014	Gridpath Solutions	Fruit- and vegetable-based juices	285
	2014	Burnbrae Farms	Egg salad, egg dips, and egg spreads	286
	2014	Lassonde	Raw fruit juices	287
	2016	Alberta Food Development	Marinated meats	288
	2016	Health Canada	HPP is no longer a novel process	289
France	2009	ND	RTE meals (plats cuisinés)	290
	2009	ND	Marinated meats (viandes marinées)	291
	2010	ND	Packed foods (aliments emballés)	292
	2011	ND	Dry-cured duck meat products (magrets de canard)	293
	2016	ND	Heat pasteurization assisted by high-pressure for fat duck liver (foie)	294
UK	2001	Orchard House	Fruit-based products	295
	2002	ATA	Fruit and vegetable products	296

HPP is a technology used for a large variety of foods having a water activity (A_w) higher than 0.8. In 2016, the total worldwide production of HPP-treated food was estimated to be about 1.2 million tons per year (2.6 billion lb per year) (Table 7.4). This estimation was performed by Hiperbaric and takes into account all industrial HPP equipment in the world, regardless of the manufacturer. They used data, including number of cycles per hour, volumetric efficiency, operating hours per day, and working days per year, provided by food companies operating high-pressure food processing lines for different types of food. It was assumed that the distribution of different types of products processed by HPP tolling equipment is identical with the distribution of the rest of the equipment owned by different food sector companies.

Table 7.4 Estimation of global HPP food production in 2016 based on global production equipment capacity and average production conditions in different food sectors (assuming that the distribution of different type of products processed by HPP tolling companies is identical with the distribution of the rest of the equipment owned by the different food sector companies).

Products	Total equipment volume/L	No. of cycles per hour	Volumetric efficiency (%)	Working hours per day	Working days per year	Production per year 10^6 kg	10^6 lb
Meat products	31 734	6	50	16	250	381	840
Fruits and vegetables	24 569	7	50	20	200	344	758
Juices and beverages	20 953	7	45	20	250	330	728
Seafood products	7185	10	45	20	125	81	178
Ready meals	3817	6	40	16	250	37	81
Dairy products	2087	7	50	16	250	29	64
Total	90 345					1201	2649

Meat products, fruit and vegetable products, and juices and beverages are each estimated to account for about 30% of HPP global food production. Seafood products represent about 7% of the production, and dairy products and RTE meals account for about 2–3% each. When these data are compared with those from the global food and juices market provided by both the Food and Agriculture Organization, and Euromonitor, HPP products can still be categorized as niche products. For example, less than 0.1% of meat products are processed under pressure, and only approximately 0.4% of processed juices commercialized in the world are pressurized products.

Fruit purées and jams were the first pressurized products to be commercialized, at the beginning of the 1990s by Meidi Ya (Japan), and were followed by juices from Wakayama and Pokka (1991, Japan) and Ulti (1994, France).[246] These products were processed at 400 MPa, a moderate pressure, which was in line with the performance of the industrial machines at that time. This pressure was sufficient to increase the product's shelf life substantially, while maintaining aroma and palatability similar to those of the fresh fruit. Today, the worldwide fastest growing market for high-pressure technology is for the fruit juice and beverage sector. In the USA alone, more than 170 different HPP beverages were listed at the end of 2016 at BevNET.com.

7.6.1 Juices and Beverages

The first international recognition of a pressurized juice occurred in 2009 when the Australian SME company Preshafood received the "Best New Juice" award and the "Best New Beverage Concept" award for their pressurized apple juices during the Beverage Innovation Awards Competition held during the Drinktec trade show, the world's leading exhibition for the beverage and liquid food industry. Since 2012, several pressurized juices have been recognized annually at the BevNET awards, one of the most important shows for beverages in the USA. American companies being recognized for their pressurized juices include Evolution Fresh, Suja, Blue Print, Bolthouse Farms, WTRMLN WTR, Tio Gazpacho, Temple Turmeric, Daily Greens, and Malk. HPP juices and beverage are considered premium products. This category has had a double-digit growth in sales in the last few years. Visiongain, a market study company, indicated in their 2015 report that HPP juices and beverages should have an average growth rate of 20% in the next 10 years.[247] Sales should increase from US$2.72 billion in 2015 to US$18.51 billion in 2025. The first juice companies to invest in HPP were small or medium sized, but large industry players are now interested in the technology. Some of the world's largest food and beverage companies have purchased the four largest US HPP juices brands in the last few years.[248] The Starbucks group purchased Evolution Fresh in 2011. In 2012, The Hain Celestial group acquired Blue Print, and Campbell group acquired Bolthouse Farms. In addition, Suja made an alliance with Coca-Cola in 2015. The key European HPP co-processing companies are Fruityline in The Netherlands, Go Fruselva in Spain, and Fruitapeel in the UK, which are producing pressurized juices for retail brands. Other European brands include Hoogesteger in The Netherlands, Frubaça in Portugal, Cruzoe and Juste Pressé in France, Cold Press, Savsé, and B'Fresh in the UK, and Beskyd and Ugo Juices in the Czech Republic.

Pressurized beverages commercialized by the above companies represent a wide range of products, including fruit and vegetable juices, blends of fruit and/or vegetable juices, smoothies, nut milks, cold brew teas and coffees, chocolate drinks, and cold soups, such as gazpacho. The shelf life of these products is between 3 weeks and 3 months or more, under refrigerated conditions.

HPP juice or beverage labeling generally contains the terms "high-pressure processed" or "HPP" or "cold pressured" or "cold pressed." It is important to point out that the use of "cold pressured" and "cold pressed" can be misleading, and is confusing to consumers. "Cold pressed" refers to juices that were extracted by hydraulic pressure and not heat processed. However, it does not include juice that has been processed under high pressure. The term "cold pressured" refers to juices subjected to HPP, but does not include cold-pressed juice. Nevertheless, cold-pressured juice could be thermally processed before HPP, but this is not usually the case. The term "cold pasteurized" is sometimes misused on HPP product labels as pasteurization is a thermal process, whereas HPP occurs at or below room temperature.

7.6.2 Non-beverage Fruit and Vegetable Products

In Japan, the company Echigo Seika has been commercializing pressurized RTE rice and cereals since 1994.[249,250] The company patented a process that consists in a pressure treatment at 200–400 MPa followed by heat sterilization. This combination of the two treatments gave the product a shelf stability of 1 year at room temperature. The process is called "high pressure-induced transformation" (Hi-Pit) and has the benefits of modifying proteins, enzymatic activities, and starch properties along with providing grain enrichment in components with health benefits (γ-aminobutyric acid).[251]

Fresherized Foods, formerly Avomex, was the first company in the USA to commercialize pressurized products with the launch of its HPP guacamole in 1997.[252] This product has revolutionized the market, and became the world leader in the fresh guacamole market. For many years, this company had the highest HPP production capacity with a production of more than 20 000 tons per year.[253] HPP guacamole's sensory quality is very close to that of fresh (non-processed) guacamole, with increased food safety and a commercial shelf life ranging from 4 to 8 weeks. Pressurized guacamole retains its green color due to the inactivation of polyphenoloxidase under pressure. It is currently the most successful HPP vegetable product. Fresherized Foods promoted the technology and established quality standards for pressurized guacamole/avocado products that were adopted by other companies. Now more than 20 companies in the world, including the early adopters, Calavo, Simplot, and Verfruco in Mexico and other companies in Australia, Chile, Dominican Republic, Guatemala, Peru, Spain, South Africa and the USA, are using HPP to preserve guacamole.[254] Pressurized guacamole packaging usually includes the term "all natural" along with "Fresherized® with High Pressure Processing" or "Cold Pasteurization *via* Ultra-High Pressure (UHP) Technology," "Ultra High Pressure Processing," and "Fresh HPP," which are now considered by the consumers as a guarantee of quality.

The variety of vegetable HPP foods has expanded in recent years with the processing of hummus, vegetable dips, and wet salads. Some of these products are commercialized in the USA by Good Foods, Hope, Cedarlane, and Sandridge. The benefits of applying pressure to these products are that chemical preservatives are not needed, improvement of sensory attributes, and increased shelf life from 1 to 2 months. SimplyFresco and Garden Fresh Salsa produce HPP tomato-based salsas. These preservative-free pressurized sauces are made with fresh spices and herbs, and maintain a fresh-like flavor for 1–3 months when stored in refrigerated conditions. Recently, several companies have launched fruit- and vegetable-based baby foods processed under pressure, including Once upon a Farm and Pure Spoon in the USA and in Born Pure in Australia.

7.6.3 Meat Products

The meat industry was the fastest growing food sector for HPP from 2004 to 2013. The technology gained acceptance by the meat industry in 2003, when the USDA Food Safety and Inspection Service recognized HPP as a valid post-lethality treatment against *L. monocytogenes* for RTE meat and poultry products.[255] In the following years, food safety authorities such as the Food

Safety Authority of Ireland (2005), Health Canada (2006), and the French food agency (AFSSA) (2010) similarly recognized HPP.

The meat industry's interest in the technology is justified by three factors: (1) the ubiquitous efficiency of the technology to kill microorganisms regardless of the volume/size/weight of packaged products, (2) the opportunity to label the product as "natural" (or additive-free), and (3) the availability of reliable industrial-sized equipment, which allows high productivity. The technology is mainly used for RTE products such as sliced and diced dry cured or cooked hams, RTE meat meals, cooked poultry cuts and sausages, and also raw ground beef products (*e.g.* hamburgers). HPP guarantees post-packaging destruction of spoilage microorganisms (coliforms, lactic acid bacteria) and pathogens (*Listeria*, *Salmonella*, *E. coli*). The formulation of pressurized meat products can be modified to include a reduced amount of salt and/or less or no chemical preservatives, without compromising the microbial safety of the products. The pressurized products maintain good organoleptic and microbial quality along with an increased shelf life, which is doubled or tripled compared with similar products not subjected to HPP. This increased shelf life provides greater opportunities for export.

In 1998, Espuña in Spain was the world pioneer in processing meat under pressure by producing cooked ham with a 2-month shelf life. The packaging of this product stated "High Pressure Pasteurized product remains fresh until eaten."[256] Currently, Hormel Foods in the USA produces the largest amount of pressurized meat products under the following brands: Hormel Natural Choice's natural lunchmeats, Bread Ready's sliced meats for food service, and Jennie-O's ground turkey.[257] In 2015, the Hormel Foods group acquired Applegate Farms, which produces pressurized organic and "natural" meat products.[258] Hormel Foods started to use the technology in 2001 for processing prosciutto to eliminate *Listeria.*[259] In 2006, they launched several natural sliced meats, which were very well received by consumers. Most of their products are subjected to a 600 MPa treatment applied for a few minutes. This allows the shelf life of refrigerated meat deli products with no artificial preservatives such as nitrates or nitrites, flavors, and colors to be extended from 60 to 90 days. The success of Hormel Foods in using HPP extends to many deli meat and poultry products. Subsequently, the technology was adopted by other large American companies, including Perdue Farms, Foster Farms, Tyson Foods, and Oscar Meyer.[260,261] West Liberty Foods is the world's second largest producer of pressurized meat-based products and acquired its first equipment in 2013; it now owns six industrial-sized HPP machines.[262]

In Europe, the food safety regulations concerning *L. monocytogenes* contamination are less strict than in other countries, such as the USA and Japan. Commercialization of an RTE product that has up to 100 *L. monocytogenes* cells per gram at the end of its shelf life is allowed in Europe. In contrast, US and Japanese regulations require less than one *L. monocytogenes* cell in 25 g of product. Dry-cured products, such as prosciutto, Serrano ham, chorizo, and dry-cured sausage, have a relatively high tolerance to *Listeria* contamination. However, some European companies, such as Abraham in Germany, Campofrio, Espuña, and Jamcal in Spain, Ferrarini in Italy, and Santa Maria in Canada, have acquired HPP equipment to pressurize these types of products for export to the USA and Japan, and complain about the less than one *L. monocytogenes* cell per 25 g requirement in these countries.[263] Other European companies interested in exporting cured meat products use the service of HPP tollers such as APA Processing in Spain and SterilParma in Italy, both of which have been USDA certified since 2015. Espuña developed an HPP sliced dry-cured ham for the European market that is shelf stable for 40 days at ambient temperature.[264,265]

The use of HPP for raw meat treatment is more challenging than for dry-cured meat, not because of a lack of microbial inactivation efficiency, but because of the textural and color modifications. This explains why only a few companies have commercialized HPP raw ground beef products and marinated meats. The first commercial HPP raw beef product, a steak tartare with spices, was commercialized in 2010 by Zwanenberg in The Netherlands.[266] This product had a shelf life that was extended from a few days to 15 days, which significantly decreased *E. coli*'s risk. In 2011, the American company Cargill introduced its "Fressure™ Ground Beef Patties" processed under pressure to extend the product's shelf life to up to 42 days.[267] However, this product is for food service only, as consumers will not accept a raw product with the significant color change produced after HPP. These changes in color disappear with cooking. Hence the food service industry is the main user of pressurized fresh ground beef because they want the extended shelf life and increased food safety. Furthermore, the sauce used in marinated products covers the color-induced changes.

The pet food industry is a growing market for pressurized raw meat. Companies commercializing pressurized pet food products include Raw Advantage, Stella & Chewy, Nature's Variety, and Primal Pet Foods, using HPP to provide microbially safe premium pet foods. They all use tolling services.

7.6.4 Seafood

HPP is used to extend the shelf life of RTE HPP fish products, raw marinated fish fillets, and fish- and shrimp-based RTE meals mainly for the European market. In 2014, Ghezzi in Italy began the commercialization of a chemical preservative-free desalted cod. Delpierre, a French company, was recognized in 2016 by the Seafood Process Show in Brussels for its line of pressurized fresh fish fillets, including seabass, cod, sea bream, mackerel, red mullet, and sardines. Delpierre has also commercialized shrimp and scallop ready-to-cook products since 2015.[268]

HPP has also two niche application in the seafood sector: mollusk shucking and pressure-assisted extraction of meat from crustaceans. The treatment triggers the opening of oyster and clam shells by unfolding the abductor muscle, which is the muscle that keeps the shell closed. The process of shucking oysters, and bivalves in general, using pressures from

100 to 400 MPa for 0.5–10 min was patented by the Japanese company Aohata in 1991. However, in 1999, the US company Motivatit Seafoods was the first to commercialize pressurized bivalve processing, with its HPP shucked "Gold Band" oysters.[269] A heat-shrinkable plastic band is placed around the pressure-shucked oyster to keep the shell closed to differentiate their products. The oysters are introduced alive and unpacked in the equipment and pressurized for 1–2 min at much lower pressure of ~300 MPa than required for microbial inactivation. The short processing time is a major advantage of the process. Consequently, the use of HPP significantly reduced the need for manual labor, reduced injury due to the use of a knife, increased yield up to 25% and increased the shelf life. Processing below 300 MPa maintains the raw meat characteristics. Higher pressures and longer processing times can be used to ensure microbial safety but denaturation of meat protein will also occur.[270] US companies such as Joey Oyster, Goose Point, and Prestige Oyster are commercializing pressurized oyster as American consumers want fresh and frozen oysters without the shell.[271] In many other countries, the more traditional in-shell seafood is preferred.

HPP can promote the extraction of meat from crustaceans by exploiting the difference in compressibility between the meat and the shell to promote meat separation after treatment. Clearwater in Canada patented the process and was the first company to process lobster and crab in 2003. The animal is placed alive in the pressure chamber without packaging. Pressurized water at 300 MPa enters under the shell and compresses the meat by about 8%, which separates the meat from the shell. HPP revolutionized the lobster market by providing raw uncooked products with superior sensory quality. HPP raw lobster meat is now produced by several companies in Canada, such as Westmorland Fisheries, Riverside Lobster, and Shediac Lobster. Other companies include Shuck Maine Lobster in the USA and Cinq Degrés Ouest in France. Pressurized lobsters are exported all over the world. They are sold frozen because the pressure applied is too low to increase the shelf life and because of their seasonality.

7.6.5 Dairy Products

Only few pressurized commercial dairy products are available. Rodilla in Spain has used HPP since 2005 to increase fresh cheese-based sandwich fillings, such as cheese mixed with nuts, shrimp, or vegetables, to ensure microbial safety and increase shelf life from one to about three weeks. A few other companies in Spain use HPP to extend the shelf life of fresh cheese, *via* pressure-inactivation of lactic acid bacteria, molds, and yeasts. Some Italian companies use HPP to increase the shelf life of buffalo mozzarella cheese for export to Japan. Non-thermally processed milk treated under pressure to increase its safety and shelf life could be a growing niche market. In 2014, the first commercialized raw cow milk preserved by high pressure was produced by Villa de Patos in Mexico.[272] The Australian company Made by Cow obtained the approval of the New South Wales and Western Australia Food

Authority in 2016 to use HPP as an alternative to conventional heat pasteurization and started the commercialization of "cold-pressed raw milk."[273] This non-homogenized milk is a premium product that is sold at a price of about A$6 or US$6 for 750 mL. This product was awarded the title of "Best New Dairy Drink" at the World Beverage Innovation Awards in Frankfurt in November 2016.

7.7 Industrial HPP Equipment

The key parameters to be considered in the development of industrial HPP equipment are safety, reliability, productivity, ease of installation, operation, and servicing and ease of cleaning. Industrial-scale HPP machines need to be compact, assure food safety, and last at least 10 years and should be able to be run daily almost continuously at pressures up to 600 MPa. A pressure of 600 MPa is the current standard maximum pressure used in the food industry to achieve extended shelf life and increased food safety. To achieve these requirements, optimum design and material selection are essential owing to the high mechanical stresses generated under high pressures.[274]

A summary of some of the HPP equipment vendors is provided in Table 7.1.

7.7.1 Design

The first HPP machines for food applications were similar in design to those used in the chemical and metallurgical industries. They were built to withstand long holding times and a few processing cycles per hour. They usually consisted of a vessel in a vertical position with a unique mobile plug.[275] As a consequence, loading and unloading of product were slow. One of the key accomplishments in the development of HPP equipment for food application was a horizontal design of the vessel, allowing easy opening of both plugs (Figure 7.3).

The horizontal design has been successfully used in the food industry since its first installation by the French company Alstom for the Spanish company Espuña in 1998. In 2009, 40% of the HPP vessels throughout the world were of horizontal design, and this number increases to 95% when industrial machines less than 5 years old are considered.

With horizontal equipment, the packaged products are placed in plastic carriers or baskets and an automatic loading system places the carriers inside the vessel. The vessel is then moved horizontally and placed inside the yoke, which is the main structure that supports the vessel and maintains the unscrewed plugs closed under pressure. Screwed plugs are commonly used in laboratory equipment but unscrewed plugs are preferred in large-volume vessels (>25 L) and large-diameter vessels (>15 cm). Unscrewed plugs are preferred because of their better mechanical resistance to fatigue under high pressure and because the opening and closing are quicker. When the plugs are in place, the vessel is filled with water at low pressure (<1 MPa). When

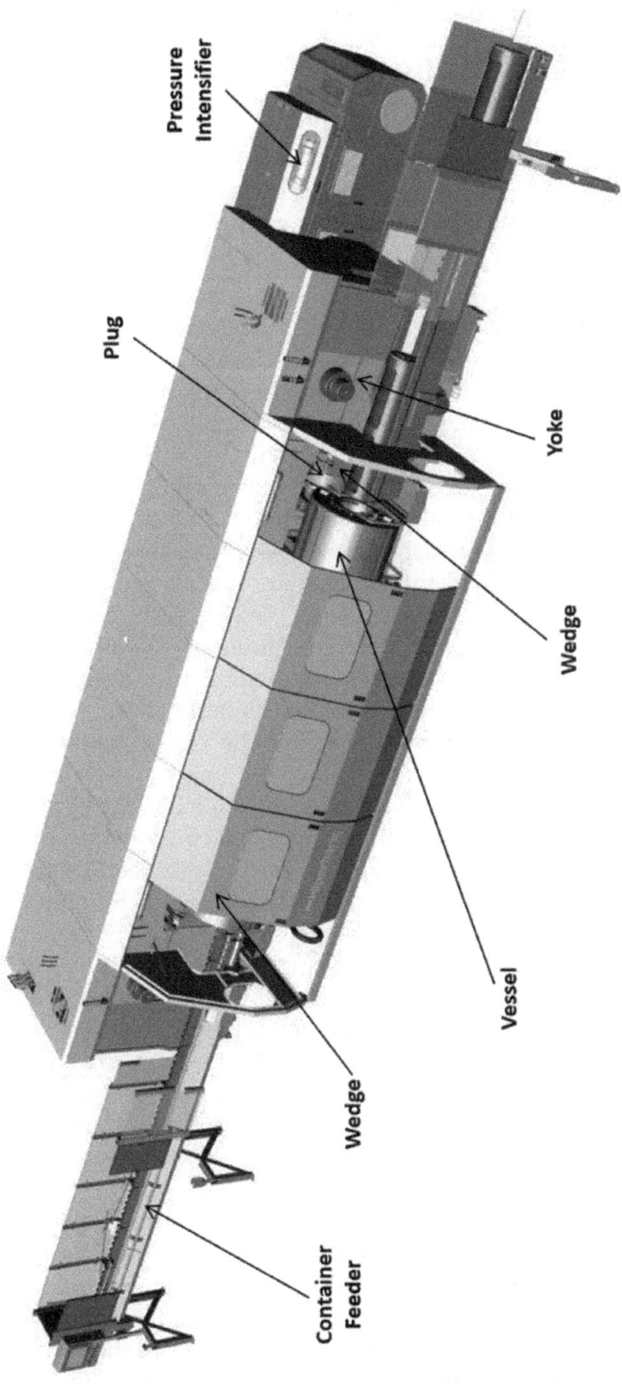

Figure 7.3 Basic components of a Hiperbaric HPP machine with pressure intensifiers on the ground floor.

the vessel has been completely filled with water at low pressure, the plugs are further secured and the high-pressure intensifiers start to pump more water into the chamber until the desired pressure is reached. This step usually takes few minutes and depends on the pressure being reached. This pressure is maintained during the desired holding time. Processing time varies depending on the time needed to reach the desired pressure, the holding time, and the time to return to atmospheric pressure. The pressure is then discharged in a few seconds. The discharge valves then open, followed by the plugs, which are put back into their initial position. The vessel moves back to the loading/unloading position. Once in this position, the processed batch carriers are pushed into the unloading line, while the next batch of carriers is automatically loaded, ready for a new processing cycle (Figure 7.4).

The horizontal design has three main advantages relative to vertical machines.[276] Loading and unloading are carried out at waist level and can be more easily and quickly performed, either manually or automatically, than with a machine that is several meters high. This operation takes less than 10 s as the machines have an automatic loading and unloading line. Therefore, the process is almost continuous. As an unprocessed batch automatically enters the vessel, the processed products are simultaneously exiting on the opposite side and the operator can place the next batch on the loading line. When a cycle is completed, the next cycle starts automatically. With vertical systems, cranes are used to load/unload samples, which is time consuming. Another successful development in the design of horizontal systems is the installation of a conveyor parallel to the loading and unloading lines for the automatic return of the empty carriers back to the loading point (Figure 7.5).

Second, the installation of a horizontal machine in a food processing plant is easier, more practical, and consequently less costly than that of a vertical machine. For example, the height of a horizontal machine (about 2.5 m) allows its transport through standard doors in any factory and no crane is required to stand the vessel up inside the factory. For a vertical vessel, special arrangements are necessary owing to its height, which is about 6 m for a 215 L vessel. The weights of horizontal industrial machines are 20 and 100 tons for a 55 and 525 L machine, respectively, and the weight is well distributed over the vessel "footprint" of 7 and 20 m^2 for a 55 and 525 L vessel, respectively. Therefore, reinforced floors are not usually required for the smaller machines. A reinforced floor with special foundations is necessary for a 40 ton, 215 L vertical machine because the footprint area of the vessel remains fairly small, 4.5 m^2.

Third, the use of a horizontal vessel improves tremendously the traceability of the product through the process. As both plugs of the vessel open at each cycle, the product can enter the vessel at one end and exit from the other end. In contrast, in vertical equipment, the unprocessed products are loaded and the processed products are unloaded from the same end, which increases the risk of misidentification of processed and unprocessed products.

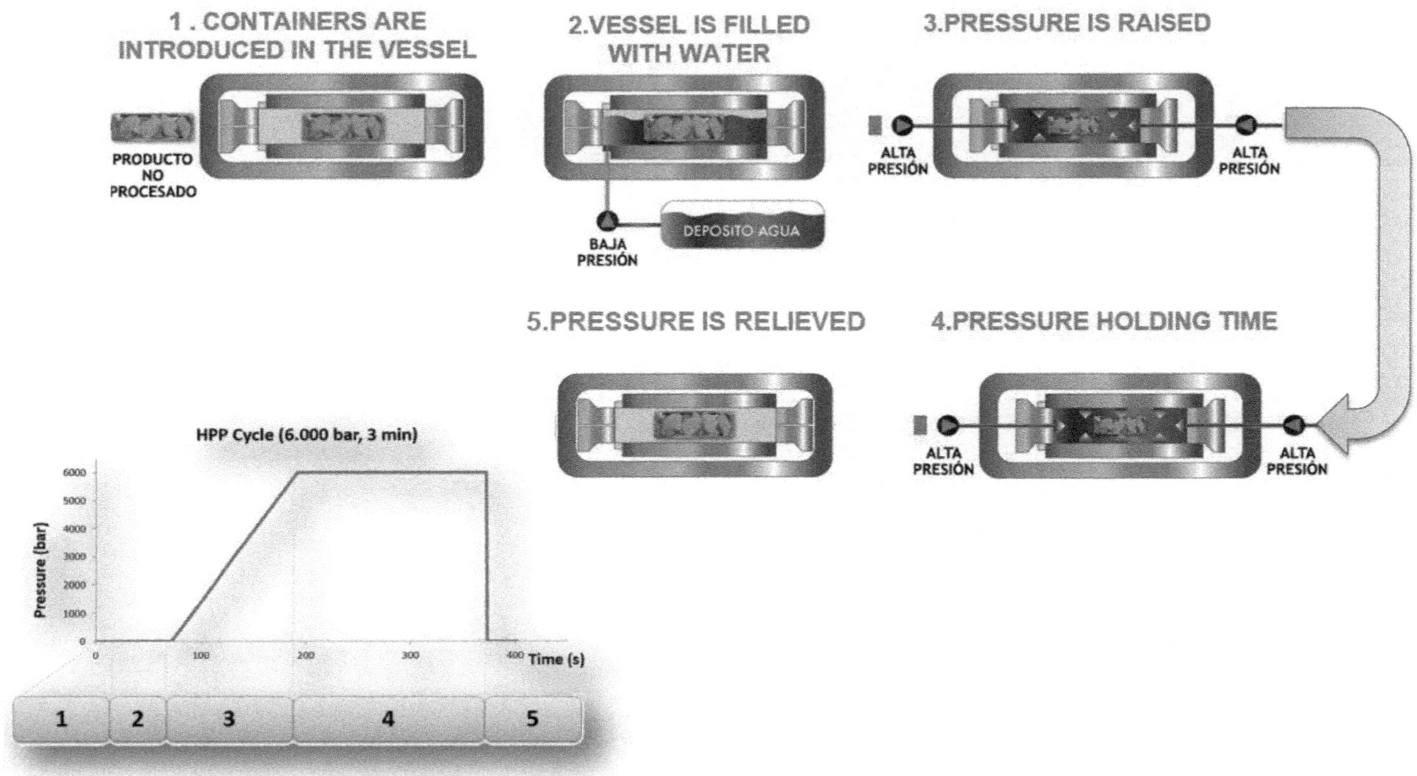

Figure 7.4 Diagram of operation of a horizontal HPP machine in five steps, and example of an HPP cycle at for a process at 6000 bar/600 MPa/87 000 psi.

A problem with the horizontal design is that when the plugs are opened, the processing water completely leaves the vessel. This release of water does not affect water consumption, as it is filtered and recycled. However, the vessel needs to be completely filled for each cycle, and this step needs to be performed quickly to assure time efficiency. Current water-filling systems allow this step to require less than 25 s for a 525 L vessel. There is one advantage of completely emptying the vessel after each cycle, namely better control of the water temperature compared with that of a vertical vessel. Another challenge with the horizontal design is the horizontal length and volume of the vessel and the yoke. To reduce the total area required, the peripherals such as a water tank, electric cabinet, operator panel, and optimized loading/unloading line are integrated into the system itself. Intensifier cabinets are designed to be as small as possible. When peripherals are integrated into the system, the total space required for a horizontal machine is similar to that for a vertical machine of the same capacity, without integrated peripherals. Hiperbaric recently improved their vessel's design by placing the machine intensifiers on the top of the vessel (Figure 7.5). This design saves floor space, without requiring a crane inside the factory because the platform supporting the high-pressure pump is easily assembled on-site and the HPP pumps (having a weight of about 2.5 tonnes) are placed on the platform with forklifts.

Each HPP vessel requires a compression system to generate pressure. The only one currently used in the food industry is an indirect compression system that uses high-pressure intensifiers, which pump water into the chamber. The pressure build-up speed is about 50–300 MPa min^{-1} and depends on the ratio between the intensifier power and vessel volume. The indirect compression system has two major disadvantages. First, it is a non-continuous process. Second, the production rate depends greatly on the vessel filling ratio, also

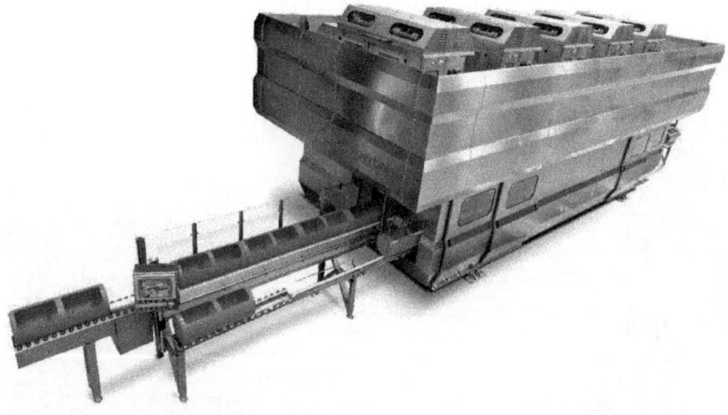

Figure 7.5 HPP equipment: Hiperbaric 525. Vessel volume 525 L, integrated design with five high-pressure pumps (including two high-pressure intensifiers each) on the top of the vessel, carrier loading and unloading lines, carrier return line, and operator panel. Maximum working pressure: 600 MPa/6000 bar/87 000 psi.

called volumetric efficiency, which is the volume or weight of products loaded in the vessel at each batch. It varies from 30 to 80%, depending on the geometry, shape, density, and the headspace of the packaged or unpacked food in each carrier placed in the vessel. In general, the larger the vessel diameter, typically 190–380 mm, the better is the volumetric efficiency. A larger diameter carrier of about 300 mm provides more opportunity to find an efficient means to load the food packages than a smaller one of 200 mm diameter or less.

When pressure is generated through a piston, reducing the internal volume of the vessel, the compression system is called a direct compression system. This system has two advantages. The pressure build-up speed is higher, with pressures up to 3000 MPa min^{-1}. Liquid products can be processed without packaging with a 100% volumetric efficiency, with the piston pressurizing the beverage directly. Direct compression of unbottled liquid requires aseptic filling after treatment. Direct compression systems are commonly used with laboratory-scale equipment but are not practical for industrial use. This is due to the risk of a valve leaking, which might contaminate pressure-treated liquid foods, such as juices, with unprocessed liquid. In addition, as the juice is in direct contact with the machine, these direct compression systems require long and complicated cleaning and maintenance operations, more particularly when the processed liquid contains particles or is acidic, which is generally the case for juices. Owing to cleaning procedures required, the time during which the equipment could be used for treatment is shortened by 25%. Moreover, the operating costs with direct compression equipment are not lower than those with indirect compression systems. With indirect systems, hermetic closure is achieved thanks to static seals between the vessel and plugs, which are more reliable and affordable than moving seals whatever the diameter of the vessel. For direct systems, hermetic closure under pressure is achieved with moving seals located between the vessel wall and piston. The affordability and reliability of the moving seals decrease significantly with increase in vessel diameter. Therefore, internal piston machines (direct systems) have limitations in terms of vessel diameter (<200 mm) and length (<5 m), which reduce the maximum vessel volume to less than 150 L. In contrast, an HPP vessel with indirect compression systems working at 600 MPa can have a volume of 525 L with an internal diameter of 380 mm or more.

7.7.2 Size and Output

The size and output of commercial HPP vessels have improved considerably in the past 15 years. The volume of a vessel that can reach a pressure of 600 MPa more than doubled between 1998 and 2014, increasing from 215 to 525 L. Similarly, the average volume of an industrial vessel in production has more than doubled compared with 15 years ago, increasing from about 90 L in the late 1990s to about 235 L in 2016. Moreover, the performance, as represented by the number of cycles per hour, has also greatly improved, mostly because of a significant reduction in the total cycle time. The duration of low-pressure operations, including vessel and plug movements, loading/unloading, and water filling the vessel at low pressure, has been shortened. In addition, the pressure build-up duration has been reduced by installing more efficient and powerful high-pressure intensifiers. Currently, industrial machines equipped with a standard configuration perform a cycle at 600 MPa, without holding time, in about 3 min. Ten cycles per hour can be performed with a 3 min holding time. With these processing conditions, and a vessel efficiency of 60%, the newer largest machines can process more than 3000 kg h^{-1} (Table 7.4).

7.7.3 Investment and Processing Costs

In the past 15 years, the cost of a treatment at 600 MPa has been reduced more than threefold. This is mainly due to technical improvements that reduced the processing cycle time. The use of HPP contributes to the overall product cost. The cost of HPP per kilogram of food processed is significantly higher (5–10 times higher) than that of a thermal treatment. However, it is an affordable processing technique for premium, high-quality, or niche products. For example, for a typical RTE product treated with a horizontal indirect compression system having a 2 min holding time at 600 MPa with a volumetric efficiency of 60%, the processing cost is €0.064–0.163 kg^{-1} or ~US$0.032–0.081 lb^{-1}, depending on the HPP equipment size (Table 7.5). The lowest operating costs are achieved with the largest equipment. Depreciation of the equipment is the most important factor in the operating costs, and the equipment cost per liter decreases with increase in vessel volume.

Table 7.5 Main characteristics and processing cost of the Hiperbaric equipment portfolio.

Model	Hiperbaric 55	Hiperbaric 135	Hiperbaric 300	Hiperbaric 420	Hiperbaric 525
Vessel volume/L	55	135	300	420	525
Vessel diameter/cm	20	30	30	38	38
Footprint/m^2	22	39	61	56	63
No. of cycles per hour	10.7	9.5	9	10	10
Output[a]/kg h^{-1} (lb h^{-1})	260 (573)	640 (1411)	1330 (2933)	2500 (5513)	3150 (6946)
Processing cost[b]/€ kg^{-1} ($ lb^{-1})	0.163 (0.081)	0.114 (0.057)	0.081 (0.040)	0.067 (0.033)	0.064 (0.032)

[a]Calculations for a process with a holding time of 2 min at 600 MPa/6000 bar/87 000 psi, considering a vessel filling efficiency of 60%, including amortization in 5 years, 300 working days per year, 16 hours per day, wear parts and utilities. Exchange rate: €1 = US$1.1.
[b]Calculations included depreciation in 5 years, 300 working days per year, 16 hours per day, wear parts and utilities, but not labor.

The cost of commercial HPP machines ranges from €500 000 to €2 500 000 or US$550 000 to US$2 750 000. The processing cost depreciation of the equipment is responsible for about 30–60% of the cycle cost, depending on moderate or intensive use of the machine.[277] Replacing parts represents 30% of the processing cost. The energy cost is about 10% and water consumption is negligible. Labor costs can increase processing costs by 20–50%. This depends on the level of automation of the processing lines, which includes loading of products in the baskets and drying after processing.

7.8 Final Remarks

There are increasing numbers of industrial high-pressure machines around the world, with about 400 at the beginning of 2017, compared with two in 1990. This illustrates the growing recognition of the benefits of HPP in the food industry. Initial applications of emerging technologies are usually in niche markets and for premium products. This is because it is difficult to compete with thermal treatment and the use of chemical preservatives, which ensure food safety at a reasonable cost. However, food poisoning cases continue to be reported and can have a dramatic economic impact on a food company. This risk may encourage the food industry to adopt HPP, especially when consumers ask for a long shelf life and chemical additive-free products (green foods) with fresh-like sensory qualities. Toll processing facilities offer the food industry an opportunity to use the technology without the challenges of owning their own equipment. The growing number of such facilities is another good indicator of the steadily increasing interest in the technology in the food industry.

In conclusion, larger quantities and a broader variety of commercially available pressurized food products are anticipated in the future. Further development of this technology could come from hurdle approaches, such as adding natural pressure-resistant antimicrobial compounds. This would allow processing at a lower pressure level to provide maximum food safety and extended shelf life, while minimizing the undesirable effects of pressure on food quality.

References

1. C. Jermann, T. Koutchma, E. Margas, C. Leadley and V. Ros-Polski, *Innovative Food Sci. Emerging Technol.*, 2015, **31**, 14.
2. A. V. Cardello, *Appetite*, 2003, **40**, 217.
3. A. V. Cardello, H. G. Schutz and L. L. Lesher, *Innovative Food Sci. Emerging Technol.*, 2007, **8**, 73.
4. H. B. Nielsen, A.-M. Sonne, K. G. Grunert, D. Banati, A. Pollak-Toth, Z. Lakner, N. V. Olsen, T. P. Zontar and M. Peterman, *Appetite*, 2009, **52**, 115.

5. J. Hjelmqwist, in *Novel Food Processing Technologies*, ed. G. V. Barbosa-Canovas, M. S. Tapia and M. Pilar Cano, CRC Press, Boca Raton, FL, 2005, pp. 361–373.
6. C. Caner, R. J. Hernandez and B. R. Harte, *Packag. Technol. Sci.*, 2004, **17**, 23.
7. Y. Lambert, G. Demazeau, A. Largeteau, J. M. Bouvier, S. Laborde-Croubit and M. Cabannes, *Packag. Technol. Sci.*, 2000, **13**, 63.
8. A. Le-Bail, N. Hamadami and S. Bahuaud, *Packag. Technol. Sci.*, 2006, **19**, 237.
9. A. Schauwecker, V. M. Balasubramaniam, G. Sadler, M. A. Pascall and C. Adhikari, *Packag. Technol. Sci.*, 2002, **15**, 255.
10. S. Balasubramaniam and V. M. Balasubramaniam, *Food Res. Int.*, 2003, **36**, 661.
11. E. Patazca, T. Koutchma and V. M. Balasubramaniam, *J. Food Eng.*, 2007, **80**, 199.
12. W. B. C. de Heij, L. J. M. M. van Schepdael, R. Moezelaar, H. Hoogland, A. M. Matser and R. W. van den Berg, *Food Technol.*, 2003, **57**, 37.
13. M. H. Lau and E. J. Turek, in *High Pressure Processing of Foods*, ed. C. Doona and F. Feeherry, Blackwell Publishing, Ames, IA, 2007, pp. 194–217.
14. G. V. Barbosa-Canovas and P. Juliano, in *Food Engineering: Integrated Approaches*, ed. G. F. Gutierrez-Lopez, G. V. Barbosa-Canovas, J. Welti-Chanes and E. P. Arias, Springer, New York, 2008, pp. 9–46.
15. US Food and Drug Administration, http://www.fda.gov/Food/ScienceResearch/ResearchAreas/SafePracticesforFoodProcesses/ucm101456.htm (updated 2009), accessed November 2009.
16. D. R. Wilson, L. Dabrowski, S. Stringer, R. Moezelaar and T. F. Brocklehurst, *Trends Food Sci. Technol.*, 2008, **19**, 289.
17. J. Ahn, V. M. Balasubramaniam and A. E. Yousef, *Int. J. Food Microbiol.*, 2007, **113**, 321.
18. H. S. Ramaswamy, H. Jin and S. Zhu, *Food Bioprod. Process.*, 2009, **87**, 1.
19. P. Juliano, S. Clark, T. Koutchma, M. Ouattara, J. W. Mathews, C. P. Dunne and G. V. Barbosa-Canovas, *J. Food Qual.*, 2007, **30**, 57.
20. P. Juliano, B. Li, S. Clark, J. W. Mathews, P. C. Dunne and G. V. Barbosa-Canovas, *J. Food Qual.*, 2006, **29**, 505.
21. P. Juliano, M. Toldra, T. Koutchma, V. M. Balasubramaniam, S. Clark, J. W. Mathews, C. P. Dunne, G. Sadler and G. V. Barbosa-Canovas, *J. Food Sci.*, 2006, **71**, E52.
22. L. Ludikhuyze, A. Van Loey, Indrawati, C. Smout and M. Hendrickx, *Crit. Rev. Food Sci. Nutr.*, 2003, **43**, 527.
23. S. Toepfl, A. Mathys, V. Heinz and D. Knorr, *Food Rev. Int.*, 2006, **22**, 405.
24. J. L. Boye and Y. Arcand, *Food Eng. Rev.*, 2013, **5**, 1.
25. O. Rodriguez-Gonzalez, R. Buckow, T. Koutchma and V. M. Balasubramaniam, *Compr. Rev. Food Sci. Food Saf.*, 2015, **14**, 536.

26. M. Mor-Mur and J. Yuste, in *Emerging Technologies for Food Processing*, ed. D.-W. Sun, Elsevier Academic Press, San Diego, CA, 2005, pp. 47–66.
27. E. Rendueles, M. K. Omer, O. Alvseike, C. Alonso-Calleja, R. Capita and M. Prieto, *LWT – Food Sci. Technol.*, 2011, **44**, 1251.
28. E. Georget, R. Sevenich, K. Reineke, A. Mathys, V. Heinz, M. Callanan, C. Rauh and D. Knorr, *Innovative Food Sci. Emerging Technol.*, 2015, **27**, 1.
29. S. Buzrul, *Innovative Food Sci. Emerging Technol.*, 2014, **26**, 1.
30. I. Baptista, S. M. Rocha, Â. Cunha, J. A. Saraiva and A. Almedia, *Innovative Food Sci. Emerging Technol.*, 2016, **36**, 128.
31. M. Gänzle and Y. Liu, *Front. Microbiol.*, 2015, **6**, 599.
32. P. Chilton, N. S. Isaacs, B. Mackey and R. Stenning, in *High Pressure Research in the Biosciences and Biotechnology*, ed. K. Heremans, Leuven University Press, Leuven, 1997, pp. 225–228.
33. D. Saucedo-Reyes, A. Marco-Celdran, M. Consuelo P-Perez, D. Rodrigo and A. Martinez-Lopez, *Innovative Food Sci. Emerging Technol.*, 2009, **10**, 135.
34. M. M. Hayman, R. C. Anantheswaran and S. J. Knabel, *Int. J. Food Microbiol.*, 2007, **115**, 220.
35. R. Pagan and B. Mackey, *Appl. Environ. Microbiol.*, 2000, **66**, 2829.
36. M. F. Patterson, *J. Appl. Microbiol.*, 2005, **98**, 1400.
37. N. Kalchayanand, A. Sikes, C. P. Dunne and B. Ray, *Food Microbiol.*, 1998, **15**, 207.
38. R. K. Simpson and A. Gilmour, *Food Microbiol.*, 1997, **14**, 567.
39. K. M. Considine, A. L. Kelly, G. F. Fitzgerald, C. Hill and R. D. Sleator, *FEMS Microbiol. Lett.*, 2008, **281**, 1.
40. M. Smiddy, L. O'Gorman, R. D. Sleator, J. P. Kerry, M. F. Patterson, A. L. Kelly and C. Hill, *Innovative Food Sci. Emerging Technol.*, 2005, **6**, 83.
41. S. Koseki and K. Yamamoto, *Int. J. Food Microbiol.*, 2007, **115**, 43.
42. R. Escriu and M. Mor-Mur, *Food Microbiol.*, 2009, **26**, 834.
43. H. S. Ramaswamy, S. U. Zaman and J. P. Smith, *J. Food Eng.*, 2008, **87**, 99.
44. C. C. Tassou, P. Galiatsatou, F. J. Samaras and C. G. Mallidis, *Innovative Food Sci. Emerging Technol.*, 2007, **8**, 478.
45. E. Ananta and D. Knorr, *Food Microbiol.*, 2009, **26**, 542.
46. A. Delgado, C. Rauh, W. Kowalczyk and A. Baars, *Trends Food Sci. Technol.*, 2008, **19**, 329.
47. US Food and Drug Administration, http://www.fda.gov/Food/ResourcesForYou/Consumers/ucm079311.htm. Accessed November 2009.
48. V. C. H. Wu, *Food Microbiol.*, 2008, **25**, 735.
49. M. F. Patterson, M. Quinn, R. Simpson and A. Gilmour, *J. Food Prot.*, 1995, **58**, 524.
50. F. Bozoglu, H. Alpas and G. Kaletunc, *FEMS Immunol. Med. Microbiol.*, 2004, **40**, 243.
51. E. P. Black, A. L. Kelly and G. F. Fitzgerald, *Innovative Food Sci. Emerging Technol.*, 2005, **6**, 286.
52. A. Jofre, M. Garriga and T. Aymerich, *Meat Sci.*, 2008, **78**, 53.

53. L. Espina, D. García-Gonzalo, A. Laglaoui, B. M. Mackey and R. Pagán, *Int. J. Food Microbiol.*, 2013, **161**, 23.
54. A. Al-Nehlawi, S. Guri, B. Guamis and J. Saldo, *LWT – Food Sci. Technol.*, 2014, **58**, 404.
55. M. Abid, S. Jabbar, B. Hu, M. M. Hashim, T. Wu, Z. Wu, M. A. Khan and X. Zeng, *LWT – Food Sci. Technol.*, 2014, **59**, 70.
56. T. J. Lopez-Pedemonte, A. X. Roig-Sagues, A. J. Trujillo, M. Capellas and B. Guamis, *J. Dairy Sci.*, 2003, **86**, 3075.
57. Y. J. Crawford, E. A. Murano, D. G. Olson and K. Shenoy, *J. Food Prot.*, 1996, **59**, 711.
58. M. K. Bull, S. A. Olivier, R. J. van Diepenbeek, F. Kormelink and B. Chapman, *Appl. Environ. Microbiol.*, 2009, **75**, 434.
59. H. S. Ramaswamy, Y. Shao and S. Zhu, *J. Food Eng.*, 2010, **96**, 249.
60. Y.-L. Gao and X.-R. Ju, *J. Food Eng.*, 2007, **82**, 68.
61. E. P. Black, P. Setlow, A. D. Hocking, C. M. Stewart, A. L. Kelly and D. G. Hoover, *Compr. Rev. Food Sci. Food Saf.*, 2007, **6**, 103.
62. T. Aymerich, P. A. Picouet and J. M. Monfort, *Meat Sci.*, 2008, **78**, 114.
63. M. Hugas, M. Garriga and J. M. Monfort, *Meat Sci.*, 2002, **62**, 359.
64. J. C. Cheftel and J. Culioli, *Meat Sci.*, 1997, **46**, 211.
65. M. Garriga, N. Grebol, M. T. Aymerich, J. M. Monfort and M. Hugas, *Innovative Food Sci. Emerging Technol.*, 2004, **5**, 451.
66. A. Jofre, T. Aymerich, N. Grebol and M. Garriga, *Lebensm.-Wiss. Technol.*, 2009, **42**, 924.
67. M. M. Hayman, I. Baxter, P. J. O'Riordan and C. M. Stewart, *J. Food Prot.*, 2004, **67**, 1709.
68. A. M. Diez, E. M. Santos, I. Jaime and J. Rovira, *Food Microbiol.*, 2008, **25**, 154.
69. J. L. Bricher, *Food Saf. Mag.*, 2008, June/July, 40.
70. A. Lopez-Malo, E. Palou, G. V. Barbosa-Canovas, J. Welti-Chanes and B. G. Swanson, *Food Res. Int.*, 1998, **31**, 549.
71. E. Palou, C. Hernandez-Salgado, A. Lopez-Malo, G. V. Barbosa-Canovas, B. G. Swanson and J. Welti-Chanes, *Innovative Food Sci. Emerging Technol.*, 2000, **1**, 69.
72. G. Aleman, D. F. Farkas, J. A. Torres, E. Wilhelmsen and S. McIntyre, *J. Food Prot.*, 1994, **57**, 931.
73. B. B. Boynton, C. A. Sims, S. Sargent, M. O. Balaban and M. R. Marshall, *J. Food Sci.*, 2002, **67**, 409.
74. F. C. Lavinas, M. A. L. Miguel, M. L. M. Lopes and V. L. Valente Mesquita, *J. Food Sci.*, 2008, **73**, M273.
75. M. Houska, J. Strhalm, K. Kocurova, J. Totusek, D. Lefnerova, J. Triska, N. Vrchotova, V. Fieldlerova, M. Holasova, D. Gabrovska and I. Paulickova, *J. Food Eng.*, 2006, **77**, 386.
76. G. Prestamo, M. Lesmes, L. Otero and G. Arroyo, *J. Agric. Food Chem.*, 2000, **48**, 2943.
77. K. Smith, A. Mendonca and S. Jung, *Food Microbiol.*, 2009, **26**, 794.

78. M. E. Barcenas, R. Altamirano-Fortoul and C. M. Rosell, *Lebensm.-Wiss. Technol.*, 2010, **43**, 12.
79. M. R. Garcia-Risco, E. Cortes, A. V. Carrascosa and R. Lopez-Fandino, *J. Food Prot.*, 1998, **61**, 735.
80. J. G. Waite, J. M. Jones, E. J. Turek, C. P. Dunne, A. O. Wright, T. C. S. Yang, R. Beckwitt and A. E. Yousef, *J. Food Sci.*, 2009, **74**, M83.
81. H. He, R. M. Adams, D. F. Farkas and M. T. Morrissey, *J. Food Sci.*, 2002, **67**, 640.
82. M. Cruz-Romero, A. L. Kelly and J. P. Kerry, *Innovative Food Sci. Emerging Technol.*, 2008, **9**, 441.
83. L. W. Murchie, M. Cruz-Romero, J. P. Kerry, M. Linton, M. F. Patterson, M. Smiddy and A. L. Kelly, *Innovative Food Sci. Emerging Technol.*, 2005, **6**, 257.
84. N. Prapaiwong, R. K. Wallace and C. R. Arias, *Int. J. Food Microbiol.*, 2009, **131**, 145.
85. L. G. Fidalgo, M. D. Santos, R. P. Queiros, R. S. Inacio, M. J. Mota, R. P. Lopes, M. S. Goncalves, R. F. Neto and J. A. Saraiva, *Food Bioprocess Technol.*, 2014, **7**, 2028.
86. A. Bermejo-Prada, M. E. Lopez-Caballero and L. Otero, *Innovative Food Sci. Emerging Technol.*, 2016, **33**, 154.
87. C. Sanchez-Moreno, B. De Ancos, L. Plaza, P. Elez-Martinez and P. Cano, *Crit. Rev. Food Sci. Nutr.*, 2009, **49**, 552.
88. F. J. Barba, N. S. Terefe, R. Buckow, D. Knorr and V. Orlien, *Food Res. Int.*, 2015, **77**, 725.
89. T. Koutchma, V. Popović, V. Ros-Polski and A. Popielarz, *Compr. Rev. Food Sci. Food Saf.*, 2016, **15**, 844.
90. A. Patras, N. P. Brunton, S. Da Pieve and F. Butler, *Innovative Food Sci. Emerging Technol.*, 2009, **10**, 308.
91. R. Doblado, J. Frias and C. Vidal-Valverde, *Food Chem.*, 2007, **101**, 918.
92. F. Sancho, Y. Lambert, G. Demazeau, A. Largeteau, J. M. Bouvier and J. F. Narbonne, *J. Food Eng.*, 1999, **39**, 247.
93. I. Oey, I. Van der Plancken, A. Van Loey and M. Hendrickx, *Trends Food Sci. Technol.*, 2008, **19**, 300.
94. S. Jung, P. A. Murphy and I. Sala, *Food Chem.*, 2008, **111**, 592.
95. F. J. Barba, N. S. Terefe, R. Buckow, D. Knorr and V. Orlien, *Food Res. Int.*, 2015, **77**, 725.
96. Q. Xia, L. Wang, C. Xu, J. Mei and Y. Li, *Food Chem.*, 2017, **214**, 533.
97. V. Briones-Labarca, G. Venegas-Cubillos, S. Ortiz-Portilla, M. Chacana-Ojeda and H. Maureira, *Food Chem.*, 2011, **128**, 520.
98. V. Briones-Labarca, C. Munoz and H. Maureira, *Food Res. Int.*, 2011, **44**, 875.
99. G. Linsberger-Martin, K. Weiglhofer, T. P. T Phuong and E. Berghofer, *LWT – Food Sci. Technol.*, 2013, **51**, 331.
100. C. Nerín, M. Aznar and D. Carrizo, *Trends Food Sci. Technol.*, 2016, **48**, 63.

101. K. de Vleeschouwer, I. Van Der Plancken, A. V. Loey and M. C. Hendrickx, *J. Agric. Food Chem.*, 2010, **58**, 11740.
102. R. Sevenich, F. Bark, C. Crews, W. Anderson, C. Pye, K. Riddellova, J. Hradecky, E. Moravcova, K. Reineke and D. Knorr, *Innovative Food Sci. Emerging Technol.*, 2013, **20**, 42.
103. R. Sevenich, F. Bark, E. Kleinstueck, C. Crews, C. Pye, J. Hradecky, K. Reineke, M. Lavilla, I. Martinez-de-Maranon, J. C. Briand and D. Knorr, *Food Control*, 2015, **50**, 539.
104. D. F. Keenan, N. Brunton, F. Butler, R. Wouters and R. Gormley, *Innovative Food Sci. Emerging Technol.*, 2011, **12**, 261.
105. W. F. Gomes, B. K. Tiwari, Ó. Rodriguez, E. S. de Brito, F. A. N. Fernandes and S. Rodrigues, *Food Chem.*, 2017, **218**, 261.
106. F. M. Rodrigues, A. Rosenthal, J. H. Tiburski and A. G. da Cruz, *Food Sci. Technol.*, 2016, **36**, 1.
107. P. E. Bouton, A. L. Ford, P. V. Harris, J. J. MacFarlane and J. M. O'shea, *J. Food Sci.*, 1977, **42**, 132.
108. H. J. Ma, D. A. Ledward, A. I. Zamri, R. A. Frazier and G. H. Zhou, *Meat Sci.*, 2004, **68**, 347.
109. A. I. Zamri, D. A. Ledward and R. A. Frazier, *J. Agric. Food Chem.*, 2006, **54**, 2992.
110. S. Jung, M. Ghoul and M. de Lamballerie-Anton, *Meat Sci.*, 2000, **56**, 239.
111. S. Ichinoseki, T. Nishiumi and A. Suzuki, *J. Food Sci.*, 2006, **71**, E276.
112. S. Jung, M. Ghoul and M. de Lamballerie-Anton, *Lebensm.-Wiss. Technol.*, 2003, **36**, 625.
113. S. Jung, M. de Lamballerie-Anton, R. G. Taylor and M. Ghoul, *J. Agric. Food Chem.*, 2000, **48**, 2467.
114. R. Buckow, B. Q. Truong and C. Versteeg, *Food Chem.*, 2010, **120**, 474.
115. A. Suzuki, M. Watanabe and Y. Ikeuchi, *Meat Sci.*, 1993, **35**, 17.
116. C. M. Crehan, D. J. Troy and D. J. Buckley, *Meat Sci.*, 2000, **55**, 123.
117. D. Hygreeva and M. C. Pandey, *Trends Food Sci. Technol.*, 2016, **54**, 175.
118. A. L. Sikes, A. B. Tobin and R. K. Tume, *Innovative Food Sci. Emerging Technol.*, 2009, **10**, 405.
119. T. Iwasaki, K. Noshitoya, N. Saitoh, K. Okano and K. Yamamoto, *Food Chem.*, 2006, **95**, 474.
120. F. Fernandez-Martin, S. Cofrades, J. Carballo and F. Jimenez-Colmenero, *Meat Sci.*, 2002, **61**, 15.
121. P. Trespalacios and R. Pla, *Food Chem.*, 2007, **104**, 1718.
122. E. Fulladosa, X. Serra, P. Gou and J. Arnau, *Meat Sci.*, 2009, **82**, 213.
123. G.-P. Hong, S.-H. Ko, M.-J. Choi and S.-G. Min, *Meat Sci.*, 2008, **79**, 236.
124. G.-P. Hong, S.-G. Min, S.-E. Ko and M.-J. Choi, *Int. J. Food Sci. Technol.*, 2008, **43**, 1484.
125. A. Grossi, J. Søltoft-Jensen, J. C. Knudsen, M. Christensen and V. Orlien, *Meat Sci.*, 2012, **92**, 481.
126. C. C. O'Flynn, M. C. Cruz-Romero, D. Troy, A. M. Mullen and J. P. Kerry, *Meat Sci.*, 2014, **96**, 1266.

127. A. Tamm, T. Bolumar, B. Bajovic and S. Toepfl, *Innovative Food Sci. Emerging Technol.*, 2016, **36**, 294.
128. M. Clariana, L. Guerrero, C. Sárraga, I. Díaz, Á Valero and J. A. García-Regueiro, *Innovative Food Sci. Emerging Technol.*, 2011, **12**, 456.
129. K. Angsupanich, M. Edde and D. A. Ledward, *J. Agric. Food Chem.*, 1999, **47**, 92.
130. R. Cheret, N. Chapleau, C. Delbarre-Ladrat, V. Verrez-Bagnis and M. de Lamballerie, *J. Food Sci.*, 2005, **70**, 477.
131. J. Gomez-Estaca, M. E. Lopez-Caballero, M. C. Gomez-Guillen, A. Lopez, de Lacey and P. Montero, *Innovative Food Sci. Emerging Technol.*, 2009, **10**, 148.
132. R. Mengden, A. Röhner, N. Sudhaus and G. Klein, *Innovative Food Sci. Emerging Technol.*, 2015, **32**, 9.
133. M. E. Lopez-Caballero, M. C. Gomez-Guillen, M. Perez-Mateos and P. Montero, *J. Food Sci.*, 2005, **70**, M166.
134. J. C. Ramirez-Suarez and M. T. Morrissey, *Innovative Food Sci. Emerging Technol.*, 2006, **7**, 19.
135. R. M. Uresti, G. Valazquez, M. Vazquez, J. A. Ramirez and J. A. Torres, *Eur. Food Res. Technol.*, 2005, **220**, 113.
136. J. Varith, G. V. Barbosa-Canovas and B. G. Swanson, in *Innovations in Food Processing*, ed. G. V. Barbosa-Canovas and G. W. Gould, Technomic Publishing, Lancaster, PA, 2000, pp. 103–110.
137. J. Varith, G. V. Barbosa-Canovas and B. G. Swanson, in *Innovations in Food Processing*, ed. G. V. Barbosa-Canovas and G. W. Gould, Technomic Publishing Company, Inc., Lancaster, PA, 2000, pp. 111–122.
138. A. Matser and R. Timmermans in *High Pressure Processing of Foods*, ed. V. M. Balasubramaniam *et al.*, Springer Science+Business Media, New York, NY, 2016, pp. 541–551.
139. R. Tangwongchai, D. A. Ledward and J. M. Ames, *J. Agric. Food Chem.*, 2000, **48**, 1434.
140. D. N. Sila, T. Duvetter, A. de Roeck, I. Verlent, C. Smout, G. K. Moates, B. P. Hills, K. K. Waldron, M. Hendrickx and A. Van Loey, *Trends Food Sci. Technol.*, 2008, **19**, 309.
141. O. Schluter, J. Foerster, M. Geyer, D. Knorr and W. B. Herppich, *Food Bioprocess Technol.*, 2009, **2**, 291.
142. N. S. Terefe, P. Tepper, A. Ullman, K. Knoerzer and P. Juliano, *Innovative Food Sci. Emerging Technol.*, 2016, **33**, 56.
143. G. I. Denoya, S. R. Vaudagna, V. C. Chamorro, M. F. Godoy, C. O. Budde and G. A. Polenta, *LWT – Food Sci. Technol.*, 2017, **78**, 367.
144. G. Gao, P. Ren, X. Cao, B. Yan, X. Liao, Z. Sun and Y. Wang, *Food Bioprod. Process*, 2016, **100**, 221.
145. F. Speroni, S. Jung and M. de Lamballerie, *J. Food Sci.*, 2010, **74**, E30.
146. E. Molina, A. B. Defaye and D. A. Ledward, *Food Hydrocolloids*, 2002, **16**, 625.
147. N. Kajiyama, S. Isobe, K. Uemura and A. Noguchi, *Int. J. Food Sci. Technol.*, 1995, **30**, 147.

148. S. Saowapark, A. Apichartsrangkoon and A. E. Bell, *Food Chem.*, 2008, **197**, 984.
149. H. Zhang, L. Li and G. S. Mittal, *J. Food Process. Preserv.*, 2009, **33**, 560.
150. S. Volk Deterres and S. Jung, Annual Meeting and Food Exposition of the Institute of Food Technologists, Anaheim, CA, June 6–10, 2009, Abstract 125-08.
151. S. Savadkoohi, A. Bannikova, N. Mantri and S. Kasapis, *Food Hydrocolloids*, 2016, **53**, 115.
152. P. Juliano in *High Pressure Processing of Foods*, ed. V. M. Balasubramaniam *et al.*, Springer Science+Business Media, New York, NY, 2016, pp. 671–686.
153. T. Huppertz, M. A. Smiddy, V. K. Upadhyay and A. L. Kelly, *Int. J. Dairy Technol.*, 2006, **59**, 58.
154. A. J. Trujillo, M. Capellas, J. Saldo, R. Gervilla and B. Guamis, *Innovative Food Sci. Emerging Technol.*, 2002, **3**, 295.
155. M. F. San Martin-Gonzalez, J. Welti-Chanes and G. V. Barbosa-Canovas, *Food Rev. Int.*, 2006, **22**, 275.
156. W. Messens, J. Van Camp and K. Dewettinck, in *Dairy Processing*, ed. G. Smit, CRC Press, Boca Raton, FL, 2003, pp. 310–332.
157. T. Considine, H. A. Patel, S. G. Anema, H. Singh and L. K. Creamer, *Innovative Food Sci. Emerging Technol.*, 2007, **8**, 1.
158. M. A. Drake, S. L. Harrison, M. Asplund, G. Barbosa-Canovas and B. G. Swanson, *J. Food Sci.*, 1997, **62**, 843.
159. M. F. San Martin-Gonzalez, J. J. Rodriguez, S. Gurram, S. Clark, B. G. Swanson and G. V. Barbosa-Canovas, *Lebensm.-Wiss. Technol.*, 2007, **40**, 697.
160. M. Linton, A. B. Mackle, V. K. Upadhyay, A. L. Kelly and M. F. Patterson, *Innovative Food Sci. Emerging Technol.*, 2008, **9**, 423.
161. V. Ferragut, V. M. Martinez, A. J. Trujillo and B. Guamis, *Milchwissenschaft*, 2000, **55**, 267.
162. D. E. Johnston, B. A. Austin and R. J. Murphy, *Milchwissenschaft*, 1993, **48**, 206.
163. E. C. Needs, M. Capellas, A. P. Bland, P. Manoj, D. Macdougal and G. Paul, *J. Dairy Res.*, 2000, **67**, 329.
164. M. N. Buffa, A. J. Trujillo, M. Pavia and B. Guamis, *Int. Dairy J.*, 2001, **11**, 927.
165. M. Schwertfeger and W. Buchheim, *Int. Dairy J.*, 1999, **9**, 487.
166. J. Van Camp and A. Huyghebaert, *Food Chem.*, 1995, **54**, 357.
167. J. Van Camp and A. Huyghebaert, *Lebensm.-Wiss. Technol.*, 1995, **28**, 111.
168. J. Van Camp, W. Messens, J. Clement and A. Huyghebaert, *Food Chem.*, 1997, **60**, 417.
169. J. Van Camp, W. Messens, J. Clement and A. Huyghebaert, *J. Agric. Food Chem.*, 1997, **45**, 1600.
170. B. Fertsch, M. Muloler and J. Hinrichs, *Innovative Food Sci. Emerging Technol.*, 2003, **4**, 143.

171. M.-H. Famelart, L. Chapron, M. Piot, G. Brule and C. Durier, *J. Food Eng.*, 1998, **36**, 149.
172. C. Kanno, T. H. Mu, T. Hagiwara, M. Ametani and N. Azuma, *J. Agric. Food Chem.*, 1998, **46**, 417.
173. W. Messens and K. Dewettinck, *Int. Dairy J.*, 1999, **9**, 569.
174. W. Messens, K. Dewettinck, J. Van Camp and H. Huyghebaert, *Lebensm.-Wiss. Technol.*, 1998, **31**, 552.
175. N. M. Rynne, T. P. Beresford, T. P. Guinee, E. Sheehan, C. M. Delahunty and A. L. Kelly, *Innovative Food Sci. Emerging Technol.*, 2008, **9**, 429.
176. G. I. Katsaros, M. N. Giannoglou and P. S. Taoukis, *J. Food Sci.*, 2009, **74**, E219.
177. D. E. Johnston and P. C. Darcy, *Milchwissenschaft*, 2000, **55**, 617.
178. D. E. Johnston, M. O'Hagan and D. W. Balmer, *Milchwissenschaft*, 2002, **57**, 198.
179. S. Garde, J. L. Arques, P. Gaya, M. Medina and M. Nunez, *Int. Dairy J.*, 2007, **17**, 1424.
180. W. Messens, D. Van de Walle, J. Arevalo, K. Dewittinck and A. Huyghebaert, *Int. Dairy J.*, 2000, **10**, 359.
181. M. K. Walker, D. F. Farkas, V. Loveridge and L. Meunier-Goddik, *Int. J. Food Sci. Technol.*, 2006, **41**, 464.
182. B. de Ancos, M. Pilar Cano and R. Gomez, *Int. Dairy J.*, 2000, **10**, 105.
183. A. Fitria Devi, R. Buckow, Y. Hemar and S. Kasapis, *J. Food Eng.*, **114**, 106.
184. A.J. Trujillo, V. Ferragut, B. Juan, A.X. Roig-Sagues and B. Guamis in *High Pressure Processing of Foods*, ed. V. M. Balasubramaniam *et al.*, Springer Science+Business Media, New York, NY, 2016, pp. 553–590.
185. G. Kresic, V. Lelas, Z. Herceg and A. Rezek, *Lait*, 2006, **86**, 303.
186. G. Kresic, V. Lelas, A. R. Jambrak, Z. Herceg and S. R. Brncic, *J. Food Eng.*, 2008, **87**, 64.
187. R. Lopez-Fandino, *Crit. Rev. Food Sci. Nutr.*, 2006, **46**, 351.
188. S. Iametti, P. Transidico, F. Bonomi, G. Vecchio, P. Pittia, P. Rovere and G. Dall'Aglio, *J. Agric. Food Chem.*, 1997, **45**, 23.
189. S. G. Anema, *Int. Dairy J.*, 2008, **18**, 228.
190. F. Speroni, M. C. Puppo, N. Chapleau, M. de Lamballerie, O. Castellani, M. C. Anon and M. Anton, *J. Agric. Food Chem.*, 2005, **53**, 5719.
191. J. Van der Plancken, A. V. Loey and M. E. Hendrickx, *J. Food Eng.*, 2007, **78**, 1410.
192. J. Ahmed, H. S. Ramaswamy, I. Alli and M. Ngadi, *Lebensm.-Wiss. Technol.*, 2003, **36**, 517.
193. R. Lakshmanan, M. De Lamballerie and S. Jung, *J. Food Sci.*, 2006, **71**, E384.
194. M. C. Puppo, F. Speroni, N. Chapleau, M. de Lamballerie, M. C. Anon and M. Anton, *Food Hydrocolloids*, 2005, **19**, 289.
195. N. Chapleau, Y. Chausse, E. Tronc and M. de Lamballerie-Anton, in *Recent Advances in Agricultural and Food Chemistry: Food Emulsions and Dispersion*, ed. M. Anton, Research Signpost, Kerala, 2002, pp. 127–141.

196. E. Molina, A. Papadopoulou and D. A. Ledward, *Food Hydrocolloids*, 2001, **15**, 263.
197. H. Wackerbarth, U. Kuhlmann, F. Tintchev, V. Heinz and P. Hildebrandt, *Food Chem.*, 2009, **115**, 1194.
198. R. Lakshmanan, D. Miskin and J. R. Piggott, *J. Sci. Food Agric.*, 2005, **85**, 655.
199. M. Capellas, M. Mor-Mur, E. Sendra and B. Guamis, *Int. Dairy J.*, 2001, **11**, 165.
200. M. P. Cano and B. de Ancos, in *Novel Food Processing Technologies*, ed. G. V. Barbosa-Canovas, M. S. Tapia and M. P. Cano, CRC Press, Boca Raton, FL, 2005, pp. 283–310.
201. P. Dervisi, J. Lamb and I. Zabetakis, *Food Chem.*, 2001, **73**, 85.
202. A. M. Matser, E. R. Knott, P. G. M. Teunissen and P. V. Bartels, *J. Food Eng.*, 2000, **45**, 11.
203. P. B. Cheah and D. A. Ledward, *Meat Sci.*, 1996, **43**, 123.
204. P. B. Cheah and D. A. Ledward, *J. Am. Oil Chem. Soc.*, 1995, **72**, 1059.
205. H. J. Ma and D. A. Ledward, *Food Chem.*, 2007, **104**, 1575.
206. V. Orlien, E. Hansen and L. H. Skibsted, *Eur. Food Res. Technol.*, 2000, **211**, 99.
207. N. Bragagnolo, B. Danielsen and L. H. Skibsted, *Innovative Food Sci. Emerging Technol.*, 2007, **8**, 24.
208. L. R. B. Mariutti, V. Orlien, N. Bragagnolo and L. H. Skibsted, *Eur. Food Res. Technol.*, 2008, **227**, 337.
209. P. A. Vazquez-Landaverde, J. A. Torres and M. C. Qian, *J. Agric. Food Chem.*, 2006, **54**, 9184.
210. X. Liu, J. R. Powers, B. G. Swanson, H. H. Hill and S. Clark, *J. Food Sci.*, 2005, **70**, C581.
211. G.-C. Yen and H.-T. Lin, *J. Agric. Food Chem.*, 1999, **47**, 2082.
212. Y. Lambert, G. Demazeau, A. Largeteau and J.-M. Bouvier, *Food Chem.*, 1999, **67**, 7.
213. I. Zabetakis, A. Koulentianos, E. Orruno and I. Boyes, *Food Chem.*, 2000, **71**, 51.
214. E. N. C. Mills, A. I. Sancho, N. M. Rigby, J. A. Jenkins and A. R. Mackie, *Mol. Nutr. Food Res.*, 2009, **53**, 963.
215. E. Penas, H. Snel, R. Floris, G. Presetamo and R. Gomez, *Int. Dairy J.*, 2006, **16**, 969.
216. E. Penas, G. Prestamo and R. Gomez, *Food Chem.*, 2004, **85**, 641.
217. M. Zeece, T. Huppertz and A. Kelly, *Innovative Food Sci. Emerging Technol.*, 2008, **9**, 62.
218. I. Lopez-Exposito, R. Chicon, J. Belloque, I. Recio, E. Alonso and R. Lopez-Fandino, *J. Agric. Food Chem.*, 2008, **56**, 11809.
219. A. Quiros, R. Chichon, I. Recio and R. Lopez-Fandino, *Food Chem.*, 2007, **104**, 1734.
220. K. Olsen, K. R. Kristiansen and L. F. Skibsted, *Food Chem.*, 2003, **80**, 255.
221. E. Penas, G. Prestamo, F. Polo and R. Gomez, *Food Chem.*, 2006, **99**, 569.

222. R. Chicon, J. Belloque, E. Alonso and R. Lopez-Fandino, *Int. Dairy J.*, 2008, **18**, 367.
223. S.-W. Yin, C.-H. Tang, Q.-B. Wen, X.-Q. Yang and L. Li, *Food Chem.*, 2008, **110**, 938.
224. T. Kato, E. Katayama, S. Matsubara, Y. Omi and T. Matsuda, *J. Agric. Food Chem.*, 2000, **48**, 3124.
225. M. Lavilla, J. Ocajo, A. Diaz-Perales and P. Gamboa, *Innovative Food Sci. Emerging Technol.*, 2016, **38**, 334.
226. S. Jung in *High Pressure Processing of Foods*, ed. V. M. Balasubramaniam *et al.*, Springer Science+Business Media, New York, NY, 2016, pp. 173–192.
227. H.-M. Bi, S.-Q. Zhang, C.-J. Liu and C.-Z. Wang, *J. Food Process Eng.*, 2009, **32**, 53.
228. Z. Shouqin, C. Ruizhan and W. Changzheng, *J. Food Eng.*, 2007, **79**, 1.
229. X. Jun, *J. Food Eng.*, 2009, **94**, 105.
230. K. N. Prasad, B. Yang, N. Ruenroengklin, M. Zhao and Y. Jiang, *J. Food Process Eng.*, 2009, **32**, 828.
231. S. Zhang, J. Xi and C. Wang, *J. Chem. Technol. Biotechnol.*, 2005, **80**, 50.
232. S. Zhang, R. Chen and C. Wang, *J. Food Eng.*, 2007, **79**, 1.
233. S. Zhang, H.-M. Bi and C.-J. Liu, *Sep. Purif. Technol.*, 2007, **57**, 277.
234. K. N. Prasad, E. Yang, C. Yi, M. Zhao and Y. Jiang, *Innovative Food Sci. Emerging Technol.*, 2009, **10**, 155.
235. K. N. Prasad, J. Hao, J. Shi, T. Liu, J. Li, X. Wei, S. Qiu, S. Xue and Y. Jiang, *Innovative Food Sci. Emerging Technol.*, 2009, **10**, 413.
236. M. Corrales, S. Toepfl, P. Butz, D. Knorr and B. Tauscher, *Innovative Food Sci. Emerging Technol.*, 2008, **9**, 85.
237. E. Roldan-Marin, C. Sanchez-Moreno, R. Lloria, D. De Ancos and M. P. Cano, *Lebensm.-Wiss. Technol.*, 2009, **42**, 835.
238. A. Fernandez Garcia, P. Butz and B. Tauscher, *J. Food Sci.*, 2001, **66**, 1033.
239. M. C. Gomez-Guillen, B. Gimenez and P. Montero, *Food Hydrocolloids*, 2005, **19**, 923.
240. S. Jung, A. Mahfuz and D. Maurer, *J. Am. Oil Chem. Soc.*, 2009, **86**, 475.
241. K. Hori, Y. Matanabe, M. Kaneko, T. Sekimoto, Y. Sugimoto and T. Yamane, in *High Pressure and Biotechnology*, ed. C. Balny, R. Hayashi, K. Heremans and P. Masson, Colloque Inserm/John Libbey Eurotext, Montrouge, 1992, Vol. 224, pp. 499–507.
242. A. Sloan, *Food Technol.*, 2015, **69**, 29.
243. US Department of Agriculture, Published January 2014, https://www.fsis.usda.gov/wps/wcm/connect/d3373299-50e6-47d6-a577-e74a1e549fde/Controlling-Lm-RTE-Guideline.pdf?MOD=AJPERES.
244. G. Eisenbrand, *Mol. Nutr. Food Res.*, 2005, **49**, 1168.
245. ANSES, https://www.anses.fr/fr/system/files/AAAT2010sa0193.pdf – Saisine n°2010-SA-0193, 2010.
246. M. Dumoulin, *Ind. Aliment. Agric.*, 1998, **115**, 15.

247. https://www.visiongain.com/Report/1406/The-Food-High-Pressure-Processing-%28HPP%29-Technologies-Market-Forecast-2015-2025). Accessed July 5, 2017.
248. N. Martinez-Belkin, *Bevnet*, 2016, September, 28.
249. A. Suzuki, in *Trends in High Pressure Bioscience and Biotechnology*, ed. R. Hayashi, *Elsevier Science*, Amsterdam, 2002, 19, 365.
250. R. Hayashi, in *Rice is Life: Scientific Perspectives for the 21st Century, Proceedings of the World Rice Research Conference*, ed. K. Toriyama, K. L. Heong and B. Hardy, International Rice Research Institute, 2005, p. 287.
251. A. Yamazaki, *Foods Food Ingredients J. Jpn.*, 2005, **210**, 29.
252. S. Ennen, *Food Process.*, 2001, January, 21.
253. J. L. Britcher, *Food Saf. Mag.* 2008, June/July, 40.
254. C. Tonello, D. Flores, D. Wilches, F. Purroy, M. Onrubia, R. Peregrina, in VIII World Avocado Congress Proceedings, 2015.
255. USDA-FSIS, in *FSIS Compliance Guideline: Controlling Listeria monocytogenes in Post-lethality Exposed Ready-to-Eat Meat and Poultry Products*, January, 2014, 143.
256. N. Grèbol, in *Pressure to Succeed Conference – An Insight into High Pressure Processing, Proceedings Book*, 2003, p. 1.
257. B. Salvage, FoodBusinessNews.net, 2014, April 3.
258. R. Gabett, Meating place – 2012, April: 37–43.
259. D. J. Castaldo, *Meat Process.*, 2003, 2.
260. J. Crews, *Meat Poult.*, 2007, 57.
261. J. Crews, *Meat Poult.*, 2007, 56.
262. B. Sims, *Meat Poult.*, 2017, 54.
263. E. Harms, *Fleischwirtsch. Int.*, 2006, **2**, 40.
264. C. Astruc, *LSA*, 2006, **1977**, 92.
265. F. Biaggi, *RIA*, 2016, **776**, 20.
266. C. Van der Plas, Distrifood, 2010, April 26, http://www.distrifood.nl/assortiment/nieuws/2010/4/zwanenberg-brengt-hpp-filet-americain-10155572.
267. M. Stone, FoodQuality News.com, 2011, February 15, http://www.foodqualitynews.com/Regulation-and-safety/HPP-technology-boosts-burgers-shelf-life-and-safety-says-Cargill.
268. M. Cérou and M. Guillon, *Process Aliment.*, 2015, **1325**, 27.
269. K. M. Muth, D. W. Anderson, S. A. Karns, B. C. Murray and J. L. Domanico, in Economic impacts of requiring post-harvest treatment of oysters, Section 3. Treatment technologies and costs, Final Report, Research Triangle Institute, North Carolina, 2000, pp. 3–11 to 3–15.
270. D. G. Hoover, *Food Technol.*, 2003, 1.
271. C. Kirkham, The Times-Picayune, 2011, October 5, http://www.nola.com/business/index.ssf/2009/11/processed_oysters_have_a_niche.html.
272. C. Barberis, *Animal Gourmet*, 2014, July 4, http://www.animalgourmet.com/2014/07/04/villa-de-patos-amor-a-primer-gusto/.
273. E. Han, The Sydney Morning Herald, 2016, June 1, http://www.smh.com.au/business/consumer-affairs/coldpressed-raw-milk-method-wins-regulatory-approval-20160601-gp8r76.html.

274. V. Balasubramaniam, V. Barbosa-Cánovas and H. Lelieveld, *High Pressure Processing of Food*, ed. Balasubramaniam, V. M., Barbosa-Cánovas, G. V., Lelieveld, H. L. 2016, 39–65.
275. D. Farkas, *High Pressure Processing of Food*, ed. Balasubramaniam, V. M., Barbosa-Cánovas, G. V., Lelieveld, H. L. 2016, 19–36.
276. A. Hernando Saiz, M. S. Tarrago, B. F. Purroy and S. C. Tonello, *Food Aust.*, 2008, **60**, 154–156.
277. H. Mujica-Paz, A. Valdez-Fragoso, C. Tonello-Samson, J. Welti-Chanes and A. Torres, *Food Bioprocess Technol.*, 2011, **4**, 969–985.
278. Government of Canada, Novel Food Information on: Applesauce and Applesauce/Fruit Blends Treated by High Hydrostatic Pressure, https://www.canada.ca/en/health-canada/services/food-nutrition/genetically-modified-foods-other-novel-foods/approved-products/applesauce-applesauce-fruit-blends-treated-high-hydrostatic-pressure.html.
279. Government of Canada, Novel Food Information on: Use of High Hydrostatic Pressure for the Control of L. monocytogenes in Ready to Eat (RTE) Meats and Poultry, https://www.canada.ca/en/health-canada/services/food-nutrition/genetically-modified-foods-other-novel-foods/approved-products/use-high-hydrostatic-pressure-control-listeria-monocytogenes-ready-meats-poultry.html.
280. Government of Canada, Novel Food Information on: Use of High Hydrostatic Pressure for Processing Ready to Eat Meat-containing Entrees, Salads and Products, https://www.canada.ca/en/health-canada/services/food-nutrition/genetically-modified-foods-other-novel-foods/approved-products/novel-food-information-use-high-hydrostatic-pressure-processing-ready-meat-containing-entrees-meat-containing-salads-meat-products-health.html.
281. Government of Canada, Novel Food Information on: Use of Multiple-cycle High Pressure Processing (HPP) Treatment of Ready To Eat (RTE) Meat and Poultry Products, https://www.canada.ca/en/health-canada/services/food-nutrition/genetically-modified-foods-other-novel-foods/approved-products/use-multiple-cycle-high-pressure-processing-treatment-ready-meat-poultry-products.html.
282. Government of Canada, Novel Food Information on: Avocado Pulp, Tomato-based Salsas, and Guacamole Treated by High Pressure Processing (HPP), https://www.canada.ca/en/health-canada/services/food-nutrition/genetically-modified-foods-other-novel-foods/approved-products/novel-food-information-avocado-pulp-tomato-based-salsas-guacamole-treated-high-pressure-processing.html.
283. Government of Canada, Novel Food Information on: High Pressure Processing (HPP)-Treated Raw Ground Beef (Fressure- fresh ground beef), https://www.canada.ca/en/health-canada/services/food-nutrition/genetically-modified-foods-other-novel-foods/approved-products/novel-food-information-high-pressure-processing-treated-ground-beef-fressure-fresh-ground-beef.html.
284. Government of Canada, Novel Food Information on: High Pressure Processing (HPP)-Treated Fruit and Vegetable-Based Smoothies,

https://www.canada.ca/en/health-canada/services/food-nutrition/genetically-modified-foods-other-novel-foods/approved-products/novel-food-information-high-pressure-processing-treated-fruit-vegetable-based-smoothies.html.
285. Government of Canada, Novel Food Information on: High Pressure Processing (HPP)-Treated Fruit and Vegetable-Based Juices, https://www.canada.ca/en/health-canada/services/food-nutrition/genetically-modified-foods-other-novel-foods/approved-products/novel-food-information-high-pressure-processing-treated-fruit-vegetable-based-juices.html.
286. Government of Canada, Novel Food Information on: High Pressure Processing (HPP)-Treated Egg Salad, Egg Dips, and Egg Spreads, https://www.canada.ca/en/health-canada/services/food-nutrition/genetically-modified-foods-other-novel-foods/approved-products/novel-food-information-high-pressure-processing-treated-egg-salad-egg-dips-egg-spreads.html.
287. Government of Canada, Novel Food Information on: High Pressure Processing (HPP)-Treated Raw Fruit Juices, https://www.canada.ca/en/health-canada/services/food-nutrition/genetically-modified-foods-other-novel-foods/approved-products/novel-food-information-high-pressure-processing-treated-fruit-juices.html.
288. R. Bergen, High Pressure Processing, Canadian Cattlemen, September 2016, p. 44, https://www.canadiancattlemen.ca/digital-edition/canadian-cattlemen-57.
289. Government of Canada, Health Canada Position - High Pressure Processing (HPP) Is No Longer a Novel Process, https://www.canada.ca/en/health-canada/services/food-nutrition/legislation-guidelines/guidance-documents/position-high-pressure-processing-no-longer-novel-process-treated-food-products-treated-food-products-2013.html.
290. Agence Française de Sécurité Sanitaire des Aliments, Avis de l'Agence française de sécurité sanitaire des aliments relatif à l'évaluation de l'impact d'un traitement de pascalisation (hautes pressions) sur des plats cuisinés, https://www.anses.fr/fr/system/files/AAAT2009sa0204.pdf.
291. Agence Française de Sécurité Sanitaire des Aliments, Avis de l'Agence française de sécurité sanitaire des aliments relatif à l'évaluation de l'impact d'un traitement par hautes pressions sur de la viande de volaille marinée et de la viande de volaille farcie, https://www.anses.fr/fr/system/files/AAAT2009sa0316.pdf.
292. Agence Nationale Sécurité Sanitaire Alimentaire Travail, Avis de l'Agence nationale de sécurité sanitaire de l'alimentation, de l'environnement et du travail, https://www.anses.fr/fr/system/files/AAAT2010sa0193.pdf.
293. Agence Française de Sécurité Sanitaire des Aliments, Avis de l'Agence française de sécurité sanitaire des aliments relatif à l'autorisation de mise sur le marché de magrets de canards séchés, ou séchés et fumés,

stabilisés par hautes pressions hydrostatiques comme nouvel aliment dans le cadre du règlement (CE) n° 258/97, https://www.anses.fr/fr/system/files/AAAT2007sa0164.pdf.
294. Agence Nationale Sécurité Sanitaire Alimentaire Travail, Avis de l'Agence nationale de sécurité sanitaire de l'alimentation, de l'environnement et du travail relatif à une demande d'avis relatif à une demande d'autorisation de mise sur le marché d'un nouvel aliment ou d'un ingrédient alimentaire : procédé de pasteurisation assisté par hautes pressions pour la transformation de foies gras, https://www.anses.fr/fr/system/files/ESPA2015SA0249.pdf.
295. Advisory Committee on Novel Foods and Processes, ACNFP Meeting minutes 4 July 2001, https://acnfp.food.gov.uk/meetings/82297/50meet/minutesofthe50thmeeting.
296. Advisory Committee on Novel Foods and Processes, Letter issued to Ata SpA regarding a request for a scientific opinion on high pressure processed (HPP) fruit and vegetable products, 2002, https://acnfp.food.gov.uk/acnfppapers/nfissues/AtaSpA.

CHAPTER 8

Ultrasonic Food Processing

THOMAS S. H. LEONG,[a,b] GREGORY J. O. MARTIN[a,c] AND MUTHUPANDIAN ASHOKKUMAR*[a,b]

[a] The ARC Dairy Innovation Hub, The University of Melbourne, Parkville, Victoria 3010, Australia; [b] School of Chemistry, The University of Melbourne, Parkville, Victoria 3010, Australia; [c] Department of Chemical Engineering, The University of Melbourne, Parkville, Victoria 3010, Australia

*Email: masho@unimelb.edu.au

8.1 Introduction

The application of ultrasonics in food processing is an area of growing interest, highlighted by a number of recent review articles and book chapters on the topic.[1–7] The applicability of ultrasound to food processing is diverse, with the capability to improve a wide range of conventional processes such as mixing, emulsification, extraction, separation/filtration, drying and pasteurisation/sterilisation. In the majority of these examples, ultrasound dramatically enhances the rate and effectiveness of the process, typically owing to increased heat and mass transfer. This often results in a significant reduction in the energy and chemical requirements compared with an equivalent conventional process, making ultrasound a promising alternative 'green' technology.[1]

Ultrasound can produce desired food processing outcomes *via* several mechanisms. Firstly, the propagation of an ultrasonic wave and its interaction with the surrounding medium can enhance mixing and other transport phenomena. Second, the application of ultrasound in a fluid will induce a phenomenon known as acoustic cavitation. The mechanical effects of the ultrasound wave, or the strong physical/chemical effects arising from

Green Chemistry Series No. 53
Alternatives to Conventional Food Processing: 2nd Edition
Edited by Andrew Proctor

Published by the Royal Society of Chemistry, www.rsc.org

acoustic cavitation, can benefit in a range of food applications. Ultrasound can accelerate a number of conventional processes such as extraction, drying and crystallisation. In addition, the unique combination of mechanisms through which ultrasound interacts with food media can be used to create novel food products, for example, products with encapsulated bioactives[8] or modified functional properties.[3] Innovative new food products with stabilised emulsions[9] or heat-stable whey protein concentrates[10,11] are examples that have been created using ultrasonic technology.

This chapter aims to provide an overview of how ultrasound can be used in specific applications for selected food systems, with a focus on recent developments and trends of interest. A brief background regarding the concepts, theory and mechanistic aspects of ultrasound is first presented. These concepts will be generally applicable to any food system.

8.2 Mechanisms Involved in Ultrasonic Food Processing

Ultrasound is, to put it simply, vibration that occurs at very high frequencies. These frequencies typically range from 20 kHz to 500 MHz, which are not audible to humans when transmitted through the air. Ultrasound can propagate through any medium with molecules that can vibrate. For food processing, ultrasound can be applied directly to fluids, transmitted into solids or transmitted through air, depending on the requirement for the specific application.

8.2.1 Acoustic Cavitation in Fluids

In food processing applications, ultrasound is most commonly delivered between 20 kHz and 3 MHz (the so-called power ultrasound region) into fluids to promote a process known as acoustic cavitation. Acoustic cavitation is the initiation, growth and collapse of a gas bubble within an oscillating pressure field[12] (Figure 8.1).

The ultrasonic pressure oscillations cause expansion and contraction of compressible nuclei present in the liquid. These are typically dissolved gas bubbles, but can also be pockets of gas trapped in crevices of reactor surfaces or on surfaces of motes/solid particles dispersed within a fluid. The bubbles expand during the negative pressure cycle and contract during the positive pressure cycle at an extremely rapid rate when subjected to oscillations at ultrasonic frequencies. The rapid oscillations are accompanied by mass transport of gas/vapour into and out of the bubbles during the expansion and contraction, respectively. If imbalanced, this results in progressive bubble growth. This process, known as rectified diffusion, is unique to bubbles oscillating within a sound field. In addition to rectified diffusion, coalescence of colliding gas bubbles[14] results in bubble growth in a sound field. These gas bubbles grow until they reach what is known as the

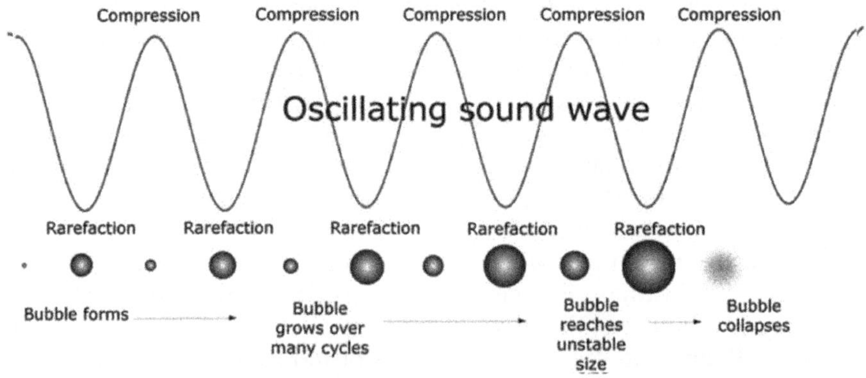

Figure 8.1 Growth and collapse of a bubble influenced by an oscillating sound wave. Reproduced with permission from T. S. H. Leong *et al.*, *Acoust. Aust.*, 2011, **39**, 54–63,[13] Copyright Acoustics Australia.

bubble resonance size, at which point they collapse. The sound field-influenced collapse of a bubble is a high-energy event that produces localised regions of extreme temperature and intense physical forces. Bubbles driven at low ultrasonic frequency (~20–100 kHz) may collapse extremely violently, releasing sufficient energy to produce temperatures up to 10 000 K within the bubble core and pressures of several hundred atmospheres within a few hundred micrometres of the bubble collapse point.[15] This can also lead to the formation of highly reactive radicals, which can be used to promote chemical reactions. In water, for example, hydrogen and hydroxyl radicals can be formed from the splitting of the water molecule by pyrolysis.[16] Both the physical shear and radical formation can be beneficially employed to process foods.

8.2.1.1 Frequency Characteristics

The effects of acoustic cavitation are dependent on the applied frequency of the ultrasound. The intensity of bubble collapse (*i.e.* amount of energy released) and the maximum bubble size prior to collapse (resonance size) are correlated and approximately inversely proportional to the applied frequency.[13] A simple but approximate relationship that relates the resonance size of the bubble with the frequency is

$$F \times R \approx 3 \tag{8.1}$$

where F is the frequency in hertz and R is the bubble radius in metres.[12,17] The resonance size for an active bubble is usually smaller than this owing to the non-linear nature of the bubble pulsation[18] and experimental conditions such as liquid viscosity and external pressure.

Based on the acoustic energy, ultrasound is usually categorised into different regions along the frequency spectrum. The *power ultrasound* region[13] spans the low-frequency range 20–100 kHz. It is characterised by bubbles that grow to a large resonance size prior to collapse. Food processes that benefit from this include emulsification,[19] homogenisation[20] and extraction.[21] This category of ultrasound is usually delivered at a high energy density in the order of 10–1000 W cm^{-2}.

The region between 100 and 1000 kHz can be labelled *intermediate ultrasound*. This region tends to produce only moderately intense bubble collapse, but importantly generates the highest 'sonochemically active' bubble population. As such, highly efficient radical production is achieved in this region. Koda *et al.*[22] and Mason *et al.*[23] showed that peak radical production occurs between 400 and 800 kHz, although it also depends on the power applied and the physical and chemical properties of the fluid system. This intermediate ultrasound region is selected when chemical modification is the primary goal. As the chemical integrity of food components is generally desired, intermediate ultrasound is not typically considered for food applications as there is a risk of promoting undesired oxidation.

Above 1000 kHz, the physical effects of bubble collapse become relatively benign owing to the small bubble resonance size prior to collapse and a reduced proportion of bubbles undergoing cavitation owing to an increased cavitation threshold.[24] The cavitation threshold refers to the pressure (or size) above which bubble nucleation, a necessary precursor to cavitation, occurs. However, despite a reduction in cavitation activity, radical formation is still possible at frequencies around 1000 kHz,[25] provided that sufficient energy intensity is employed. This regime, typically labelled the *diagnostic* or *megasonic* region, is used if only gentle physical effects are desired. It is particularly useful for applications such as selective particle separation and product recovery,[26] where the ultrasound does not negatively impact upon the product integrity.

8.2.1.2 Radical Formation and Sonochemistry

Localised temperature increases caused by acoustic cavitation can be sufficient to split solvent/solute molecules that have diffused into the bubbles, forming radical species. In foods, water is usually the solvent present, resulting in the following reaction:[16]

$$H_2O \leftrightarrow 2H^{\bullet} + {}^{\bullet}OH \tag{8.2}$$

that is, water molecules can be split into highly reactive hydrogen and hydroxyl radicals. These radical species can be used to induce or speed up redox-type reactions. In food processing, ultrasound can be used to speed up oxidation reactions such as those that occur in alcohol fermentation.[27] The radical species can also be used to promote crosslinking of proteins to form larger networks and structures.[28] These protein networks can be used to coat bubbles, droplets or other templates, forming rigid spheres that can be useful for stable encapsulation of bioactive materials in foods.[8] The formation of protein microspheres by application of ultrasound results from a combination of shear-induced emulsification and radical formation.[28,29] The high shear and temperature can partially unfold the proteins,[30] rendering them more hydrophobic so that they can accumulate at the air or organic phase boundary and undergo crosslinking.[28] The protein crosslinking can be reversible, for instance through hydrophobic interactions and hydrogen bonding, or irreversible if covalent links are produced, for example, disulfide bonds. In the latter case, this can potentially be facilitated by free radicals generated through ultrasonic cavitation.[31]

8.2.2 Physical Effects of Ultrasound

The shear forces generated during bubble collapse occur in the form of pressure shockwaves, liquid micro-jets and acoustic streaming. Acoustic streaming is the propagation of disturbances in the fluid caused by the ultrasonically induced oscillation of gas bubbles. This results in localised shear forces in the immediate vicinity of the bubbles. Liquid micro-jets result from asymmetric and extremely rapid collapse of bubbles during cavitation, leading to the unidirectional expulsion of high-velocity jets into the surrounding fluid. Naudé and Ellis[32] hypothesised that the observed pitting of solid surfaces and particle size reduction of colloids on exposure to ultrasound was in fact due to the formation of these micro-jets. It has since been shown that micro-jets with velocities in the order of 100 m s^{-1} can be formed,[33] and that these can create pitting and erosion of surfaces.[34] Micro-jetting is also the primary cause of ultrasound-induced bulk mixing.

Symmetric bubble collapse generates shockwaves that propagate radially outwards from the collapse point into the surrounding fluid. These shockwaves can be useful for increasing the rates of mass transfer across interfacial boundaries, enhancing the efficiency of multi-phase reactions.[34] All of these physical effects are commonly utilised in, for example, laboratory ultrasonic baths to facilitate the cleaning of surfaces and increase the rate of dissolution of solids.

Although collapsing bubbles may reach temperatures of many thousands of degrees Celsius, these extreme temperatures are confined to small areas at the core of the collapsing bubble and near the bubble surface.[35] In the bulk solution, the increase in temperature resulting from a single bubble collapse is insignificant. Nevertheless, the dissipation of heat from many cavitation bubbles, vibrating transducer surfaces, shockwave propagation and acoustic streaming can all contribute to incremental increases to the surrounding bulk temperature. As foods are thermally sensitive, a form of temperature control (*e.g.* a cooling jacketed reactor) should be used to prevent undesired temperature effects, *e.g.* denaturation of proteins.

8.2.2.1 Acoustic Radiation Forces and Acoustic Streaming

The physical effects of ultrasound are not confined to those that result from bubble collapse. The propagation of the sound field itself through a fluid can induce forces as it becomes attenuated and absorbed by the medium. One category of forces useful in food processing applications is known as acoustic radiation forces.[36]

When a sound wave is reflected upon itself, the reflected wave can superimpose to form what is known as an acoustic standing wave. Such waves are characterised by regions of high local pressure, where constructive superimposition occurs, known as pressure antinodes, and regions of minimum local pressure where destructive interference occurs, known as pressure nodes.

When a standing wave field is sustained in a medium containing a suspension of food particles or emulsion droplets, the wave will be partially scattered by the particles if there is an acoustic impedance mismatch between the fluid and the material of the particles. This scattering of the wave will produce what is known as the primary acoustic radiation force.[37] This force acts on the particles and droplets, causing them to migrate towards either the node or antinode of the standing wave. Where they move is determined by the acoustic contrast factor.

The acoustic contrast factor ϕ depends on the density and compressibility of the particles and their surrounding medium and can be calculated using the equation

$$\phi = \frac{5\rho_p - 2\rho_l}{2\rho_p + \rho_l} - \frac{\beta_p}{\beta_l} \tag{8.3}$$

where ρ is the density, β is the compressibility and the subscripts l and p refer to the liquid medium and the particles, respectively.[38] Particles move towards the pressure nodes of the standing wave when the contrast factor has a positive sign and towards the pressure antinodes when the contrast factor has a negative sign. Generally, food particles that have a density less than that of the surrounding fluid will have a contrast factor that drives them towards the pressure maxima. Those with a density higher than that of the surrounding fluid will be driven towards the pressure minima. The positioning of individual droplets or particles on pressure nodes or antinodes within the standing wave will cause them to aggregate or coalesce rapidly into larger entities. Figure 8.2 shows a schematic representation of particles moving towards pressure nodal planes and pressure antinodal planes.

Another important fluid flow effect caused by a propagating sound wave is known as acoustic streaming.[40–42] This fluid flow is generated from the spatial or temporal variations in a pressure field. In acoustic streaming, the whole fluid is set into motion, typically in a spinning/curling motion that drives mixing and enhanced mass transfer. In general, the higher the acoustic energy input into the system, the stronger these flows become.

Three main types of acoustic streaming phenomena are often observed: (i) Eckart streaming, (ii) Rayleigh streaming and (iii) Schlichting streaming. Eckart streaming[43] refers to the type of streaming that is generated in a

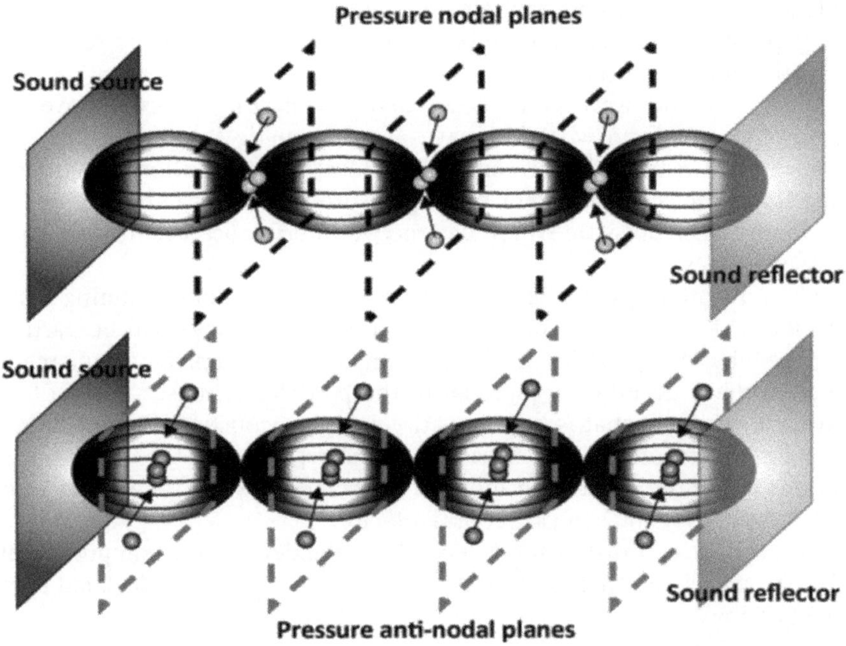

Figure 8.2 Migration of food materials within an acoustic standing wave field. Reprinted with permission from ref. 39. Copyright 2013 American Chemical Society.

large, non-uniform sound field, and occurs due to intense ultrasound being attenuated with distance from the sound source. Rayleigh streaming[44] refers to vortex-like streaming that is formed in a standing wave field, and develops from a streaming vorticity within the boundary layer at the solid vessel wall, driving streaming in the bulk fluid. Schlichting streaming[45] is the vortex-like motion that occurs around an object placed in the sound field, and is usually on a scale much smaller than the wavelength of the sound.

The complex relationships between these mechanisms of power ultrasound and their effects and relevance to food processing applications are described schematically in Figure 8.3.

8.3 Delivery of Ultrasound into Food

The mechanical vibrations that produce ultrasonic waves are generated by transducers and can be delivered into food systems *via* direct or indirect contact with emitter devices. The ultrasound acts on the food through some or all of the mechanisms described in Figure 8.3 to impart the desired improvement (*e.g.* increased mass transfer kinetics during extraction). Successful application of ultrasound in food processing requires the correct combination of equipment and parameters specific to the application. This section aims to provide an understanding of how this can be achieved.

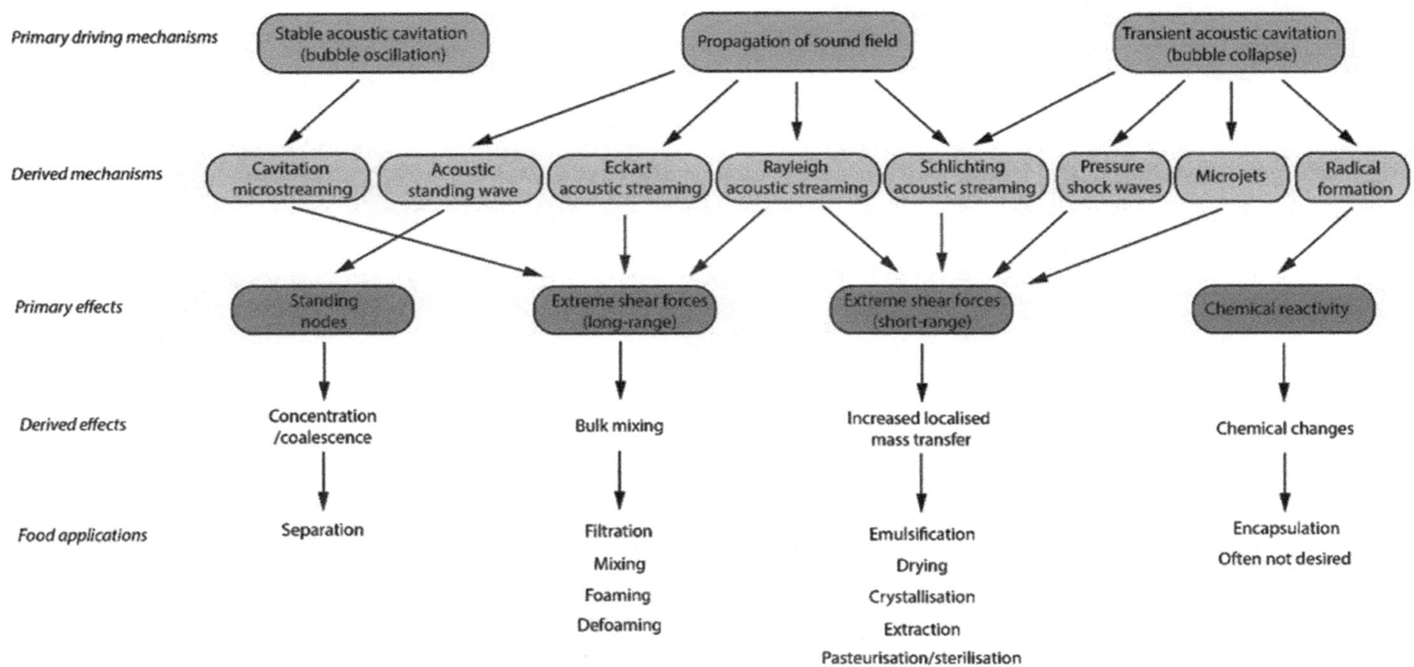

Figure 8.3 Schematic representation of the interconnections between the mechanisms and effects of power ultrasound.

8.3.1 Generation and Delivery of Ultrasonic Waves

As with all longitudinal waves, the defining characteristics of ultrasound are the frequency, wavelength, amplitude and velocity (speed and direction). The frequency of the ultrasound is determined by the frequency of mechanical vibration in a transducer. These vibrations are propagated at increased amplitude by an emitter. The emitter mechanically vibrates at the desired frequency to create a longitudinal pressure wave that emanates into the medium, creating an approximately conical volume of medium subjected to ultrasonic waves. A typical conical bubble cloud structure created from an ultrasonic horn is depicted in Figure 8.4.

Ultrasonic processing devices are usually specified on the basis of power (W or J s^{-1}). The power of the devices determines the amplitude of the pressure difference between rarefaction and compression and the volume of material into which the vibration is directly translated (*i.e.* the surface area of the probe). Two emitters can have the same power (and frequency), but different surface areas and therefore amplitudes (*i.e.* small surface area and high amplitude, or large surface area and low amplitude). The surface area of the probe will determine how much of the fluid is exposed to the pressure oscillations (*i.e.* the diameter of the cylindrical volume of medium

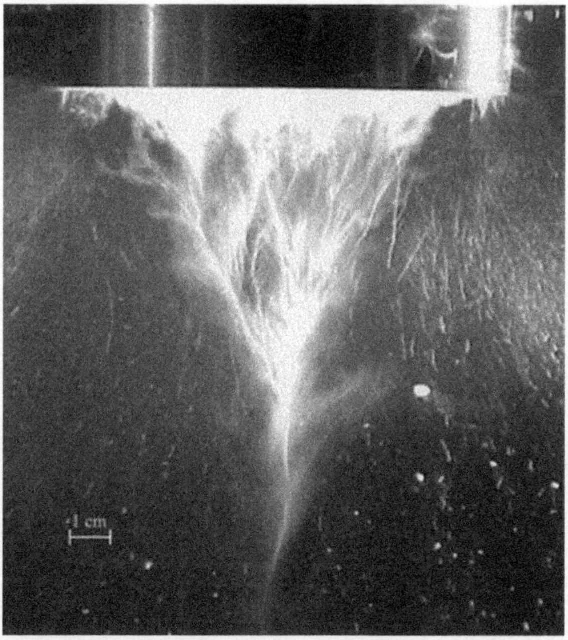

Figure 8.4 A conical bubble cloud structure generated from a typical ultrasonic horn/probe.[46]
Reprinted from *Ultrasonics Sonochemistry*, **17**, B. Dubus *et al.*, On the physical origin of conical bubble structure under an ultrasonic horn, 810–818, Copyright (2010), with permission from Elsevier.

experiencing ultrasonic vibrations), the amplitude represents the magnitude of the pressure oscillations and the frequency represents how rapidly those pressure oscillations occur.

The pressure oscillations (sound waves) propagate away from the emitter at a speed dependent on the elastic bulk modulus and density of the medium (according to the Newton–Laplace equation). The amplitude of the pressure oscillations decreases with increasing distance away from the emitter as energy is dissipated as heat (due to inelastic vibration of the medium), kinetic energy (bulk movement of the media through streaming), interfacial energy (in the case of emulsification), light (in the case of sonoluminescence) or chemical energy (in the case of radical formation). Beyond a certain distance from the emitter (primarily dependent on the viscosity of the medium), the sound will have completely dissipated (*i.e.* no further pressure oscillations will be present in the medium) due to attenuation. If a surface is present before this point, the sound can reflect or be partially transmitted into the new material.

8.3.2 Ultrasonic Parameters for Food Processing

As explained in Section 8.2, the physical and chemical effects of ultrasound depend on the frequency and intensity of ultrasound and also the properties of the medium (see the summary in Table 8.1). The desired ultrasonic effects, and therefore the required ultrasonic frequency, intensity and exposure, are specific to different applications (see Table 8.1).

The frequency of ultrasound is simple to define and control. In theory, the intensity of ultrasound can be simply defined as the amplitude of the ultrasound (magnitude of the pressure fluctuations) and the exposure as the time for which the medium has been subjected to these fluctuations. In practice, however, this is less straightforward as elements of fluid closer to the emitter will experience sound at a higher amplitude (*i.e.* greater pressure oscillations) than those further away. This is further complicated by bulk mixing of the material in relation to the emitter, resulting in dynamic exposure of different elements of the material to ultrasound and its physical and chemical effects. Achieving reasonably consistent exposure of all elements of the food material to ultrasound and all its physical and chemical effects therefore requires consideration of the bulk mixing in relation to processing time and ultrasound intensity.

A potentially useful measure that is often reported when applying ultrasound to food processing is the so-called energy intensity. This is the total energy drawn by ultrasonic device per unit volume of material processed (*e.g.* $\mathrm{J\,mL^{-1}}$) and is the product of ultrasound power intensity (*e.g.* $\mathrm{W\,mL^{-1}}$) and exposure duration (*e.g.* s) (where the power density is a product of the number and surface area of the emitters and the amplitude of the ultrasound generated). For instance, a given energy density can be achieved by long exposure at low intensity or a short exposure at high intensity. Whether or not similar results are achieved at different intensity–time combinations

Table 8.1 General effects of ultrasound as a function of frequency, intensity and medium properties and examples of food processing applications.

Frequency (kHz)	Intensity ($W\,cm^{-2}$)	Medium properties	Dominant effects	Food processing examples
20–100	10–1000	Liquid	Transient cavitation Acoustic streaming	Emulsification of oil in water[19] Extraction of saponins from ginseng[47] Reduction of viscosity in heated whey protein concentrate[48]
20–100	10–1000	Solid	Acoustic vibration Transient cavitation	Enhancing thawing and freezing of fish[49] Enhancing drying of fruits[50]
20–100	10–1000	Air/foam	Acoustic attenuation Soundwave propagation and absorbance	Defoaming of frothy beverages during bottling[51]
100–1000	1–100	Liquid	Sonochemistry/radicals Acoustic streaming Stable cavitation	Degradation of polysaccharides, *e.g.* chitosan, starch[52]
100–3000	1–100	Liquid	Acoustic standing waves/ pressure nodes Acoustic streaming Stable cavitation	Separation of cream from milk[53] Recovery of oil from palm oil milling[54]

for a given energy density depends on the system and application (see Table 8.1). Regardless, application of ultrasound at a given energy density will produce a characteristic amount of heat that will increase the temperature of the material unless removed from the system.

8.3.3 Ultrasonic Processing Equipment

Delivering the right dose of ultrasound for a particular food application relies on having the correct equipment and setup. This requires consideration of the spatial and temporal contact arrangement between the emitter(s) and the food, and also how the acoustic wave penetrates the food medium. Typically, the acoustic wave is of highest intensity near the surface of the emitter (Figure 8.5). The conical 'bubble cloud,' shown in Figure 8.4, is formed within the high-intensity region near the surface, and represents a region of maximum cavitation activity. The wave becomes attenuated and

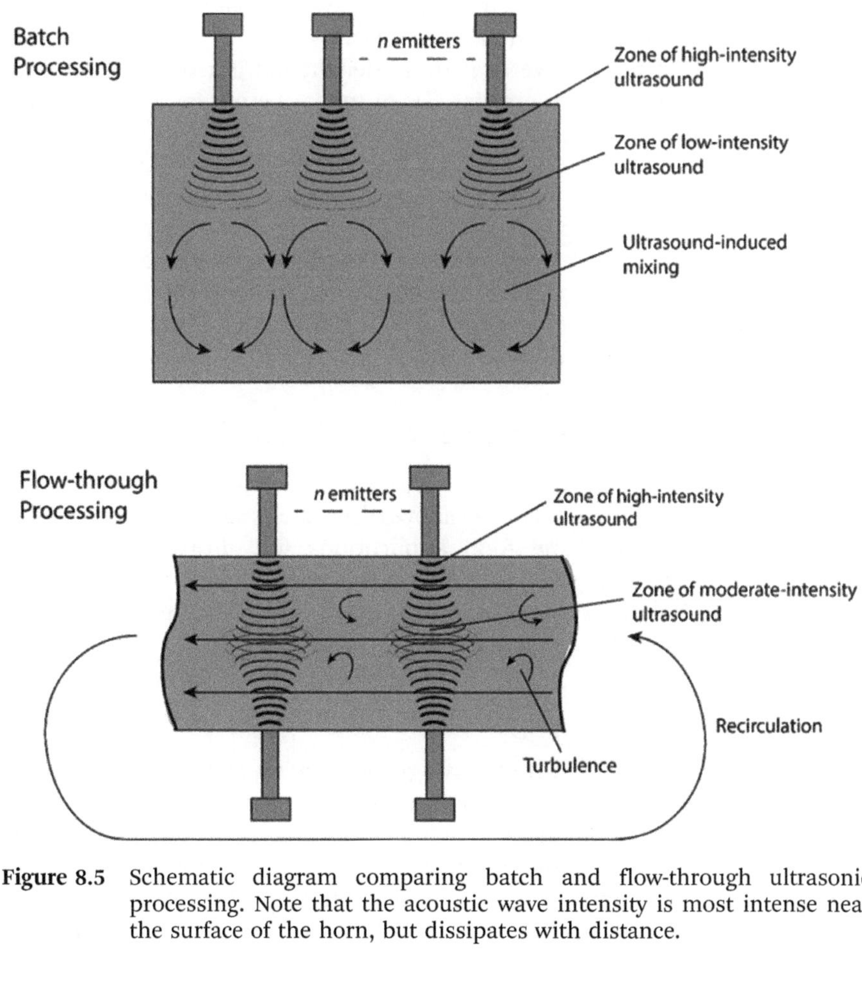

Figure 8.5 Schematic diagram comparing batch and flow-through ultrasonic processing. Note that the acoustic wave intensity is most intense near the surface of the horn, but dissipates with distance.

reduced in intensity with distance from the emitter, but will influence the food medium by imparting shear forces and acoustic streaming that drives mixing.

One of the most important distinctions is between batch and flow-through processing (see Figure 8.5). Batch processing relies on having adequate mixing to ensure that all elements of product receive approximately equal exposure to ultrasound. Achieving the desired average energy intensity is simply a matter of processing the given batch volume for the required time given the total energy intensity of the emitters. If properly designed, flow-through processing can ensure that all elements of product receive fairly consistent ultrasonic processing. As the residence time of an element of fluid within the ultrasonic exposure zone is typically short, achieving high degrees of processing requires either long banks of emitters and/or recirculation of the product to provide multiple passes through the ultrasound. Other considerations in the design and selection of ultrasonic equipment include consideration of the orientation and spacing of the emitters in relation to the geometry of the processing vessel and the reflection and dissipation of ultrasonic wave.

In batch processing, the average energy density of ultrasound applied is a product of the power of each emitter, the number of emitters and the time of treatment over the total volume. Even treatment is dependent on providing sufficient mixing (ultrasound-induced or mechanical) in relation to processing time and volume. In flow-through processing, the average energy density of ultrasound applied is a product of the power of each emitter, the number of emitters, the residence time in the chamber and the number of passes through the chamber over the total volume. Reasonably even treatment is provided by ensuring that all fluid passes through a zone of ultrasound, and can be improved further by turbulence and more passes.

8.3.4 Ultrasound Propagation in Liquid Foods

Ultrasound has been most widely applied to and studied in liquids, particularly aqueous systems. All of the above discussions can be directly related to liquids. There are many food applications for ultrasound involving liquids, such as emulsification and processing of milk, juice, wine and other beverages. The following two subsections briefly address important aspects of applying ultrasound to solid foods and foams.

8.3.5 Ultrasound Propagation in Solid Foods

Ultrasound can be transmitted through a solid by stimulating vibration through the object. Food materials are not homogeneous, and contain a number of different components, such as moisture, plant cells and protein. These will interact with the applied ultrasound in different ways. For example, different components will have differing absorption and compression behaviours. The reflection of sound waves from different components can be used for the analysis of defects and faults within food materials.[55] The rapid and successive compressions and rarefactions that the solid food material experiences can be useful in, for example, enhancing drying applications[1] (see Section 8.4.1.1).

8.3.6 Ultrasound Propagation in Foams

Ultrasound, especially high-frequency ultrasound, does not transmit through air particularly effectively, owing to strong attenuation.[56] Many ultrasonic devices are not suitable for operation directly in air, because a large amount of energy may become reflected back into the transducer, leading to rapid overheating and potential mechanical failure of the piezo-ceramic material in the ultrasound device, particularly if operated at very high energies.[57] Typically, plate- and bath-type ultrasonic transducers should not be operated without being in contact with a liquid or solid interface. Focused ultrasound applied using a horn-type transducer can be usefully operated at short distances for specific applications. In foods, this is useful for targeting foams, which are primarily made up of air. Defoaming is an important process in food processing, and ultrasound has been shown to be a useful non-contact technique that can be used to break up foams in carbonated soft drink manufacture[1,51] (see Section 8.4.1.2).

8.4 Ultrasonic Food Processing Applications

Ultrasonics can be used in a diverse range of food processing applications. The applications presented here are categorised according to the main mechanism by which ultrasound is effective in the process.

8.4.1 Applications Due to Mechanical Vibration Caused by Ultrasound

8.4.1.1 Ultrasonic Enhancement of Drying/Dehydration

A traditional method still employed for preserving fruits and vegetables is by drying or dehydration, *i.e.* removal of moisture. This is typically performed using convective hot-air drying, and involves simultaneous heat and mass transfer to remove moisture from the food product. Drying rates are often slow, and the high temperatures to which foods can be exposed within the hot air stream may result in a reduction in the quality of the food, such as product shrinkage, undesirable off-flavours and product discoloration.[58] Ultrasound can be used to accelerate the drying rate (Figure 8.6) by enhancing the mass transfer rate at which moisture is expelled from solids. This can also enable food drying to be affectively achieved using lower temperatures, thereby preventing degradation of food at high temperatures.

There are a number of mechanisms responsible for ultrasonically enhanced drying of solid foods. Primarily, the drying enhancement is due to

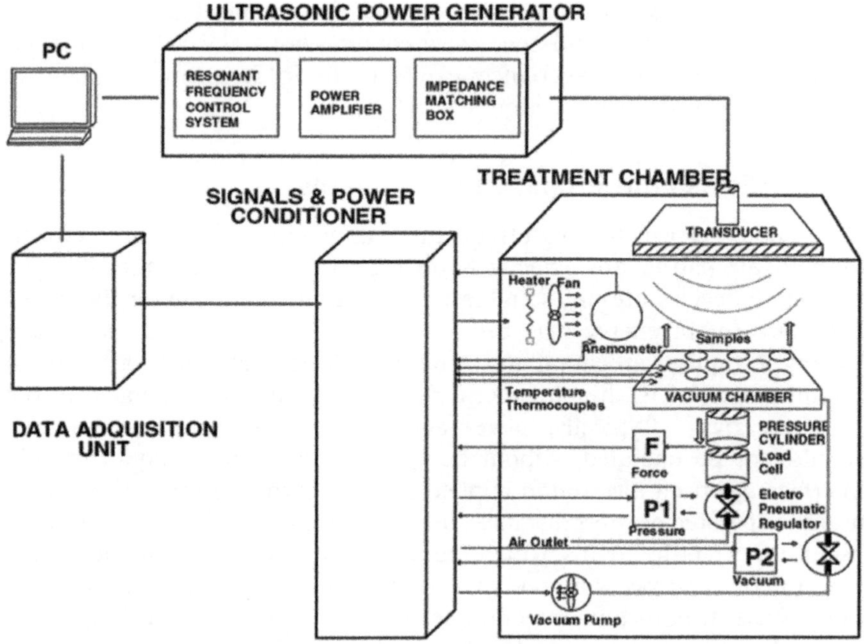

Figure 8.6 Schematic diagram of an ultrasonic dehydration system.[59] Reprinted from *Ultrasonics*, **44**, S. de la Fuente-Blanco *et al.* Food drying process by power ultrasound, e523–e527, Copyright (2006), with permission from Elsevier.

the oscillations of ultrasonic waves that causes solid materials to vibrate rapidly. The alternate compressions and expansions that occur as sound is transmitted through the solid promote the formation of microscopic channels inside the material that enhance the diffusion of moisture locked within the material to the outside. This is analogous to the squeezing of a sponge, but at a very rapid rate. Cavitation can also be induced within the aqueous phase trapped inside the solid material as a secondary mechanism, causing localised heating that makes removal of the moisture easier.[6]

For the drying of solid foods, an ultrasonic transducer can be coupled in direct contact with the food or applied *via* air-borne ultrasound[59] (see Figure 8.6). In some studies, the ultrasound is applied as a pretreatment step so that the diffusivity of moisture is enhanced during convective drying,[60,61] whereas in others, the ultrasound can be applied simultaneously to the drying process.[59] Ultrasonic compression and expansion tend to be more intense in highly porous materials and with greater acoustic energy adsorption. It should be noted that an acoustic intensity threshold must be exceeded before enhancement of drying occurs. Little enhancement of drying rate is achieved if ultrasound is applied below this threshold. de la Fuente-Blanco *et al.*[59] showed how increasing the acoustic power delivered to the system resulted in faster dehydration rates of carrots. The application of

100 W acoustic power for 90 min reduced the sample mass by 70% compared with controls without ultrasonic application, which reduced the sample mass by only 10%. Other variables that can influence the drying rate include the drying temperature used and the product thickness, which need to be optimised on a case-by-case basis. There are many examples in which ultrasound has been used successfully to accelerate food drying. Fruits and vegetables such as carrots,[59] plums,[62] apples[50] and papaya[61] can be dried significantly faster, with further advantages including the ability to dry at lower temperatures, thereby improving product quality.

8.4.1.2 Ultrasonic Defoaming and Degassing

Foaming can have negative consequences in food processing, as it significantly decreases the useful volume of processing equipment. It can also lead to product loss and a decline in production efficiency.[1] Conventionally, foaming is controlled either physically using mechanical breakers, or chemically by using antifoams. When bottling carbonated drinks, for example, foaming can slow the overall process[7] and hence reduce production rate.

As mentioned, ultrasound can be applied through air to interact with foams, provided that suitable working distances (up to a few centimetres) are used.[56] The foam bubbles break almost instantly when the applied ultrasonic energy is dissipated within the elastic foam layer, causing it to collapse. The primary mechanism is the acoustic streaming in the liquid phase of the foam, which leads to a reduction in the apparent viscosity (due to shear thinning behaviour) that permits rapid fluid draining, merging of bubbles and collapse.[51] The stability of the foam, however, determines whether or not ultrasound can be used effectively for defoaming. Foams that are stabilised by proteins are generally not able to be defoamed effectively by ultrasound because the proteins are denatured into stable structures at the aqueous liquid–gas interfaces.[63] For example, Mawson *et al.*[51] reported the use of airborne ultrasound to knock down foams of carbonated soft-drink beverages and premixed alcoholic beverages that have relatively weak foam structures. However, beer foams are less effectively reduced, owing to the presence of residual proteins from barley that can stabilise the foams. Dairy milk foams and foams heavily laden with fats stabilised by fatty acid surfactants in waste treatment tanks also cannot be defoamed effectively by ultrasound. Nevertheless, ultrasound provides a unique, contactless and hence clean method by means of which foaming can be controlled in selected food processing applications.[64] Increased throughput and reduction or elimination of antifoam chemicals[65] can be achieved.

Alternatively, ultrasonics can be used to promote foaming. This can be achieved by positioning the sonication device at the air–fluid interface. This promotes the entrainment of air bubbles into the mixture as sonication proceeds. This can be used, for example, to produce aerated gelatin gels or for the entrapment of volatile aromas and flavours.[7]

Ultrasound can also be used to initiate degassing of solutions.[64] The mechanisms of bubble growth, rectified diffusion and coalescence lead to the formation of large bubbles that rise rapidly to the surface and hence leave the solution. This is a useful method to degas carbonated beverages (*e.g.* beer), and can be used to control the gas content and size of gas bubbles remaining in a liquid product.

8.4.1.3 Ultrasonic Separation of Food Materials

The separation of components from food mixtures is an important unit operation in food processing. A number of conventional food separation processes exist, including centrifugation, sedimentation or clarification, chemical-induced flocculation and membrane filtration units, which are well established and robust unit operations. Nevertheless, these techniques still bear a number of potential issues including, but not limited to, (1) high energy consumption, (2) excessive shear that may damage the integrity of the product, (3) product fouling that limits throughput or requires extensive cleaning such as in the case of filtration, (4) slow separation rates and/or (5) excessive use of chemicals.[57] Ultrasonic separation involves the application of acoustic standing waves to promote rapid gravitational sedimentation of dispersed materials (see Section 8.2.1.4), and offers a non-contact, clean and rapid mechanism that can be used alone or complementary to these conventional methods for the purpose of enhancing the recovery or purification of food materials.

Ultrasonic separation technology has recently been commercialised for the palm oil milling industry in Malaysia, where a typical milling operation may experience a loss of recoverable product in the effluent with a potential revenue of >US$200 000 per year. The success of the ultrasonic application is due to the ultrasound (applied at 600 kHz frequency) generating physical 'rubbing' of the plant material. This activity assists the extraction of oil from the vegetable matter. The oil in the bulk liquid is then coalesced into larger droplets by the acoustic standing waves, resulting in faster oiling-off by gravity settling in downstream processing. Thus, ultrasonic application in this case improves the total recoverable oil and the rate of separation. Juliano and co-workers[54,66] successfully demonstrated the ability to enhance oil recovery from the extract obtained from pressed oil palm fruit mesocarp at laboratory and pilot scales. This success was extended to full-scale commercial operation, capable of a processing throughput of up to 45 tonnes per hour (Figure 8.7). An additional recovery of between 1% and 10% of palm oil product can be achieved in the process, which prevents virtually any oil being lost to the effluent stream.

Another food industry where ultrasonic separation shows promise is dairy processing. The separation of cream from milk is typically achieved using a centrifuge. Since milk fat is a valuable commodity, a common practice involves removing all of the fat from the milk and recombining the separated cream as required back into the skim milk to make products with standardised fat concentrations. This also evens out seasonal fluctuations in milk

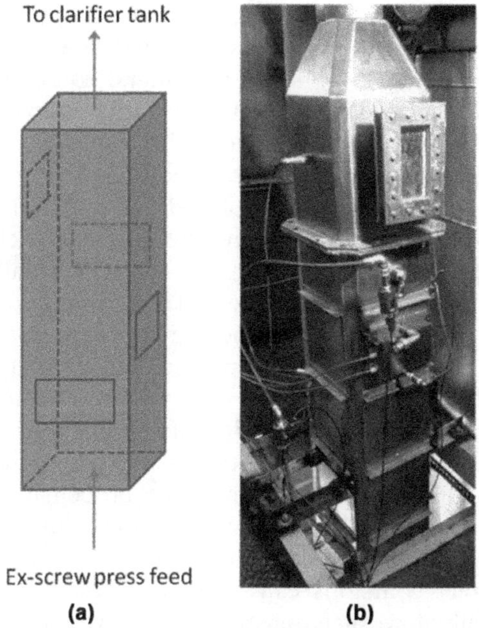

Figure 8.7 (a) Schematic representation of the megasonics palm oil separation reactor; the rectangles represent a transducer/transmission plate setup. (b) Photograph of the commercially operating megasonics reactor (45 $t\,h^{-1}$; 4 × 600 kHz, 1 kW each).[57]
Reprinted from *Food Engineering Reviews*, Megasonic separation of food droplets and particles: design considerations, 7, 2015, 298–320, T. S. H. Leong, © Springer Science+Business Media, New York, 2015, with permission of Springer.

composition, enhancing product consistency. Ultrasonic separation offers an alternative method of cream separation and refining. In an acoustic standing wave field applied to a volume of milk, the cream will gather at the pressure antinodes and experience an enhanced probability of aggregating into larger floccules. These larger floccules then begin to rise more rapidly to the surface owing to the increased hydrodynamic radius as described by Stokes law.[67] Leong and co-workers[68–70] recently applied these concepts to achieve the separation and fractionation of fat globules from milk and showed that higher frequency ultrasound, in this case >1 MHz, was more effective in manipulating and separating the fat globules present in natural whole milk, owing to stronger acoustic radiation forces.

8.4.2 Applications Due to Physical Effects of Acoustic Cavitation

8.4.2.1 Ultrasonic Emulsification

Emulsification is the process of dispersing two or more immiscible liquids, typically oil and water, together to form a semi-stable mixture. Common food emulsions include mayonnaise, sauces and spreads. Emulsions are thermodynamically unstable mixtures, and usually require the addition of surface-active materials, known as emulsifiers, to stabilise the interface formed between the oil and water.[9] Ultrasonication is an effective method by which stable emulsions can be formed.[19,71]

One of the attributes that govern the stability, appearance and taste of a food emulsion is the emulsion droplet size (EDS) and size distribution. Whereas emulsions with droplet sizes >100 nm are characterised by a 'milky' opaque appearance, emulsions that have EDS smaller than ~100 nm can appear translucent and almost clear. The formation of nanoemulsions, although not specific to ultrasonic processing, is of great interest in the food industry as it is recognised as an efficient delivery method for lipophilic bioactives,[72] such as omega 3 fatty acids, carotenoids and phytosterols.

In addition to an improvement in the visual appearance, emulsions smaller than ~100 nm become kinetically stable.[73] That is, the Brownian motion of the small droplets overcomes the propensity of the emulsified droplets to rise as individual droplets to the surface and form a cream layer. Eventually, colloidal forces will cause phase separation by Ostwald ripening[74] and droplet–droplet collisions can lead to coalescence. These processes, however, are usually slow, such that an emulsion with a small EDS will have a very long shelf life, *i.e.* several months. The rheological properties of an emulsion, such as its viscosity, are also influenced by the EDS, which plays a role in the sensory characteristics.[75] In general, emulsions containing smaller droplets tend to have a higher viscosity.[75] This can be usefully applied to give products such as cheese[76] and creams[77] an improved taste sensation such as a 'creamier' mouthfeel.[9]

Various high-energy shearing techniques are available for emulsification, with common methods being rotor–stator-type devices, high-pressure homogenisation, microfluidisation and ultrasonication. It is commonly recognised that the Microfluidizer® is the most effective nanoemulsion preparation technique currently available,[78] with relatively high energy efficiency for producing emulsions with very small and narrowly distributed EDS. Microfluidisation is, however, fairly expensive to operate, and so is less suited for the production of high-volume, low-value products such as food emulsions. Jafari *et al.*[78] compared emulsion preparation using ultrasonication at matched specific energies with microfluidisation, and found comparable performances. It was found that when using a matched 20 kJ kg^{-1} energy input, particle size reduction by the Microfluidizer achieved a mean volume-weighted particle size of 0.83 μm compared with 1.02 μm for ultrasonication at 20 kHz. A significant advantage of ultrasonication over microfluidisation is its relative ease of use with regard to cleaning and maintenance.[73]

The mechanism of ultrasonic emulsification is proposed to be due to two processes.[9] First, the application of the sound field produces interfacial waves that become unstable as they propagate through the fluid. This results in the eruption of the oil phase into the continuous water phase in the form of mid- to large-sized droplets. Second, acoustic cavitation generates high shear and turbulence in the vicinity of collapsing bubbles, which act to break up these initially formed droplets of dispersed oil into droplets of sub-micrometre size.[34] The first mechanism is primarily relevant when the continuous and dispersed phases are separate prior to the beginning of emulsification. Once a coarse emulsion has formed, continual droplet size reduction occurs by acoustic cavitation.

Competing mechanisms are also at play during sonication that may result in droplet re-coalescence and hence an increase in the EDS. Ostwald ripening, Brownian motion and droplet collisions all contribute to an increased likelihood of droplet coalescence. Coalescence during emulsification can be reduced by ensuring fast stabilisation of the new interfaces formed from disruption by sufficient and appropriate surfactant molecules or by increasing hydrodynamic effects such as retardation of liquid drainage between two colliding droplets.[79]

One of the advantages of ultrasonication is the potential to reduce or even eliminate the requirement for using synthetic emulsifiers to stabilise an emulsion. Shanmugam and Ashokkumar[80] showed that ultrasonic emulsification can be used to create stable food-based emulsions of flax seed oil directly in skim milk without the requirement for additional surfactants. The native milk proteins are partially denatured ($<1\%$) by the ultrasound, allowing them to coat the formed oil droplets effectively, stabilising the emulsion for at least 7 days. Stable emulsification could not be achieved in the absence of ultrasound even when using matched applied energies in a rotor–stator system, suggesting the importance of acoustic cavitation to the stabilisation process. Leong *et al.*[81] recently demonstrated the ability to form what are known as double emulsions using ultrasonication, again using the native milk proteins to stabilise the external interface. These double emulsions (Figure 8.8) were able to encapsulate an inner aqueous phase within the oil droplets, and could be a viable delivery vehicle for encapsulating aqueous-based bioactives in foods.

Although ultrasonic emulsification is typically achieved using low-frequency ultrasound (*i.e.* 20–100 kHz), higher frequency ultrasound (midsonic to megasonic) has also been reported to be useful at forming nanosized emulsion droplets when applied following low-frequency ultrasound through a process known as *tandem acoustic emulsification*.[82,83] The emulsions formed in this process were stable for 1–2 years, and were notably formed without requiring any additional surfactant. Oleic acid–water nanoemulsions were prepared by Kamogawa *et al.*[83] using this technique, and Nakabayashi *et al.*[82] also reported the production of transparent emulsions of ethylenedioxythiophene (EDOT) monomer formed by sequential emulsification with 20 kHz, 1.6 MHz and 2.4 MHz ultrasound.

It was proposed that the small droplets achieved in tandem acoustic emulsification upon application of higher frequency ultrasound was not due to destructive shear forces such as micro-jets and shockwaves prominent with low-frequency ultrasound. Instead, it is due to the enhanced

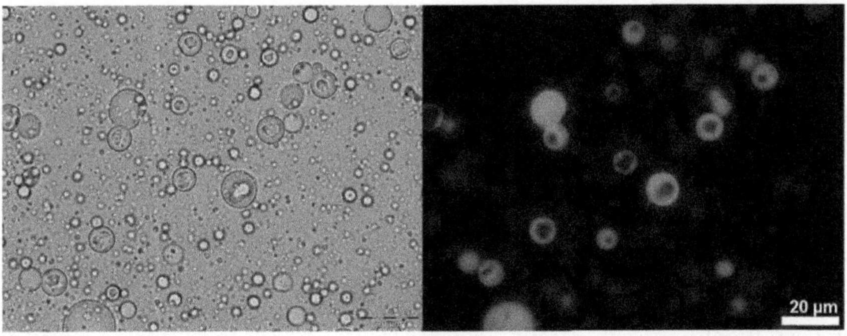

Figure 8.8 Double emulsion of skim milk encapsulated within sunflower oil dispersed within skim milk, formed using ultrasonication.[81]
Reprinted from *Food Hydrocolloids*, **63**, T. S. H. Leong *et al.*, Preparation of water-in-oil-in-water emulsions by low frequency ultrasound using skim milk and sunflower oil, 685–695, Copyright (2017), with permission from Elsevier.

acceleration of solvent and the emulsion droplets caused by acoustic radiation forces and acoustic streaming[83] such that they collide and break apart into smaller droplets. These acceleration forces become stronger with increased frequency, and the sequence in which the different frequencies of ultrasound are applied is noted to be important, with reversal of the order (*i.e.* high frequency followed by low frequency) resulting in ineffective emulsification.

Recent research has highlighted the capability of using ultrasonically produced nanoemulsions to deliver essential oils into food- or pharmaceutical-based products. Essential oils have a wide range of beneficial properties such as high nutritional value[84] and antimicrobial properties,[85] but can be difficult to incorporate stably into aqueous-based food products due to the hydrophobic nature of the oils. Ultrasonication has been used as an effective method to create stable nanoemulsions using basil,[86] orange peel,[87,88] annatto seed[84] and *Thymus daenensis*[85] essential oils. In addition to being shelf stable for many months without phase separation, these nanoemulsions were generally found to possess good to excellent antimicrobial activity. The antimicrobial activity of some of the nanoemulsions of these essential oils were found to be superior to that of the native state.[85] It was speculated that the reason for this increased activity was because the nano-emulsions enabled the essential oils to come nearer to the bacterial cell membrane interface, allowing more effective disruption of the phospholipid bilayer. The activity of bioactive components may be negatively affected by the formation of radicals during ultrasonication.[65] However, the treatment times required for emulsion formation are typically short, which limits the amount of degradation that occurs.

Another interesting application of ultrasonics for food emulsions reported recently is improving meat-based emulsions *via* an indirect sonication approach. Meat emulsions consist of ground meat, comprising primarily water, protein and fat, salt and other ingredients.[89] Krasulya *et al.*[89] reported that sonication of the brining solution used to treat the meat emulsions leads to improved textural properties of the meat and hence can be used as a technique to control the viscosity of sausage meat.

8.4.2.2 Ultrasound-assisted Extraction

Natural foods are a key source of essential oils, aromas, pigments, antioxidants and other bioorganic compounds, often locked within the cell walls of the material. These compounds can be selectively extracted from foods to produce highly valuable nutraceutical concentrates. Conventionally, such extractions are performed using polar or non-polar solvents, depending on the nature of the material being extracted. For oil-based extractions, hexane and mixtures of hexane with 2-propanol are commonly used.[6] Polar compounds are typically extracted using solvents such as water, ethanol, methanol or acetone. Conventional solvent extractions can take many hours or even days to complete, and often large amounts of solvents are required, making such processes potentially very expensive. The solvents used must eventually be disposed of or regenerated using high temperatures, creating a large environmental footprint. Supercritical fluid extraction is another technique that can be used for such extractions.[1] Typically, CO_2 is the most widely used supercritical extraction solvent, because it is non-toxic, non-flammable, cheap and easily removed after extraction. It also has a high solvating capability for non-polar molecules[1] Although supercritical CO_2 extraction is highly effective, it is also an expensive process requiring pressurised equipment.

Ultrasound can be applied to both solvent and supercritical extraction processes to enhance and accelerate dramatically the extraction of food compounds. The mechanism involves enhancement of the heat and mass transfer between the solvent and the solid boundaries of the food material by the physical shearing effects resulting from cavitation, such as micro-jetting and pressure shockwaves. The release of extractable compounds is further enhanced by the disruption of the cell walls of the food material, accelerating the movement of compounds into the solvent phase. Power ultrasound in the 20–100 kHz range is known to be most effective for extraction, and the result is often higher extraction yields and reduced extraction time and solvent consumption.[6] In some cases, extractions requiring many hours can be completed in a manner of minutes using ultrasonication, with high reproducibility. Ultrasonic extraction can also enhance the effectiveness of solvents, so that generally recognised as safe (GRAS) solvents (*e.g.* water) can be used instead of organic-based solvents. This can be used as a strategy to reduce environmental impact and costs.[90]

There are a vast number of food materials, including herbs, crushed fruit pomace and oil seeds, from which products such as polyphenolics, anthocyanins, aromatic compounds, polysaccharides and functional compounds can be extracted, with ultrasound[91] able to accelerate extraction and improve yields. Wu *et al.*[47] used ultrasound to achieve a threefold increase in the extraction rate of saponins from ginseng. Vinatoru[92] achieved increased yields of up to 34% compared with conventional stirred reactors in the extraction of extracts from fennel, hops, marigold and mint. Balachandran *et al.*[93] employed ultrasonication as a pretreatment for supercritical CO_2 extraction, improving the extraction yield by 30% and reducing the extraction time required. Valuable compounds such as polyphenols can be recovered from waste materials. Grape marc, which is a waste stream from the wine industry, was used by Vilkhu *et al.*[91] to extract phenolics. A reported increase in yield of 17–35% was achieved when ultrasound was used. Mechanistically, in all of these examples the application of ultrasound interacts with the plant materials to alter its physical and, in some cases, chemical properties. Some example applications and methods of application are reported in Table 8.2.

The optimum temperature used for extraction is application dependent. In general, an increased temperature increases the softening and swelling of the raw materials and also increases the solubility of the desired compounds to be extracted in the solvent. For ultrasonication, temperatures cannot be raised too high, however, as the vapour pressure of the cavitation bubbles may also increase, resulting in a significant decline in the cavitation intensity. Ultrasound in general has a greater influence on extraction at lower temperatures where the energy generated upon cavitation bubble collapse is greater.[90] Careful temperature control should also be considered to protect against thermal degradation of sensitive components.

8.4.2.3 Sonocrystallisation and Freezing

The crystallisation of dissolved solids from supersaturated fluids is an important process in many food manufacturing applications as a method of production, purification and recovery.[101] Crystallisation is also a key mechanism in the freezing of food products.[1] Important crystallisation processes in food processing include the removal of lactose from dairy whey,[102] tempering of fat crystals in chocolate to optimise appearance and texture[103] and control of ice crystal size in frozen products such as ice-cream.[104]

Supersaturation is a prerequisite state for initiating crystallisation. Such a state is often achieved by evaporation, cooling or the addition of an antisolvent. Currently, the rates of crystallisation processes employed in industrial applications tend to be slow and expensive in terms of energy usage since high supersaturation concentrations are often required.[105] Careful control of variables such as temperature, concentration, agitation rates and time must be made to ensure that crystals with the correct size and morphology are formed.[105,106] The formation of excessively large crystals or incorrect shapes leads to products that are less desirable in taste, texture and appearance in, for example, chocolate products.[103,107]

Table 8.2 Ultrasonic extraction applications.[91] Adapted from *Innovative Food Science & Emerging Technologies*, **9**, K. Vilkhu *et al.*, Applications and opportunities for ultrasound assisted extraction in the food industry—A review, 161–169, Copyright (2008), with permission from Elsevier.

Product	Ultrasound delivery	Solvent	Performance outcomes	Ref.
Almond oils	Batch, 20 kHz	Supercritical CO_2	30% increased yield or extraction time reduction	Riera *et al.*, 2004[94]
Herbal extracts (fennel, hops, marigold, mint)	Stirred batch, 20–2400 kHz	Water and ethanol	Up to 34% increased yield over stirred	Vinatoru, 2001[92]
Ginseng saponins	Batch, 38.5 kHz	Water, methanol and *n*-butanol	Threefold increase of extraction rate	Wu *et al.*, 2001[47]
Ginger	Batch, 20 kHz	Supercritical CO_2	30% increased yield or extraction time reduction	Balachandran *et al.*, 2006[93]
Soy protein	Continuous, 20 kHz, 3 W g^{-1}	Water and sodium hydroxide	53% and 23% yield increase over equivalent ultrasonic batch conditions	Moulton and Wang, 1982[95]
Soy isoflavones	Batch, 24 kHz	Water and solvent	Up to 15% increase in extraction efficiency	Rostagno *et al.*, 2003[96]
Rutin from Chinese scholar trees	Batch, 20 kHz	Water and methanol	Up to 20% increase in 30 min	Paniwnyk *et al.*, 2001[97]
Carnosic acid from rosemary	Batch, 20 and 40 kHz	Butanone and ethyl acetate	Reduction in extraction time	Albu *et al.*, 2004[98]
Polyphenols, amino acids and caffeine from green tea	Batch, 40 kHz	Water	Increased yield at 65 °C compared with 85 °C	Xia *et al.*, 2006[99]
Pyrethrins from flowers	Batch, 20 and 40 kHz	Hexane	Increased yield at 40 °C compared with 66 °C	Romdhane and Gourdon, 2002[100]

Ultrasound can be used to enhance and control the growth of crystals from supersaturated solutions. This process is also known as sonocrystallisation. The collapsing bubbles created during acoustic cavitation can act as primary nucleation sites for crystallisation processes.[106] Rapid cooling also occurs after bubble collapse, creating localised regions of very high supersaturation that are not achieved under quiescent conditions.

Many studies have demonstrated beneficial outcomes from sonocrystallisation, including a reduction in the induction time and supersaturation concentrations required prior to crystal nucleation and growth.[106,108–110] The crystallisation of lactose, if performed by conventional means such as stirring, for example, is slow and often requires anywhere from 12 to 72 h to complete. The application of ultrasound can shorten this to within 2 min.[108]

Sonocrystallisation also permits improved control of crystal size and shape with high reproducibility.[103,107] The crystal size and shape are important aspects that influence taste, appearance and texture in many food products. This is particularly important in chocolate manufacture, where crystals of cocoa butter can form up to six different polymorphic forms depending on how the melted chocolate is solidified. Polymorphic form V is the most desirable, imparting the glossy appearance, characteristic snap and 'cooling' sensation as it melts in the mouth of high-quality chocolate.[103,107]

Chow and co-workers[104,111,112] were able to visualise the nucleation and subsequent growth of an ice crystal from a bubble undergoing acoustic cavitation in a sound field (Figure 8.9). Mechanistically, it implies that the collapsing bubbles act as primary nucleation sites for crystallisation. Once nucleated, these crystals grow rapidly and continue to propagate. Acoustic cavitation can therefore be used to control the rate at which ice crystals form, which is an important phenomenon for frozen food products.

The quality of frozen foods is highly dependent on the size of the ice crystals that form during freezing. Rapid freezing is usually employed as it produces smaller ice crystals that are more desirable and produces a higher quality product. Ultrasound can be used to induce more rapid freezing, thereby improving freezing logistics and maintaining higher quality frozen foods. Ultrasound mainly enhances the freezing rate by creating cavitation bubbles that act as nucleation sites where ice crystals develop, increasing heat and mass transfer, and by controlling the crystal shape and size distribution within the frozen food.[6] An example is the freezing of solid vegetables such as potatoes[113] and carrots,[114] where freezing rates can be significantly enhanced when ultrasound is applied.

For the purpose of creating nucleation sites, very short duration ultrasound is required. In fluid samples, 1 s of ultrasound is usually sufficient to initiate the nucleation process.[6] Ultrasound can also be used to regulate the nucleation temperature of ice crystals. Typically, nucleation temperatures are higher when ultrasound is used compared with its absence.

For food preservation and ice-cream manufacture, small-sized crystals are always desired, owing to the benefit to the final product quality. However, there have been some conflicting reports on the usefulness of

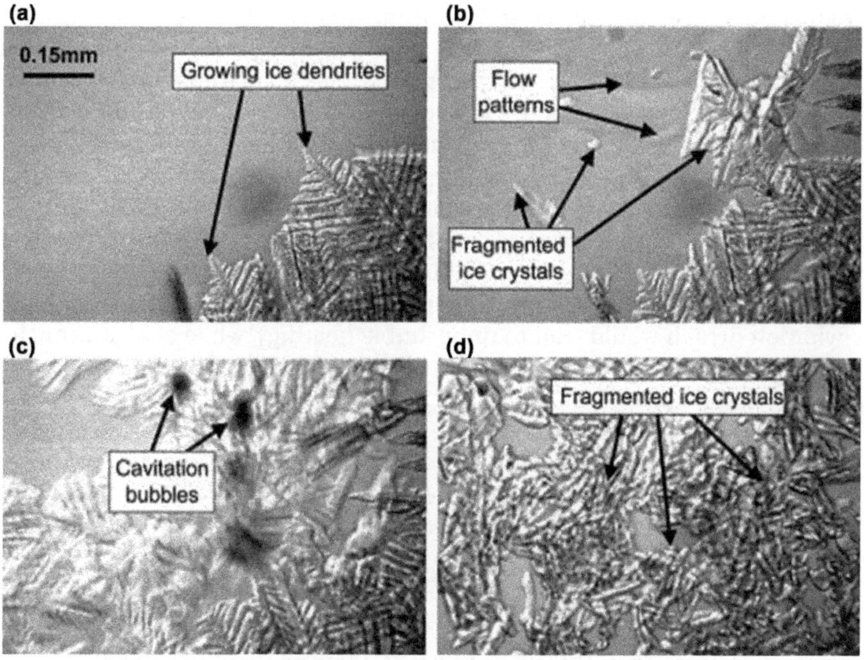

Figure 8.9 Effect of ultrasound on the secondary nucleation of ice in a 15 wt% sucrose solution. (a) Ice dendrite (no ultrasound); (b) flow patterns and fragments of ice dendrites after 1.36 s of ultrasound; (c) fragmentation of ice crystals by cavitation bubbles after 2.38 s of ultrasound; (d) ice crystals produced after 17.38 s of sonication.[112]
Reprinted from *Ultrasonics*, **41**, R. Chow *et al.*, The sonocrystallisation of ice in sucrose solutions: primary and secondary nucleation, 595–604, Copyright (2003), with permission from Elsevier.

sonocrystallisation to create ice-creams with more uniform ice crystals. This is because sonication can, as mentioned previously, remove air from solutions. A good ice-cream is considered to be one in which the air content is maximised.[7] In a study by Mortazavi and Tabatabaie,[115] ultrasound was shown to increase the freezing rate by 35% and enhance sensory characteristics relative to controls when applied for durations <20 min. Negative sensory effects were observed in ice-creams made with extended sonication durations of >20 min, likely due to more extensive degassing of the ice-cream that counteracts the positive sensory effects of smaller ice crystal formation.

Thawing can also be significantly enhanced by the application of sound waves. Thawing is typically much slower than freezing and is a very time-consuming process. By applying acoustic waves to frozen foods, the ice crystals within the food responds to the applied energy through molecular reorientation and stress/strain.[116] The system becomes thermodynamically unbalanced and a thermal current develops that increases the energy absorption and hence melting rate of the ice crystals. This process is termed thermoelastic relaxation. By applying sound waves at the relaxation frequency range of ice crystals within foods, thawing rates can be enhanced.[116] This frequency range is temperature and material dependent.

For frozen fish, Kissam *et al.*[116] reported that a sub-ultrasonic frequency range between 500 and 2000 Hz would provide maximal relaxation response by the ice crystals in the food. A frequency of 1500 Hz (60 W power) resulted in a 71% reduction in the thawing time required relative to controls. Miles *et al.*[49] investigated the response of thawing frozen meat and fish subjected to ultrasonic frequencies in the range 220 kHz–3.3 MHz. They found an 'optimal' frequency at ~500 kHz that maximised penetration of the ultrasound through the frozen product without causing unnecessary cavitation (which would lead to undesirable heating), while also minimising the degree of surface heating due to attenuation of the ultrasound that occurs at higher frequencies.

It should be noted also that the rate of thawing increases with the intensity of the acoustic energy applied.[49] The acoustic energy absorbed by the ice crystals increases the rate at which they melt. Furthermore, since this energy is absorbed throughout the product, thawing can be enhanced without problems of creating excessive surface heating and hence dehydration of the food product.

8.4.2.4 Ultrasonic Enhancement of Membrane Filtration Processes

Membrane filtration is widely used for the concentration or separation of food components.[117,118] Membrane fouling is a significant issue for food processing, with many food components, in particular proteins, acting as foulants. Fouling reduces the flux through the membrane, meaning that filtration processes need to be stopped periodically for cleaning to return the flux to an acceptable level. Sometimes an irreversible fouling may occur, meaning that flux never returns to the same level as preoperation even after cleaning. Concentration polarisation is the build-up of solute particles at the surface of the membrane that are left behind by the permeating solvent. Concentration polarisation represents a resistance to flux, the effect of which can be reduced by employing crossflow filtration in which the fluid flows parallel to the membrane surface.

Ultrasound can be used to reduce the effects of membrane fouling and also the effects of concentration polarisation, thereby enhancing filtration efficiency.[6] It achieves this primarily by the physical removal of debris and particles from the fouled or polarised surface. Typically, low-frequency ultrasound has superior performance, since cavitation is more intense at low frequency, as demonstrated by Muthukumaran and co-workers.[118,119] High-frequency ultrasound (1 MHz) was found to be significantly less effective than low frequency ultrasound (50 kHz) for the purpose of flux enhancement and membrane cleaning for the ultrafiltration of whey protein concentrates. The reason for this is the stronger physical effects from acoustic cavitation at lower frequency ultrasound.

The application of ultrasound to filtration processes is potentially energy intensive. One strategy to reduce the energy requirements is to use intermittent application of ultrasound,[6] *i.e.* pulsed operation, so that cake layer is removed at timed intervals. Muthukumaran *et al.*,[119] however, found that continuous application resulted in more effective cleaning. Continuous application is viable on a cost/energy perspective, since low powers of the order of 2 W L^{-1} are used.

8.4.3 Applications Due to a Combination of Physical and Chemical Effects of Ultrasound Arising from Cavitation

8.4.3.1 Viscosity and Rheology Modification

Proteins and polysaccharides in foods can be subject to aggregation, particularly by heating, which can increase the viscosity of the food materials, making them more difficult to process. Shear forces induced by cavitation have been shown to be highly effective in reducing the viscosity of a range of food products.[7,120] This is achieved by break-up of the protein aggregates formed by disrupting hydrophobic interactions or by reducing the polymer chain length. It has been shown in dairy systems that once whey protein aggregates have been broken up by ultrasound[10] they do not re-form during subsequent thermal treatment.

Polysaccharide-rich fluids, such as those containing starch, dextran and pectin, can also undergo gelatinisation due to heating in the presence of water, transforming the aqueous suspension into an amorphous gel phase. Iida *et al.*[121] extensively studied the effects of ultrasonication on starch solutions after gelatinisation. It was found that sonication for 30 min could reduce the viscosity of starch solutions (5–10 wt%) by two orders of magnitude, to ~100 mPa s. High concentration starch gels (20–30 wt%) could even become liquefied after treatment with ultrasonication. This liquefaction and viscosity reduction can be useful as a pretreatment of viscous starch solutions to increase the efficiency of spray drying.[121]

Another application is in the control of dairy fluid viscosity. In the processing of whey proteins, high temperatures will cause protein aggregation, leading to an increase in the viscosity that makes downstream processing more challenging. Zisu *et al.*[122] showed in a pilot-scale reactor that sonication at 20 kHz with an energy density of 260 J mL^{-1} could significantly reduce the size of whey protein aggregates formed in a 33 wt% solids solution. This process allows the production of a heat-stable dairy powder or fluid, which can be of great importance in downstream food manufacture.

Ultrasonic cavitation can also be employed beneficially for promoting the formation of dairy gels. The high energy of collapsing bubbles can partially disrupt casein micelles (particularly when applied at elevated pH), which increases the surface area for gelation, enhancing the rennet and acid gelation kinetics.[123,124]

8.4.3.2 Enhancement of Wine and Rice Wine Maturation

Post-fermentation maturation processes for wine and rice wine are slow, requiring months to years for completion. Maturation is a mild oxidation process that can be accelerated by application of ultrasound,[6] which can promote the many oxidative reactions responsible for flavour development. The application of ultrasound at 20 kHz to rice wine over the course of 1 week produced a rice wine with quality identical with that of a market product that had been aged naturally for 1 year.[125] Similarly, ultrasound has been shown to have the potential to accelerate the ageing of red wine.[126] Zhang *et al.*[127] recently showed that ultrasound applied at 80 kHz leads to significant changes in the physicochemical properties of red wine in terms of colour, total phenolic content and electrical conductivity, but had no effect in changing the pH or titratable acidity. Ultrasonication is also useful in its ability to reduce the activity of spoilage enzymes present in beverages and foods. For example, enzymes such as polyphenyloxidase, peroxidase and protease, which are responsible for producing negative coloration to wine by browning,[128,129] can be deactivated by the intense physical affects and chemical radicals generated during cavitation. This capability may be useful in reducing the amount of preservatives required to be added to wines, such as sulfur dioxide (SO_2).[128] A decrease in SO_2 is desirable as it can induce allergic responses in some consumers.[130]

High-frequency ultrasound, which produces more radicals, is not as effective as power ultrasound in enhancing wine maturation, as it produced wines that were deemed unacceptable by sensory analysis.[131] One of the mechanisms possibly responsible for this is that the radicals can degrade some of the phenolic compounds[3] that contribute to wine's organoleptic characteristics. Careful consideration must therefore be made as to the duration of treatment and frequency/intensity of application, since wine quality is very sensitive to these processing conditions.

8.4.3.3 Effects of Ultrasound on Microorganisms and Enzymes

Maintaining food safety and preventing food spoilage require the effective destruction of pathogenic bacteria and other microorganisms.[7] Conventionally, thermal treatment is used to pasteurise and/or sterilise foods to make them safe for human consumption. Although well established, thermal processing is known to lead to undesirable loss of vitamins, nutrients and flavours.[132] Owing to consumer demand, there is great interest in the development of non-thermal processing methods to preserve better the nutritional and sensory properties of foods.[133]

Ultrasound can be used as a processing aid to inactivate microbes for the purpose of preserving food. Both transient cavitation and stable cavitation can lead to microorganism and/or enzyme deactivation. Localised temperature hot spots, shockwaves, micro-jets, formation of free radicals and microstreaming all contribute to the deactivation of microorganisms by damaging cell walls and membranes that result in cell death.

The use of ultrasound for microbe deactivation is considered a non-thermal food processing technique,[7] since deactivation occurs without requiring the heating of bulk food material to high temperatures. Although ultrasound has some effect in deactivating microorganisms when applied alone, it is most effective when applied in conjunction as a complementary treatment to heat (thermosonication), pressure (manosonication), combinations of heat and/or chemical treatment.[133] For example, Sams and Feria[134] showed that ultrasonication alone could not deactivate bacteria located on chicken broiler drum sticks. The likely explanation for this was that it was due to the irregular surface of the skin, which limited the effectiveness of the ultrasonication applied. By contrast, the application of ultrasonication by Lillard[135] showed ~1 log reduction in viable bacteria when applied alone on chicken breasts. However, when applied together with a 0.5 ppm chlorine solution, a synergistic reduction of up to 3.9 log could be achieved. The reasoning was that the ultrasound was able to dislodge bacteria that were firmly attached on the surface of the chicken breast, enhancing the sanitisation effect of the chlorine. Similarly, ultrasonication alone was noted to be ineffective in sanitising whole leaf lettuce, achieving a maximum reduction of 90%.[136]

Gao *et al.*[132] investigated the influence of different bacterial properties, such as Gram positivity/Gram negativity and the size and shape of bacteria cells, on the effectiveness of ultrasonication deactivation. It was concluded that none of these properties actually has any effect on the inactivation capability of ultrasound, although studies of these effects have resulted in conflicting reports.[137,138] Rather, the relative stiffness and thickness of the bacterial cell walls were considered to be the most important factors. Bacteria with thin, stiff cell walls were more susceptible to critical failure when subjected to ultrasonic vibrations, whereas those comprising thicker more flexible cell wall materials can dampen the mechanical effects due to cavitation. Ultrasound was not effective when used alone to deactivate certain types of bacteria such as *Staphyloccus* spp.[132] or bacterial spores,[133] both of which have thick cell walls that are resistant to the effects of ultrasound.

Thermosonication and manosonication are noted to be particularly interesting deactivation techniques in dairy systems, as they permit pasteurisation of milk at reduced temperatures and cost. Bermuddez-Aguirre demonstrated the capability of achieving a 5 log reduction of *Listeria* bacteria after 10 min of sonication. Notably, the thermosonicated milks had better colour and similar physicochemical characteristics compared with conventional pasteurised milk. Villamiel and de Jong[20] reported not only synergistic effects of ultrasound and heat on the deactivation of spoilage enzymes, but also a size reduction of the fat globules in milk. Thermosonication could potentially be used to pasteurise and homogenise milk simultaneously.

8.5 Future Outlook and Conclusion

The application of ultrasonics in food processing has been extensively investigated and proven to be useful in laboratory and pilot scale studies in the past few decades. The food industry is typically slow to adopt new technologies, so it has only been in the last few years that commercialised outcomes have been achieved. The current trends in the food industry towards the development of functional foods and non-thermal processing have largely been driven by consumer demands for foods with improved health benefits and 'fresh-like' qualities. These trends have increased interest in the application of ultrasonics for various food processing applications.

Ultrasonics will play an important role in fulfilling niche applications where significant value addition can be achieved. The versatility of ultrasound to provide a range of beneficial effects makes it an intriguing alternative processing technology for food processors. As the continuing development of ultrasonic equipment serves to bring down the required capital cost investment, we expect to see ultrasonics being used in more applications to enhance traditional food processing methodologies.

Acknowledgements

This research was supported under The Australian Research Council's Industrial Transformation Research Program (ITRP) funding scheme (project number IH120100005). The ARC Dairy Innovation Hub is a collaboration between The University of Melbourne, The University of Queensland and Dairy Innovation Australia Ltd.

References

1. F. Chemat and M. K. Khan, Applications of ultrasound in food technology: Processing, preservation and extraction, *Ultrason. Sonochem.*, 2011, **18**, 813–835.
2. M. J. W. Povey and T. J. Mason, *Ultrasound in Food Processing*, Blackie Academic, London, 1998.
3. M. Ashokkumar, D. Sunartio and S. Kentish, *et al.*, Modification of food ingredients by ultrasound to improve functionality: A preliminary study on a model system, *Innovative Food Sci. Emerging Technol.*, 2008, **9**, 155–160.
4. J. Chandrapala and T. Leong, Ultrasonic processing for dairy applications: Recent advances, *Food Eng. Rev.*, 2015, 1–16.
5. H. Feng, G. V. Barbosa-Cánovas and J. Weiss, *Ultrasound Technologies for Food and Bioprocessing*, Springer, New York, 2011.
6. Y. Tao and D.-W. Sun, Enhancement of food processes by ultrasound: A review, *Critic Rev Food Sci Nutr*, 55 2015 570–594.
7. S. Kentish and H. Feng, Applications of power ultrasound in food processing, *Annu. Rev. Food Sci Technol.*, 2014, **5**, 263–284.

8. T. S. Leong, G. J. Martin and M. Ashokkumar, Ultrasonic encapsulation–A review, *Ultrason. Sonochem.*, 2016, **35**, 605–614.
9. S. Kentish, T. Wooster and M. Ashokkumar, *et al.*, The use of ultrasonics for nanoemulsion preparation, *Innovative Food Sci. Emerging Technol.*, 2008, **9**, 170–175.
10. M. Ashokkumar, J. Lee and B. Zisu, *et al.*, Hot topic: Sonication increases the heat stability of whey proteins, *J. Dairy Sci.*, 2009, **92**, 5353–5356.
11. J. Chandrapala, B. Zisu, M. Palmer, *et al.*, A possible mechanism to understand the ultrasound induced heat stability of whey protein concentrates, in *Int Non Thermal Workshop, Melbourne*, 2012.
12. T. G. Leighton, *The Acoustic Bubble*, Academic Press, San Diego, 1994.
13. T. Leong, M. Ashokkumar and S. Kentish, The fundamentals of power ultrasound – A review, *Acoust. Aust.*, 2011, **39**, 54–63.
14. M. Ashokkumar, J. Lee and S. Kentish, *et al.*, Bubbles in an acoustic field: An overview, *Ultrason. Sonochem.*, 2007, **14**, 470–475.
15. E. B. Flint and K. S. Suslick, The temperature of cavitation, *Science*, 1991, **253**, 1397–1399.
16. A. Weissler, Formation of hydrogen peroxide by ultrasonic waves: free radicals, *J. Am. Chem. Soc.*, 1959, **81**, 1077–1081.
17. F. R. Young, *Cavitation*, McGraw-Hill, London, 1989.
18. K. Yasui, T. Tuziuti and J. Lee, *et al.*, The range of ambient radius for an active bubble in sonoluminescence and sonochemical reactions, *J. Chem. Phys.*, 2008, **128**, 184705–184712.
19. T. S. H. Leong, T. J. Wooster and S. E. Kentish, *et al.*, Minimising oil droplet size using ultrasonic emulsification, *Ultrason. Sonochem.*, 2009, **16**, 721–727.
20. M. Villamiel and P. de Jong, Influence of high-intensity ultrasound and heat treatment in continuous flow on fat, proteins, and native enzymes of milk, *J. Agric. Food Chem.*, 2000, **48**, 472–478.
21. G. Cravotto, L. Boffa and S. Mantegna, *et al.*, Improved extraction of vegetable oils under high-intensity ultrasound and/or microwaves, *Ultrason. Sonochem.*, 2008, **15**, 898–902.
22. S. Koda, T. Kimura and T. Kondo, *et al.*, A standard method to calibrate sonochemical efficiency of an individual reaction system, *Ultrason. Sonochem.*, 2003, **10**, 149–156.
23. T. J. Mason, A. J. Cobley and J. E. Graves, *et al.*, New evidence for the inverse dependence of mechanical and chemical effects on the frequency of ultrasound, *Ultrason. Sonochem.*, 2011, **18**, 226–230.
24. R. E. Apfel, Acoustic cavitation prediction, *J. Acoust. Soc. Am.*, 1981, **69**, 1624–1633.
25. L. Johansson, T. Singh and T. Leong, *et al.*, Cavitation and non-cavitation regime for large-scale ultrasonic standing wave particle separation systems–In situ gentle cavitation threshold determination and free radical related oxidation, *Ultrason. Sonochem.*, 2016, **28**, 346–356.

26. T. Leong, L. Johansson and P. Juliano, *et al.*, Ultrasonic separation of particulate fluids in small and large scale systems: a review, *Ind. Eng. Chem. Res.*, 2013, **52**, 16555–16576.
27. K. Matsuura, M. Hirotsune and Y. Nunokawa, *et al.*, Acceleration of cell growth and ester formation by ultrasonic wave irradiation, *J. Ferment. Bioeng.*, 1994, **77**, 36–40.
28. F. Cavalieri, M. Ashokkumar and F. Grieser, *et al.*, Ultrasonic synthesis of stable, functional lysozyme microbubbles, *Langmuir*, 2008, **24**, 10078–10083.
29. M. W. Grinstaff and K. S. Suslick, Air-filled proteinaceous microbubbles: Synthesis of an echo-contrast agent, *Proc. Natl. Acad. Sci. U. S. A.*, 1991, **88**, 7708–7710.
30. A. Shanmugam, J. Chandrapala and M. Ashokkumar, The effect of ultrasound on the physical and functional properties of skim milk, *Innovative Food Sci. Emerging Technol.*, 2012, **16**, 251–258.
31. F. Cavalieri, M. Zhou and F. Caruso, *et al.*, One-pot ultrasonic synthesis of multifunctional microbubbles and microcapsules using synthetic thiolated macromolecules, *Chem. Commun.*, 2011, **47**, 4096–4098.
32. C. F. Naudé and A. T. Ellis, On the mechanism of cavitation damage by nonhemispherical cavities collapsing in contact with a solid boundary, *J. Fluids Eng.*, 1961, **83**, 648–656.
33. K. S. Suslick, Sonochemistry, *Science*, 1990, **247**, 1439–1445.
34. L. Thompson and L. Doraiswamy, Sonochemistry: Science and engineering, *Ind. Eng. Chem. Res.*, 1999, **38**, 1215–1249.
35. D. J. Flannigan and K. S. Suslick, Plasma formation and temperature measurement during single-bubble cavitation, *Nature*, 2005, **434**, 52–55.
36. H. Bruus, Acoustofluidics 7: The acoustic radiation force on small particles, *Lab Chip*, 2012, **12**, 1014–1021.
37. F. J. Trujillo, P. Juliano and G. Barbosa-Cánovas, *et al.*, Separation of suspensions and emulsions via ultrasonic standing waves – A review, *Ultrason. Sonochem.*, 2014.
38. K. Yosioka and Y. Kawasima, Acoustic radiation pressure on a compressible sphere, *Acustica*, 1955, **5**, 167.
39. T. Leong, L. Johansson and P. Juliano, *et al.*, Ultrasonic separation of particulate fluids in small and large scale systems: A review, *Ind. Eng. Chem. Res.*, 2013, **52**(57), 16555–16576.
40. J. Lighthill, Acoustic streaming, *J. Sound Vib.*, 1978, **61**, 391–418.
41. H. Mitome, The mechanism of generation of acoustic streaming, *Electron. Comm. Jpn.*, 1998, **81**, 1–8.
42. N. Riley, Steady streaming, *Annu. Rev. Fluid Mech.*, 2001, **33**, 43–65.
43. C. Eckart, Vortices and streams caused by sound waves, *Phys. Rev.*, 1948, **73**, 68–76.
44. N. Riley, Acoustic streaming, *Theor. Comput. Fluid Dyn.*, 1998, **10**, 349–356.
45. M. Wiklund, R. Green and M. Ohlin, Acoustofluidics 14: Applications of acoustic streaming in microfluidic devices, *Lab Chip*, 2012, **12**, 2438–2451.

46. B. Dubus, C. Vanhille and C. Campos-Pozuelo, *et al.*, On the physical origin of conical bubble structure under an ultrasonic horn, *Ultrason. Sonochem.*, 2010, **17**, 810–818.
47. J. Wu, L. Lin and F.-t. Chau, Ultrasound-assisted extraction of ginseng saponins from ginseng roots and cultured ginseng cells, *Ultrason. Sonochem.*, 2001, **8**, 347–352.
48. M. Ashokkumar, R. Bhaskaracharya and S. Kentish, *et al.*, The ultrasonic processing of dairy products – An overview, *Dairy Sci. Technol.*, 2010, **90**, 147–168.
49. C. Miles, M. Morley and M. Rendell, High power ultrasonic thawing of frozen foods, *J. Food Eng.*, 1999, **39**, 151–159.
50. H. Sabarez, J. Gallego-Juarez and E. Riera, Ultrasonic-assisted convective drying of apple slices, *Drying Technol.*, 2012, **30**, 989–997.
51. R. Mawson, J. Tongaonkar, S. S. Bhagwat, *et al.*, 13 – Airborne ultrasound for enhanced defoaming applications, in *Innovative Food Processing Technologies*, Woodhead Publishing, 2016, pp. 347–359.
52. R. Czechowska-Biskup, B. Rokita and S. Lotfy, *et al.*, Degradation of chitosan and starch by 360-kHz ultrasound, *Carbohydr. Polym.*, 2005, **60**, 175–184.
53. T. Leong, L. Johansson and P. Juliano, *et al.*, Design parameters for the separation of fat from natural whole milk in an ultrasonic litre-scale vessel, *Ultrason. Sonochem.*, 2014.
54. P. Juliano, P. Swiergon and K. H. Lee, *et al.*, Effects of pilot plant-scale ultrasound on palm oil separation and oil quality, *J. Am. Oil Chem. Soc.*, 2013, 1–8.
55. T. Awad, H. Moharram and O. Shaltout, *et al.*, Applications of ultrasound in analysis, processing and quality control of food: A review, *Food Res. Int.*, 2012, **48**, 410–427.
56. E. Riera, J. A. Gallego-Juárez and T. J. Mason, Airborne ultrasound for the precipitation of smokes and powders and the destruction of foams, *Ultrason. Sonochem.*, 2006, **13**, 107–116.
57. T. Leong, K. Knoerzer and F. J. Trujillo, *et al.*, Megasonic separation of food droplets and particles: Design considerations, *Food Eng. Rev.*, 2015, **7**, 298–320.
58. H. Qing-Guo, Z. Min and A. S. Mujumdar, *et al.*, Effects of different drying methods on the quality changes of granular edamame, *Drying Technol.*, 2006, **24**, 1025–1032.
59. S. de la Fuente-Blanco, E. R.-F. de Sarabia and V. Acosta-Aparicio, *et al.*, Food drying process by power ultrasound, *Ultrasonics*, 2006, **44**, e523–e527.
60. M. Nowacka, A. Wiktor and M. Śledź, *et al.*, Drying of ultrasound pretreated apple and its selected physical properties, *J. Food Eng.*, 2012, **113**, 427–433.
61. F. A. Fernandes, F. I. Oliveira and S. Rodrigues, Use of ultrasound for dehydration of papayas, *Food Bioprocess Technol.*, 2008, **1**, 339–345.
62. H. Sabarez, Ultrasonic drying of horticultural food products, (2016).

63. J. M. R. Patino, M. D. N. Delgado and J. L. Fernández, Stability and mechanical strength of aqueous foams containing food proteins, *Coll. Surf., A*, 1995, **99**, 65–78.
64. A. Patist and D. Bates, Ultrasonic innovations in the food industry: From the laboratory to commercial production, *Innovative Food Sci. Emerging Technol.*, 2008, **9**, 147–154.
65. A. C. Soria and M. Villamiel, Effect of ultrasound on the technological properties and bioactivity of food: a review, *Trends Food Sci. Technol.*, 2010, **21**, 323–331.
66. P. Juliano, P. Swiergon and R. Mawson, *et al.*, Application of ultrasound for oil separation and recovery of palm oil, *J. Am. Oil Chem. Soc.*, 2013, **90**, 579–588.
67. H. Lamb, *Hydrodynamics*, Cambridge university press, 1993.
68. T. Leong, L. Johansson and P. Juliano, *et al.*, Design parameters for the separation of fat from natural whole milk in an ultrasonic litre-scale vessel, *Ultrason. Sonochem.*, 2014, **21**, 1289–1298.
69. T. Leong, P. Juliano and L. Johansson, *et al.*, Temperature effects on the ultrasonic separation of fat from natural whole milk, *Ultrason. Sonochem.*, 2014, **21**, 2092–2098.
70. T. Leong, L. Johansson and R. Mawson, *et al.*, Ultrasonically enhanced fractionation of milk fat in a litre-scale prototype vessel, *Ultrason. Sonochem.*, 2016, **28**, 118–129.
71. T. Leong, 16 – High-power ultrasonication for the manufacture of nanoemulsions and nanodispersions, in *Innovative Food Processing Technologies*, Woodhead Publishing, 2016, pp. 413–428.
72. D. J. McClements, E. A. Decker and J. Weiss, Emulsion-based delivery systems for lipophilic bioactive components, *J. Food Sci.*, 2007, **72**, R109–R124.
73. S. M. Jafari, Y. He and B. Bhandari, Production of sub-micron emulsions by ultrasound and microfluidization techniques, *J. Food Eng.*, 2007, **82**, 478–488.
74. C. Solans, P. Izquierdo and J. Nolla, *et al.*, Nano-emulsions, *Curr. Opin. Colloid Interface Sci.*, 2005, **10**, 102–110.
75. D. Kilcast and S. Clegg, Sensory perception of creaminess and its relationship with food structure, *Food Qual. Prefer.*, 2002, **13**, 609–623.
76. H. Goudédranche, J. Fauquant and J.-L. Maubois, Fractionation of globular milk fat by membrane microfiltration, *Le lait*, 2000, **80**, 93–98.
77. M. Akhtar, J. Stenzel and B. S. Murray, *et al.*, Factors affecting the perception of creaminess of oil-in-water emulsions, *Food Hydrocolloids*, 2005, **19**, 521–526.
78. S. M. Jafari, Y. He and B. Bhandari, Nano-emulsion production by sonication and microfluidization—A comparison, *Intern. J. Food Prop.*, 2006, **9**, 475–485.
79. S. M. Jafari, E. Assadpoor and Y. H. He, *et al.*, Re-coalescence of emulsion droplets during high-energy emulsification, *Food Hydrocolloids*, 2008, **22**, 1191–1202.

80. A. Shanmugam and M. Ashokkumar, Ultrasonic preparation of stable flax seed oil emulsions in dairy systems–Physicochemical characterization, *Food Hydrocolloids*, 2014, **39**, 151–162.
81. T. S. Leong, M. Zhou and N. Kukan, *et al.*, Preparation of water-in-oil-in-water emulsions by low frequency ultrasound using skim milk and sunflower oil, *Food Hydrocolloids*, 2017, **63**, 685–695.
82. K. Nakabayashi, F. Amemiya and T. Fuchigami, *et al.*, Highly clear and transparent nanoemulsion preparation under surfactant-free conditions using tandem acoustic emulsification, *Chem. Commun.*, 2011, **47**, 5765–5767.
83. K. Kamogawa, G. Okudaira and M. Matsumoto, *et al.*, Preparation of oleic acid/water emulsions in surfactant-free condition by sequential processing using midsonic-megasonic waves, *Langmuir*, 2004, **20**, 2043–2047.
84. E. K. Silva, M. T. M. Gomes and M. D. Hubinger, *et al.*, Ultrasound-assisted formation of annatto seed oil emulsions stabilized by biopolymers, *Food Hydrocolloids*, 2015, **47**, 1–13.
85. R. Moghimi, L. Ghaderi and H. Rafati, *et al.*, Superior antibacterial activity of nanoemulsion of *Thymus daenensis* essential oil against E. coli, *Food Chem.*, 2016, **194**, 410–415.
86. V. Ghosh, A. Mukherjee and N. Chandrasekaran, Ultrasonic emulsification of food-grade nanoemulsion formulation and evaluation of its bactericidal activity, *Ultrason. Sonochem.*, 2013, **20**, 338–344.
87. A. M. Hashtjin and S. Abbasi, Optimization of ultrasonic emulsification conditions for the production of orange peel essential oil nanoemulsions, *J. Food Sci. Technol.*, 2015, **52**, 2679–2689.
88. A. M. Hashtjin and S. Abbasi, Nano-emulsification of orange peel essential oil using sonication and native gums, *Food Hydrocolloids*, 2015, **44**, 40–48.
89. O. Krasulya, V. Bogush and V. Trishina, *et al.*, Impact of acoustic cavitation on food emulsions, *Ultrason. Sonochem.*, 2016, **30**, 98–102.
90. J. Chandrapala, C. M. Oliver and S. Kentish, *et al.*, Use of power ultrasound to improve extraction and modify phase transitions in food processing, *Food Rev. Int.*, 2013, **29**, 67–91.
91. K. Vilkhu, R. Mawson and L. Simons, *et al.*, Applications and opportunities for ultrasound assisted extraction in the food industry – A review, *Innovative Food Sci. Emerging Technol.*, 2008, **9**, 161–169.
92. M. Vinatoru, An overview of the ultrasonically assisted extraction of bioactive principles from herbs, *Ultrason. Sonochem.*, 2001, **8**, 303–313.
93. S. Balachandran, S. Kentish and R. Mawson, *et al.*, Ultrasonic enhancement of the supercritical extraction from ginger, *Ultrason. Sonochem.*, 2006, **13**, 471–479.
94. E. Riera, Y. Golas and A. Blanco, *et al.*, Mass transfer enhancement in supercritical fluids extraction by means of power ultrasound, *Ultrason. Sonochem.*, 2004, **11**, 241–244.

95. K. Moulton, L. Wang and A. Pilot-Plant, Study of continuous ultrasonic extraction of soybean protein, *J. Food Sci.*, 1982, **47**, 1127–1129.
96. M. A. Rostagno, M. Palma and C. G. Barroso, Ultrasound-assisted extraction of soy isoflavones, *J. Chromatogr. A*, 2003, **1012**, 119–128.
97. L. Paniwnyk, E. Beaufoy and J. Lorimer, *et al.*, The extraction of rutin from flower buds of *Sophora japonica*, *Ultrason. Sonochem.*, 2001, **8**, 299–301.
98. S. Albu, E. Joyce and L. Paniwnyk, *et al.*, Potential for the use of ultrasound in the extraction of antioxidants from *Rosmarinus officinalis* for the food and pharmaceutical industry, *Ultrason. Sonochem.*, 2004, **11**, 261–265.
99. T. Xia, S. Shi and X. Wan, Impact of ultrasonic-assisted extraction on the chemical and sensory quality of tea infusion, *J. Food Eng.*, 2006, **74**, 557–560.
100. M. Romdhane and C. Gourdon, Investigation in solid–liquid extraction: Influence of ultrasound, *Chem. Eng. J.*, 2002, **87**, 11–19.
101. R. W. Hartel, Advances in food crystallization, *Annu. Rev. Food Sci. Technol.*, 2013, **4**, 277–292.
102. S. R. Patel and Z. Murthy, Lactose recovery processes from whey: A comparative study based on sonocrystallization, *Sep. Purif. Rev.*, 2012, **41**, 251–266.
103. N. Deora, N. Misra and A. Deswal, *et al.*, Ultrasound for improved crystallisation in food processing, *Food Eng. Rev.*, 2013, **5**, 36–44.
104. R. Chow, R. Blindt and R. Chivers, *et al.*, A study on the primary and secondary nucleation of ice by power ultrasound, *Ultrasonics*, 2005, **43**, 227–230.
105. R. S. Dhumal, S. V. Biradar and A. R. Paradkar, *et al.*, Ultrasound assisted engineering of lactose crystals, *Pharm. Res.*, 2008, **25**, 2835–2844.
106. M. L. De Castro and F. Priego-Capote, Ultrasound-assisted crystallization (sonocrystallization), *Ultrason. Sonochem.*, 2007, **14**, 717–724.
107. S. Martini, *Sonocrystallization of Fats*, Springer, 2013.
108. P. R. Gogate and A. B. Pandit, Sonocrystallization and its application in food and bioprocessing, in *Ultrasound Technologies for Food and Bioprocessing*, Springer, 2011, pp. 467–493.
109. N. Lyczko, F. Espitalier and O. Louisnard, *et al.*, Effect of ultrasound on the induction time and the metastable zone widths of potassium sulphate, *Chem. Eng. J.*, 2002, **86**, 233–241.
110. Z. Guo, M. Zhang and H. Li, *et al.*, Effect of ultrasound on anti-solvent crystallization process, *J. Cryst. Growth*, 2005, **273**, 555–563.
111. R. Chow, R. Mettin, B. Lindinger, *et al.*, The importance of acoustic cavitation in the sonocrystallisation of ice-high speed observations of a single acoustic bubble, in *Ultrasonics, 2003 IEEE Symposium on,* IEEE, 2003, pp. 1447–1450.
112. R. Chow, R. Blindt and R. Chivers, *et al.*, The sonocrystallisation of ice in sucrose solutions: Primary and secondary nucleation, *Ultrasonics*, 2003, **41**, 595–604.
113. B. Li and D.-W. Sun, Effect of power ultrasound on freezing rate during immersion freezing of potatoes, *J. Food Eng.*, 2002, **55**, 277–282.

114. C. Dai, X. Zhou and S. Zhang, *et al.*, Influence of ultrasound assisted nucleation on freeze drying of carrots, *Drying Technol.*, 2015.
115. A. Mortazavi and F. Tabatabaie, Study of ice cream freezing process after treatment with ultrasound, *World Appl. Sci. J.*, 2008, **4**, 188–190.
116. A. Kissam, R. Nelson and J. Ngao, *et al.*, Water-thawing of fish using low frequency acoustics, *J. Food Sci.*, 1982, **47**, 71–75.
117. J. Maubois, Membrane microfiltration: A tool for a new approach in dairy technology, *Aust. J. Dairy Technol.*, 2002, **57**, 92.
118. S. Muthukumaran, S. E. Kentish and M. Ashokkumar, *et al.*, Mechanisms for the ultrasonic enhancement of dairy whey ultrafiltration, *J. Membr. Sci.*, 2005, **258**, 106–114.
119. S. Muthukumaran, S. E. Kentish and G. W. Stevens, *et al.*, The application of ultrasound to dairy ultrafiltration: the influence of operating conditions, *J. Food Eng.*, 2007, **81**, 364–373.
120. S. Karaman, M. T. Yilmaz and M. F. Ertugay, *et al.*, Effect of ultrasound treatment on steady and dynamic shear properties of glucomannan based salep dispersions: Optimization of amplitude level, sonication time and temperature using response surface methodology, *Ultrason. Sonochem.*, 2012, **19**, 928–938.
121. Y. Iida, T. Tuziuti and K. Yasui, *et al.*, Control of viscosity in starch and polysaccharide solutions with ultrasound after gelatinization, *Innovative Food Sci. Emerging Technol.*, 2008, **9**, 140–146.
122. B. Zisu, R. Bhaskaracharya and S. Kentish, *et al.*, Ultrasonic processing of dairy systems in large scale reactors, *Ultrason. Sonochem.*, 2010, **17**, 1075–1081.
123. Z. Liu, P. Juliano and R. P. Williams, *et al.*, Ultrasound effects on the assembly of casein micelles in reconstituted skim milk, *J. Dairy Res.*, 2014, **81**, 146–155.
124. Z. Liu, P. Juliano and R. P. Williams, *et al.*, Ultrasound improves the renneting properties of milk, *Ultrason. Sonochem.*, 2014, **21**, 2131–2137.
125. A. C. Chang and F. C. Chen, The application of 20 kHz ultrasonic waves to accelerate the aging of different wines, *Food Chem.*, 2002, **79**, 501–506.
126. C. G. Leonhardt and J. A. Morabito, *Wine aging method and system, U.S. Pat.*, US 7220439 B2, 2007.
127. Q.-A. Zhang, Y. Shen and X.-H. Fan, *et al.*, Preliminary study of the effect of ultrasound on physicochemical properties of red wine, *CyTA – J. Food*, 2016, **14**, 55–64.
128. M. C. Santos, C. Nunes and J. A. Saraiva, *et al.*, Chemical and physical methodologies for the replacement/reduction of sulfur dioxide use during winemaking: review of their potentialities and limitations, *Eur. Food Res. Technol.*, 2012, **234**, 1–12.
129. C. O'donnell, B. Tiwari and P. Bourke, *et al.*, Effect of ultrasonic processing on food enzymes of industrial importance, *Trends Food Sci. Technol.*, 2010, **21**, 358–367.
130. H. Vally, N. L. Misso and V. Madan, Clinical effects of sulphite additives, *Clin. Exp. Allergy*, 2009, **39**, 1643–1651.

131. A. C. Chang, Study of ultrasonic wave treatments for accelerating the aging process in a rice alcoholic beverage, *Food Chem.*, 2005, **92**, 337–342.
132. S. Gao, G. D. Lewis and M. Ashokkumar, *et al.*, Inactivation of microorganisms by low-frequency high-power ultrasound: 1. Effect of growth phase and capsule properties of the bacteria, *Ultrason. Sonochem.*, 2014, **21**, 446–453.
133. P. Piyasena, E. Mohareb and R. McKellar, Inactivation of microbes using ultrasound: A review, *Int. J. Food Microbiol.*, 2003, **87**, 207–216.
134. A. Sams and R. Feria, Microbial effects of ultrasonication of broiler drumstick skin, *J. Food Sci.*, 1991, **56**, 247–248.
135. H. Lillard, Bactericidal effect of chlorine on attached Salmonellae with and without sonification, *J. Food Prot.*, 1993, **56**, 716–717.
136. B. Kim, O. Kim, D. Kim, *et al.*, *Development of a surface sterilization system combined with a washing process technology for leafy lettuce*, in International Symposium on Vegetable Quality of Fresh and Fermented Vegetables **483**, 1997, pp. 311–318.
137. T. Monsen, E. Lövgren and M. Widerström, *et al.*, In vitro effect of ultrasound on bacteria and suggested protocol for sonication and diagnosis of prosthetic infections, *J.Clin. Microbiol.*, 2009, **47**, 2496–2501.
138. G. Scherba, R. Weigel and W. D. O'Brien, Quantitative assessment of the germicidal efficacy of ultrasonic energy, *Appl. Environ. Microbiol.*, 1991, **57**, 2079–2084.

CHAPTER 9

High-intensity Pulsed Light Processing

JAQUELINE O. DE MORAES AND CARMEN I. MORARU*

Department of Food Science, Cornell University, Ithaca, NY 14853, USA
*Email: cim24@cornell.edu

9.1 Introduction

Pulsed light (PL) treatment is a non-thermal method of microbial inactivation, which was approved by the US Food and Drug Administration (FDA) as an indirect additive in food production, processing, and handling (Code of Federal Regulation, CFR:21CFR179.41).[1] PL uses high-intensity light emitted by gas discharge lamps that generate radiation in the wavelength range 200–1000 nm, containing the ultraviolet (UV) (200–400 nm), visible (380–780 nm), and near-infrared (NIR) (700–1100 nm) spectral ranges.

It is generally accepted that PL inactivates microorganisms mainly due to the absorption of UV light by microbial DNA, but other mechanisms may also contribute to inactivation. According to the FDA regulation, PL treatments that are used in the food industry must use xenon flashlamps, typically operated at 1–20 pulses per second, with a pulse duration no longer than 2 ms, and the total cumulative treatment shall not exceed 12.0 $J\,cm^{-2}$.

PL efficiency has been demonstrated for a large number of microorganisms in clear or translucent liquids, and also on food and food contact surfaces, and under certain conditions has the potential to replace thermal treatments and chemical sanitizers (*e.g.* hydrogen peroxide, peracetic acid).

Green Chemistry Series No. 53
Alternatives to Conventional Food Processing: 2nd Edition
Edited by Andrew Proctor

Published by the Royal Society of Chemistry, www.rsc.org

PL has also shown some efficacy in inactivating enzymes and toxins and reducing various food allergens. However, few industrial applications are currently available, and more investigations are needed to define the best uses and requirements for the application of PL on an industrial scale.

This chapter provides an overview of the principles of PL technology and its current and potential applications in food safety and preservation.

9.2 Fundamentals of Pulsed Light Technology

Pulsed light (PL), also known as intense pulsed light (IPL), has emerged in recent years as a promising non-thermal technology that has the capability of increasing food safety while maintaining many food quality aspects that are typically altered by thermal processes. PL technology uses high-intensity pulses of broad-spectrum light (200–1100 nm), with UV-C being the most relevant part of the spectrum, to inactivate bacteria, yeasts, molds, and viruses in liquids and on solid food and food contact surfaces.[2–8] As with any other light-based treatment, the effectiveness of PL is highly dependent on the properties of the substrate being treated, particularly those that affect light absorption, reflection, and transmission. PL most efficiently inactivates microorganisms in clear liquids or on smooth solid substrates, which are less limited by light penetration or microbial hiding effects.

Compared with UV-C treatment, which uses low-pressure mercury lamps that emit germicidal monochromatic light with a peak at 254 nm,[9] PL uses gas discharge lamps filled with xenon gas (or another inert gas). This avoids the use of mercury, which could be an environmental and health hazard if the lamp is broken. Furthermore, UV-C light typically has poorer penetration depth and lower emission power than PL, resulting in longer treatment times for comparable microbial inactivation effectiveness. The short pulse duration and cooling period between pulses during PL treatment minimize temperature increases due to heat build-up.[10]

9.2.1 Pulsed Light Dose

The PL treatment dose is quantified by "fluence," expressed in $J\,cm^{-2}$, which represents the total radiant energy through an infinitesimal area dA. There are some difficulties associated with the accurate quantification of PL fluence. Some of the most frequently used methods to measure the fluence delivered by PL systems include biodosimetry[11,12] and pyroelectric joulemeters.[13,14] Jin *et al.*[15] compared different methods of fluence measurement and concluded that although the use of a radiometer coupled with a germicidal filter matched a biodosimeter in describing fluence supplied by a collimated beam system, the radiometer cannot accurately measure light from all directions, as it measures only the radiation incident normal to the detector surface. Ideally, the total fluence in addition to the number of pulses should be reported to allow direct comparisons of different treatments, regardless of the experimental setup.

9.2.2 Components of Pulsed Light Systems

A PL unit is typically comprised of the following components: a high-voltage power supply, a storage capacitor, a pulse-forming network that determines the pulse shape and spectrum characteristics, a gas discharge flashlamp, and a trigger that initiates discharge of the electrical energy to the flashlamp.

PL generation requires direct current. Therefore, the alternating current from the outlet is converted into direct current by a transformer. The electrical energy is then stored in a dense, high-energy electrical storage capacitor over relatively long periods. The electricity released by the capacitor is discharged in fractions of a second in the flashlamp. Owing to the absorption of electrical energy, electrons in the xenon atoms are elevated to an excited state, which is a temporary, higher energy state. The electrons then release a discrete amount of energy as photons, and return to the most stable and lowest energy state, the ground state. The wavelength spectrum of the released photons depends upon the elemental nature of the inert gas and the power supplied. In PL systems that use xenon gas, broad-spectrum light ranging from the UV to the NIR region is emitted. An example of an emission spectrum for a xenon flashlamp is shown in Figure 9.1.

The key element of any PL unit is the flashlamp, filled with an inert gas such as xenon or krypton. Xenon is more commonly used because of its higher conversion efficiency. Xenon is also the gas of choice for most microbial inactivation applications. The main structural components of the flashlamp are the envelope, the seals, and the electrodes. The envelope is the jacket that surrounds the electrodes and physically contains the filling gas. Typically, lamp envelopes for flashlamps are made of clear fused quartz of about 1 mm thickness.[16] The electrodes are connected to a capacitor charged with a high voltage, and are responsible for delivering the electric

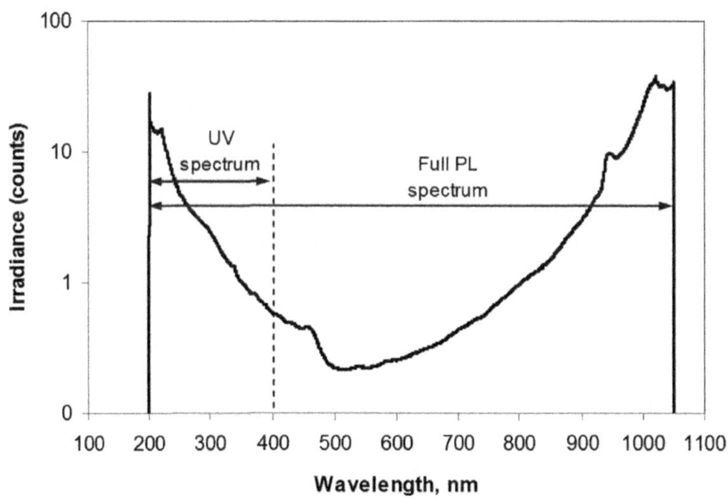

Figure 9.1 Spectral output of a pulsed xenon discharge lamp.

current into the gas. The cathode must be able to supply an adequate amount of electrons without sputtering, since sputtering can strip off significant amounts of cathode material, drastically reducing the lamp lifetime. This effect is caused by the formation of hot spots that occurs at peak power. The most important consideration for the anode is that it has sufficient mass or surface area to satisfy the power loading due to electron bombardment from the arc.[16] In order for flashlamps to be hermetic, the quartz envelope and the electrode assembly must have a gas-tight seal.[16]

According to lamp manufacturers, the conversion of electrical power into radiated optical power for xenon flashlamps is approximately 50%. Generally, the efficiency improves with increasing current density and gas fill pressure.[16,17]

Overheating is a significant issue encountered during the operation of flashlamps. When operated at high power and flash rates, additional cooling of the lamp is required in order to ensure reliable operation and a long lamp lifetime. This can be achieved by using forced air cooling by means of a cooling fan incorporated in the PL equipment, which reduces the heat build-up in the treatment area. Reflectors are sometimes incorporated in the treatment chamber to help redirect the light to the sample.

9.3 Microbial Inactivation Using Pulsed Light

9.3.1 Mechanisms of Inactivation

The mechanisms of microbial inactivation by PL are not yet fully elucidated. Since a significant portion of the PL spectrum includes UV light, it is generally accepted that UV light plays an important role in the microbial cell death by PL. Nonetheless, some studies have also presented evidence of non-UV-related death mechanisms, such as physical microbial-membrane destruction caused by cell overheating. The relative importance of these inactivation mechanisms likely depends on the fluence, wavelength, and target microorganism, and need more studies to be well understood.

Since the UV range is a significant portion of the PL spectrum, it is generally accepted that UV-B and UV-C light play an important role in microbial cell death by PL. The UV spectrum consists of the long-wave UV-A (320–400 nm), medium-wave UV-B (280–320 nm), and short-wave UV-C (200–280 nm) regions.[9] The antimicrobial effects of UV light are caused by the absorption of radiation by the conjugated carbon–carbon double bonds in proteins and nucleic acids and subsequent structural changes in DNA or RNA,[15–21] and also abnormal ion flow, increased cell membrane permeability, and depolarization of the cell membrane.[22] The high-energy UV-C range is credited with carrying most of the bactericidal effects.[21] In the 250–260 nm range, alteration of the cellular DNA occurs due to pyrimidine dimer formation and/or other photochemical products; the UV-A and UV-B ranges are believed to impart lethal effects *via* membrane damage and the formation of peroxides.[9]

To quantify the contributions of different UV spectral ranges to cell death, Wang *et al.*[23] directed a xenon flashlamp into a monochromator at 14 different wavelengths in the range 230–360 nm. The results showed that *Escherichia coli* inactivation depended on wavelength; the germicidal efficiency, expressed by the log reduction in *E. coli* activity per unit fluence, peaked at 270 nm. However, no *E. coli* inactivation was observed at wavelengths higher than 300 nm. An earlier study by Takeshita *et al.*[24] also found that no inactivation took place in PL experiments in which a filter removed UV light below 320 nm. Woodling and Moraru[25] conducted a detailed study on the effect of wavelength on inactivation. Three filters were used to select different portions of the PL spectrum: (1) a cold mirror allowed partial transmission of wavelengths between 200 and 400 nm and near total transmission for wavelengths above 700 nm; (2) The UV-A-transmitting filter transmitted wavelengths between 300 and 400 nm; and (3) the UV-blocking filter transmitted wavelengths above 400 nm. A quartz filter that transmitted wavelengths between 200 and 2000 nm was used as a control. The authors treated stainless-steel coupons inoculated with *Listeria innocua* with PL at 6 $\mathrm{J\,cm^{-2}}$. The full spectrum treatment resulted in a 4 log reduction of *L. innocua* activity, whereas the removal of $\lambda < 200$ nm decreased the reduction to less than 2 log. When all UV light ($\lambda < 400$ nm) was filtered out, no lethal or sublethal effects of the treatment on *L. innocua* were observed. The results demonstrated that the portions of the PL spectrum responsible for bacterial inactivation were located in the range $\lambda < 300$ nm (UV-B and UV-C ranges), with some bacterial death taking place as a result of exposure to λ between 300 and 400 nm (UV-A range).

In addition to nucleic acid damage, several studies indicated other types of microbial cell injury following exposure to PL. Wekhof[26] reported the physical destruction of *Aspergillus niger* spores as a result of structural collapse, and Takeshita *et al.*[24] reported enlarged vacuoles in *Saccharomyces cerevisiae* yeast cells after PL treatment. These effects were attributed to intracellular heating resulting from exposure to PL. Takeshita *et al.*[24] reported that continuous UV treatment resulted in a larger amount of DNA damage in *S. cerevisiae* than PL, while PL resulted in expanded vacuoles and compromised cell membranes.

There is still no consensus as to whether PL is more effective than continuous UV radiation. Although some authors reported that PL is more effective for microbial inactivation than continuous UV radiation,[27] contrary findings have also been reported by Cheigh *et al.*[28] and Takeshita *et al.*,[24] who carried out electrophoresis studies of DNA extracted from cells treated with either UV light or PL, and reported that cells treated with continuous UV radiation exhibited greater amounts of DNA damage. On the other hand, Uesugi *et al.*[29] investigated the response of *Listeria monocytogenes* to PL and continuous UV radiation at treatment levels that yielded comparable levels of inactivation (3.20 $\mathrm{J\,cm^{-2}}$ for PL and 33 $\mathrm{mJ\,cm^{-2}}$ for UV). Whole-genome microarray analysis was used to identify differential gene expression after the treatments. Compared with untreated cells, PL- and UV-treated cells had higher transcription levels for 80 and 39, respectively, multiple stress-related proteins, motility genes, and transcriptional regulators. Exposure to UV-blocked PL resulted in 131 motility-related genes and cell membrane-related genes with lower transcription levels, and no genes with higher transcription levels, relative to untreated cells. However, no significant differences in gene expression between PL- and UV-treated cells were found, which suggested that microbicidal effects in both situations were primarily due to UV light.

Some studies have also claimed that although cells are able to repair DNA damage after continuous UV treatment, these repair mechanisms are inactivated by PL. For instance, Gómez-López *et al.*[30] suggested that the broader spectrum might inactivate photolyase, the enzyme responsible for light repair, which would make PL-mediated microbial DNA damage irreversible.

Heating effects produced by PL treatment have also been considered responsible for structural damage in cells. However, most studies showing structural damage of microbial cells have used high doses of PL, much higher than the 12 $J\,cm^{-2}$ established by the FDA as the upper limit for PL treatment (in the USA).

Wekhof[31] argued that there must be an effect of PL treatment beyond the UV effect on microbial DNA that inactivates UV-resistant microorganisms, such as *Cryptosporidium*, *Aspergillus niger*, or *Bacillus subtilis*. He suggested that the high-intensity light pulses delivered by PL may cause rapid overheating in microorganisms due to the absorption of UV light. By estimating the size, absorption coefficient, and cooling properties, Wekhof approximated that *E. coli* bacteria lose only 1–10% of the heat gained during PL treatment to cooling. If over 90% of the energy received by the bacteria is retained as heat, the internal temperature in the cell could quickly rise to over 100 °C during longer PL treatments. This hypothesis was tested for *E. coli* cells, and the results showed that little temperature increase occurred until a threshold fluence was reached, after which the temperature increased dramatically to over 120 °C.

Other researchers have also attempted to quantify the temperature changes associated with PL treatments. A study on *Staphylococcus aureus* by Krishnamurthy *et al.*[32] reported temperature increases for PL treatments longer than 5 s, which corroborated Wekhof's findings. Negligible temperature increases were seen for PL treatments shorter than 5 s in a subsequent study on *S. aureus.*[33] Although bulk heating of the sample was not observed in the latter study, it is still possible that localized heating of bacterial cells occurred, causing the structural damage that was observed by transmission electron microscopy. Generally, minimal heating is found at the substrate level after PL treatment and the magnitude of the heating effect depends on the substrate being treated and the PL dose.[33–35] However, heating that occurs at a cellular level, which would be directly responsible for cell damage, is difficult to quantify, hence the proposed overheating mechanisms are difficult to verify.

Some studies suggest some effectiveness of PL against bacterial spores, but the mechanisms responsible for these purported inactivation effects have not yet been elucidated. Considering the effect of UV-C treatment, PL could cause nucleic acid damage by the formation of 5-thyminyl-5,6-dihydrothymine and single-strand and double-strand breaks, and cyclobutane–pyrimidine dimers.[36,37]

9.3.2 Critical Factors That Affect Microbial Inactivation by Pulsed Light

The antimicrobial effectiveness of PL is influenced by a range of factors, including: optical and surface properties of microorganisms and substrate, type of microorganism and cell population density, and process characteristics, among others. Some factors are co-dependent, *e.g.* the light absorption and scattering properties of the treated substrate and/or microorganisms affect treatment fluence and spectral make-up. A brief overview of the main factors that influence PL treatment is presented below.

9.3.2.1 Optical Properties of Microbial Cells and the Substrate

Knowledge of the optical properties of microbial cells and the substrate is important in understanding the microbicidal effects of PL treatment, as they can significantly diminish irradiance and alter the spectral makeup of the incident light.

Waltham *et al.*[38] measured light scattering and absorption from a population of *E. coli* in water. They observed a significant decrease in these properties in response to radiation UV to visible range. Similar results were reported by Arakawa *et al.*[39] for *Erwinia herbicola*. Tuminello *et al.*[40] reported an increase in the UV to NIR transmission by *Bacillus* spores. Such findings support the idea that during PL treatment the UV radiation affects the microbial cells mostly, since this component of the electromagnetic spectrum is absorbed the most.

The interaction between light and the treated substrate is critically important in determining the outcome of PL treatments. For instance, Montgomery and Banerjee[41] reported activity reductions in *L. monocytogenes* and *E. coli* O157:H7 biofilms of 2.7 and 3.9 log cfu mL^{-1}, respectively (the total fluence received by the samples was not reported). The lower inactivation levels for these microorganisms when entrapped in a dense biofilm matrix, compared with suspended in media, show the importance of light–substrate interactions during PL inactivation. The incident light can be reflected, refracted, scattered, or absorbed to varying extents, depending on the composition and structure of the substrate media and the light wavelength. *Refraction*, which occurs when light enters a substrate with a different optical density than the surrounding air, is particularly relevant to transparent and colored materials. *Reflection* tends to be the dominant phenomenon for opaque surfaces. When the incident light bounces on a smooth surface and comes out at the same angle as the incident beam, with the same spectral distribution of energy, the phenomenon is termed *specular reflection*. For rough surfaces, light travels through the outer layers of the material, where the incident light is partly absorbed, a phenomenon called diffuse reflection. Since absorption is uneven at different wavelengths, the spectral distribution of the incident light changes and the diffused light comes out in all directions.[42] Such effects are very important for PL treatment, as reflection by the substrate could significantly decrease the efficiency of the treatment.

Extinction phenomena such as absorption and scattering are particularly important light interactions in the case of in-depth treatments, as they significantly diminish the irradiance and alter the spectral makeup of the incident light. In the case of translucent materials, a portion of the incident light interacts with the internal structure and undergoes multiple internal reflections and redirections, resulting in *scattering*. Light propagation in biological tissues thicker than several tens of micrometers is typically characterized by multiple scattering events, due to structural heterogeneity and relative refractive index mismatch at the boundaries between cell membranes or intracellular structures.[43,44] Tissues that show strong scattering cause light to be backscattered towards the tissue surface and increase the delivered irradiance, resulting in a higher fluence rate and greater penetration.[45] This can be particularly relevant in the PL treatment of tissue foods, as it can affect the actual fluence delivered to microbial cells present in their internal structure or on their surface. Scattering can also occur in opaque substrates that have surface irregularities. When the distance between two points of irregularity is larger than the wavelength (λ) of the incident light, scattering effects are negligible. However, when this distance is comparable to λ, substantial scattering occurs.[46] This effect is particularly important for the UV region of the PL spectrum, as it could affect the UV dose available for microorganism inactivation.

The decrease in the intensity of the light traversing the substrate due to light *absorption* can significantly diminish the amount of light that reaches the microbial cells during PL treatment. Absorption of photons by the substrate molecules results in an increase in the energy content of those molecules, which causes an increased level of vibration or rotation.[47]

For biological systems, including foods, absorption of UV light is mainly due to proteins that contain the amino acids tyrosine, phenylalanine, and tryptophan, and to nucleic acids, since these molecules absorb strongly in the near-UV region. Visible light absorption is due to tissue chromophores, such as hemoglobin, deoxyhemoglobin, myoglobin, melanin, and cytochrome *c* oxidase from meat[44,48] or carotene, chlorophyll, and xanthophyll from plant tissues. In the NIR region, most of the absorption by food substrates is due to water, which has an absorption peak at 970 nm,[49] and fat.[50] UV absorption by the substrate is the most relevant in PL treatments, as it can decrease the irradiance responsible for antimicrobial effects. Hence UV-absorbing substrates diminish the antimicrobial effectiveness of the treatment.

For practical applications of light-based treatments, it is very important to be able to quantify the overall effect of the substrate on the light dose received by the microorganisms. The optical property that describes quantitatively the decrease in light dose inside a substrate is the *optical penetration depth* (δ), which represents the distance over which light decreases in fluence rate to 1/e or 37% of its initial value. It is important to note that optical penetration varies with wavelength, with shorter wavelengths providing deeper penetration into the food than longer wavelengths.[51] Sauer and Moraru,[52] and Uesugi and Moraru[53] determined the optical penetration depth of PL for several substrates. For Butterfield's phosphate buffer, apple juice, and apple cider, the optical penetration depth of PL was calculated to be 166.7, 41.7, and 15.9 mm, respectively. For Vienna sausages, the optical penetration depth of PL was estimated as 2.3 mm. This demonstrates that PL treatments will be effective as a surface treatment only for solid food substrates, but could be used as an in-depth treatment for clear liquid substrates.

9.3.2.2 *Surface Topography*

The surface properties of the treated substrate/product plays an important role in the effectiveness of PL surface treatments. In particular, microorganisms may lodge in surface irregularities, which can shield them from PL exposure. For instance, *E. coli* O157:H7 cells were found to be capable of penetrating up to 20 μm into cut lettuce leaves.[54] This means that the use of PL for surface decontamination of lettuce, or any rough food surface, may not be very effective. Nonetheless, a smoother surface does not necessarily guarantee a high level of inactivation. Surface imperfections can lead to shading effects, which significantly limit the efficiency of the treatment, and induce a great level of variability. Woodling and Moraru[35] studied the effectiveness of PL in inactivating *L. innocua* on solid substrates with different degrees of surface roughness, using stainless-steel surfaces with varying degrees of surface roughness and reflectivity. They expected the roughest surface to have the lowest level of inactivation, since surface roughness can provide locations that shield bacterial cells from PL. However, in this study, the smoother surfaces actually had lower levels of inactivation. This was attributed to the hydrophobicity of these surfaces, which induced clumping and layering of cells, thus shielding some cells from direct exposure to PL. Surfaces of high reflectivity were also found to have a negative impact on inactivation levels by PL.[55]

9.3.2.3 *Effect of Microbial Cell Attributes on PL*

The properties of the treated microorganisms also play a role in the efficacy of PL treatments. Some studies have reported a difference between the susceptibility of Gram-negative bacteria, Gram-positive bacteria, and fungal spores to PL, with Gram-negative bacteria being the most susceptible and fungal spores the least susceptible of these three groups.[34,56] *L. monocytogenes* and its surrogate *L. innocua* inoculated into different products have been reported to be among the most resistant bacteria to PL.[57–60] Other studies found that microorganisms differed in sensitivity to PL, but no pattern was observed among different groups;[57] another stated that susceptibility must be determined on an individual organism basis.[61]

Gómez-López *et al.*[57] extensively evaluated the lethal effect of PL on a wide variety of microorganisms, using a light source of 100 W (ST100-IE, Sysmat Industrie (St. Thibault des Vignes, France). They used a pulse duration of 30 μs, pulse intensity of 7 J. with a distance between the light and the surface of inoculated Petri dishes of 8.4 cm, 50 flashes per treatment, at a rate of one pulse per second. Initial counts for Gram-negative spoilage bacteria ranged from 4.8 to 5.6 log cfu cm^{-2}, and the most resistant cell was *Shewanella putrefaciens*, with a reduction of 3.9 log cfu cm^{-2}. Among five species of Gram-positive spoilage bacteria, the most resistant was *Lactobacillus sakei*, which showed a reduction of 2.5 log cfu cm^{-2} from an initial count of 5.0 log cfu cm^{-2}. Two species of Enterobacteriaceae, *Enterobacter aerogenes* and *Klebsiella oxytoca*, were evaluated with initial counts of about 5.0 log cfu cm^{-2} and reductions of 2.4 and 4.2 log cfu cm^{-2} were achieved, respectively. Among the Gram-negative pathogens *Aeromonas hydrophila*, *E. coli*, *Salmonella typhimurium*, *Shigella flexnii*, and *Yersinia enterocolitica*, the most resistant to PL treatment was *A. hydrophila*, which showed a reduction of 2.3 log cfu cm^{-2}. No countable survivors were found after treating the Gram-positive pathogens *Bacillus cereus* and *Clostridium perfringens* with initial counts of about 3.3 log cfu cm^{-2} and *S. aureus* with initial counts of 5.5 log cfu cm^{-2}. A reduction in *L. monocytogenes* activity of 2.5 log cfu cm^{-2} (initial count 5.0 log cfu cm^{-2}) was achieved. The yeasts *Candida lambica* and *Rhodotorula mucilaginosa* showed a reduction of 2.8 log cfu cm^{-2} (initial count ~3.3 log cfu cm^{-2}). Conidia were very resistant to PL treatment, showing a reduction of 2.2 log cm^{-2} for *Aspergillus flavus* (initial count 5.2 log cm^{-2}) and 1.2 log cm^{-2} for *Botrytis cinerea* (initial count 4.1 log cm^{-2}). When bacterial spores were tested (*Alicyclobacillus acidoterrestris*, *Bacillus circulans*, and *Bacillus cereus*), PL was most effective on *B. cereus*, which showed complete inactivation, whereas *B. circulans* showed a reduction from an initial 5.7 log spores cm^{-2} to 3.7 log spores cm^{-2}. In general, spores are known to be more resistant to PL than vegetative cells.[62,63] Artíguez and de Marañón[62] reported a 5 log reduction of vegetative cells and only a 3 log reduction for spores of *B. subtilis* with a cell density of 10^7 mL^{-1} treated with a fluence of 0.5 J cm^{-2}.

Inactivation is affected by the microbial population density, which determines light penetration. Several studies on vegetative cells have shown that PL inactivation efficiency decreased on increasing the initial microbial population.[14,57,62,64,65] Artíguez and Marañón[62] investigated the effectiveness of PL inactivation of vegetative cells with different population densities. Suspensions of *Geobacillus stearothermophilus* (10^5–10^7 cfu mL^{-1}), *B. subtilis* (10^7–10^9 cfu mL^{-1}) and their spores (10^5–10^8 mL^{-1} for *G. stearothermophilus* and 10^7–10^9 mL^{-1} for *B. subtilis*) were treated with a fluence in the range 0.14–12 J cm^{-2}. The antimicrobial effect of PL decreased when the population density increased, for both vegetative cells and spores.

Uesugi *et al.*[14] found that this correlation depends on the substrate being treated. For instance, the initial microbial population density did not influence the reduction by PL in clear liquid substrates, but for stainless-steel surfaces a higher level of inactivation was observed for higher initial microbial population densities compared with lower microbial cell densities.[14]

The substrate in which the microorganism is present also affects the effectiveness of PL treatment. Huang *et al.*[66] inactivated about 8 log cfu mL^{-1} of *E. coli* O157:H7, 8 log cfu mL^{-1} of *Salmonella enterica*, and 6 log pfu mL^{-1} of murine norovirus virus suspended in phosphate-buffered saline, after PL treatment with 4.94 J cm^{-2}. However, when the same microorganisms were inoculated on strawberries and blueberries, the inactivation decreased significantly.

For laboratory studies, the method of inoculation can also affect the effectiveness of PL treatments. Xu *et al.*[67] spot inoculated and dip inoculated green onion leaves to simulate contamination of green onions by workers or equipment during processing and during washing, respectively. When using spot-inoculated green onion leaves, a 5 J cm^{-2} dose resulted in a 4.8 log cfu g^{-1} reduction in *E. coli* O157:H7. In contrast, the 5 J cm^{-2} dose yielded only a 0.6 log cfu g^{-1} reduction in *E. coli* for dip-inoculated green onion leaves. This is possibly due to deeper penetration of the bacterial cells within the substrate during dip inoculation.

9.3.2.4 Process Characteristics

Treatment dose (fluence) is a critical treatment factor, since it affects both microbicidal results and the quality of the treated product. As already discussed, although many studies have reported the number of pulses, pulse characteristics, and distance from the lamp, ideally the total fluence should be reported in order to allow direct comparisons of different PL treatments.[68,69]

Hsu and Moraru[69] studied in detail the dependence of fluence on the location of the sample in relation to the light source. Their results revealed that fluence varies significantly in all directions within a PL chamber (Figure 9.2). This has a significant influence on treatment effectiveness and uniformity. Therefore, PL treatments should be designed with this in mind, and direct exposure of the sample to the light source should be ensured. Knowledge of the spatial distribution of fluence within the PL chamber and within different liquids within the PL chamber would help processors determine where to place their substrates relative to the lamp and how to locate multiple lamps in order to ensure homogeneous treatment and maximize PL effectiveness. The design of the PL chamber/equipment also affects the outcome of the treatment. A strategic placement of light reflectors can be used to improve treatment effectiveness and efficiency.

The effect of sublethal treatment temperature on PL inactivation was investigated by Hilton *et al.* for three bacterial strains (*L. innocua*, *E. coli* ATCC 25922, and *Pseudomonas fluorescens*) suspended in a thin layer of clear, liquid phosphate buffer.[60] In the range 5–40 °C, temperature did not affect the PL inactivation of *E. coli* or *P. fluorescens*, but a modest synergistic effect

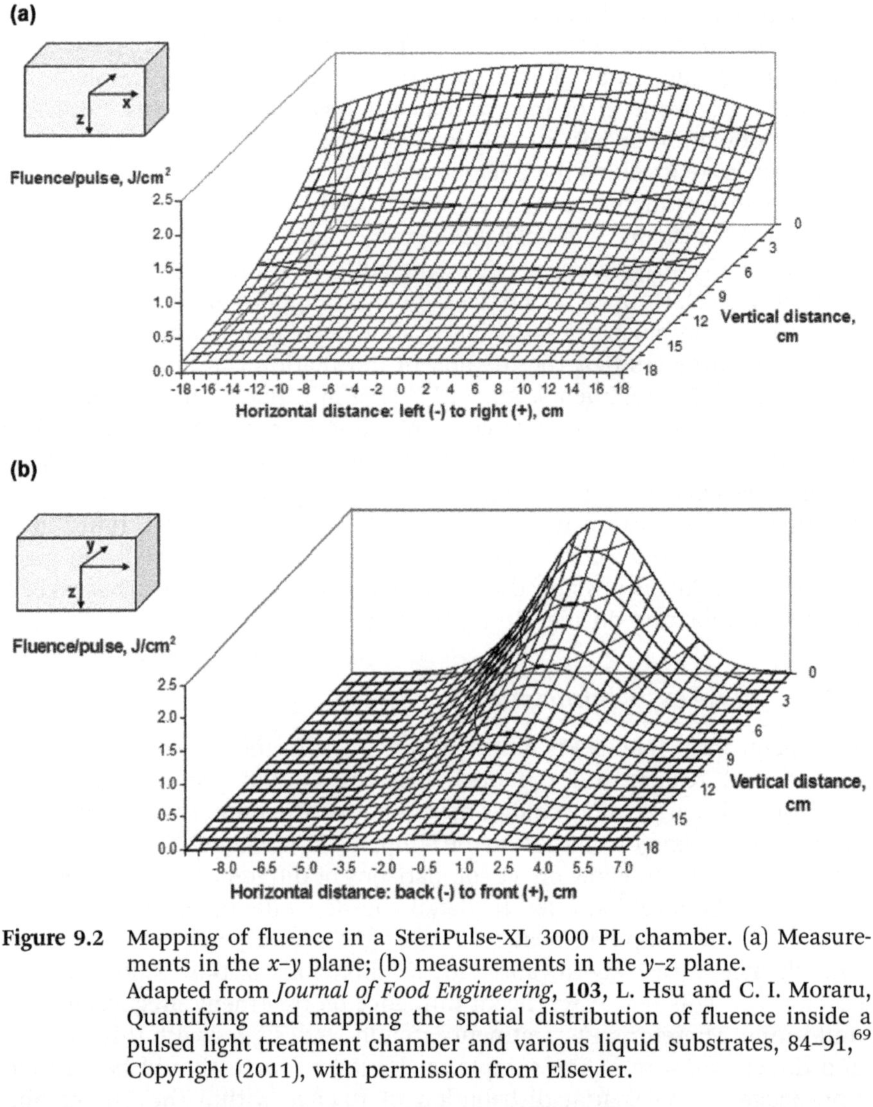

Figure 9.2 Mapping of fluence in a SteriPulse-XL 3000 PL chamber. (a) Measurements in the *x*–*y* plane; (b) measurements in the *y*–*z* plane.
Adapted from *Journal of Food Engineering*, **103**, L. Hsu and C. I. Moraru, Quantifying and mapping the spatial distribution of fluence inside a pulsed light treatment chamber and various liquid substrates, 84–91,[69] Copyright (2011), with permission from Elsevier.

between PL inactivation and temperature was observed for *L. innocua* treated above 40 °C. The practical implication of this finding is that PL treatments can be conducted in a wide range of environmental temperatures without any change in treatment outcome.

9.3.3 Microbial Inactivation Kinetics in Pulsed Light Treatment

Microbial survivor curves in PL treatment show a rapid decrease in survivors at low fluence, but the curve reaches a plateau as fluence increases (Figure 9.3). The plateau could be due to the resistance of the treated

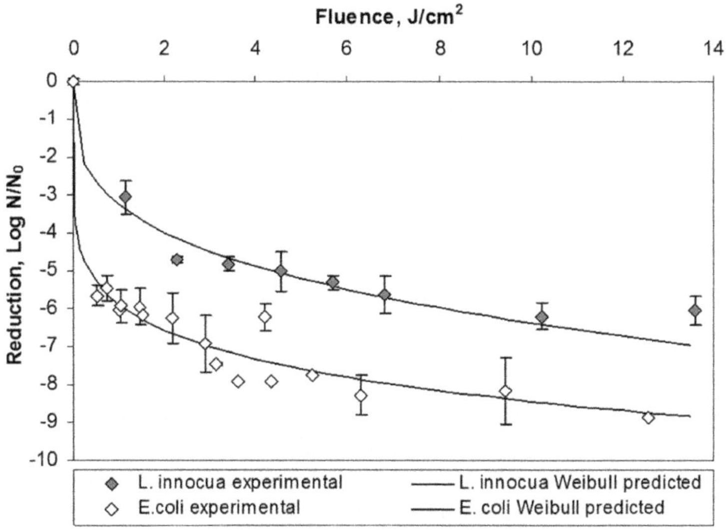

Figure 9.3 Experimental and Weibull-calculated survival ratios ($S = N/N_0$) for *L. innocua* and *E. coli* treated with PL in Butterfield's phosphate buffer. Adapted with permission of the International Association for Food Protection, from Kinetics and factors of variability in the pulsed light treatment of *Listeria innocua* cells, *Journal of Food Protection*, A. R. Uesugi, S. E. Woodling and C. I. Moraru, **70**, 2518–2525,[14] © 2007, and Inactivation of *Escherichia coli* ATCC 25922 and *Escherichia coli* O157:H7 in apple juice and apple cider, using pulsed light treatment, *Journal of Food Protection*, A. Sauer and C. I. Moraru, **72**, 937–944,[52] © 2009; permission conveyed through Copyright Clearance Center, Inc.

microbial species and/or the features of the treated substrate, which causes inefficient exposure of some cells to PL.[70]

First-order kinetics are not applicable to describe the PL microbial inactivation kinetics quantitatively, owing to the non-linear shape of survivor curves. The Weibull model has been successfully used to describe the kinetics of microbial inactivation by PL because of its ability to model non-linear inactivation successfully.[14,52] The Weibull model, an empirical model used for situations when a period of stress exposure causes failure, has often been used to model microbial inactivation kinetics.[71]

For PL treatments, the following form of the Weibull model was proposed by Uesugi *et al.*:[14]

$$\frac{\log N}{N_0} = -\frac{1}{2.303}\left(\frac{F}{\alpha}\right)^{\beta} \tag{9.1}$$

where N = survivors after PL treatment; N_0 = initial number of microorganisms; α = scale parameter; β = shape parameter; and F = pulsed light fluence (J cm^{-2}). The parameter β indicates the shape of the survivor curve, with $\beta > 1$ indicating concave up survival curves, $\beta < 1$ concave down survival curves, and $\beta = 1$ linear survival curves.

An example of Weibull-predicted survivor curves for PL treatment of *E. coli* and *L. innocua* is shown in Figure 9.3.

Yi *et al.*[72] modeled the non-linear relationship between fluence and the amount of *Pseudomonas aeruginosa* inactivation using a double Weibull model according to eqn (9.2). This model assumes that microbial populations can be divided into two subpopulations that exhibit different resistances towards the lethal source, with subpopulation 1 being more sensitive to stress than subpopulation 2 ($\delta_1 < \delta_2$).

$$\frac{N}{N_0} = \frac{1}{1 + 10^{\alpha}} \left[10^{-\left(\frac{F}{\delta_1}\right)^{\alpha}_{+\alpha}} + 10^{-\left(\frac{F}{\delta_2}\right)^{\beta}} \right] \tag{9.2}$$

where F = fluence (J cm^{-2}); β = shape parameter ($\beta > 1$, convex; $\beta < 1$, concave; and $\beta = 1$, linear); α = log proportion between the sensitive fraction (f, subpopulation 1) and the resistant fraction ($1 - f$, subpopulation 2); and δ_1 and δ_2 = fluence values (J cm^{-2}) for the first log reduction of subpopulations 1 and 2, respectively.

Yi *et al.*[72] also introduced some interrelated factors, including the F_R value (fluence needed to obtain the first log reduction), V_F (the increase in voltage required for a 1 J cm^{-2} increase in the fluence), and the z_{prr} and z_{pw} values (the increases in repetition rate and pulse width, respectively, that result in a 1 unit increase in the V_F value). These terms allow an evaluation of the effects of pulse repetition rate and width on PL fluence. They also suggested that it would be interesting to investigate the fate of the emitted light energy that is not involved in PL inactivation.

9.4 Applications of Pulsed Light Treatment

9.4.1 Pulsed Light Treatment of Liquids

PL has the potential to be an effective alternative to traditional thermal treatment for increasing the safety and shelf life of foods, particularly liquid food products. The efficacy of PL treatments in liquids depends on multiple factors, including fluence level, turbidity, the depth of the liquid, and its optical properties, particularly absorptivity, transmissivity, and light scattering properties.[73,74] It is difficult to ensure antimicrobial effectiveness and treatment uniformity in liquids with high absorptivity and turbidity for PL, or any other light-based technology.[75,76] The use of turbulent flow and thin flow channels can improve the contact area between non-clear liquids and light, and thus improve the effectiveness of PL treatment.

Huffman *et al.*[77] examined the inactivation of bacteria, viruses, and parasites suspended in water and treated with PL at 0.25 J cm^{-2} using a PureBright water treatment unit. For the bacteria *Klebsiella terrigena*, two pulses resulted in a reduction of >7.4 log cfu mL^{-1}. For viruses (Poliovirus type 1 and Rotavirus SA11), human viral surrogates (bacteriophages MS-2 and PRD-1) and a parasite (*Cryptosporidium parvum*), two pulses achieved an inactivation of >4 log reduction. Huang *et al.*[66] proposed the use of PL to inactivate a murine norovirus and Tulane virus. The viruses were suspended in phosphate-buffered saline to final populations of <6 log pfu mL^{-1}. The samples were irradiated with PL at 0.31–4.93 J cm^{-2} at a distance of 16 cm from the quartz window. Complete inactivation with higher fluence was achieved for murine norovirus and Tulane virus.

Palgan *et al.*[78] found that PL inactivation of bacteria and yeast decreased with decreasing transparency of the liquid. They reported maximum reductions for *E. coli* of 4.7, 1, and <1 log cfu for apple juice, orange juice, and milk, respectively. Hwang *et al.*[76] determined the inactivation effects of the PL on *P. aeruginosa* in liquids with different transparency and color, namely mineral water, an isotonic beverage, two types of apple juice, orange juice, grape juice, plum juice, three types of carbonated drink with different color, milk, and a coffee beverage without milk. In mineral water and the isotonic beverage, a 7.0 log reduction was obtained at a fluence of 0.97 J cm^{-2}. At 12.17–24.35 J cm^{-2}, 7.0 log reductions were observed for the apple juices, carbonated beverages, and plum juice, while only 0.5–2.0 log reductions were obtained in the other beverages, even after applying a total fluence of 29.21 J cm^{-2}. Takeshita *et al.*[24] noted a decrease of 5.8 log cfu mL^{-1} of *Saccharomyces cerevisiae* cells suspended in potassium phosphate buffer in a 110 mm diameter watch-glass treated with a total of 3.5 J cm^{-2} from a flashlamp and reflector positioned above the sample. The SteriPulse-XL 3000 (XENON Corporation) benchtop unit, which offers a pulse width of 360 μs, 3800 V input power, and a pulse rate of three flashes per second, has been used in many of PL inactivation studies. Krishnamurthy *et al.*[32] suspended *S. aureus* in phosphate buffer and treated the suspension with PL for 5 s at a distance of 80 mm from the lamp. They reported an inactivation level of 7.5 log cfu mL^{-1}. *L. innocua* in Butterfield's phosphate buffer in a thin layer (about 1.2 mm), exposed to 13.3 J cm^{-2} at a distance of 50 mm from the quartz face of the lamp, by Uesugi *et al.*[14] was reduced by 6.04 log cfu mL^{-1}. *E. coli* suspended in apple cider and apple juice was reduced by 5.5 log cfu ml and over 7 log cfu mL^{-1}, respectively, when samples were subjected to agitation and treated with PL at a distance of 50 mm from the lamp, at fluence levels below 12 J cm^{-2}.[52]

Milk samples treated with 25.1 J cm^{-2} of pulsed UV light from a UV laser showed reductions of >2 log cfu mL^{-1} of *Serratia marcescens*.[79] Subsequent plating of treated milk samples did not show signs of growth after 21 days of storage. Innocente *et al.*[80] used PL to treat raw milk. A fluence of 26.25 J cm^{-2} produced a 3.2 log cfu mL^{-1} reduction in total microbial counts. However, this fluence is more than twice the FDA-approved limit. The penetration depth of UV light in milk was determined to be 0.17 mm, indicating that liquid samples could be treated in a thin layer or in a turbulent flow to improve microbial inactivation. Miller *et al.*[81] performed static and turbulent treatments of 1.3 mm thick layers of skim and 2% fat milk inoculated with *E. coli* ATCC 25922 with PL doses ranging from 2.14 to 14.85 J cm^{-2}.

The maximum reduction for either milk type was 0.35 log cfu mL^{-1} after static treatments, while turbulent flow resulted in maximum reductions of 3.36 and 2.89 log cfu mL^{-1} for skim milk and 2% milk, respectively. It was hypothesized that inactivation in 2% milk was decreased due to scattering of light by milk fat globules, which in homogenized milk are approximately 1 μm in diameter.

Sauer and Moraru[52] showed that PL applied under turbulence and in a thin layer was able to inactivate *E. coli* ATCC 25922 and *E. coli* O157:H7 effectively in both filtered and unfiltered apple juice, which exceeds the FDA-required 5 log reduction. Subsequently, Pataro *et al.*[82] used a continuous-flow unit (tubes of 1 mm diameter) to inactivate *E. coli* and *L. innocua* in commercially available apple and orange juices. A fluence of 4 J cm^{-2} resulted in *E. coli* activity reductions of 4.00 and 2.90 log cfu mL^{-1} and *L. innocua* reductions of 2.98 and 0.93 log cfu mL^{-1} in apple juice and orange juice, respectively. Kasahara *et al.*[84] inoculated pasteurized goat milk with *E. coli* and subsequently treated the inoculated goat milk with PL doses ranging from 1.3 to 10 J cm^{-2}. A 6 log cfu mL^{-1} decrease in *E. coli* activity was achieved with the 10 J cm^{-2} dose.

The effect of PL on liquid food quality has been evaluated in a few studies. Elmnasser *et al.*[83] explored the effect of PL treatments on the quality of milk, and found that minimal changes were induced. The changes that were observed included the formation of disulfide bond-linked dimers in *β*-lactoglobulin, increased polarity of the tryptophan environment within milk proteins, and the destruction of vitamin A. Kasahara *et al.*[84] reported no difference in the protein or lipid content of goat milk treated with PL compared with the untreated product. Sensory panelists detected a difference in aroma between untreated and PL-treated samples, but not in flavor.

9.4.2 Pulsed Light Treatment of Solid Foods

9.4.2.1 Foods of Animal Origin

9.4.2.1.1 Meat Products. Several studies have looked at the potential of PL to inactivate microorganisms on the surface of meat products. PL treatments of meat and fish products are limited in their efficacy owing to product surface roughness and the ability of the microorganisms to penetrate the product surface. Inactivation of *L. innocua* on the surface of Vienna sausages by PL treatment at 9.4 J cm^{-2} was investigated by Uesugi and Moraru.[53] A reduction of 1.39 log cfu was observed immediately after PL treatment, and no significant growth occurred until after 8 days of refrigerated storage at 4 °C. When PL was used in combination with a nisin dip, a decrease in *L. innocua* activity of 4–5 log cfu was observed, and the number of survivors did not change until after 28 days of refrigerated storage. The survivors did not reach the initial level of inoculum until 40 days of storage.

Ozer and Demirci[85] examined the reduction of *E. coli* O157:H7 and *L. monocytogenes* activity on the muscle and skin side of raw salmon fillets using a SteriPulse-XL 3000 at 180 pulses and at a distance of 80 mm from the lamp. For *E. coli* O157:H7 activity reductions of 0.30 and 1.09 log cfu g^{-1} and for *L. monocytogenes* 74 and 1.02 log cfu g^{-1} were observed on the salmon muscle and skin, respectively. Surface temperatures increased after 60 s of PL treatment from a distance of 80 mm by 28 and 51 °C on the muscle and skin side, respectively. For treatments shorter than 5 s, the muscle and skin surface temperatures did not increase.

Hierro *et al.*[86] treated vacuum-packed ham and bologna inoculated with *L. monocytogenes* before packaging. After treatment with a maximum fluence of 8.4 J cm^{-2}, activity reductions of 1.78 and 1.11 log cfu cm^{-2}, respectively, were obtained. This treatment extended the shelf life of ham by 30 days beyond that of the vacuum-packed control samples, and did not affect the sensory properties of the ham. In contrast, bologna samples treated with doses above 4.2 J cm^{-2} exhibited negative changes in sensory properties. Wambura and Verghese[87] also observed negative quality changes in sliced ham exposed to PL (no fluence values were reported). Ganan *et al.*[88] examined ready-to-eat (RTE) cured meats, namely salchichón (dry, fermented sausage) and dry, cured loin. A dose of 11.9 J cm^{-2} yielded 1.5–1.8 log cfu cm^{-2} activity reductions in *L. monocytogenes* and *Salmonella*, respectively. Sensory tests detected little difference between PL-treated and untreated cured meats. Hierro *et al.*[89] studied the efficacy of PL doses in the range 0.7–11.9 J cm^{-2} in improving the safety of carpaccio. Beef and tuna slices were surface inoculated with approximately 3 log cfu cm^{-2} of *L. monocytogenes*, *E. coli*, *Salmonella* Typhimurium (*Salmonella enterica* serovar Typhimurium), and *Vibrio parahaemolyticus*. Treatments at 8.4 and 11.9 J cm^{-2} inactivated the selected pathogens by approximately 1 log cfu cm^{-2}, although they also induced changes in the product color. Tuna carpaccio treated with fluences $\leq$4.2 J cm^{-2} showed no significant changes in either color or odor.

Rajkovic *et al.*[90] applied a fluence of 3 J cm^{-2} to the surface of dry, fermented salami to reduce *L. monocytogenes*, *E. coli* 0157:H7, *S.* Typhimurium, and *S. aureus*. This resulted in inactivation of slightly greater than 2 log cfu g^{-1} for all products. Increasing the fluence to 15 J cm^{-2} did not increase the level of microbial inactivation. After PL treatment, fermented salami slices were packed in ES-Plastic thin plates (PP/EVOH/PP; polypropylene–ethylene vinyl acetate copolymer) under vacuum or in an 80% CO_2–20% N_2-modified atmosphere. They were then stored at 4 °C and the effect of PL treatment on protein and lipid oxidation was investigated. Lipid oxidation for PL-treated vacuum and modified-atmosphere packed fermented salami slices fell within the acceptable threshold for rancid odor. However, an exception was the sample treated with the highest fluence (15 J cm^{-2}) packed in a modified atmosphere and kept in cold storage for 9 weeks. A significant rise in protein oxidation for PL-treated fermented salami slices (28% increase in carbonyl content compared with untreated samples) was observed only after 9 weeks of cold storage in both vacuum and modified-atmosphere packed samples. Keklik *et al.*[91] compared the inactivation of *Salmonella* on unpackaged and vacuum-packed boneless chicken breast over a wide range of fluence (2.9–67 J cm^{-2}). A 1.2 log cfu cm^{-2} reduction in *Salmonella* activity was achieved for unpackaged chicken treated at a fluence of 8.7 J cm^{-2}, and a 0.9 log cfu cm^{-2} reduction was achieved on packaged chicken treated at 8.2 J cm^{-2}. These samples were treated for the same period of time at the same distance from the lamp, but the packaging absorbed a portion of the light. This resulted in a net fluence exposure for the packaged chicken that was slightly lower than for the unpackaged chicken. Lipid oxidation of the PL-treated samples was not significantly different from that of untreated samples for treatments below the FDA-approved limit (12 J cm^{-2}). Paskeviciute *et al.*[92] also found that PL treatment of unpackaged skinless chicken breast did not result in significant increases in lipid oxidation. Haughton *et al.*[93] observed color changes in packaged chicken skin and chicken breast, but the treatment doses were substantially higher than 12 J cm^{-2}. Similarly, Nicorescu *et al.*[94] found that lipid oxidation in PL-treated raw pork roast, roasted pork, and raw salmon was significant only in samples treated with a dose of 30 J cm^{-2}. These studies highlight the importance of using a fluence level that provides adequate levels of microbial inactivation while also maintaining product quality.

9.4.2.1.2 Eggs. Macias-Rodriguez *et al.*[95] reported a 5 log cfu reduction of *E. coli* K12 activity on hard-cooked peeled eggs without inducing any heat, color, or texture changes in the eggs, but fluence values were not provided. Eggshells have also been decontaminated by PL. A fluence of 11 J cm^{-2} yielded a 3.5 log activity reduction in *Salmonella* inoculated on eggshells.[96]

Manzocco *et al.*[97] inoculated egg pasta with *Salmonella* by adding it to the surface of the finished pasta and by incorporating it into the dough mixture before the pasta was formed. Significant differences in inactivation were observed for the two types of samples. A dose of 3.50 J cm^{-2} resulted in a 3.3 log reduction of *Salmonella* for the surface-inoculated pasta, but only a 1 log reduction for the dough-inoculated pasta. The rate of lipid oxidation was slower for both PL-treated samples than for untreated samples, which was hypothesized to be due to the PL-induced formation of antioxidant non-enzymatic browning products.

9.4.2.1.3 Dairy Products. Several studies have shown that PL has some potential for the microbial decontamination of cheese surfaces, even after packaging in transparent packaging materials. The effectiveness of the treatment was limited by light absorption and scattering by the cheese. Proulx *et al.*[59] examined the effectiveness of PL on *P. fluorescens*, *E. coli* ATCC 25922, and *L. innocua* on the surface of white Cheddar and processed cheese slices, using PL doses below the FDA limit of 12 J cm^{-2}. *E. coli* was the most sensitive to PL, with a maximum reduction of

5.41 log. *L. innocua* was the least sensitive to PL treatment, with a maximum inactivation level of 3.37 log, followed by *P. fluorescens*, with a maximum inactivation level of 3.74 log.

PL treatment of cheese through UV-transparent packaging resulted in similar inactivation levels compared with unpackaged cheese. Fernández *et al.*[100] treated two cheese varieties with different surface features (Gouda and Manchego) by PL with fluences between 0.9 and 8.4 J cm^{-2}. The treatment was less effective in Manchego than in Gouda, in which a 3 log cfu cm^{-2} reduction was obtained with a fluence of 0.9 J cm^{-2}. Can *et al.*[98] studied the inactivation of *Penicillium roqueforti* and *L. monocytogenes* inoculated onto packaged and unpackaged white American cheeses. A 10 J cm^{-2} PL dose resulted in 0.81 and 1.0 log cfu cm^{-2} reductions of *P. roqueforti* on unpackaged and packaged cheeses, respectively. The treatment was more effective for *L. monocytogenes*, with reductions of 2.3 and 2.6 log cfu cm^{-2} achieved for packaged and unpackaged cheeses, respectively. The color and lipid oxidation in PL-treated cheeses were not significantly different from those of untreated controls. Fernández *et al.*[99,100] examined the effects of PL on protein oxidation in processed cheese. The level of protein oxidation was significantly higher in samples treated with fluence values above 8.4 J cm^{-2} than in control samples, while still within the normal range for cheese.

PL treatment of milk has already been described in Section 9.4.1.

9.4.2.1.4 Honey. Honey naturally has low levels of microorganisms. However, it has been known to contain spores of yeast, mold, and bacteria that can lead to spoilage. *Clostridium botulinum* is also commonly found in honey, and is known to cause botulism and infant botulism. After treatment of clover honey inoculated with 6.25 log g^{-1} of *Clostridium sporogenes* spores with PL using a SteriPulse-XL 3000 unit at different honey depths, treatment times, and distances from the lamp, spore reductions from 0.0 to 5.65 log were obtained.[101] The best conditions reported for treating clover honey were 540 pulses (treatment time 180 s) and an 8 mm honey depth.[101] No treatment combination could completely remove spores from the honey.

9.4.2.2 Foods of Plant Origin

9.4.2.2.1 Fruit and Vegetable Products. Decontamination of fruits and vegetables is the most studied food application of PL. Surface irregularities in some fruits and vegetables lower inactivation levels, and both negative and positive effects of PL on produce quality have been reported.

Gómez-López *et al.*[102] examined the PL treatment of celeriac, green bell peppers, iceberg lettuce, radicchio, soybean sprouts, spinach, and white cabbage with up to 2700 pulses. A xenon flashlamp with a pulse duration of 30 μs and an intensity of 7 J, at a distance of 128 mm from the lamp was used, with samples spread over a sterile 14×21 cm tray. Microbial reductions ranged from 0.56 to 2.04 log for these vegetables, which were naturally contaminated and not laboratory inoculated. It was hypothesized that the variability among samples was due to: varying resistances of the naturally occurring microbial population, the location of the microorganisms, shadowing effects, or protective substances found in the vegetables. Processing (shredding, grating, chopping, or whole), shape, and sample size did not lead to any observable patterns in microbial reduction. Although the initial microbial load on the minimally processed vegetables was reduced, their shelf life was not increased compared with untreated controls.

Kramer *et al.*[103] studied endive salad and mung bean sprouts treated with PL to inactivate both the naturally occurring microbiota, and also *E. coli* (DSM 498) and *L. innocua* (DSM 20649) that were artificially inoculated onto the samples, and 2 log reductions were reached with one flash in the case of fresh-cut salad whereas the reduction on mung bean sprouts was limited to ~1.6 log, irrespective of whether the natural flora or laboratory-inoculated organisms were involved. PL treatments improved the color and general appearance of mung bean sprouts slightly, whereas respiration was almost unaffected during storage.

Strawberries inoculated with *Botrytis cinerea* conidia and treated with PL with a pulse duration of 30 μs and an intensity of 7 J within a treatment chamber with reflective inner walls showed no reduction of storage rot.[104] Combining PL treatment with additional UV-C exposure at $\lambda = 254$ nm and 1.0 kJ m^{-2} for 120 s delayed the appearance of fungal mycelia by 1 day. Previous studies had shown that similar treatment of *B. cinerea* conidia in phosphate buffer resulted in a 3–4 log reduction,[105] but this level of reduction could not be achieved on strawberries. Luksiene *et al.*[106] observed reductions of 1.0–2.2 log cfu g^{-1} for several microorganisms after treating inoculated strawberries with a PL dose of 3.9 J cm^{-2}. The PL treatment extended the shelf life of strawberries by 2 days and did not cause any significant changes to their color, firmness, or ascorbic acid, anthocyanin, and phenolic contents.

Bialka and Demirci[107] reported reductions of 1.1 and 4.3 log cfu g^{-1} of *E. coli* O157:H7 on blueberries treated at a fluence of 1.9 and 22.6 J cm^{-2}, respectively, at a distance of 80 mm from the lamp. Treatments at a distance of 30 mm resulted in slightly greater reductions, but also in damaged fruit. When treating *Salmonella* on blueberries, increasing the fluence did not significantly increase the activity reduction at 30 and 80 mm from the lamp. At a treatment distance of 130 mm, however, the reduction of *Salmonella* activity increased significantly with fluence. The color and sensory properties of treated blueberries were not significantly affected by the treatment.

Agüero *et al.*[108] reported the bactericidal effect of PL fluence level (1.0–12.0 J cm^{-2}) against *L. innocua* and *E. coli* inoculated on spinach leaves. Reductions of 1.85 and 1.72 log cfu g^{-1} were obtained for *L. innocua* and *E. coli*, respectively, after an application of 1.0 J cm^{-2}. Increasing the fluence level did not produce a proportional increase in treatment effectiveness. Samples treated with PL had lower microbial growth rates than untreated samples during storage. Spinach treated with PL showed an enhanced total phenolic concentration and antioxidant capacity, but also a greater degradation rate of phytochemical compounds during refrigerated storage.

PL treatment with 4.8 J cm^{-2} reduced the native yeast population in mushrooms, sufficiently to extend shelf life by 2–3 days.[109] Higher fluences changed the mushroom texture and promoted enzymatic browning. Ignat *et al.*[110] found that PL treatment of apple slices at a fluence of 1.75 J cm^{-2} resulted in reductions in *L. monocytogenes* and *Lactobacillus brevis* of 2.7 and 3 log cfu cm^{-2}, respectively. Quality changes were similar to those for untreated cut slices at this fluence, but higher fluence levels caused dehydration and browning in apple slices. Gómez *et al.*[111] dipped apple slices in an antibrowning solution (ascorbic acid in $CaCl_2$ solution) before PL treatments, which inhibited browning for PL doses up to 71.6 J cm^{-2}.

Fresh-cut cantaloupes were subjected to repetitive PL treatment using fluence levels of up to 1.2 J cm^{-2} every 2 days, which resulted in a total cumulative fluence of up to 15.6 J cm^{-2} at the end of storage. Repetitive PL treatment at 0.9 J cm^{-2} every 2 days was the most effective treatment for extending the shelf life of fresh-cut cantaloupes. Microbiological shelf life was extended by 20 days compared with untreated controls, while firmness, color, and total soluble solids did not change. The effect of repetitive PL treatments relative to a single PL treatment, at the same dose, was also evaluated. Fresh-cut cantaloupes subjected to a single treatment at 11.7 J cm^{-2} had intense cell-wall breakage, leading to alterations in cell shape and plasmolysis, but these negative effects were minimized by repetitive PL treatment.[112]

Rodov *et al.*[113] focused on the quality effects of mild PL treatments (0.1–1.8 J cm^{-2}) on pale fig fruit. The anthocyanin and phenolic content increased and the fruit darkened during storage, which improved their marketability. The treatment of red-ripe tomatoes by PL induced both positive and negative quality changes.[114] Three days after treatment, PL-treated tomatoes lost firmness and developed a wrinkled exterior; however, lycopene became more bioavailable and the carotenoid content increased slightly as a result of PL treatment.

Lopes *et al.*[115] treated peel and pulp from physiologically mature mangoes with 0.6 J cm^{-2} of PL before storage for 7 days at 20 °C and 80% relative humidity. Oxidative stress occurred in the mango pulp due to overproduction of reactive oxygen species, whereas in mango peel the antioxidant defense system was stimulated without mediation by reactive oxygen species. The observed changes in mango tissue physiology suggest that PL may promote higher antioxidant levels, without any permanent serious damage to the plant tissue or cells, a condition referred to as hormesis. Charles *et al.*[116] also treated mangoes with a PL dose of 8 J cm^{-2}. After 3 days of storage, the firmness, color, and carotenoid content were better preserved in the treated mangoes than in control samples. Furthermore, the phenolic and ascorbic acid contents remained the same as in the control samples. UV-C hormesis has been shown to induce disease resistance against a wide range of pathogens in horticultural crops.[117] Postharvest hormetic treatment of mature green tomato fruit with 16 pulses of PL significantly delayed ripening to levels comparable to those achieved using 3.7 kJ m^{-2} of continuous UV-C.[117] The total duration of the treatment times for the PL and continuous UV sources were 10 and 370 s, respectively. This showed a clear advantage of PL relative to UV treatment.

Generally, PL inactivation of microorganisms on the surfaces of vegetables and fruits has been applied in the dry state, but a few studies have been conducted in which PL was applied during a simulated washing process. Carrot slices inoculated with *S. cerevisiae* prior to PL treatment using two flashlamps located above and below the sample were investigated by Kaack and Lyager.[118] After two pulses with a fluence of 0.7 J cm^{-2} per pulse, a reduction of 3.07 log cfu g^{-1} was obtained. After exposure to 24 pulses, inactivation levels increased to 4.93 log cfu g^{-1}. It was suggested that PL may reduce the level of yeast on sliced carrots by up to 6 log cfu when combined with washing. Blueberries inoculated with *E. coli* O157:H7 were submerged in turbulent water and then treated with PL.[119] The dry treatment with a fluence of 5.0 J cm^{-2} resulted in a 3.8 log cfu g^{-1} reduction of *E. coli*, but negative quality changes were observed in the blueberries. Water-assisted treatments at the same fluence resulted in a 4.5 log cfu g^{-1} reduction of *E. coli*, without inducing negative quality changes. Huang and Chen[120] used the same water-assisted technique to treat strawberries and raspberries. The treatment was less effective than with blueberries, which is likely due to the much higher degree of surface irregularities on raspberries and strawberries. Kramer *et al.*[121] evaluated the use of PL treatment to reduce the microbial loads on mung bean sprouts and endive salad during a simulated wash process lasting for up to 60 s. PL was more efficient than electrolyzed water (40 ppm free chlorine) and chlorine dioxide (15 ppm) for the same treatment time. PL with an incident fluence on the water surface of 0.32 and 0.58 J cm^{-2} per pulse, at a distance of 5 cm between the water surface and the flashlamp, was applied during a simulated washing process. Reductions of microbial counts of up to 2.5 log were found for PL-treated endive salad, whereas both sanitizers caused maximum reductions of only about 1.5 log. An additional benefit of PL in such combined treatments is that it also keeps the microbial loads in the washing water at a low level.

9.4.2.2.2 Grains and Dry Foods. A few studies have investigated the effect of PL on grains. Sharma and Demirci[122] treated alfalfa seeds inoculated with 5 log cfu g^{-1} of *E. coli* O157:H7 with 270 pulses of light at a distance of 30–130 mm from the lamp. Changing the distance from the lamp decreased the *E. coli* O157:H7 reduction from 4.89 to 0.07 log, but did not significantly change the germination rate of the seeds.[122] Aron Maftei *et al.*[123] treated both sides of wheat grains with PL in sterile polypropylene bags and obtained a reduction in the mold load of the wheat grains of approximately 4 log cfu g^{-1} at the highest energy treatment used. Hwang *et al.*[124] treated sesame seeds using a pilot-scale PL unit and achieved a maximum inactivation of 1.46 log at a total fluence of 44.46 J cm^{-2}. Another interesting observation in this study was that under the same conditions, the fluence range of PL measured in a stainless-steel chamber was 64.77–235.23 mJ cm^{-2} per pulse, compared with 51.38–187.35 mJ cm^{-2} per pulse in an acrylic chamber. This indicates that the stainless-steel chamber generated light reflection effects that increased the net fluence.

PL treatment of powdered foods has been investigated in a few studies.[122,124–126] A problem encountered with powdered foods is the shadowing effect, which could only be resolved by the development of treatment chambers that can minimize this effect and the use of very high PL energy levels.

Fine and Gervais[126] achieved reductions in *S. cerevisiae* of 2.93 and 0.7 log in the PL treatment of inoculated black pepper powder and wheat flour, respectively, after exposure to 31.12 J cm^{-2} at a distance of 20 mm from the lamp. The difference in inactivation between the two products was attributed to the difference in color. The visual and flavor quality of the black pepper and the wheat flour decreased prior to reaching an acceptable yeast reduction, as both products became progressively burned with increased exposure. Nicorescu *et al.*[127] examined the efficacy of PL treatments of 10 J cm^{-2} on ground caraway, ground red pepper, and ground black pepper inoculated with *B. subtilis*. Only modest inactivation levels, of 0.8 log (ground caraway and ground black pepper) and 1.0 log (ground red pepper), were obtained. Shadowing of microorganisms by spice particles and also light absorption likely limited the efficacy of the PL treatment.

Aspergillus niger spores were significantly inactivated in corn meal by Jun *et al.*[128] Corn meal with an initial inoculum of 5 log cfu g^{-1} was exposed to a 100 s treatment at a distance of 8 cm from the lamp and a reduction of 4.95 log cfu g^{-1} was achieved. The reduction changed to 3.26 and 2.95 log cfu g^{-1} when the distance from the lamp was increased from 30 to 130 mm. A sample temperature of >120 °C was noted at the maximum treatment level, leading to changes in product quality.

Choi *et al.*[125] treated infant powdered food with PL inoculated with 10^5 cfu g^{-1} of freeze-dried *L. monocytogenes* culture. The infant powdered milk was treated in a layer of 1 or 2 mm thickness on a Petri dish at 15 kV triggering voltage pulse, 10 Hz pulse frequency, 1.5 μs pulse width, and 0–900 s lamp operating time at a distance of 6 mm from the lamp. The *L. monocytogenes* population was reduced by 1, 2, and 3 log for 2300, 4700, and 9500 μs of PL treatment, respectively.

9.4.2.3 Pulsed Light Treatment of Packaging Materials

Post-processing contamination is a recurrent contributory factor to foodborne illness outbreaks, and some of it can occur during packaging, cutting, transportation, and storage.[129] Therefore, decontamination of already prepackaged food is highly desirable, because it prevents product recontamination, in contrast to a system that treats the food and packaging material separately.[130,131] PL can be used to inactivate microorganisms on the surface of food packaging materials and potentially on the surface of products packaged in clear, "UV-transparent" packaging. The use of PL could lead to a decrease in the use of chemical sterilizing agents such as hydrogen peroxide, propylene oxide, or peracetic acid. Additionally, PL has the advantage of not leaving undesirable residues after the treatment. One aspect that needs to be considered when using PL to treat packaged products is that the treatment is limited to the product surface and is restricted by the lack of ability of light to penetrate opaque and irregular surfaces.

PL inactivation of *L. innocua* cells on food packaging materials was investigated by Ringus and Moraru.[132] *L. innocua* activity reductions of 7.2, 7.1, 4.4, 4.5, and 3.5 log cfu were obtained after treatment with up to 8.0 J cm^{-2} of low-density polyethylene (LDPE), high-density polyethylene (HDPE), polyethylene-laminated ultra-metalized poly(ethylene terephthalate) (AF), polyethylene-coated paperboard (TR), and polyethylene-coated aluminum foil paperboard laminate (EP) coupons, respectively. The lowest inactivation was obtained for EP, which was the packaging material with the highest reflectance. These results demonstrate the potential of PL as an effective method for decontamination of food packaging materials, before or after filling, and further substantiate how different surface properties affect PL inactivation.

Castillo *et al.*[133] investigated the migration of toxic components from PL-treated polymers into food. Two commercial polypropylene (PP) films were treated with PL (no fluence values were reported) and the it was concluded that the formation of potentially health-relevant migrant substances after PL treatment cannot be ruled out, and it may depend on the nature of the additives, their concentration in the polymer, other additives present, and the treatment conditions. These results suggest that further studies are needed to assess the potential toxicity of polymeric food packaging treated with PL.

9.4.3 Other Applications of Pulsed Light Treatment

PL has also been used in foods for other purposes besides microbial inactivation. Kalaras *et al.*[134] reported the formation of vitamin D_2, previtamin D_2, lumisterol$_2$, and tachysterol$_2$ from ergosterol in white button mushrooms. In another study, Kalaras *et al.*[135] determined that a 2.4 J cm^{-2} dose of PL increased the vitamin D_2 content of mushrooms to more than 100% the recommended daily allowance (RDA) per serving. Moreau *et al.*[136] studied the deactivation of the mycotoxin aflatoxin B_1 (AFB1) by PL, and found that a PL dose of 8 J cm^{-2} caused 92.7% AFB1 destruction by fragmentation of the AFB1 structure. The mutagenic activity in AFB1 samples treated with PL has also been destroyed.[137]

Janve *et al.*[138] explored the possibility of using PL to deactivate soybean lipoxygenase (LOX). PL treatments of 16 s (fluence values were not reported) resulted in 95–99.95% inactivation of LOX, depending on the distance of the sample from the lamp. Manzocco *et al.*[139] studied the effect of PL treatment on polyphenoloxidase (PPO), which is responsible for enzymatic browning. They found that fluences >8.75 J cm^{-2} resulted in complete inactivation of PPO.

PL has also been investigated as a means of modifying the structural and functional properties of proteins and obtaining protein derivatives with targeted functionality. Siddique *et al.*[140] treated a whey protein isolate (WPI) solution at PL fluence levels from 4 to 16 J cm^{-2}. The treatments induced changes in WPI conformational and functional properties in a fluence-dependent manner. Exposure to light treatment increased the concentration of SH groups and formation of carbonyl groups. Overall, the study suggested the modification of protein tertiary, secondary, and primary structures. The solubility and functional properties were significantly improved by applying PL treatments at fluence levels from 8 to 12 J cm^{-2}.

A reduction in allergenic compounds by PL has been reported for peanut products,[141,142] soybeans,[143] shrimp,[144] and almonds,[145] although practical applications of PL for this purpose are severely impaired by the very low penetration depth of light in such products. Meinlschmidt *et al.*[146] treated soy protein isolate with PL, and their sodium dodecyl sulfate poyacrylamide gel electrophoresis (SDS-PAGE) analysis revealed reduced protein intensity bands corresponding to the major soy allergens β-conglycinin and glycinin. However, very high fluence levels of around 100 J cm^{-2} were required to achieve these effects, and such high treatment doses are likely to induce overheating and significant quality changes to the treated foods. Funes *et al.*[147] found that PL is capable of reducing patulin, a heat-resistant fungal toxin known to contaminate apples, with the reduction increasing with increase in PL dose. Despite the expected limitations, the ability of PL to cause inactivation of undesirable or dangerous chemical components in foods is very intriguing and is worth further investigation.

Additionally, Uslu *et al.*[148] suggested the use of PL technology as an alternative to chlorine and conventional UV light for the treatment of municipal wastewater effluents. Using a flow-through one-pass pulsed UV treatment, they obtained complete inactivation of *E. coli* at a flow rate of 10 L min^{-1} and of *B. subtilis* at 6 L min^{-1}. The treatment resulted in significant reductions of the chemical oxygen demand and total organic carbon.

9.5 Combination of Pulsed Light with Other Treatments

Many researchers have explored the possibility of combining PL with other antimicrobial treatments that have complementary microbial inactivation mechanisms in order to obtain additive or even synergistic effects. The premise is that sublethal damage caused to microbial cells by the light exposure may increase their sensitivity to stressful environmental factors and other injurious agents. Such a strategy could minimize the dose of each of the treatments used, and thus better preserve the quality of the treated foods.

Ferrario *et al.*[149] found that the combination of ultrasound and PL worked well for the inactivation of *S. cerevisiae* cells suspended in commercial and fresh squeezed apple juices. Furthermore, thermosonication before or after PL exposure improved the inactivation of *E. coli* in apple juice compared with thermosonication or PL treatment alone.[150]

Most combination treatments involving PL have used antimicrobial chemicals as the secondary treatment. For instance, Uesugi and Moraru[53] combined PL with nisin, a natural antimicrobial, for the disinfection of ready-to-eat sausages inoculated with *L. innocua*. The combination of PL and nisin dip yielded a reduction of 4.03 log cfu, whereas PL and nisin dip treatments alone resulted in reductions of only 1.37 and 2.35 log cfu, respectively. Ramos-Villarroel *et al.*[151] compared the efficacies of PL, malic acid, and their combination for the inactivation of *L. innocua* and *E. coli* on avocado, watermelon, and mushrooms. The malic acid treatment alone was the least effective, producing reductions of 0.96–1.08 log cfu g^{-1} in *E. coli* immediately after treatment. PL alone resulted in reductions of 2.58–2.97 log cfu g^{-1}. PL combined with malic acid treatment was the most effective, giving reductions of 3.14–3.48 log cfu g^{-1}. One important consideration when using PL in conjunction with antimicrobial treatments is that the order of the individual treatments can be very important. If the antimicrobial compound used absorbs light in the microbicidal UV range, this may diminish the effectiveness of PL, rather than enhancing it.[152]

9.6 Sensory Effects on Products Treated with PL

Although PL can be effective for the microbial decontamination of certain food products, its potential impact on the sensory properties could deter commercial adoption. Sensory analysis data for foods treated with PL are scarce. A few examples of PL-treated products that have been evaluated for sensory characteristics include goat milk,[84] Cheddar cheese,[153] Gouda and Manchego cheese,[100] apple juice,[154] fresh-cut apple,[110] orange and carrot juice blend,[155] beef and tuna carpaccio,[89] and ready-to-eat cooked meat products.[86]

Kasahara *et al.*[84] treated goat milk with PL doses ranging from 1.3 to 10 J cm^{-2}. Sensory panelists detected a difference in aroma between untreated and PL-treated samples, but not in flavor. Proulx *et al.*[153] examined the effect of PL doses from 1.02 to 12.29 J cm^{-2} on Cheddar cheese packaged in transparent polyethylene bags and stored at 6 °C. Sixty untrained panelists performing triangle tests were unable to detect significant differences between untreated and PL-treated samples, although PL had a significant effect on overall liking, flavor, and appearance. Fernández *et al.*[100] treated two cheese varieties with different surface features (Gouda and Manchego) with PL, and assessed the effect of PL on the sensory quality of the cheese after 0,

15, and 30 days of storage at 4 °C, using a panel of 11 tasters. Differences in odor and flavor were reported immediately in both cheeses after treatment with fluence levels higher than 4.2 J cm^{-2}. The development of specific sulfur notes was observed, but the sulfur volatiles and the corresponding sensory notes disappeared during refrigerated storage.

Hierro *et al.*[86] reported that PL at 8.4 J cm^{-2} did not affect the sensory quality of vacuum-packaged cooked ham, whereas treatments above 2.1 J cm^{-2} negatively influenced the sensory properties of bologna slices. Hierro *et al.*[89] also treated carpaccio from beef and tuna at fluences of 8.4 and 11.9 J cm^{-2}. These treatments inactivated selected pathogens, but at certain doses the treatment induced color changes and had a negative effect on the sensory quality of the products. The paler tuna color was hypothesized to be a consequence of changes in the myoglobin oxidation state. PL treatment of beef carpaccio at 11.9 J cm^{-2} gave a slightly lighter color, although the samples were considered acceptable. On the other hand, the treatments at 8.4 and 11.9 J cm^{-2} significantly changed the product odor. Although panelists reported a loss of freshness, they considered the samples to be acceptable.

Caminiti *et al.*[155] processed a blend of orange and carrot juice by a combination of PL (3.3 J cm^{-2}) and manothermosonication (MTS) (400 kPa, 35 °C, 1000 W, 20 kHz). A sensory evaluation of the product by 37 untrained panelists showed that the color of the product processed by hurdle combinations was preferred to that of a heat-pasteurized sample (72 °C, 26 s), while the flavor was adversely affected by the combination treatment. Panelists did not perceive differences in the odor, sweetness, or acidity of the product. Ferrario and Guerreiro[154] investigated the effect of a continuous flow-through PL system at a maximum fluence of 0.73 J cm^{-2} and a flow rate of 155 mL min^{-1}, alone or in combination with ultrasound, on commercial and freshly pressed apple juices. The combination treatment prevented juices from browning during storage. The freshly pressed, PL-treated apple juice was well accepted by a group of consumers, who noted the fresh natural apple taste.

Ignat *et al.*[110] studied the effect of PL at fluences between 1.75 and 15.75 J cm^{-2} on apple slices. The treatments induced some moderate heating effects, and the sample temperature ranged from 6 to 26 °C after processing. PL also resulted in significant weight loss and changes in color and sensory attributes after storage at 6 °C. PL treatment at 1.75 J cm^{-2} resulted in apple slices comparable to the untreated samples, with limited quality changes. In contrast, at higher fluences the apple slices underwent dehydration and browning during storage, due to loss of cell integrity. Negative flavor changes were also observed, and were attributed to alterations in the composition of volatile components, including esters, aldehydes, and alcohols. Apple slices submitted to 15.75 J cm^{-2} of PL developed an anomalous, "cooked" flavor.

These data highlight the fact that some products are more prone to negative sensory impacts of PL treatment than others. Therefore, treatment optimization is needed for each type of product in order to identify the PL parameters that result in the required level of microbial reduction, while minimizing any negative changes to the product quality and sensory attributes.

9.7 Pulsed Light Systems

The application of PL technology in the food industry is still modest. PL treatment of food packaging is the most developed current application. PL systems for either batch or continuous treatments are available, with the batch systems being predominant at present. In batch configurations, samples are placed within a treatment chamber, with lamps located along the walls of the chambers. The simplest designs include a single lamp located above the sample and an adjustable tray to hold the samples. More complex designs incorporate several lamps within a chamber along with a quartz stand to hold the sample and allow a 360° exposure and treatment.

The US company XENON Corporation (Wilmington, MA, USA) and the French company Claranor (Manosque, France) dominate the market of PL equipment for the treatment of food.

XENON is commercializing a wide range of laboratory-scale systems and modular units that can be easily configured inline for larger scale, industrial applications. Some of their commercially available systems are used for non-food applications, *e.g.* the S-Series is used for sintering solutions for printed electronics production, while the RC-Series comprises modular systems designed for curing applications. The XENON X-1100 is a low-cost benchtop compact PL system that delivers up to 9 $J cm^{-2}$ radiant energy per pulse, and allows adjustment of the pulse peak radiant power and pulse duration. Their Z-series systems are designed for research, and consist of a controller and separate sterilization chamber, able to deliver high-intensity (1.27 $J cm^{-2}$), short (360 μs), broadband pulses. The configuration includes a stainless-steel treatment chamber, a linear flashlamp in a fully enclosed air-cooled lamp housing, and a light intensity monitor. The Z-1000 system for pilot- or small-scale production includes a benchtop controller and an air-cooled, standalone enclosed lamp housing with a flashlamp. The lamp housing may be located up to 10 m from the controller, and the system offers operator control *via* front-panel controls or a remote PLC.[156]

Claranor is commercializing a range of PL equipment for the food processing industry, including static equipment for laboratories and industrial systems for the disinfection of caps, cups, preformed packaging, or films. The latter are continuous systems, and the treated products are placed on flat conveyor belts, on spool-bars, or in tunnels. A flashlamp, or several flashlamps, are located above the moving conveyor belt. Other components include a hood to contain the lamp, reflectors to redirect light to the sample, and a quartz panel that protects both the flashlamp and the workers from exposure. Claranor have also built a reactor for the in-line treatment of clear liquids and water, in which liquids travel through a cylindrical reactor equipped with a lamp placed in the center of the reactor. The system can be integrated into the manufacturing line as a single-treatment reactor, or several reactors can be coupled in series to increase PL exposure and thus treatment efficiency.

Most of the PL studies reported in the literature were conducted using static equipment, but a few studies have also investigated continuous systems. The bactericidal effects of PL on liquid foods in continuous-flow PL devices were studied by Artíguez *et al.*[157] and Pataro *et al.*[82] Artíguez *et al.*[157] highlighted the importance of identifying all the critical factors that affect microbial inactivation in a continuous flow-through unit. PL treatments were performed using a flow-through pilot unit (Maria PUD system; Claranor). The impact of several flow-through process parameters, such as voltage input (1000 and 3000 V), flow rate (1 and 5 L min^{-1}), liquid thickness (2.15 and 6.23 mm), number of pulses, and total fluence (up to 10 J cm^{-2}), on water inoculated with *L. innocua* was evaluated. A fluence of 10 J cm^{-2} induced a >5 log reduction in *L. innocua* activity, with no significant rise in water temperature (<7 °C). A greater microbial inactivation was observed in treatments performed at 1000 V than at 3000 V. These results are consistent with the results obtained using static systems.

Yi *et al.*[158] studied microbial inactivation in untreated groundwater using a self-designed pilot-scale, continuous-flow PL unit. This pilot-scale PL system comprised a power source, a pump to draw water, filters to remove impurities, flow-rate controller, treatment chamber, and an air compressor to circulate the sample in the chamber. The cylindrical lamps used in this pilot-scale system were 145 mm long with an external diameter of 7.14 mm (Type NL 4006; Heraeus Noblelight, Cambridge, UK), protected by quartz tubes. The cylindrical treatment chamber was made of stainless steel, with an inner diameter of 31 cm and a height of 93 cm. The chamber volume capacity was 66 L and flow rates of 40, 32, 20, and 12 L min^{-1} were tested. This system has a capacity of over 17 t with 24 h of operation per day. This pilot-scale continuous-flow PL system was proven to inactivate a diverse range of microorganisms, including *E. coli* (maximum 4.79 log reduction), aerobic and facultative anaerobic heterotrophs (maximum 2.91 log reduction), and murine norovirus (maximum 3.35 log reduction), in a large amount of water.

Hwang *et al.*[124] proposed the design of a pilot-scale continuous-flow PL system to treat 3 kg of sesame seeds. The PL device consists of a power supply, treatment chamber, and a xenon lamp (NL9553) with length 271 mm and diameter 11 mm. The maximum voltage and frequency were 2400 V and 15 Hz, respectively. The pulse duration in this equipment could be adjusted from 0.5 to 3.0 ms. The xenon lamp was installed in the middle of the treatment chamber (made of either acrylic or stainless steel) and placed in a quartz pipe filled with circulating distilled water for cooling. As mentioned earlier, the highest microbial reduction was achieved by using the stainless-steel rather than the acrylic chamber. Further modifications to the system to improve the treatment effectiveness were suggested, such as control of cyclonic flow, a conveyor belt-type chamber, or the addition of photocatalytic effects by using pitch-coated lamps coated with TiO_2.

For both batch and continuous designs, process validation using commercial-scale equipment should be carried out before commercial adoption of the technology, whether the equipment is static or continuous. Since the number of lamps, and their shape, size, and orientation, and the material and configuration of the treatment chamber have a direct impact on the energy delivered and the outcome of the process, and direct extrapolation of results obtained with small-scale, benchtop equipment can lead to significant inaccuracies.

9.8 Conclusions

The research data currently available in the literature and discussed in this chapter clearly demonstrate the potential of PL technology to inactivate pathogenic and spoilage microorganisms in food products and on food contact materials. Since the effectiveness of PL treatment is strongly influenced by the substrate characteristics, particularly optical properties and topography, it is important to consider these limitations when developing PL applications. Arguably, the best way to take advantage of the unique benefits of PL technology, which include fast inactivation and limited undesirable changes to the treated substrates, is to focus on applications where no other antimicrobial treatment can be used. An interesting prospect is the use of PL as a terminal, post-process treatment for surface decontamination of various products, including fresh-cut fruits and vegetables and ready-to-eat meats or cheeses. PL treatment of food packaging is currently the most significant application of this technology. Although PL is attracting increasing attention from scientists and the food industry, much work still needs to be done, particularly to identify possible undesirable side effects, commercial scale-up, and process validation. Furthermore, the development of more energy-efficient light sources in the future may also increase the utilization of this technology for both food and non-food applications.

References

1. Food and Drug Administration, 1996. Code of Federal Regulations, 21 CFR, part 179, 41.
2. J. Dunn, W. Clark and T. Ott, Pulsed-light treatment of food and packaging, *Food Technol.*, 1995, **49**(9), 95–98.
3. J. E. Dunn, T. M. Ott and R. W. Clark, 1997. Prolongation of shelf life in perishable pod products. *U. S. Pat.* 5,489,442.
4. S. J. MacGregor, J. G. Anderson, R. A. Fouracre, O. Farish, L. McIlvaney and N. J. Rowan, Light inactivation of food-related pathogenic bacteria using a pulsed power source, *Lett. Appl. Microbiol.*, 1998, **27**(2), 67–70.
5. USFDA/CFSAN. 2002. Kinetics of microbial inactivation for alternative food processing technologies: pulsed light technology. USFDA Center

for Food Safety and Applied Nutrition. Available from: http://www.cfsan.fda.gov/~comm/ift-puls.html. Last accessed: May 1, 2008.
6. B. F. Ozen and J. D. Floros, Effects of emerging food processing techniques on the packaging materials, *Trends Food Sci. Technol.*, 2001, **12**, 60–67.
7. K. McDonald, R. Curry and P. Hancock, Comparison on pulsed and CW ultraviolet light sources to inactivate bacterial spores on surfaces, *IEEE Trans. Plasma Sci.*, 2002, **30**(5), 1986–1989.
8. C. I. Moraru and A. Uesugi 2009. Pulsed Light Treatment: Principles and Applications, in *Ultraviolet Light in Food Technology: Principles and Applications*, ed. T. Koutchma, L. Forney and C. I. Moraru. pp. 235–265.
9. T. Bintsis, E. Litopoulou-Tzanetaki and R. K. Robinson, Existing and potential application of ultraviolet light in the food industry – a critical review, *J. Sci. Food Agric.*, 2000, **80**, 637–645.
10. G. Oms-Oliu, O. Martín-Belloso and R. Soliva-Fortuny, Pulsed Light Treatments for Food Preservation. A Review, *Food Bioprocess Technol.*, 2010, **3**, 13.
11. M. Otaki, A. Okuda, K. Tajima, T. Iwasaki, S. Kinoshita and S. Ohgaki, Inactivation differences of microorganisms by low pressure UV and pulsed xenon lamps, *Water Sci. Technol.*, 2003, **47**, 185–190.
12. E. Weber and K. Scheible, Pulsed-UV unit may inactivate biological agents, *J. – Am. Water Works Assoc.*, 2003, **95**, 34–46.
13. A. A. Mofidi, H. Baribeau, P. A. Rochelle, R. De Leon, B. M. Coffey and J. F. Green, Disinfection of Cryptosporodium parvum with polychromatic UV light, *J. Am. Water Works Assoc.*, 2001, **93**, 95–109.
14. A. R. Uesugi, S. E. Woodling and C. I. Moraru, Inactivation kinetics and factors of variability in the pulsed light treatment of *Listeria innocua* cells, *J. Food Prot.*, 2007, **70**, 2518–2525.
15. S. Jin, A. A. Mofidi and K. G. Linden, Polychromatic UV fluence measurement using chemical actinometry, biodosimetry, and mathematical techniques, *J. Environ. Eng.*, 2006, **132**, 831–841.
16. Perkin Elmer Optoelectonics, High Performance Flash and Arc Lamps, Technical document available online at www.optoelectronics.perkinelmer.com. Accessed July 2008.
17. XENON Corporation, 2005, AN-104 Application note, Achieving faster cure time with Pulsed UV. Available online at http://www.globalspec.com.
18. R. M. Tyrrell, Induction of pyrimidine dimmers in bacterial DNA by 365 nm radiation, *Photochem. Photobiol.*, 1973, **17**, 69–73.
19. B. S. Rosenstein and J. M. Ducore, Induction of DNA strand breaks in normal human fibroblasts exposed to monochromatic ultraviolet and visible wavelengths in the 240–546 nm range, *Photochem. Photobiol.*, 1983, **38**, 51–55.
20. J. Farkas, Physical methods of food preservation, in *Food Microbiology, Fundamentals and Frontiers*, ed., M. P. Doyle, L. R. Beauchat and T. J. Montville, ASM Press, Washington, 1997, pp. 497–519.

21. J. M. Jay, *Modern Food Microbiology*, Gaithersburg, MD: Aspen Publishers, Inc. 2000, p. 679.
22. E. Y. Wuytack, L. D. Phuong, A. Aertsen, K. M. Reyns, D. Marquenie, B. De Ketelaere, B. Masschalck, I. Van Opstal, A. M. Diels and C. W. Michiels, Comparison of sublethal injury induced in *Salmonella enterica* serovar Typhimurium by heat and by different nonthermal treatments, *J. Food Prot.*, 2003, **66**(1), 31–37.
23. T. Wang, S. J. MacGregor, J. G. Anderson and G. A. Woolsey, Pulsed ultraviolet inactivation spectrum of *Escherichia coli*, *Water Res.*, 2005, **39**, 2921–2925.
24. K. Takeshita, J. Shibato, T. Sameshima, S. Fukunaga, S. Isobe, K. Arihara and M. Itoh, Damage of yeast cells induced by pulsed light irradiation, *Int. J. Food Microbiol.*, 2003, **85**, 151–158.
25. S. E. Woodling and C. I. Moraru, Effect of spectral range in surface inactivation of *Listeria innocua* using broad spectrum Pulsed Light, *J. Food Prot.*, 2007, **70**(4), 906–916.
26. A. Wekhof, 2003, Sterilization of packaged pharmaceutical solutions, packaging and surgical tools with Pulsed UV Light, July 9–11, 2003, Vienna, Austria.
27. M. Grapperhaus, R. B. Schaefer and K. Linden, Pulsed UV lamp performance and comparison with UV mercury lamps, *J. Environ. Eng. Sci.*, 2007, **6**, 303–310.
28. C.-I. Cheigh, M.-H. Park, M.-S. Chung, J.-K. Shin and Y.-S. Park, Comparison of intense PL and UV (UVC)-induced cell damage in *Listeria monocytogenes* and *Escherichia coli* O157:H7, *Food Control*, 2012, **25**, 654–659.
29. A. R. Uesugi, L. C. Hsu, R. W. Worobo and C. I. Moraru, Gene expression analysis for *Listeria monocytogenes* following exposure to pulsed light and continuous ultraviolet light treatments, *LWT – Food Sci. Technol.*, 2016, **68**, 579–588.
30. V. M. Gómez-López, P. Ragaert, J. Debevere and F. Devlieghere, Pulsed light for food decontamination: a review, *Trends Food Sci. Technol.*, 2007, **18**, 464–473.
31. A. Wekhof, Disinfection with flash lamps, *PDA J. Pharm. Sci. Technol.*, 2000, **54**, 264–276.
32. K. Krishnamurthy, A. Demirci and J. Irudayaraj, Inactivation of *Staphylocloccus aureus* by pulsed UV-light sterilization, *J. Food Prot.*, 2004, **67**, 1027–1030.
33. K. Krishnamurthy, J. C. Tewari, J. Irudayaraj and A. Demirci, Microscopic and spectroscopic evaluation of inactivation of *Staphylococcus aureus* by pulsed UV light and infrared heating, *Food Bioprocess Technol.*, 2010, **3**, 93–104.
34. N. J. Rowan, J. G. Anderson, S. J. MacGregor and R. A. Fouracre, Inactivation of food-borne enteropathogenic bacteria and spoilage fungi using pulsed-light, *IEEE Trans. Plasma Sci.*, 2000, **28**, 83–88.

35. S. E. Woodling and C. I. Moraru, Influence of surface topography on the effectiveness of Pulsed Light treatment for the inactivation of *Listeria innocua* on stainless steel surfaces, *J. Food Sci.*, 2005, **70**(7), 345–351.
36. W. L. Nicholson, N. Munakata, G. Horneck, H. J. Melosh and P. Setlow, Resistance of *Bacillus* endospores to extreme terrestrial and extraterrestrial environments, *Microbiol. Mol. Biol. Rev.*, 2000, **64**(3), 548–572.
37. P. Setlow, Spores of *Bacillus subtilis*: their resistance to and killing by radiation, heat and chemicals, *J. Appl. Microbiol.*, 2006, **101**(3), 514–525.
38. C. Waltham, J. Boyle, B. Ramey and J. Smit, Light scattering and absorption caused by bacterial activity in water, *Appl. Opt.*, 1994, **33**(31), 7536–7540.
39. E. T. Arakawa, P. S. Tuminello, B. N. Khare and M. E. Milham, Optical properties of *Erwinia herbicola* bacteria at 0.190–2.50 μm, *Biopolym. (Biospectroscopy)*, 2003, **72**, 391–398.
40. P. S. Tuminello, E. T. Arakawa, B. N. Khare, J. M. Wrobel, M. R. Querry and M. E. Milham, Optical properties of *Bacillus subtilis* spores from 0.2 to 2.5 μm, *Appl. Opt.*, 1997, **36**(13), 2818–2824.
41. N. L. Montgomery and P. Banerjee, Inactivation of *Escherichia coli* O157:H7 and *Listeria monocytogenes* in biofilms by pulsed ultraviolet light, *BMC Res. Notes*, 2015, **8**, 235.
42. L. Durán and C. Calvo, 2002, Optical properties of foods, in *Encyclopedia of Life Support Systems*. Available online at www.eolss.net.
43. S. T. Flock, B. C. Wilson and M. S. Patterson, Total attenuation coefficients and scattering phase functions of tissues and phantom materials at 633 nm, *Med. Phys.*, 1987, **14**(5), 835–841.
44. V. Hollis, 2002, Non-invasive monitoring of brain tissue temperature by near-infrared spectroscopy, PhD Thesis, University College London, 237 (pp. 28–52, Ch.2). Available online at http://www.medphys.ucl.ac.uk/.
45. S. Xie, H. Li and B. Li, Measurement of optical penetration depth and refractive index of human tissue, *Chin. Opt. Lett.*, 2003, **1**(1), 44–46.
46. K. H. Guenther, J. A. McCandless and F. D. Orazio Jr., Correlation of light-scattering measurements and visual ranking of optical surfaces, *Appl. Opt.*, 1993, **32**(19), 3425–3432.
47. N. Sakai and T. Hanzawa, Applications and advances in far-infrared heating in Japan, *Trends Food Sci. Technol.*, 1994, **5**, 357–362.
48. G. Marquez, L. V. Wang, S. Lin, J. A. Schwartz and S. L. Thomsen, Anisotropy in the absorption and scattering spectra of chicken breast tissue, *Appl. Opt.*, 1998, **37**(4), 798–804.
49. C. E. Cooper, E. E. Elwell, J. H. Meek, S. J. Matcher, J. S. Wyatt, M. Cope and D. T. Delpy, The Noninvasive Measurement of Absolute Cerebral Deoxyhaemoglobin Concentration and Mean Optical Path Length in the Neonatal Brain by Second Derivative Near Infrared Spectroscopy, *Pediat. Res.*, 1996, **39**, 32–38.

50. J. M. Conway, K. H. Norris and C. E. Bodwell, A new approach for the estimation of body composition: infrared interactance, *Am. J. Clin. Nutr.*, 1984, **40**, 1123–1130.
51. M. Dagerskog and L. Österström, Infra-red radiation for food processing. I. A study of the fundamental properties of infra-red radiation, *Lebensm.-Wiss. Technol.*, 1979, **12**, 237–242.
52. A. Sauer and C. I. Moraru, Inactivation of *E. coli* ATCC 25922 and *E. coli* O157:H7 in apple juice and apple cider using Pulsed Light treatment, *J. Food Prot.*, 2009, **72**(5), 937–944.
53. A. Uesugi and C. I. Moraru, Reduction of Listeria on ready-to-eat sausages after exposure to a combination of Pulsed Light and nisin, *J. Food Prot.*, 2009, **72**(2), 347–353.
54. K. H. Seo and J. F. Frank, Attachment of *Escherichia coli* O157: H7 to lettuce leaf surface and bacterial viability in response to chlorine treatment as demonstrated by using confocal scanning laser microscopy, *J. Food Prot.*, 1999, **62**, 3–9.
55. D. L. Ringus and C. I. Moraru, Pulsed Light inactivation of *Listeria innocua* on food packaging materials of different surface roughness and reflectivity, *J. Food Eng.*, 2013, **114**, 331–337.
56. J. G. Anderson, N. J. Rowan, S. J. MacGregor, R. A. Fouracre and O. Farish, Inactivation of food-borne enteropathogenic bacteria and spoilage fungi using pulsed-light, *IEEE Trans. Plasma Sci.*, 2000, **28**, 83–88.
57. V. M. Gómez-López, F. Devlieghere, V. Bonduelle and J. Debevere, Factors affecting the inactivation of micro-organisms by intense light pulses, *J. Appl. Microbiol.*, 2005a, **99**, 460–470.
58. A. Rajkovic, I. Tomasevic, N. Smigic, M. Uyttendaele, R. Radovanovic and F. Devlieghere, Pulsed UV light as an intervention strategy against *Listeria monocytogenes* and *Escherichia coli* O157:H7 on the surface of a meat slicing knife, *J. Food Eng.*, 2010, **100**, 446–451.
59. J. Proulx, L. C. Hsu, B. M. Miller, G. Sullivan, K. Paradis and C. I. Moraru, Pulsed light inactivation of pathogenic and spoilage bacteria on cheese surface, *J. Dairy Sci.*, 2015, **98**, 5890–5898.
60. S. T. Hilton, J. O. de Moraes and C. I. Moraru, Effect of sublethal temperatures on pulsed light inactivation of bacteria, *Innovative Food Sci. Emerging Technol.*, 2017, **39**, 49–54.
61. A. Lasagabaster and I. M. de Marañón, Sensitivity to pulsed light technology of several spoilage and pathogenic bacteria isolated from fish products, *J. Food Prot.*, 2012, **75**, 2039–2044.
62. M. L. Artíguez and I. M. de Marañón, Improved process for decontamination of whey by a continuous flow-through pulsed light system, *Food Control*, 2015, **47**, 599–605.
63. C. Levy, X. Aubert, B. Lacour and F. Carlin, Relevant factors affecting microbial surface decontamination by pulsed light, *Int. J. Food Microbiol.*, 2012, **152**, 168–174.
64. N. Elmnasser, M. Federighi, A. Bakhrouf and N. Orange, Effectiveness of pulsed ultraviolet light treatment for bacterial inactivation on agar

surface and liquid medium, *Foodborne Pathog. Dis.*, 2010, 7(11), 1401–1406.

65. H. P. Farrell, M. Garvey, J. G. Laffey and N. J. Rowan, Investigation of critical inter-related factors affecting the efficacy of pulsed light for inactivating clinically relevant bacterial pathogens, *J. Appl. Microbiol.*, 2009, **108**, 1494–1508.
66. Y. Huang, M. Ye, X. Cao and H. Chen, Pulsed light inactivation of murine norovirus, Tulane virus, *Escherichia coli* O157: H7 and *Salmonella* in suspension and on berry surfaces, *Food Microbiol.*, 2017, **61**, 1–4.
67. W. Q. Xu, H. Q. Chen, Y. X. Huang and C. Q. Wu, Decontamination of *Escherichia coli* O157:H7 on green onions using pulsed light (PL) and PL-surfactant-sanitizer combinations, *Int. J. Food Microbiol.*, 2013, **166**, 102–108.
68. M. L. Artíguez and I. M. de Marañón, Process parameters affecting Listeria innocua inactivation by pulsed light, *Food Bioprocess Technol.*, 2014, 7, 2759–2765.
69. L. Hsu and C. I. Moraru, Quantifying and mapping the spatial distribution of fluence inside a pulsed light treatment chamber and various liquid substrates, *J. Food Eng.*, 2011, **103**, 84–91.
70. D. Marquenie, A. H. Geeraerd, J. Lammertyn, C. Soontjens, J. F. Van Impe, C. W. Michiels and B. M. Nicolai, Combinations of pulsed white light and UV-C or mild heat treatment to inactivate conidia of *Botrytis cinerea* and *Monilia fructigena*, *Int. J. Food Microbiol.*, 2003, **85**, 185–196.
71. M. A. J. S. van Boekel and A. J. S. Martinus, On the use of the Weibull model to describe thermal inactivation of microbial vegetative cells, *Int. J. Food Microbiol.*, 2002, **74**, 139–159.
72. J. Y. Yi, Y.-K. Bae, C.-I. Cheigh and M.-S. Chung, Microbial inactivation and effects of interrelated factors of intense pulsed light (IPL) treatment for *Pseudomonas aeruginosa*, *LWT – Food Sci. Technol.*, 2017, 77, 52–59.
73. T. Koutchma, S. Keller, S. Chirtel and B. Parisi, Ultraviolet disinfection of juice products in laminar and turbulent flow reactors, *Innovative Food Sci. Emerging Technol.*, 2004, 5(2), 179–189.
74. H. Liltved and S. J. Cripps, Removal of particle-associated bacteria by prefiltration and ultraviolet irradiation, *Aquac. Res.*, 1999, **30**(6), 445–450.
75. E. Gayán, M. J. Serrano, S. Monfort, I. Álvarez and S. Condón, Combining ultraviolet light and mild temperatures for the inactivation of *Escherichia coli* in orange juice, *J. Food Eng.*, 2012, **113**(4), 598–605.
76. H.-J. Hwang, C.-I. Cheigh and M.-S. Chung, Relationship between optical properties of beverages and microbial inactivation by intense pulsed light, *Innovative Food Sci. Emerging Technol.*, 2015, **31**, 91–96.
77. D. E. Huffman, T. R. Slifko, K. Salisbury and J. B. Rose, Inactivation of bacteria, virus and *Cryptosporidium* by a point-of-use device using pulsed broad spectrum white light, *Water Res.*, 2000, **34**, 2491–2498.

78. I. Palgan, I. M. Caminiti, A. Muñoz, F. Noci, P. Whyte, D. J. Morgan, D. A. Cronin and J. G. Lyng, Effectiveness of High Intensity Light Pulses (HILP) treatments for the control of *Escherichia coli* and *Listeria innocua* in apple juice, orange juice and milk, *Food Microbiol.*, 2011, **28**, 14–20.
79. W. L. Smith, M. C. Lagunas-Solar and J. S. Cullor, Use of pulsed ultraviolet laser light for the cold pasteurization of bovine milk, *J. Food Prot.*, 2002, **65**, 1480–1482.
80. N. Innocente, A. Segat, L. Manzocco, M. Marino, M. Maifreni, I. Bortolomeoli, A. Ignat and M. C. Nicoli, Effect of pulsed light on total microbial count and alkaline phosphatase activity of raw milk, *Int. Dairy J.*, 2014, **39**, 108–112.
81. B. M. Miller, A. Sauer and C. I. Moraru, Inactivation of *Escherichia coli* in milk and concentrated milk using pulsed-light treatment, *J. Dairy Sci.*, 2012, **95**, 5597–5603.
82. G. Pataro, A. Muñoz, I. Palgan, F. Noci, G. Ferrari and J. G. Lyng, Bacterial inactivation in fruit juices using a continuous flow Pulsed Light (PL) system, *Food Res. Int.*, 2011, **44**, 1642–1648.
83. N. Elmnasser, M. Dalgalarrondo, N. Orange, A. Bakhrouf, T. Haertle, M. Federighi and J. M. Chobert, Effect of pulsed-light treatment on milk proteins and lipids, *J. Agric. Food Chem.*, 2008, **56**, 1984–1991.
84. I. Kasahara, V. Carrasco and L. Aguilar, Inactivation of *Escherichia coli* in goat milk using pulsed ultraviolet light, *J. Food Eng.*, 2015, **152**, 43–49.
85. N. P. Ozer and A. Demirci, Inactivation of *Escherichia coli* O157: H7 and *Listeria monocytogenes* inoculated on raw salmon fillets by pulsed UV-light treatment, *Int. J. Food Sci. Technol.*, 2006, **41**, 354–360.
86. E. Hierro, E. Barroso, L. de la Hoz, J. A. Ordonez, S. Manzano and M. Fernandez, Efficacy of pulsed light for shelf-life extension and inactivation of *Listeria monocytogenes* on ready-to-eat cooked meat products, *Innovative Food Sci. Emerging Technol.*, 2011, **12**, 275–281.
87. P. Wambura and M. Verghese, Effect of pulsed ultraviolet light on quality of sliced ham, *LWT – Food Sci. Technol.*, 2011, **44**, 2173–2179.
88. M. Ganan, E. Hierro, X. F. Hospital, E. Barroso and M. Fernandez, Use of pulsed light to increase the safety of ready-to-eat cured meat products, *Food Control*, 2013, **32**, 512–517.
89. E. Hierro, M. Ganan, E. Barroso and M. Fernandez, Pulsed light treatment for the inactivation of selected pathogens and the shelf-life extension of beef and tuna carpaccio, *Int. J. Food Microbiol.*, 2012, **158**, 42–48.
90. A. Rajkovic, I. Tomasevic, B. De Meulenaer and F. Devlieghere, The effect of pulsed UV light on *Escherichia coli* O157:H7, *Listeria monocytogenes*, Salmonella Typhimurium, Staphylococcus aureus and staphylococcal enterotoxin A on sliced fermented salami and its chemical quality, *Food Control*, 2017, **73**, 829–837. Part B.
91. N. M. Keklik, A. Demirci and V. M. Puri, Decontamination of unpackaged and vacuum-packaged boneless chicken breast with pulsed ultraviolet light, *Poult. Sci.*, 2010b, **89**, 570–581.

92. E. Paskeviciute, I. Buchovec and Z. Luksiene, High-power pulsed light for decontamination of chicken from food pathogens: a study on antimicrobial efficiency and organoleptic properties, *J. Food Safety*, 2011, **31**, 61–68.
93. P. N. Haughton, J. G. Lyng, D. J. Morgan, D. A. Cronin, S. Fanning and P. Whyte, Efficacy of High-Intensity Pulsed Light for the Microbiological Decontamination of Chicken, Associated Packaging, and Contact Surfaces, *Foodborne Pathog. Dis.*, 2011, **8**, 109–117.
94. I. Nicorescu, B. Nguyen, S. Chevalier and N. Orange, Effects of pulsed light on the organoleptic properties and shelf-life extension of pork and salmon, *Food Control*, 2014, **44**, 138–145.
95. B. Macias-Rodriguez, W. D. Yang, K. Schneider and C. Rock, Pulsed UV light as a postprocessing intervention for decontamination of hard-cooked peeled eggs, *Int. J. Food Sci. Technol.*, 2014, **49**, 2472–2480.
96. N. M. Keklik, A. Demirci, P. H. Patterson and V. M. Puri, Pulsed UV light inactivation of *Salmonella enteritidis* on eggshells and its effects on egg quality, *J. Food Prot.*, 2010a, **73**, 1408–1415.
97. L. Manzocco, M. Maifreni, M. Anese, M. Munari, I. Bartolomeoli, S. Zanardi, M. Suman and M. C. Nicoli, Effect of pulsed light on safety and quality of fresh egg pasta, *Food Bioprocess Technol.*, 2014, **7**, 1973–1980.
98. F. O. Can, A. Demirci, V. M. Puri and H. Gourama, Decontamination of Hard Cheeses by Pulsed UV Light, *J. Food Prot.*, 2014, **77**, 1723–1731.
99. M. Fernández, M. Ganan, C. Guerra and E. Hierro, Protein oxidation in processed cheese slices treated with pulsed light technology, *Food Chem.*, 2014, **159**, 388–390.
100. M. Fernández, X. F. Hospital, K. Arias and E. Hierro, Application of pulsed light to sliced cheese: effect on *Listeria* inactivation, sensory quality and volatile profile, *Food Bioprocess Technol.*, 2016, **9**(8), 1335–1344.
101. S. L. Hillegas and A. Demirci 2003. Inactivation of Clostridium sporogenes in clover honey by pulsed UV-light treatment. In *Agricultural Engineering International: The CIGR Journal of Scientific Research and Development*, vol. V, Manuscript FP03 009.
102. V. M. Gómez-López, F. Devlieghere, V. Bonduelle and J. Debevere, Intense light pulses decontamination of minimally processed vegetables and their shelf-life, *Int. J. Food. Microbiol.*, 2005b, **103**, 79–89.
103. B. Kramer, J. Wunderlich and P. Muranyi, Pulsed light decontamination of endive salad and mung bean sprouts and impact on color and respiration activity, *J. Food Prot.*, 2015, **78**(2), 340–348.
104. D. Marquenie, C. W. Michiels, J. F. Van Impe, E. Schrevens and B. N. Nicolai, Pulsed white light in combination with UV-C and heat to reduce storage rot of strawberry, *Postharvest Biol. Technol.*, 2003, **28**, 455–461.
105. D. Marquenie, A. H. Geeraerd, J. Lammertyn, C. Soontjens, J. F. Van Impe, C. W. Michiels and B. M. Nicolai, Combinations of

pulsed white light and UV-C or mild heat treatment to inactivate conidia of *Botrytis cinerea* and *Monilia fructigena*, *Int. J. Food Microbiol.*, 2003, **85**, 185–196.

106. Z. Luksiene, I. Buchovec and P. Viskelis, Impact of high-power pulsed light on microbial contamination, health promoting components and shelf life of strawberries, *Food Technol. Biotechnol.*, 2013, **51**, 284–292.
107. K. L. Bialka and A. Demirci, Decontamination of *Escherichia coli* O157:H7 and *Salmonella enterica* on blueberries using ozone and pulsed UV-light, *J. Food Sci.*, 2007, **72**, M391–M396.
108. M. V. Agüero, R. J. Jagus, O. Martín-Belloso and R. Soliva-Fortuny, Surface decontamination of spinach by intense pulsed light treatments: impact on quality attributes, *Postharvest Biol. Technol.*, 2016, **121**, 118–125.
109. G. Oms-Oliu, I. Aguiló-Aguayo, O. Martín-Belloso and R. Soliva-Fortuny, Effects of pulsed light treatments on quality and antioxidant properties of fresh-cut mushrooms (*Agaricus bisporus*), *Postharvest Biol. Technol.*, 2010, **56**, 216–222.
110. A. Ignat, L. Manzocco, M. Maifreni, I. Bartolomeoli and M. C. Nicoli, Surface decontamination of fresh-cut apple by pulsed light: Effects on structure, colour and sensory properties, *Postharvest Biol. Technol.*, 2014, **91**, 122–127.
111. P. L. Gómez, A. Garcia-Loredo, A. Nieto, D. M. Salvatori, S. Guerrero and S. M. Alzamora, Effect of pulsed light combined with an antibrowning pretreatment on quality of fresh cut apple, *Innovative Food Sci. Emerging Technol.*, 2012, **16**, 102–112.
112. P. C. Koh, M. A. Noranizan, R. Karim and Z. A. N. Hanani, Repetitive pulsed light treatment at certain interval on fresh-cut cantaloupe *(Cucumis melo* L. *reticulatus cv. Glamour)*, *Innovative Food Sci. Emerging Technol.*, 2016, **36**, 92–103.
113. V. Rodov, Y. Vinokur and B. Horev, Brief postharvest exposure to pulsed light stimulates coloration and anthocyanin accumulation in fig fruit (*Ficus carica* L.), *Postharvest Biol. Technol.*, 2012, **68**, 43–46.
114. I. Aguiló-Aguayo, F. Charles, C. Renard, D. Page and F. Carlin, Pulsed light effects on surface decontamination, physical qualities and nutritional composition of tomato fruit, *Postharvest. Biol. Technol.*, 2013, **86**, 29–36.
115. M. M. A. Lopes, E. O. Silva, K. M. Canuto, L. M. A. Silva, M. I. Gallão, L. Urban, J. F. Ayala-Zavala and M. R. A. Miranda, Low fluence pulsed light enhanced phytochemical content and antioxidant potential of 'Tommy Atkins' mango peel and pulp, *Innovative Food Sci. Emerging Technol.*, 2016, **33**, 216–224.
116. F. Charles, V. Vidal, F. Olive, H. Filgueiras and H. Sallanon, Pulsed light treatment as new method to maintain physical and nutritional quality of fresh-cut mangoes, *Innovative Food Sci. Emerging Technol.*, 2013, **18**, 190–195.

117. G. Scott, M. Rupar, A. G. D. Fletcher, M. Dickinson and G. Shama, A comparison of low intensity UV-C and high intensity pulsed polychromatic sources as elicitors of hormesis in tomato fruit, *Postharvest Biol. Technol.*, 2017, **125**, 52–58.
118. K. Kaack and B. Lyager, Treatment of slices from carrot (*Daucus carota*) using high intensity white pulsed light, *Eur. Food Res. Technol.*, 2007, **224**, 561–566.
119. Y. X. Huang and H. Q. Chen, A novel water-assisted pulsed light processing for decontamination of blueberries, *Food Microbiol.*, 2014, **40**, 1–8.
120. Y. X. Huang and H. Q. Chen, Inactivation of *Escherichia coli* O157: H7, Salmonella and human norovirus surrogate on artificially contaminated strawberries and raspberries by water-assisted pulsed light treatment, *Food Res. Int.*, 2015, **72**, 1–7.
121. B. Kramer, J. Wunderlich and P. Muranyi, Pulsed light decontamination of endive salad and mung bean sprouts in water, *Food Control*, 2017, **73**, 367–371. Part B.
122. R. R. Sharma and A. Demirci, Inactivation of *Escherichia coli* O157: H7 on inoculated alfalfa seeds with pulsed ultraviolet light and response surface modeling, *J. Food Sci.*, 2003, **68**, 1448–1453.
123. N. Aron Maftei, A. Y. Ramos-Villarroel, A. I. Nicolau, O. Martin-Belloso and R. Soliva-Fortuny, Pulsed light inactivation of naturally occurring moulds on wheat grain, *J. Sci. Food Agric.*, 2014, **94**, 721–726.
124. H.-J. Hwang, C.-I. Cheigh and M.-S. Chung, Construction of a pilot-scale continuous-flow intense pulsed light system and its efficacy in sterilizing sesame seeds, *Innovative Food Sci. Emerging Technol.*, 2017, **39**, 1–6.
125. M.-S. Choi, C.-I. Cheigh, E.-A. Jeong, J.-K. Shin and M.-S. Chung, Nonthermal sterilization of *Listeria monocytogenes* in infant foods by intense pulsed-light treatment, *J. Food Eng.*, 2010, **97**, 504–509.
126. F. Fine and P. Gervais, Efficiency of pulsed UV light for microbial decontamination of food powders, *J. Food Prot.*, 2004, **67**, 787–792.
127. I. Nicorescu, B. Nguyen, M. Moreau-Ferret, A. Agoulon, S. Chevalier and N. Orange, Pulsed light inactivation of *Bacillus subtilis* vegetative cells in suspensions and spices, *Food Control*, 2013, **31**, 151–157.
128. S. Jun, J. Irudayaraj, A. Demirci and D. Geiser, Pulsed UV-light treatment of corn meal for inactivation of *Aspergillus niger* spores, *Int. J. Food Sci. Technol.*, 2003, **38**, 883–888.
129. J.-W. Ha, K.-H. Back, Y.-H. Kim and D.-H. Kang, Efficacy of UV-C irradiation for inactivation of food-borne pathogens on sliced cheese packaged with different types and thicknesses of plastic films, *Food Microbiol.*, 2016, **57**, 172–177.
130. F. Leipold, N. Schultz-Jensen, Y. Kusano, H. Bindslev and T. Jacobsen, Decontamination of objects in a sealed container by means of atmospheric pressure plasmas, *Food Control*, 2011, **22**, 1296–1301.

131. H. I. Yong, H.-J. Kim, S. Park, K. Kim, W. Choe, S. J. Yoo and C. Jo, Pathogen inactivation and quality changes in sliced cheddar cheese treated using flexible thin-layer dielectric barrier discharge plasma, *Food Res. Int.*, 2015, **69**, 57–63.
132. D. L. Ringus and C. I. Moraru 2009. Inactivation of Listeria innocua by pulsed light treatment on various food packaging materials. Annual Meeting of IFT, Anaheim, CA, June 2009.
133. R. Castillo, M. Biedermann, A.-M. Riqueta and K. Grob, Comprehensive on-line HPLC-GC for screening potential migrants from polypropylene into food: The effect of pulsed light decontamination as an example, *Polym. Degrad. Stab.*, 2013, **98**(9), 1679–1687.
134. M. D. Kalaras, R. B. Beelman, M. F. Holick and R. J. Elias, Generation of potentially bioactive ergosterol-derived products following pulsed ultraviolet light exposure of mushrooms (*Agaricus bisporus*), *Food Chem.*, 2012, **135**, 396–401.
135. M. D. Kalaras, R. B. Beelman and R. J. Elias, Effects of postharvest pulsed UV light treatment of White Button Mushrooms (*Agaricus bisporus*) on Vitamin D2 content and quality attributes, *J. Agric. Food Chem.*, 2012, **60**, 220–225.
136. M. Moreau, G. Lescure, A. Agoulon, P. Svinareff, N. Orange and M. Feuilloley, Application of the pulsed light technology to mycotoxin degradation and inactivation, *J. Appl. Toxicol.*, 2013, **33**, 357–363.
137. B. N. Ames, J. McCann and E. Yamasaki, Methods for detecting carcinogens and mutagens with *Salmonella*-mammalian-microsome mutagenicity test, *Mutat. Res.*, 1975, **31**, 347–363.
138. B. A. Janve, W. Yang, M. R. Marshall, J. I. Reyes-De-Corcuera and T. M. Rababah, Nonthermal Inactivation of Soy (*Glycine max* sp.) Lipoxygenase by Pulsed Ultraviolet Light, *J. Food Sci.*, 2014, **79**, C8–C18.
139. L. Manzocco, A. Panozzo and M. C. Nicoli, Inactivation of polyphenoloxidase by pulsed light, *J. Food Sci.*, 2013, **78**, E1183–E1187.
140. M. A. B. Siddique, P. Maresca, G. Pataro and G. Ferrari, Effect of pulsed light treatment on structural and functional properties of whey protein isolate, *Food Res. Int.*, 2016, **87**, 189–196.
141. S. Y. Chung, W. Yang and K. Krishnamurthy, Effects of pulsed UV-light on peanut allergens in extracts and liquid peanut butter, *J. Food Sci.*, 2008, **73**(5), C400–C404.
142. W. W. Yang, N. R. Mwakatage, R. Goodrich-Schneider, K. Krishnamurthy and T. M. Rababah, Mitigation of major peanut allergens by pulsed ultraviolet light, *Food Bioprocess Technol.*, 2012, **5**(7), 2728–2738.
143. W. W. Yang, S. Y. Chung, O. Ajayi, K. Krishnamurthy, K. Konan and R. Goodrich-Schneider, Use of Pulsed Ultraviolet Light to Reduce the Allergenic Potency of Soybean Extracts, *Int. J. Food Eng.*, 2010, **6**, 13.
144. S. K. Shriver, Effect of selected emerging nonthermal processing methods on the allergen reactivity of Atlantic white shrimp (*Litopenaeus*

setiferus), MS thesis, Department of Food Science and Human Nutrition, University of Florida, Gainesville, FL, USA, 32611, 2011.
145. Y. Q. Li, W. D. Yang, S. Y. Chung, H. Q. Chen, M. Ye, A. A. Teixeira, J. F. Gregory, B. A. Welt and S. Shriver, Effect of pulsed ultraviolet light and high hydrostatic pressure on the antigenicity of almond protein extracts, *Food Bioprocess Technol.*, 2013, **6**(2), 431–440.
146. P. Meinlschmidt, E. Ueberham, J. Lehmann, K. Reineke, O. Schlüter, U. Schweiggert-Weisz and P. Eisner, The effects of pulsed ultraviolet light, cold atmospheric pressure plasma, and gamma-irradiation on the immunoreactivity of soy protein isolate, *Innovative Food Sci. Emerging Technol.*, 2016, **38**, 374–383. Part B.
147. G. J. Funes, P. L. Gómez, S. L. Resnik and S. M. Alzamora, Application of pulsed light to patulin reduction in McIlvaine buffer and apple products, *Food Control*, 2013, **30**, 405–410.
148. G. Uslu, A. Demirci and J. M. Regan, Disinfection of synthetic and real municipal wastewater effluent by flow-through pulsed UV-light treatment system, *J. Water Process. Eng.*, 2016, **10**, 89–97.
149. M. Ferrario, S. M. Alzamora and S. Guerrero, Study of the inactivation of spoilage microorganisms in apple juice by pulsed light and ultrasound, *Food Microbiol.*, 2015, **46**, 635–642.
150. A. Muñoz, I. M. Caminiti, I. Palgan, G. Pataro, F. Noci, D. J. Morgan, D. A. Cronin, P. Whyte, G. Ferrari and J. G. Lyng, Effects on *Escherichia coli* inactivation and quality attributes in apple juice treated by combinations of pulsed light and thermosonication, *Food Res. Int.*, 2012, **45**, 299–305.
151. A. Y. Ramos-Villarroel, O. Martin-Belloso and R. Soliva-Fortuny, Combined effects of malic acid dip and pulsed light treatments on the inactivation of *Listeria innocua* and *Escherichia coli* on fresh-cut produce, *Food Control*, 2015, **52**, 112–118.
152. J. Proulx, G. Sullivan, L. F. Marostegan, S. VanWees, L. C. Hsu and C. I. Moraru, Pulsed Light and antimicrobial combination treatments for surface decontamination of cheese: favorable and antagonistic effects, *J. Dairy Sci.*, 2017, **100**(3), 1664–1673.
153. J. Proulx, M. Agustin, G. Sullivan, S. VanWees, J. Jian, S. T. Hilton and C. I. Moraru, Short communication: influence of pulsed light treatment on the quality and sensory characteristics of Cheddar cheese, *J. Dairy Sci.*, 2017, **c100**(2), 1004–1008.
154. M. Ferrario and S. Guerrero, Effect of a continuous flow-through pulsed light system combined with ultrasound on microbial survivability, color and sensory shelf life of apple juice, *Innovative Food Sci. Emerging Technol.*, 2016, **34**, 214–224.
155. I. M. Caminiti, F. Noci, D. J. Morgan, D. A. Cronin and J. G. Lyng, The effect of pulsed electric fields, ultraviolet light or high intensity light pulses in combination with manothermosonication on selected physico-chemical and sensory attributes of an orange and carrot juice blend, *Food Bioprod. Process.*, 2012, **90**(3), 442–448.

156. XENON Corporation. Products. Technical document. Available online at http://www.xenoncorp.com/. Accessed February 2017.
157. M. L. Artíguez, A. Lasagabaster and I. M. de Marañón, Factors affecting microbial inactivation by Pulsed Light in a continuous flow-through unit for liquid products treatment, *Procedia Food Sci.*, 2011, **1**, 786–791.
158. J. Y. Yi, N.-H. Lee and M.-S. Chung, Inactivation of bacteria and murine norovirus in untreated groundwater using a pilot-scale continuous-flow intense pulsed light (IPL) system, *LWT – Food Sci. Technol.*, 2016, **66**, 108–113.

CHAPTER 10

Infrared Food Processing Technology: Fundamentals and Case Studies of Recent Advances in Grain Processing

GRIFFITHS G. ATUNGULU

Department of Food Science, University of Arkansas Division of Agriculture, 2650 N. Young Avenue, Fayetteville, AR 72704, USA
Email: atunglu@uark.edu

10.1 Engineering Principles of Infrared Heating

10.1.1 Fundamentals

Any object that has a temperature higher than 0 K emits radiation, in the form of electromagnetic waves. An electromagnetic wave can be imagined as a self-propagating, transverse oscillating wave of electric and magnetic fields. Two important aspects that characterize waves are their wavelength or distance between peaks and amplitude or distance between peaks and valleys. Figure 10.1a illustrates a wave oscillation using a 3D sketch of a linearly polarized wave propagating from left to right such that the electric and magnetic fields in such a wave are in phase with each other, reaching minima and maxima together. Infrared waves are part of the electromagnetic spectrum. Based on wavelength classification, the waves that produce radiant infrared energy lie between the visible and microwave bands (Figure 10.1b). An ideal (black) body

Green Chemistry Series No. 53
Alternatives to Conventional Food Processing: 2nd Edition
Edited by Andrew Proctor

Published by the Royal Society of Chemistry, www.rsc.org

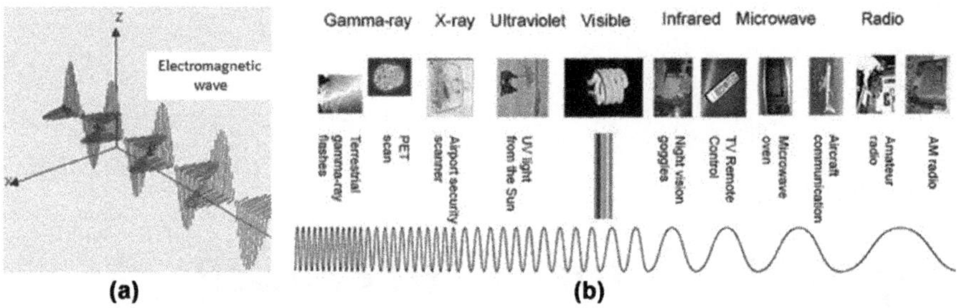

Figure 10.1 (a) Self-propagating transverse oscillating wave of electric and magnetic fields; (b) the infrared waves lie between the visible and the microwave bands.

absorbs all incident radiation, and also emits the maximum radiant energy. According to the Stefan–Boltzmann Law, radiant energy emitted from a heat source is proportional to the fourth power of the emitter's temperature T (K):[1]

$$q'' = \sigma T^4 \tag{10.1}$$

where q'' (W m^{-2}) is the radiant energy and $\sigma = 5.669 \times 10^{-8}$ W m^{-2} K^{-4} is the proportionality constant, also known as the Stefan–Boltzmann constant.

The emissivity of a non-ideal (gray) body, ε, is the ratio of the energy flux emitted by the real body to the flux emitted by an ideal (black) body at the same temperature. The energy absorbed, reflected, or transmitted depends on the incident wavelength, material type, and surface conditions. For a black body, the spectral distribution of the energy emitted is described by Planck's law:[1]

$$E_{\lambda,\mathrm{b}} = \frac{C_1 \lambda^{-5}}{\mathrm{e}^{C_2/\lambda T} - 1} \tag{10.2}$$

where the radiation constants C_1 and C_2 are 3.742×10^8 W μm^4 m^{-2} and 1.439×10^4 μm K, respectively. The spectral distribution of energy and radiant intensity are a function of wavelength and emitter temperature (Figure 10.2). The spectral distribution peak at a specified wavelength is defined by Wien's law:[1]

$$\lambda_{\mathrm{peak}} = \frac{C_3}{T} \tag{10.3}$$

where the peak wavelength λ_{peak} (μm) is inversely proportional to the emitting body's temperature T (K) with $C_3 = 2897.8$ μm K.[1] The phenomenon associated with Planck's law produces the Wien displacement curve, which is an important feature in emitter selection. Das and Das, in our book on infrared heating for food processing,[2] provided a detailed description of emitter types, design, selection, and characteristics.

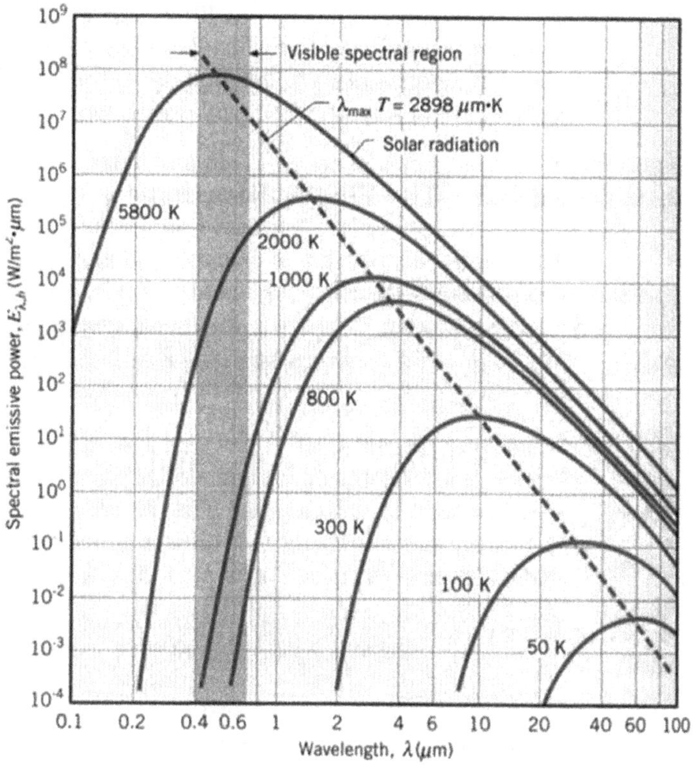

Figure 10.2 The spectral distribution peak at a specified infrared wavelength as defined by Wien's law.
Republished with permission of Taylor & Francis from *Infrared Heating for Food and Agricultural Processing*, Pan, Z. and Atungulu, G. G., ISBN 9781420090970, 2010; permission conveyed through Copyright Clearance Center, Inc.

10.1.2 Infrared Emitters

Infrared emitters can be broadly classified into three categories, namely short-, medium-, and long-wave emitters. However, the boundary between them is not very clear. Short-wave emitters emit infrared rays of wavelength ranging from 0.7 to 1.4 μm with corresponding radiator temperature around 1300–2600 K (1027–2327 °C). These type of emitters are often considered the most powerful among their counterparts since they achieve the highest power densities of up to or more than 300 kW m^{-2}. They are used extensively for industrial processes such as preheating, metal castings, powder coating, and adhesive bonding. Medium-wave emitters emit a radiation spectrum with wavelengths ranging from 1.4 to 3.0 μm with a power density of up to 90 kW m^{-2}. Medium-wave infrared emitters are used for drying and curing of food products. Long-wave emitters emit a radiation spectrum with wavelengths of more than 3.0 μm and attain power densities of up to 40 kW m^{-2}.

Long-wave emitters create a stream of hot air (due to water vapor) that is useful for processes that demand a combination of convection and infrared heating. While most materials absorb long (far) infrared wavelengths uniformly, many materials selectively absorb short (near) infrared energy in bands. In process heating applications, this selective absorption could be very critical to uniform and effective heating. New generations of carbon twin infrared emitters with a unique combination of a carbon infrared emitter and a short-wave infrared emitter in twin quartz glass tubes have been developed. Such combinations allow applications that require both short- and medium-wave infrared radiation and greater depths of penetration, for example, simultaneous cooking and browning of foodstuffs. Carbon twin emitters have a maximum power density that is twice that of conventional medium-wave carbon emitters.

Industrially, common infrared emitters are of the electric and gas-fired types. The electric type of infrared emitters have a metal filament placed inside a sealed enclosure filled with either inert gas or evacuated; energy is generated by passing an electric current through a high-resistance wire (nichrome wire, iron–chromium wire, or tungsten filament). The element and the material surrounding the element in the electric emitters are heated to an incandescent temperature; these emitters could be operated up to 2600 K. The spectrum of wavelength and emissive powers of electric infrared emitters could be varied by moderating the output power. Examples of electric infrared radiant emitters include reflector-type incandescent lamp, quartz tube, and resistance elements such as a metallic tube, ceramic tube, and non-metallic rod. The incandescent lamps are classified, based on wavelength and temperature, as short-wave emitters whereas quartz tubes and resistance elements are classified as medium- and long-wave emitters, respectively. Ready availability, ease of control, instrumentation, and possibility of a wide selection of heating elements make the electric infrared emitters very attractive. Electric infrared emitters generate zero emissions. However, electric-powered infrared heat generation is expensive in operation owing to the electricity cost.

The direct flame infrared radiator, ceramic burner, metal fiber burner, high-intensity porous burner, and catalytic burner are some of the examples of gas-fired emitters. Natural gas or liquefied gas (propane) can be used for heating a perforated ceramic surface (luminous heaters) or steel tube (tube heater) that emit radiation. In the gas-fired infrared emitters, combustion of air and the fuel stream takes place on the burner surface, which raises the surface temperature to about 1000–1200 K and emits radiation in the wavelength range 1.6–10 μm (Figure 10.3). New versions of gas-fired shortwave infrared emitters are capable of attaining surface temperatures of around 1600–1700 K. The gas-fired infrared emitters are independent of electricity, cheaper in terms of operating costs, capable of saving energy, and more reliable and durable than electric infrared emitters. However, the initial cost of gas-fired infrared emitters is higher than that of the electric infrared emitters. Operating gas-fired infrared emitters with flames requires thorough ventilation of the burning area to reduce hazards of evolved carbon monoxide.

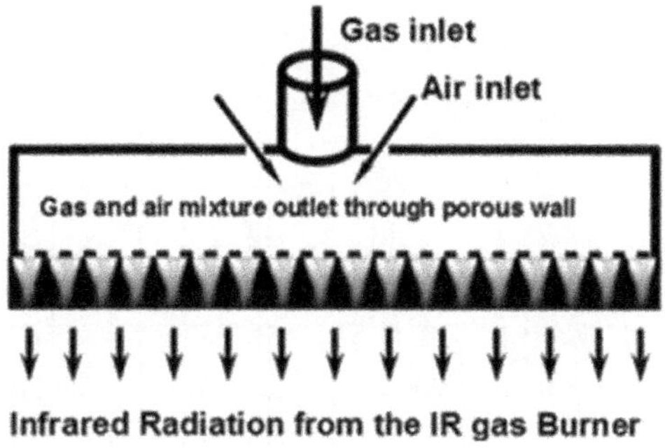

Figure 10.3 Illustration of a gas-fired infrared emitter in which combustion of air and the fuel stream on the burner surface raises the surface temperature to emit infrared radiation.
Republished with permission of Taylor & Francis from *Infrared Heating for Food and Agricultural Processing*, Pan, Z. and Atungulu, G. G., ISBN 9781420090970, 2010; permission conveyed through Copyright Clearance. Center, Inc.

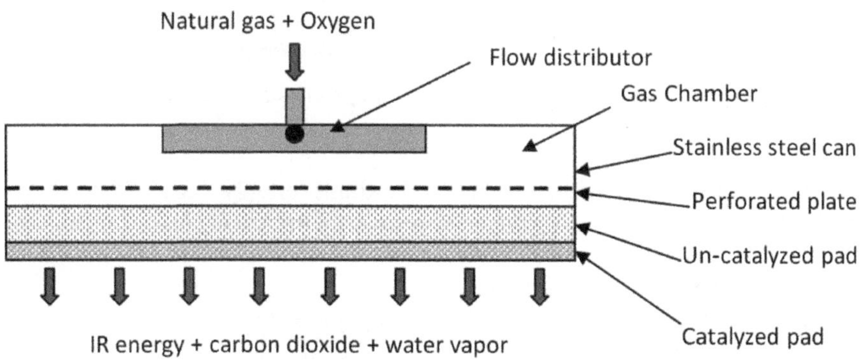

Figure 10.4 Schematic structure of a catalytic infrared emitter.

Gas-fired emitters with catalytic burners are flameless. This type releases infrared energy through exothermic catalytic chemical reactions of natural gas. The catalytic infrared emitter has gained acceptance for some applications owing to the safety concerns about open flames and emitters made of glass. The first catalytic emitters were built during World War I to prevent airplane fuel lines from freezing. Today, they are primarily used to prevent freezing of natural gas lines at junctions when the gas pressure drops.

10.1.3 Characteristics of Catalytic Infrared Emitters

In a catalytic infrared emitter, natural gas or propane gas combines with air across a platinum catalyst and reacts by oxidation–reduction to yield a controlled bandwidth of infrared energy, small amounts of carbon dioxide, and water vapor (Figure 10.4). The radiation emitted from the catalytic

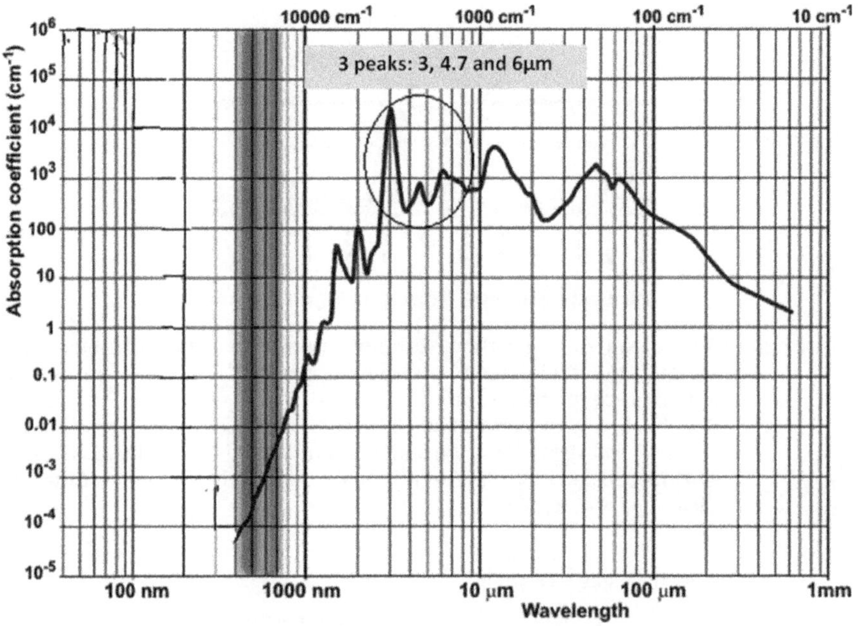

Figure 10.5 Absorption spectrum of liquid water in the infrared waveband.

infrared emitter has the bulk of the radiant energy bandwidth in the medium- and far-infrared range, with wavelengths of 3–6 μm. In this wavelength range, liquid water shows three key absorption peaks of infrared energy at 3, 4.7, and 6 μm (Figure 10.5).

The catalytic infrared emitter has many advantages over conventional gas-fired emitters or other gas-fired dryers. Since the oxidation–reduction reaction of natural/propane gas in the catalytic infrared emitter occurs at a temperature below the ignition temperature of the gas, no flame is produced. This is an important feature and a desired safety factor for food manufacturers. A recent concern for food processors is that they are facing pressure from and regulations of environmental protection agencies to limit polluted air emissions from gas-fired ovens. The catalytic infrared emitter generates virtually no harmful chemicals, such as NO_x or CO, and is environmentally friendly. Switching from regular open-flame gas heaters to catalytic infrared emitters may reduce polluted air emissions.

10.2 Interaction of Infrared Radiation with Foods

A comprehensive review of the radiative properties of food and interactions of infrared radiation with food was presented in our book on infrared heating for food processing.[2]

10.2.1 Effect of Water Content, Product Thickness, and Physicochemical Nature

The radiation absorption of foodstuffs depends mainly on the water content, thickness, and physicochemical nature of the product. During infrared heating, radiation energy is absorbed in foods by organic materials at discrete frequencies corresponding to intramolecular transitions between energy levels. The mechanisms of energy absorption are associated with the wavelength range of the incident radiation energy:[3,4]

1. Changes in the electronic state correspond to wavelengths in the range 0.2–0.7 μm (UV and visible radiation).
2. Changes in vibration state correspond to wavelengths in the range 2.5–100 μm (medium- and far-infrared radiation).
3. Changes in the rotational state correspond to wavelengths above 100 μm.

Foodstuffs absorb medium- to far-infrared energy most efficiently through stretching vibrations, which leads to the radiative heating process. Approximate values for the strong absorption bands of major food constituents are proteins at 3–4 and 6–9 μm, lipids at 3–4, 6, and 9–10 μm, and sugars at 3 and 7–10 μm. The four principal absorption bands of liquid water are at 3, 4.7, 6, and 15.3 μm.[5] Superimposing the infrared absorption bands of the principal food constituents on that of liquid water shows that the absorption spectra of food components overlap with one another.[2] Since the strong absorption bands of major food constituents are close to the water absorption peaks, it remains a challenge for practical applications to use differential or selective heating efficiently for targeting water without heating other components in the food material. It has been concluded that most foodstuffs show high transmissivity at wavelengths <2.5 μm.[2] Studies also showed that the transmissivity of foodstuffs increases abruptly when their water content is lowered in the near-infrared range while fresh and dry apple showed similar spectral absorptivity at wavelengths above 3 μm.[2]

Ginzberg[6] also showed that as the thickness of a foodstuff increases, a simultaneous decrease in transmittance and increase in absorptivity occur. For different applications, the optimal thickness of a foodstuff and the selected radiation wavelength could be different based on varying transmissivity and absorptivity. Thin products could be preferred for processes using infrared energy since the high transmissivity would result in higher heating rates. On the other hand, near-infrared radiation has advantages over medium- and far-infrared radiation owing to its superior transmissivity. However, the temperature of near-infrared radiation could be too high for processing food and agricultural products to maintain high quality of the products. In addition, the decrease in absorptivity and increase in transmissivity of near-infrared radiation during drying could also be a problem for thin materials. As materials are dried, the shrinkage of thin materials can result in low absorption of near-infrared energy since most of the radiation energy can be reflected and transmitted through the thin layer.

As the food material exposed to electromagnetic radiation undergoes changes in temperature, moisture content, structures, and other properties, the main characteristics of radiation also change.[7,8] In general, the changes in the radiative properties are influenced by various transformations in structure and food chemistry during processing. For instance, heating potato with microwaves or boiling in water sets in a starch gelatinization process which is likely to influence radiative properties such as reflectance. Similarly, such phenomena can be noted, for instance, in crumb and crust portions during bread making, which are due to structure and surface property differences resulting from the two states of gluten and the starch matrix.

10.2.2 Selective Infrared Heating and Absorption of Foods

Limited research has been focused on radiative property control for the selective heating of food materials. An excellent report by Jun and Irudayaraj[9,10] described the design and evaluation of infrared heating systems for selective heating. Subsidiary information may also be found in Chapter 1 of our book on infrared heating for food processing.[2] Some studies that have emphasized the significance of considering spectral selectivity for infrared heating include research by Lentz *et al.*,[11] which noted the importance of infrared-emitting wavelengths in the thermal processing of dough to avoid crust formation. Excessive heating of the dough surface and poor heating of the interior were observed when the infrared spectral emission was not consistent with the wavelengths absorbed best for dough. A study by Shuman and Staley[12] clearly demonstrated the importance of spectral control of the infrared source to manipulate the delivery of the correct amounts heat to specific food materials. They indicated that orange juice has a minimum absorption at the range 3–4 μm, whereas dried orange solids have a maximum absorption in the same region, hence the infrared source was controlled to emit the spectral range between 5 and 7 μm to obtain the desirable absorption of orange juice. Other studies of similar interest include those by Bolshakov *et al.*[13] and Dagerskog.[14] Overall, there still seems to be a lack of consistent methods to explore the intrinsic selective heating process in the area of food engineering. Dagerskog and Osterstrom[15] first used a bandpass filter (Optical Coating Laboratory, type L-01436-7) in their experiments on frying of pork to transmit only wavelengths above 1.5 μm, which turned out to be a good example for the design of selective infrared heating systems to emit the spectral regions of interest.

Some advances have been made Osterstrom to output directly substantial quantities of infrared radiation at selected wavelengths for the purpose of replacing broadband-type devices with narrowband irradiation sources.

Advances in semiconductor processing technology have resulted in the availability of direct electron-to-photon solid-state emitters that operate in the general mid-infrared range above 1 μm (1000 nm). These solid-state devices operate analogously to common light-emitting diodes (LEDs), except that they do not emit visible light but emit thermal infrared energy at longer mid-infrared wavelengths. In one form, they are an entirely new class of devices that utilize quantum dot technology. To distinguish this new class of devices from the conventional shorter wavelength devices (LEDs), they are more appropriately described as radiance or radiation-emitting diodes (REDs). The devices have the property of emitting radiant electromagnetic energy in a tightly limited wavelength range. Furthermore, through proper semiconductor processing operations, REDs can be tuned to emit at specific wavelengths that are most advantageous to a particular radiant treatment application. REDs may take a variety of forms, including diode forms and laser diode forms, and, in some cases, laser forms. In addition, innovations in RED technology have evolved related to the formation of a doped planar region in contact with an oppositely doped region formed as a randomly distributed array of small areas of material or quantum dots for generating photons in the targeted infrared range and potentially beyond. This fabrication technique and others such as the development of novel semiconductor compounds, if adequately applied, would yield suitable pseudo-monochromatic, solid-state mid-infrared emitters. RED devices and the technology with which to make them are subjects of a patent application.[16]

10.2.3 Radiation Model

It is crucial to understand changes in optical properties when a food material is subjected to thermal radiation. Jun and Irudayaraj[9,10] developed a novel selective infrared heating system and demonstrated the importance of optical properties in addition to thermal properties when infrared radiation is used for processing (Figure 10.6). Our contributions, including the approaches taken to simulate and validate a selective infrared heating model, are extensively reviewed in Chapter 1 of our book.[2]

The generalized approach of modeling infrared heating for agricultural products was described by Pan and Atungulu.[2] A transient simulation model was derived for the strongly coupled solution of three-dimensional conduction, convection, and radiation that combines a Monte Carlo infrared model with convention–diffusion airflow and heat transfer models to solve the complex heat transfer phenomenon. Generally, numerical methods are applied to solve sets of governing equations with the appropriate initial and boundary conditions. The numerical methods include finite elements, finite difference, and finite volume or control volume methods. Time discretization can be accomplished by using first- or second-order implicit methods. Computational fluid dynamics (CFD) commercial software, which includes the Monte Carlo ray-tracing technique[17] for estimating infrared

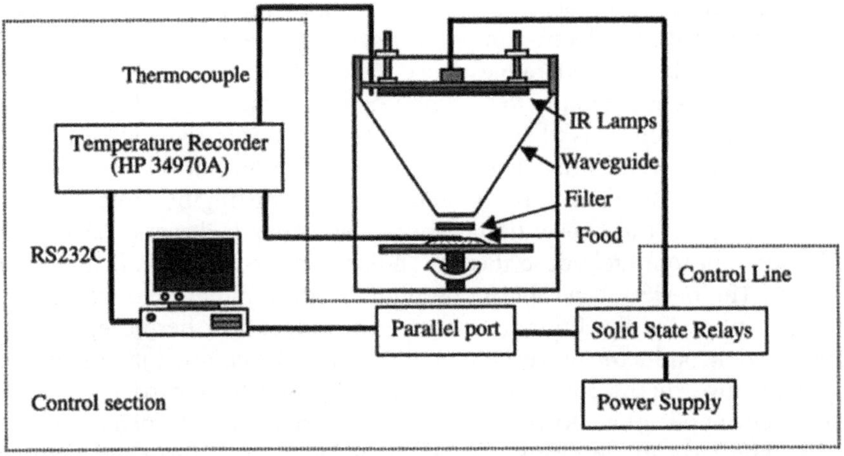

Figure 10.6 Schematic diagram of the far-infrared heating system with control part. Republished with permission of Taylor & Francis from *Infrared Heating for Food and Agricultural Processing*, Pan, Z. and Atungulu, G. G., ISBN 9781420090970, 2010; permission conveyed through Copyright Clearance. Center, Inc.

heat transfer, is used in the food processing field. Some software can solve a time-dependent scalar anisotropic conduction–convection–radiation problem with the mass transfer phenomenon.

10.3 Engineering Continuous Drying and Decontamination of Grains Using Infrared Heating

10.3.1 Description of a Newly Built Pilot-Scale Catalytic Infrared System

Innovative research at the Food Science Department, University of Arkansas System Division of Agriculture, has led to design and construction of a scaled-up infrared heating unit with modular processing parameters for food and feed processing (Figure 10.7). The newly built pilot-scale catalytic infrared system consists of a belt drive, heating chamber, gas/fuel supply lines, and control panel as shown in Figure 10.7a. The belt drive conveys the grains from the hopper through the heating chamber to discharge at the outlet. The belt has dimensions of $0.5\times0.2\times1.2$ m (Figure 10.7b) and is made of white butyl, designed for extremely high temperatures of up to 150 °C with not more than a 1% shrinkage in length. The movement of the belt is controlled by a variable-speed controller powered by an adjustable-frequency AC drive. The belt drive speed can be varied between 0.03 and 0.11 m s^{-1}. The belt assembly is equipped with an adjustable-frequency drive to vibrate the belt in case that is desirable. The vibration intensity can be varied using a

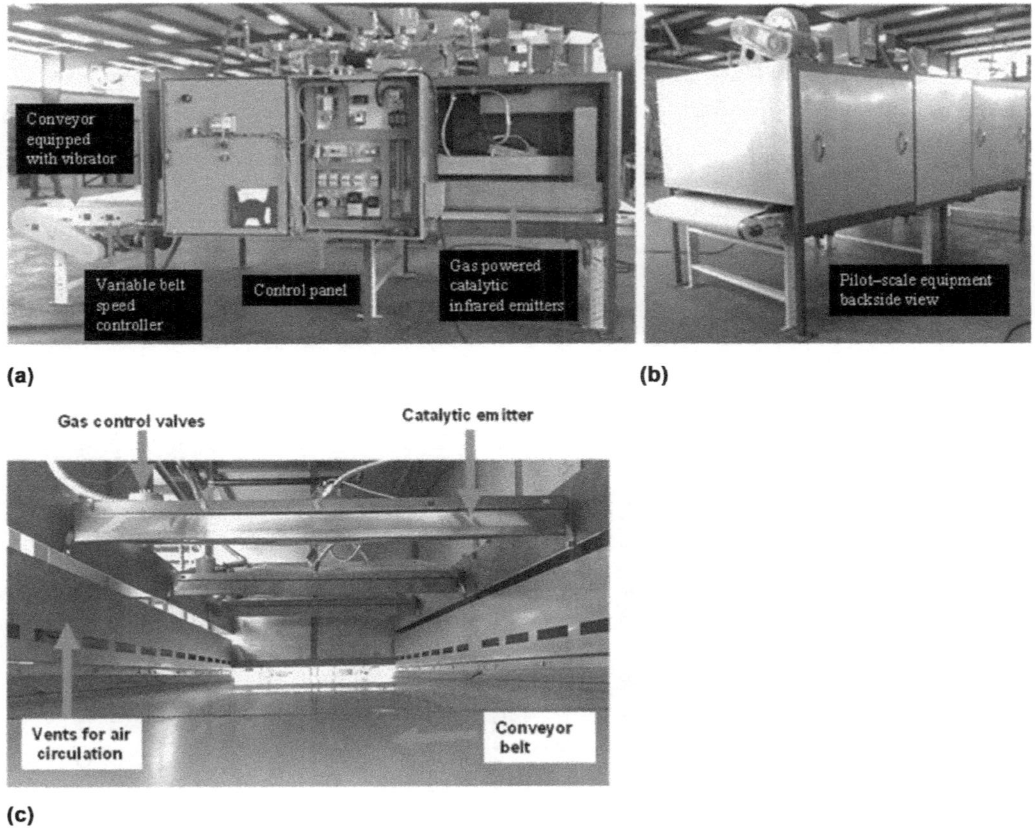

Figure 10.7 (a) Pilot-scale equipment designed and built with modular characteristics to achieve infrared drying and decontamination processes for grains and other food materials; (b) rear side view of the pilot-scale equipment; (c) arrangement of infrared heating emitters and air circulation vents inside the heating zone.

vibrator controller between 5 and 2000 Hz (Figure 10.7a). The arrangement of emitters in the heating zone is shown in Figure 10.7c: the dimensions are 0.08×0.13×0.2 m and houses four catalytic infrared heaters with dimensions 0.6×1.2 m, inclined at 30° and located above the belt at a predetermined height that is designated the product-to-emitter gap size. Each of the catalytic infrared emitters has a heat generation capacity of 57 600 BTUH; the total system energy supply is 230 400 BTUH. The catalytic infrared emitters generate medium- and far-infrared radiant energy with wavelengths from 3.3 to 8 μm by catalyzing oxidation–reduction reactions of natural gas or propane gas with a platinum catalyst. The heating chamber is enclosed using infrared heat insulators.

10.3.2 Case Study of Shelled Corn Drying and Decontamination

In the United States, corn is harvested with moisture contents ranging from 18 to 24% wet basis (w.b.); occasionally, under poor weather conditions, the moisture content could be higher. The freshly-harvested corn must be dried within a short duration to a moisture content of about 12–14% (w.b.) in order to prevent excessive mold proliferation and preserve the overall corn quality. Conventional practices use natural air or heated air for corn drying. Unfortunately, these conventional drying methods are not metered to ensure inactivation of all microbial spores that may contaminate the corn kernels. As a result, corn dried using the conventional methods may contain high levels of heat-resistant types of molds, some of which are toxigenic. Surviving spores of the highly heat-resistant molds may not germinate on corn at moisture contents of 12–14%. However, adding contaminated corn to high moisture-containing foods provides a suitable environment for the mold spores to grow. Under suitable conditions, toxigenic types of mold spores may produce secondary metabolites known as mycotoxins. Some mycotoxins, such as aflatoxins, are known carcinogens and pose health hazards to consumers.

In recent years, our group[18] has undertaken fundamental and applied research to demonstrate the feasibility of infrared heating to improve the drying rate of corn while simultaneously decontaminating mycotoxin-producing fungi that could infest the grain. These studies refined infrared processing parameters, including heating intensity, heating duration, and product-to-emitter gap size, *versus* corn initial moisture content to reduce the deterioration of the dried corn color and formation of stress cracks post-treatment. In this section, an effort is made to document the findings from our group. Practical highlights include the following:

- Scaled-up infrared treatment of corn with efficacy to reduce drying times significantly.
- Capability of using intermittent infrared heating to combine corn drying and decontamination in a single step while maintaining the quality of the finished product.

- Potential of infrared heating to inactivate heat-resistant, mycotoxin-producing mold spores that are liable to survive conventional drying methods.

Freshly harvested corn with a moisture content of 24% (w.b.) was intermittently dried with infrared radiation at intensities of 2.39, 3.78, and 5.55 kW m^{-2} for 30, 50 and 180 s and at a product-to-emitter gap size of 450 mm to achieve a safe storage moisture content of 13% (w.b). Microbial load reduction, energy use, and corn sensory and pasting qualities were evaluated. Infrared treatment of corn significantly reduced the drying time (9 to 22 min, depending on the infrared intensity and duration of intermittent infrared drying) to drive the freshly harvested corn moisture content from 23 to 13% (w.b.) (Figure 10.8). Drying the corn to a suitable storage moisture content with infrared intensities of 2.39 and 5.55 kW m^{-2} required 7.5 and 15.0 MJ kg^{-1} of water removed, respectively (Figure 10.9). Increasing the infrared intensity from 2.39 to 5.55 kW m^{-2} resulted in average microbial load reductions of 2.6 and 2.9 log cfu g^{-1} with 180 and 30 s of intermittent infrared heating, respectively (Figure 10.10). Infrared treatments reduced corn paste viscosity and stress cracks ($p<0.05$), but insignificantly impacted corn color ($p>0.05$). Hence accelerated, scaled-up infrared drying and decontamination were achieved without affecting the corn quality.

10.3.3 Case Study of Rice Drying

Head rice yield is used by the rice industry as a standard to assess commercial rice quality. Head rice yield is defined as the weight of rice that

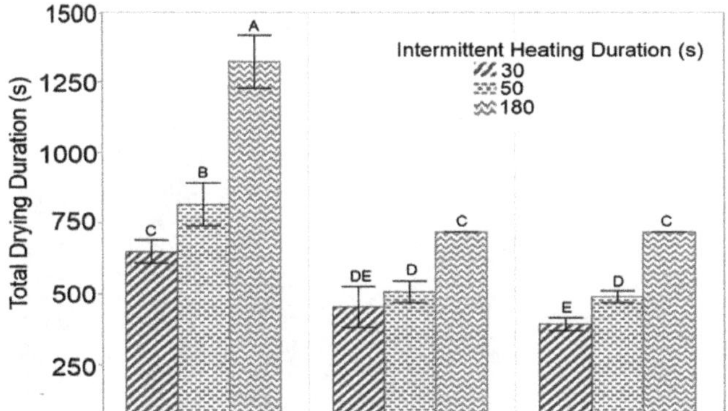

Figure 10.8 Effect of intermittent infrared heating of corn kernels and infrared intensity (low, medium, and high intensities, which correspond to 2.39, 3.78, and 5.55 kW m^{-2}) on total drying durations of corn from an initial moisture content of 24% to a final storage moisture content of 13% (w.b.). Means with the same capital letter are not significantly different at $\alpha=0.05$.

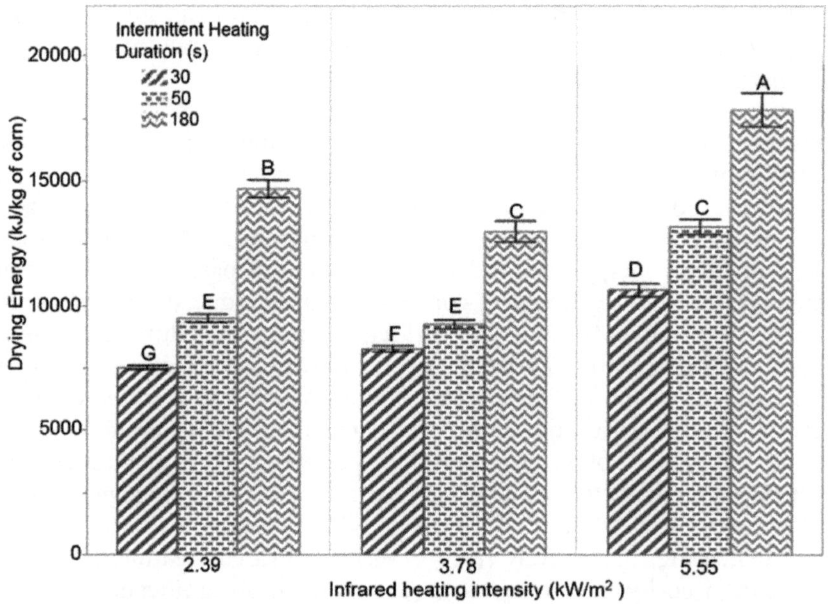

Figure 10.9 Effect of intermittent infrared heating of corn kernels and infrared intensity (low, medium and high intensities, which correspond to 2.39, 3.78, and 5.55 kW m^{-2}) on the energy required to remove unit mass of water from corn kernels that were dried from an initial moisture content of 24% to a final storage moisture content of 13% (w.b.). Means with the same capital letter are not significantly different at $\alpha = 0.05$.

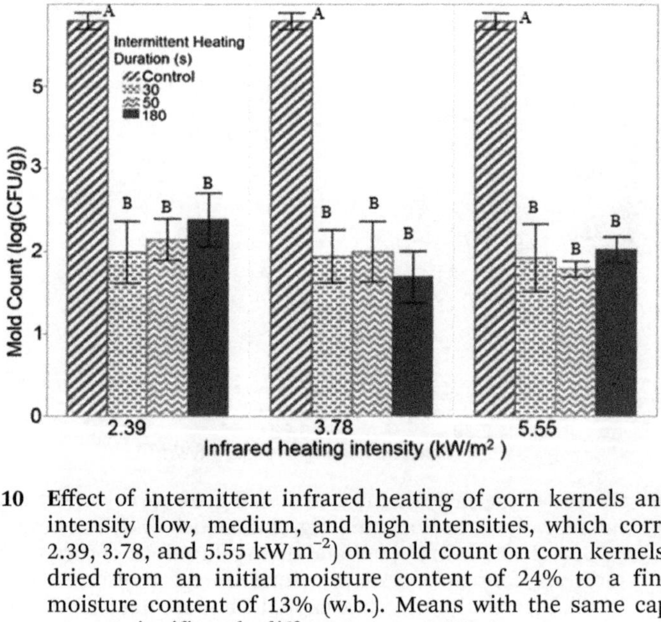

Figure 10.10 Effect of intermittent infrared heating of corn kernels and infrared intensity (low, medium, and high intensities, which correspond to 2.39, 3.78, and 5.55 kW m^{-2}) on mold count on corn kernels that were dried from an initial moisture content of 24% to a final storage moisture content of 13% (w.b.). Means with the same capital letter are not significantly different at $\alpha = 0.05$.

remains as head rice (kernels of at least three-quarters of the original kernel length) after complete milling.[19] Maximizing the proportion of head rice attained after milling is important to rice processors from a commercial point of view. The concept of using infrared drying of rice while maintaining the dried product quality has been studied using laboratory-based batch processing equipment.[20] The impact of radiation intensities (4685 and 5348 W m^{-2}), exposure time (15–120 s), and bed thickness (from a thin layer to 10 mm) for one-pass heating of rice was investigated and it was found that infrared treatments under the tested conditions showed no adverse effects on milled rice yields and quality.

The functionality of rice-based ingredients, especially pasting characteristics (viscosity profiles), are affected by the processing methods.[21,22] Rice contains primarily starch, which plays a major role in the production of thickeners, stabilizers, and gelling, water retention, and adhesive agents.[23] Rice drying temperatures and durations have been associated with decreased rice whiteness.[24] Longer drying durations and high initial moisture contents of the rice may accelerate the Maillard reaction, which may lead to discoloration.[25] In cases where high-temperature drying is employed to accelerate the drying process, tempering has been shown to help minimize kernel damage, especially rice fissuring, which leads to a decrease in head rice yield.[26] During tempering, dried rice kernels are held in non-drying conditions in order to allow the transfer of moisture from the center to the surface of the grain before another cycle of heating;[27] this eliminates the moisture gradient inside the grain imposed during the previous drying stage and reduces fissuring. Studies have shown that combining an infrared heating process with tempering stages may lead to the removal of significant amounts of moisture from a product while maintaining the dried product quality.[2,28] Intermittent heating and tempering stages, as practiced in conventional, multiple-pass driers, have been reported to have a positive influence on drying rates and the quality of the final product.[29,30] Moreover, it is also possible to use the sensible heat from the tempered product to remove more moisture in a natural cooling process after tempering.

Beyond the reported batch-type research on rice drying, there has been a need in the food industry to design, build, and test a scaled-up infrared drying system for continuous rice drying operation. In this section, research performed to assess the effectiveness of scaled-up, non-batch, infrared operations for the drying of rice is reported. Rice differing in cultivar types typically dry and mill differently,[31] hence infrared drying performance for multiple cultivars is reviewed. The infrared unit described earlier (Figure 10.6) was used to dry freshly harvested rough rice of long-grain pureline (cv. Cheniere), long-grain hybrid (cv. 6XP 756), and medium-grain (cv. CL 271) cultivars at initial moisture contents of 23, 23.5, and 24% w.b., respectively. Samples at a loading rate of 1.61 kg m^{-2} were heated with infrared radiation of intensity 5.55 kW m^{-2} for 30, 50, 90, and 180 s followed by tempering at 60 °C for 4 h, with a product-to-emitter gap size of 450 mm, in one- and two-pass drying operations. Control samples were gently dried using natural air in an equilibrium moisture content (EMC) chamber set at a relative humidity of 65% and a temperature of 26 °C to a moisture content of 12.5% (w.b.). The effects of infrared treatments followed by tempering on the percentage points of moisture removed, head rice yield, energy use, rice color, and pasting characteristics were evaluated. For all the cultivars, the percentage points of moisture removed increased with increase in infrared drying duration (Table 10.1).

For all rice cultivars, one-pass infrared treatment for 180 s resulted in a head rice yield that was significantly lower than that of rice dried using natural air in controlled environment conditions ($p<0.05$) (Table 10.2). The energy required to dry rice increased with increase in drying duration (Table 10.3). The viscosity values of all the experimental samples were significantly greater ($p<0.05$) than that of the control samples for all the cultivars except those treated with infrared radiation for 180 s. There was a significant difference ($p<0.05$) in the color index (ΔE) of treated milled samples and the controls.

The study provided information crucial to understanding the effects of scaled-up radiant heating and tempering of rough rice on drying rates and rice quality. The findings suggested that scaled-up radiant treatment of rice followed by tempering could be optimized to remove significant amounts of moisture from rice with a few passes while at the same time maintaining desirable head rice yields and other desirable quality indices such as rice color and paste viscosity. However, results from the treatments are dependent on the rice cultivar. The treatment conditions that gave the

Table 10.1 Tukey's HSD (honest significant difference) test for the effect of cultivar and infrared drying duration on the percentage points moisture removal from rough rice after heating for 30, 50, 90, and 180 s, followed by tempering at 60 °C for 4 h in one- and two-pass treatments.[a]

		One-pass treatment		Two-pass treatment	
Cultivar	IR heating duration/s	Mean moisture removal/% points	SE	Mean moisture removal/% points	SE
6XP 756	30	5.37^{F}	0.15	7.45^{C}	0.24
	50	7.03^{E}	0.14	10.65^{A}	0.16
	90	9.12^{D}	0.14	—	—
	180	14.02^{A}	0.17	—	—
Cheniere	30	4.24^{G}	0.10	8.67^{B}	0.12
	50	6.63^{E}	0.06	11.06^{A}	0.09
	90	10.98^{C}	0.08	—	—
	180	13.02^{B}	0.18	—	—
CL 271	30	5.07^{FG}	0.14	8.47^{B}	0.17
	50	7.56^{E}	0.31	10.48^{A}	0.15
	90	9.24^{D}	0.28	—	—
	180	13.76^{AB}	0.27	—	—

[a]Values are compared within columns. Values marked with the same superscript letter are not significantly different ($p>0.05$). SE, standard error.

Table 10.2 Tukey's HSD test for the effect of cultivar and infrared drying duration on the head rice yield (HRY) after heating for 30, 50, 90, and 180 s, followed by tempering at 60 °C for 4 h in one- and two-pass treatments.[a]

Cultivar	IR heating duration/s	One-pass treatment Mean HRY/%	SE	Two-pass treatment Mean HRY/%	SE
6XP 756	Control	64.4^{A}	0.61	64.4^{B}	0.61
	30	63.1^{BCD}	0.64	60.2^{C}	0.15
	50	61.7^{CDE}	0.36	59.2^{CD}	0.63
	90	60.1^{EFD}	0.08	—	—
	180	53.8^{B}	0.73	—	—
Cheniere	Control	63.6^{CB}	0.64	63.6^{B}	0.64
	30	61.8^{CDE}	0.72	56.6^{D}	0.49
	50	60.4^{EF}	0.3	58.4^{CD}	1.56
	90	60.5^{EF}	0.48	—	—
	180	47.1^{H}	0.73	—	—
CL 271	Control	69.9^{A}	0.14	69.9^{A}	0.14
	30	69.1^{A}	0.17	67.9^{A}	0.19
	50	68.8^{A}	0.09	66.5^{AB}	0.55
	90	67.8^{A}	0.19	—	—
	180	58.5^{F}	0.05	—	—

[a]Values are compared within columns. Values marked with the same superscript letter are not significantly different ($p > 0.05$). SE, standard error.

Table 10.3 Tukey's HSD test for the effect of cultivar and infrared drying duration on the energy required to remove unit mass of water from rough rice after heating for 30, 50, 90, and 180 s, followed by tempering at 60 °C for 4 h in one- and two-pass treatments.[a]

Cultivar	IR heating duration (s)	One-pass treatment Mean energy used/kJ kg^{-1} H_2O	SE	Two-pass treatment Mean energy used/kJ kg^{-1} H_2O	SE
6XP 756	30	1927.7^{H}	53.10	6478.7^{A}	184.22
	50	2454.9^{EF}	47.39	5564.2^{B}	100.9
	90	3404.8^{C}	54.03	—	—
	180	4426.2^{B}	52.71	—	—
Cheniere	30	2240.5^{FGH}	44.65	4775.3^{C}	69.09
	50	2599.1^{E}	24.94	6236.7^{A}	52.54
	90	2981.0^{D}	54.38	—	—
	180	4769.4^{A}	67.13	—	—
CL 271	30	2042.4^{GH}	55.76	4890.1^{C}	98.81
	50	2287.4^{EFG}	95.01	6579.5^{A}	91.78
	90	3361.8^{C}	98.18	—	—
	180	4510.9^{AB}	88.82	—	—

[a]Values are compared within columns. Values marked with the same superscript letter are not significantly different ($p > 0.05$). SE, standard error.

highest percentage points moisture removal, maintained head rice yield, and resulted in low energy usage were as follows:

- 6XP 756 cultivar infrared heated for 90 s followed by tempering at 60 °C for 4 h in a one-pass treatment. This treatment resulted in a head rice yield of 60.1% and energy use of 3404.8 kJ kg^{-1} H_2O.

- CL 271 cultivar infrared heated for 90 s followed by tempering at 60 °C for 4 h in a one-pass treatment. This treatment resulted in a head rice yield of 67.8% and energy use of 3361.8 kJ kg^{-1} H_2O.
- Cheniere cultivar infrared heated for 90 s followed by tempering at 60 °C for 4 h in a one-pass treatment. This treatment resulted in a head rice yield of 60.5% and energy use of 2981.0 kJ kg^{-1} H_2O.

10.4 The Future of Infrared Heat Treatments

Commercialization of any new technology often involves many unknowns. Although infrared heating technology has demonstrated many advantages, the major obstacles to is adoption in the agricultural and food industries are the lack of understanding of the fundamentals of the technology and subsequent extension past the conventional manufacturing industries such as the textile and automotive sectors. Research on the control and optimization of infrared drying processes, and also on the use of different infrared heating systems for different purposes, is still fairly limited. Optimization of infrared heating technology with respect to processed product sensory and functional characteristics, quality changes, energy savings, and process costs could place a great spotlight on the technology. Industrial implementations of the infrared processes need careful design that integrate variable heat fluxes and wavelengths to avoid product overheating and overprocessing on the surface, as this could compromise the product quality. Already efforts in research fields to utilize the existing knowledge to improve equipment design are being made by researchers across the food processing fields (Table 10.4).

In this chapter, we have succinctly covered case studies of novel infrared applications relevant to rice and corn drying and decontamination. However, not discussed are a myriad of other infrared applications that are still at the laboratory or pilot scale. Researchers at the University of California Davis and the USDA Western Region Research Center in Albany, NY, have developed the following infrared heating processes:

1. Simultaneous infrared dry blanching and dehydration of fruits and vegetables.
2. Combined and sequential infrared freeze-drying of fruits.
3. Sequential infrared and hot-air roasting of almonds.
4. Infrared pasteurization of raw almonds.
5. Simultaneous infrared disinfestation of insects from stored products.
6. Infrared heating for tomato peeling.

All the foregoing new infrared-based technologies have the potential to have a positive influence on the food processing industry.

The future of the developed infrared heating technologies will depend on how the industry moves ahead in technology adoption and commercial

Table 10.4 Commercial food applications of infrared heating equipment. Republished with permission of Taylor & Francis, from *Infrared Heating for Food and Agricultural Processing*, Pan, Z. and Atungulu, G. G., ISBN: 9781420090970, 2010; permission conveyed through Copyright Clearance Center, Inc.

Equipment	Type of application	Foods processed
Baking/heating		
Gas oven	Batch	Bread
Tunnel oven	Continuous	Fish paste, oyster, egg, laver
Drum oven	Continuous	Coffee
Cooker	Batch	Sweet potato, chestnut
Drying		
Tunnel drier	Continuous	Vegetables, flaked dried bonito, semi-dried noodle
Drum drier	Batch	Green tea
Freeze drier	Continuous	Liquid foods
Pneumatic drier	Continuous	Particulate foods
Thawing		
Tunnel thawing equipment	Continuous	Frozen fish paste
Thawing device	Batch	Frozen sushi and other frozen foods, tuna
Pasteurization		
Conveyor pasteurizer	Continuous	Packed foods
Blanching		
Infrared dry blancher	Continuous and Batch	Strawberry carrots, *etc.*

equipment manufacture. Moving the technology from research and pilot-scale applications to the commercial level and educating relevant stakeholders in the merits of these new technologies over the conventional counterparts should be given great priority; such an approach will speed up the delivery of the novel infrared-based food processing applications, with their inherent technical advantages, to the marketplace.

Acknowledgements

The author is extremely grateful to students Shantae Wilson and Anne Okeyo, whose research contributions were invaluable to the research for and preparation of this chapter.

References

1. R. Siegel and J. R. Howell, *Thermal Radiation Heat Transfer*, Taylor and Francis, Philadelphia, 4th edn, 2001, pp. 419–429.

2. Z. Pan and G. G. Atungulu, *Infrared Heating for Food and Agricultural Processing*, CRC Press, 2010.
3. I. Rosenthal, Electromagnetic radiations in food science, *Adv. Ser. Agric. Sci.*, 1992, 19.
4. N. Sakai and T. Hanzawa, Applications and advances in far-infrared heating in Japan, *Trends Food Sci. Technol.*, 1994, **5**, 357–362.
5. C. Sandu, Infrared radiative drying in food engineering: a process analysis, *Biotechnol. Prog.*, 1986, **2**, 109–119.
6. A. S. Ginzberg, *Application of Infrared Radiation in Food Processing*, Leaonard Hill Books, London, 1969.
7. V. V. Ivanov, *Optics of Scattering Media*, Nauka i Tekhnika, Minsk, 1969, p. 592.
8. S. G. Ilyasov, Theoretical principles of infrared radiation of foodstuffs. *Summary of doctoral dissertation*. Moscow, 1977, p. 47.
9. S. Jun and J. Irudayaraj, Selective far infrared heating system - design and analysis (Part 1), *J. Drying Technol.*, 2003, **21**(1), 51–67.
10. S. Jun and J. Irudayaraj, Selective far infrared heating system – spectral manipulation (Part 2), *J. Drying Technol.*, 2003, **21**(1), 69–82.
11. R. R. Lentz, P. S. Pesheck, G. R. Anderson, J. DeMars, T. R. Peck *Method of Processing Food Utilizing Infra-Red Radiation. U. S. Pat.* 5382441; 1995.
12. A. C. Shuman and C. H. Staley, Drying by infra-red radiation, *Food Technol.*, 1950, **4**, 481–484.
13. A. S. Bolshakov, V. G. Borreskov, G. N. Kasulin, F. A. Rogov, U. P. Skryabin, N. N. Zhukov 22nd *European meeting of meat research workers*, Paper 15, (*Cited by Dagerskog, 1979*), 1976.
14. M. Dagerskog, Infra-red radiation for food processing II. Calculation of heat penetration during infra-red frying of meat products, *Lebensm.-Wiss. Technol.*, 1979, **12**(5), 252–257.
15. M. Dagerskog and L. Osterstrom, Infra-red radiation for food processing. A study of fundamental properties of infrared radiation, *Lebensm.-Wiss. Technol.*, 1979, **12**(4), 237–242.
16. S. Sinharoy, D. Wilt *Quantum Dot Semiconductor Device*. Application Ser. No. 60/628,330; 2004. Availble from: http://www.faqs.org/patents/app/20090102083.
17. F. Tanaka, P. Verboven, N. Scheerlinck, K. Morita, K. Iwasaki and B. M. Nicolaï, Investigation of far infrared radiation heating as an alternative technique for surface decontamination of strawberry, *J. Food Eng.*, 2007, **79**(2), 445–452.
18. S. Wilson, G. Atungulu, A. Couch and S. Sadaka, Radiant Heating and Tempering Treatments for Improving Rate of Moisture Removal during Drying of Shelled Corn, *Appl. Eng. Agric.*, 2015, **31**, 799–808.
19. J. Aquerreta, A. Iguaz, C. Arroqui and P. Virseda, Effect of high temperature intermittent drying and tempering on rough rice quality, *J. Food Eng.*, 2007, **80**, 611–618.
20. Z. Pan, R. Khir, K. Bett-Garber, E. Champagne, J. Thompson, A. Salim, B. Hartsough and S. Mohamed, Drying characteristics and quality of

rough rice under infrared radiation heating, *Trans. ASABE*, 2011, **54**, 203–210.
21. A. Gunaratne, S. Ranaweera and H. Corke, Thermal, pasting, and gelling properties of wheat and potato starches in the presence of sucrose, glucose, glycerol, and hydroxypropyl Î²-cyclodextrin, *Carbohydr. Polym.*, 2007, **70**, 112–122.
22. F. Zhu, Y.-Z. Cai, M. Sun and H. Corke, Effect of phytochemical extracts on the pasting, thermal, and gelling properties of wheat starch, *Food Chem.*, 2009, **112**, 919–923.
23. N. Singh, J. Singh, L. Kaur, N. S. Sodhi and B. S. Gill, Morphological, thermal and rheological properties of starches from different botanical sources, *Food Chem.*, 2003, **81**, 219–231.
24. P. Bunyawanichakul, G. Walker, J. Sargison and P. Doe, Modelling and simulation of paddy grain (rice) drying in a simple pneumatic dryer, *Biosyst. Eng.*, 2007, **96**, 335–344.
25. C. Inprasit and A. Noomhorm, Effect of drying air temperature and grain temperature of different types of dryer and operation on rice quality, *Drying Technol.*, 2001, **19**, 389–404.
26. A. Cnossen and T. Siebenmorgen, The glass transition temperature concept in rice drying and tempering: Effect on milling quality, *Trans. ASAE*, 2000, **43**, 1661–1668.
27. Y. Li, C. Cao, Q. Yu and Q. Zhong, Study on rough rice fissuring during intermittent drying, *Drying Technol.*, 1998, **17**, 1779–1793.
28. N. Meeso, A. Nathakaranakule, T. Madhiyanon and S. Soponronnarit, Influence of FIR irradiation on paddy moisture reduction and milling quality after fluidized bed drying, *J. Food Eng.*, 2004, **65**, 293–301.
29. S. Kowalski and A. Pawlowski, Intermittent drying: energy expenditure and product quality, *Chem. Eng. Technol.*, 2011, **34**, 1123–1129.
30. A. K. Thakur and A. Gupta, Water absorption characteristics of paddy, brown rice and husk during soaking, *J. Food Eng.*, 2006, **75**, 252–257.
31. S. Lanning and T. Siebenmorgen, Comparison of milling characteristics of hybrid and pureline rice cultivars, *Appl. Eng. Agric.*, 2011, **27**, 787–795.

CHAPTER 11

Membrane Separations

KOEN DEWETTINCK,*[a] THIEN TRUNG LE[b] AND VIET BAO NGUYEN[c]

[a] Laboratory of Food Technology and Engineering, Department of Food Safety and Food Quality, Faculty of BioScience Engineering, Ghent University, Coupure Links 653, 9000 Ghent, Belgium; [b] Department of Food Engineering, Faculty of Food Science and Technology, Nong Lam University, Quarter 6, Linh Trung Ward, Thu Duc District, Ho Chi Minh City, Vietnam; [c] Department of Chemical Engineering, Nong Lam University, Quarter 6, Linh Trung Ward, Thu Duc District, Ho Chi Minh City, Vietnam

*Email: koen.dewettinck@ugent.be

11.1 Introduction

The use of membrane technology as a processing and separation method is gaining importance, with wide applications not only in the food processing industry but also in other fields such as the biotechnological, chemical and pharmaceutical industries, and in water and wastewater treatment. In the food industry, membrane separations can be used either as novel technology for processing new ingredients and foods or as alternatives to conventional techniques. Pressure-driven membrane separations, namely microfiltration, ultrafiltration, nanofiltration and reverse osmosis, find their major applications in the concentration of liquid ingredients (as an alternative to thermal evaporation), removal of bacteria and yeasts (cold pasteurization and sterilization), rejection of minute particle impurities and separations based on particle size and/or charges.

Green Chemistry Series No. 53
Alternatives to Conventional Food Processing: 2nd Edition
Edited by Andrew Proctor

Published by the Royal Society of Chemistry, www.rsc.org

In many cases, membrane processes are more desirable than traditional technologies. For example, the use of cold pasteurization and sterilization with suitable membranes instead of high-temperature treatment for the removal of microorganisms is more economical in terms of energy consumption. Avoiding the use of additives and preservatives also projects a "green" image for the processed foods and the membrane processing method. Membrane concentration preserves the natural flavors of food products and the nutritional value of heat-sensitive components as it conducted at ambient temperature. The recovery of valuable components in diluted effluents and wastewater treatment applications are among the most useful and currently active aspects of membrane technology.

One major obstacle towards wider acceptance of membrane technology is the problem of concentration polarization and membrane fouling. This chapter describes these and discusses their different mechanisms and the conditions that trigger these phenomena. An extensive discussion on the state-of-the-art techniques for enhancing membrane process performance is also presented. This includes optimization of operating conditions, pre-treatment of feeds, modification of membrane modules, flow manipulation and application of external forces such as ultrasound and electric fields. Membrane cleaning, a very important issue in membrane technology, is also discussed. This chapter also presents a comparison of the technological and economic advantages and disadvantages of membrane processes with those of traditional food processing methods. The technology's potential applications in the food industry such as in the processing of milk, beer, wine, juice, sugar, soy products and other food products are very important and are included.

11.2 Types of Membrane Separation Processes

11.2.1 Pressure-driven Membrane Separations

Filtration is a technique that allows the separation of two or more components present in a fluid stream based primarily on size differences. This technique is conventionally used to separate solid immiscible particles from a liquid or gas. Membrane filtration can be defined as "finer" filtration that can be applied even to separate dissolved solutes in a liquid stream. Four pressure-driven processes, microfiltration (MF), ultrafiltration (UF), nanofiltration (NF) and reverse osmosis (RO), can be distinguished as shown in Figure 11.1. The products of interest can be either in the retentate side or in the permeate side or in both streams.

MF membrane pore sizes range from 20 to 0.1 μm.[2] MF is normally used to separate suspended particles and microorganisms from other soluble components in the feed. The transmembrane pressure (TMP), which is the driving force applied in MF, is rarely higher than 3 bar.

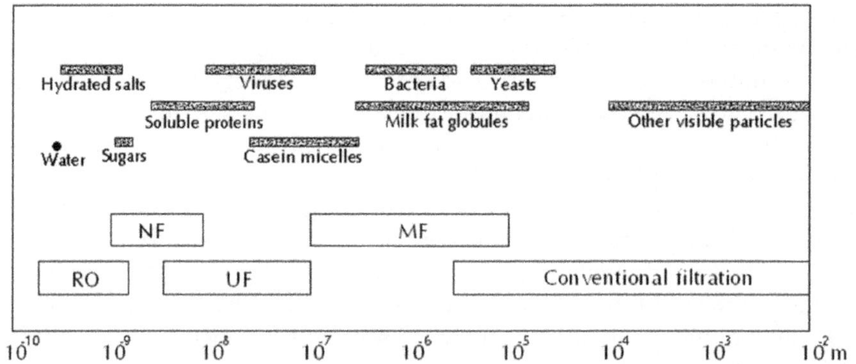

Figure 11.1 Approximate particle sizes and pressure-driven separation processes Republished with permission of Taylor and Francis Group, LLC, from Dairy Science and Technology, P. Walstra, J. T. M. Wouters and T. J. Geurts, 2nd edn,[1] © 2006; permission conveyed through Copyright Clearance Center, Inc.

In UF, the applied pressure ranges from about 3 to 7 bar.[3] The separation principle is based on the diameters of molecules and partially on their charges and affinity for the membrane.[4]

Salt rejection in NF membranes is controlled by molecular size and charge.[5,6] When negatively charged NF membranes are used, membrane rejection is greater for salts having anions of higher charge.[7] Normally, applied TMPs in NF processes are in the range 10–30 bar.[5,8–11]

The separation principle of RO, on the other hand, is based on the solubility of water and the very poor solubility of other components in the membrane.[1] Theoretically, RO membranes allow only water to be transmitted. This means that RO processes have to work against the chemical potential difference, namely osmotic pressure. This requires high operational TMPs of about 10–75 bar.[12–15] RO membranes, generally, reject ≥95% NaCl.

11.2.2 Other Types of Membrane Separation Processes

Dialysis is a concentration discrepancy-driven process. A solution is confined within a semipermeable membrane bag, which is placed in a large volume of a solvent. Owing to the gradient between the solution and the solvent, the molecules in the solution will attempt to migrate from the dialysis bag to the outside solvent and only species that are smaller than the membrane pore size can leave the solution. The process continues as long as there is a difference in concentration of the diffusible solutes between the two sides of the membrane. This technique allows the separation of smaller molecules from larger molecules or of dissolved substances from colloidal particles in a solution. Applications of dialysis are mostly found in laboratories rather than in industrial-scale facilities.

Electrodialysis (ED) utilizes selective ion-exchange membranes and an electric field as the driving force to separate charged species. ED membranes comprise monopolar membranes (cation-exchange and anion-exchange membranes) and bipolar-membranes. Applications in the food industry include demineralization of whey, skim milk and sugar, deacidification/acidification in the production of fruit juice, caseins and soy protein concentrate, inhibition of enzymatic browning in cloudy apple juice, separation of amino acids and proteins, tartaric stabilization in wine, *etc.*[16–22]

On the other hand, the driving force of *membrane osmotic distillation* (MOD) is the vapor pressure difference on the two sides of the membrane. The source of the driving force can be a temperature difference or an osmotic pressure difference on the two sides. MOD employs microporous hydrophobic membranes where water evaporates at the feed–membrane interface and migrates to the permeate side, where it is adsorbed by a colder solution or a solution with higher osmotic pressure, condenses in cold air or is swept away by a gas stream. The most promising food-related application of MOD is in the demineralization of water. MOD can also be used in combination with RO in the production of concentrated fruit juice[23,398–400] or low-alcohol beer/wine.[418–421]

Pressure-driven membrane processes, namely MF, UF, NF and RO, facilitate the separation of components with a large range of particle sizes. It is for this reason that they find a wide range of applications in the food processing industry both as alternative methods to conventional separation techniques and as innovative processing methods for the development of new foods and specialty ingredients with tailored functionalities. This chapter presents practical information concerning the pressure-driven membrane processes used in the food industry. It also describes the latest developments and innovations being made to improve the system so as to stimulate wider use of the technology.

11.3 Separation Characteristics

11.3.1 Filtration Modes

Membrane separation can be operated in *dead-end* or *crossflow* (CF) modes. The difference in fluid flow between these two modes is illustrated in Figure 11.2. In CF mode, which is also called tangential filtration, the feed is pumped over the surface of the membrane, resulting in two streams: permeate and retentate. The retentate can be recirculated back to the feed tank, where it is mixed with the fresh feed to offer some operational advantages. By tangential movement of the feed, CF filtration has a lower extent of concentration polarization and membrane fouling. On the other hand, the feed for dead-end filtration is pumped to the membrane. This mode of separation is mostly applied in laboratories for filtration of low solid-loading solutions or for the investigation of the membrane fouling characteristics of feeds. Stirrers are normally used in dead-end units to enhance the permeate flux.[24,25]

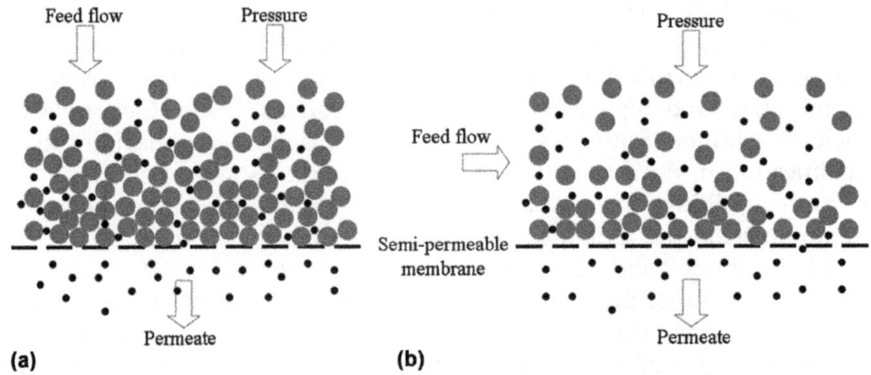

Figure 11.2 Illustration of (a) dead-end and (b) crossflow filtration.

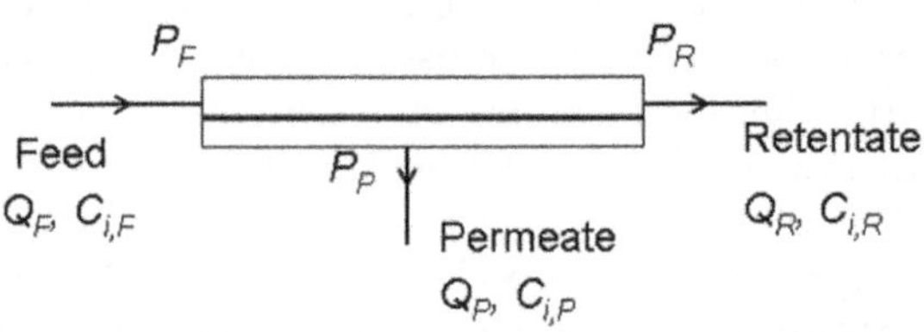

Figure 11.3 Schematic diagram of a crossflow membrane separation unit with key parameters.

11.3.2 Membrane Separation Parameters

To enable readers to follow this chapter easily, several common terms that are used in membrane filtration are summarized in this section. A schematic diagram of a membrane separation unit with key parameters is illustrated in Figure 11.3.

Feed is the fluid introduced to the system to be filtered or fractionated. *Permeate* is that portion of the feed that has passed through the filter. *Retentate* is the stream of components that is retained by the membrane.

The *permeate flux*, J, is the rate of withdrawal of permeate and is determined using eqn (11.1), where A_e is effective surface area of the membrane. J has dimensions of volume/area×time, *e.g.* $\mathrm{L\,m^{-2}\,h^{-1}}$ or $\mathrm{m^3\,m^{-2}\,s^{-1}}$ in SI units. A typical permeate flux profile during CF filtration of a constant-concentration feed (*e.g.* both permeate and retentate are completely circulated back to the feed tank) is illustrated later in Figure 11.5.

$$J = \frac{Q_P}{A_e} \tag{11.1}$$

The *volume concentration ratio* (VCR) is the degree of reduction of the feed volume achieved by concentration and is determined as the ratio of feed volume at the beginning of operation ($V_{F,i}$) to the retentate volume at time t ($V_{R,t}$):

$$\mathrm{VCR} = \frac{V_{F,i}}{V_{R,t}} = \frac{V_{F,i}}{V_{F,i} - V_{P,t}} \tag{11.2}$$

where $V_{P,t}$ is permeate volume at time t.

The *transmembrane pressure* (TMP) is the pressure discrepancy between the retentate and the permeate sides of the membrane at a particular point along the membrane. In CF membrane filtration, the decrease of TMP is linear along the membrane so an average TMP can be calculated as in eqn (11.3):

$$\mathrm{TMP} = \frac{P_F + P_R}{2} - P_P \tag{11.3}$$

where P_F, P_R and P_P are pressures at the feed inlet, retentate outlet and permeate outlet, respectively.

The *transmission* of component i, Tr_i, can be calculated as follows:[26]

$$\mathrm{Tr}_i(\%) = \frac{C_{i,P}}{C_{i,R}} \times 100 \tag{11.4}$$

where $C_{i,R}$ and $C_{i,P}$ are the concentration of component i in the retentate and at the outlet of the membrane pores, respectively. On the other hand, the *apparent rejection*, which represents the *separation efficiency*, of a compound can be determined by means of the following expression:

$$R_i(\%) = \left(1 - \frac{c_{i,P}}{c_{i,R}}\right) \times 100 \tag{11.5}$$

The yield recovery of component i, Y_i, is then simply determined as

$$Y_i(\%) = \frac{V_R c_{i,R}}{V_F c_{i,F}} \times 100 \tag{11.6}$$

and the *selectivity* of the membrane between two components i and j is

$$\phi = \frac{Tr_i}{Tr_j} \tag{11.7}$$

11.4 Concentration Polarization and Membrane Fouling

11.4.1 Concentration Polarization

During the separation process, the withdrawal of permeated solutes at the permeate side of the membrane cause a build-up of retained solutes at the

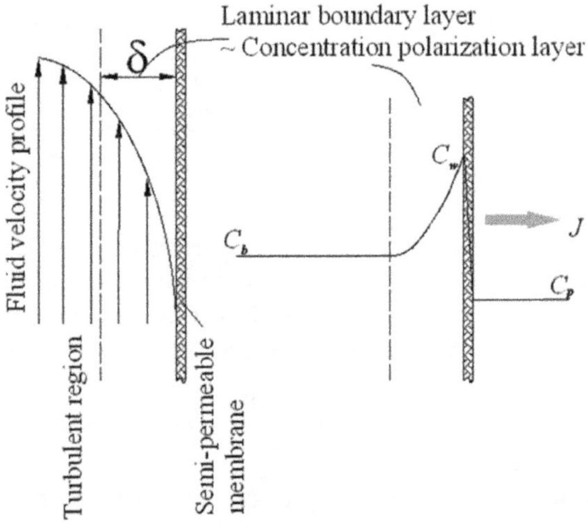

Figure 11.4 Fluid flow velocity profile (left) and schematic representation of the concentration gradient caused by concentration polarization (right). J is the permeate flux and C_b, C_w and C_p are the concentrations of a rejected solute in the bulk feed solution, at the membrane surface and in the permeate, respectively.

surface of the membrane. This creates a difference in concentration of retained solutes in the bulk phase and at the adjacent side of the membrane. Such a phenomenon is called concentration polarization (CP) (Figure 11.4). Figure 11.4 also shows a profile of fluid velocity in the feed channel of a membrane module. An average CFV is normally taken as one of the operating parameters during a separation process. However, the velocity profile of a feed in its channel is not uniform. Owing to friction, the fluid velocity is reduced to essentially zero at the membrane–fluid interface and it increases with increasing distance from the membrane surface to a maximum at the center of the channel. There exists a layer at the surface of the membrane where the flow is laminar. This layer is called the laminar boundary layer, δ. The mixing of the fluid in this layer is caused by diffusion. In contrast, the flow regime in the middle of the channel is often turbulent because of the much higher flow velocity and here the fluid is well mixed. In the boundary layer film model, the concentration gradient due to CP is assumed to be confined to the boundary layer.

11.4.2 Membrane Fouling

Membrane fouling is a process in which solutes or particles of the feed stream are deposited on the membrane surface, which is so-called *external fouling*, or in the membrane pores, which is called *internal fouling*, resulting in a decrease in membrane performance. Owing to membrane fouling, the flux declines dramatically and the selectivity of the membrane is modified (Figure 11.5). Both the internal and surface fouling materials act as

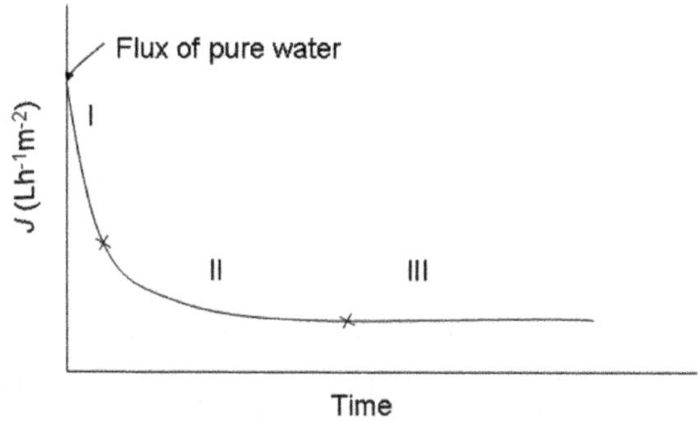

Figure 11.5 Typical flux profile during CF membrane filtration of a constant-concentration feed: (I) linear and fast decrease of flux from that of pure water, (II) gradual flux decrease and (III) time-independent steady-state permeate flux. The flux of the third stage is mainly controlled by foulants and gradually decreases during the concentration process.[27]

additional layers of resistance to the transmission of solutes through the membrane. CP and membrane fouling are two major obstacles for industrial applications of membrane separations. CP is part of reversible membrane fouling, which can be removed by physical methods such as applying a high shear force or backflushing. Severe irreversible membrane fouling may require intense chemical cleaning or even membrane replacement, and these requirements limit the economic efficiency of applications of membrane processes. The components causing membrane fouling are called *foulants.*

Three fouling mechanisms, namely adsorption, pore blocking (superficial pore blocking and deep pore clogging) and cake or gel formation, could be distinguished in an MF process.[28,29] These mechanisms are based on the differences in particle sizes of feed components and membrane pore size (Figure 11.6). Figure 11.7 represents a real fouled membrane after beer filtration.

Adsorption and adhesion of components such as proteins to membranes depends on their affinity.[30] Pore blocking is influenced by the morphology of the membrane, pore opening geometry and nature of the membrane. Fouling can develop very quickly when the process starts. The characteristic time for protein adsorption on solid surfaces can be as short as a few seconds.[3] Under some operating conditions, components that have particle sizes much smaller than the membrane pore size could aggregate (*e.g.* protein aggregation) or form complexes with other components (*e.g.* protein–polyphenolic compound complexation in MF of wine) within the pores or in the bulk phase, therefore leading to much faster blocking of the membrane pores than merely with adsorption.[31,32]

Proteins and polysaccharides (*e.g.* pectin in fruit juice) have a tendency to form a gel layer on the surface of UF, NF and RO membranes.[33] Feed properties such as pH and ionic strength are important factors influencing the thickness, porosity and compressibility of the deposited gel layer of proteins on the surface of UF membranes.[34–36] The main type of fouling of

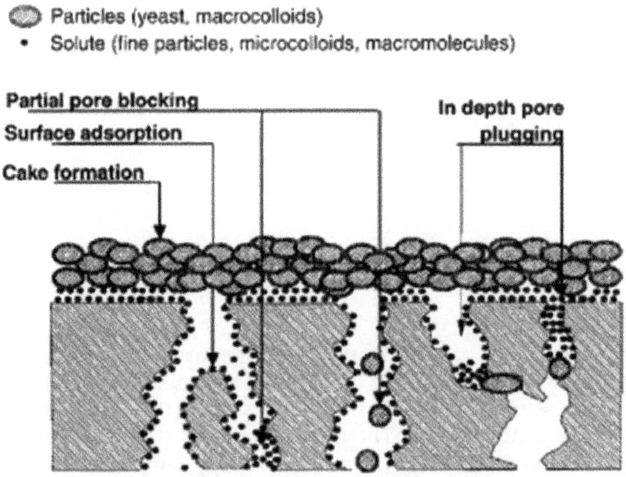

Figure 11.6 Fouling mechanisms in crossflow microfiltration.
Reprinted from *Food and Bioproducts Processing*, **76**, L. Fillaudeau and M. Lalande, A practical method to predict steady-state flux and fouling in the crossflow microfiltration of rough beer with 1.40 μm tubular ceramic membranes, 217–223,[29] Copyright (1998), with permission from Elsevier.

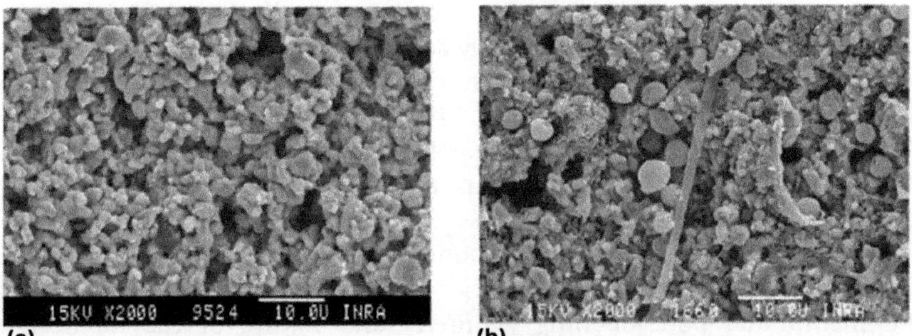

Figure 11.7 SEM images at ×2000 of (a) a clean and (b) a fouled 1.40 μm ceramic membrane after rough beer clarification.
Reprinted from *Journal of Membrane Science*, **196**, L. Fillaudeau and H. Carrère, Yeast cells, beer composition and mean pore diameter impacts on fouling and retention during cross-flow filtration of beer with ceramic membranes, 39–57,[42] Copyright (2002), with permission from Elsevier.

NF and RO membranes is surface fouling and both organic and inorganic materials are responsible. A cake layer formed from several components is more resistant to (synergistic effects in flux reduction) the transmission of solutes than the cake formed from one component.[25,37] The crystallization of salts at the membrane wall due to oversaturation causes scaling while deposition of salts that are already precipitated in the bulk phase causes caking.[38–40] Image analyzing video microscopy can be used to measure the fouling layer thickness.[41]

11.5 Membrane Characteristics and Membrane Modules

11.5.1 Membrane Characteristics

11.5.1.1 Membrane Materials

The membrane is a key component in membrane separation processes. The selection of suitable membranes and modules is a determinant in the success of their applications. Section 11.6.3 provides some basic tips regarding membrane selection.

11.5.1.1.1 Organic Polymeric Membranes. The organic polymers widely used for manufacture of membranes are cellulose acetate (CA), polyacrylonitrile (PAN), polysulfone (PS), polyethersulfone (PES), poly(vinylidene fluoride) (PVDF), polyamide (PA), polypropylene (PP) and polytetrafluoroethylene (PTFE). The first two are *hydrophilic* and the rest are, in nature, *hydrophobic* materials. CA is the "original" material for manufacturing membranes. They can be manufactured in a wide range of pore sizes and are therefore used in all four types of pressure-driven processes. Compared with other organic materials, CA membranes are normally cheaper. Moreover, owing to their hydrophilic nature, they are more resistant to fouling. However, CA membranes are not suitable for applications with harsh conditions such as extreme pH and temperature. CA also has poor resistance to chlorine, making the membranes difficult to clean for reuse.[43]

11.5.1.1.2 Ceramic Membranes. Ceramic membranes are normally fabricated in the form of tubes (see Figure 11.10). An anisotropic structure is formed inside the tubes from several slip-coated layers of progressively finer particles. Ceramic membranes cannot be manufactured with small pores, so currently they are commercially available only in MF and UF applications. Ceramic membranes are fairly hydrophilic and can tolerate a wide range of pH, cleaning and disinfecting chemicals, high temperatures and high operating pressures. Therefore, the life of ceramic membranes can be 2–10 years, whereas that of organic membranes, such as PS membranes, is only 1 year.[4] Ceramic membranes are, however, about an order of magnitude more expensive than polymeric membranes.[45] Ceramic membranes can withstand temperatures of >100 °C. However, it is advisable to change the temperature gradually (<10 C min^{-1}) to avoid cracking of the membrane.[46]

11.5.1.1.3 Graphene Oxide Membranes. Research on graphene oxide (GO) membranes has increased dramatically during the last decade. Unlike graphene, which is an expensive material, GO is fairly cheap and easy to produce on a large scale.[345] Stacked GO membranes allow water to flow freely while controlling the transport of other gas-based components through the channel width between the GO sheets and the pore size distribution.[346,347] The selectivity of the GO membrane towards electrolyte and non-electrolyte molecules in aqueous solution is closely related to their interaction with hydroxyl and carboxyl groups on the GO sheets.[348,349] The interlayer distance in these membranes is the result of competition between hydrogen bonds and electrostatic repulsion of GO lattices, which is strongly dependent on the environmental pH.[345] Coating GO membranes with reductants is an advanced method to improve their barrier properties and their stability in aqueous solution in addition to decreasing the interlayer space.[348] High mechanical strength, fast water permeation and excellent selectivity contributed to making GO membranes ideal materials for pressure-driven membrane separation.[349] Fabricating GO on an organic membrane[350,351] led to a higher permeation flux and better antifouling properties in comparison with commercial nanofiltration materials. Another advantage of GO membranes in food industry applications is their potential antimicrobial activity.[345,352–354] GO sheets can penetrate cell membranes[353] and increase oxidative stress on bacteria.[354]

11.5.1.2 Membrane Structure

Membranes can be classified into two broad groups: *symmetric* (isotropic) and *asymmetric* (anisotropic) types. Symmetric membranes have uniform composition and morphology. Asymmetric membranes, on the other hand, are composed of several layers of different morphology and permeability. An asymmetric membrane can be made from layers of a single material with different porosities and pore sizes or from layers of different materials. A typical organic membrane of this type is comprised of two layers, a support layer and a skin layer. The support or porous layer has high porosity, no selectivity and a thickness of approximately 50–200 μm. The skin has a thickness of approximately 0.1–2 μm and is responsible for membrane selectivity.[47] Since the latter is very thin, the flux is high. Most membranes are asymmetric in structure.

11.5.2 Membrane Modules

Membranes modules show how membranes are packed or arranged in a housing. In industrial applications, hundreds of square meters of membrane surface area are commonly needed. To minimize the footprint, the unit has to be configured in such a way that it is compact yet provides a large membrane area. The most common modules used for CF separations include flat-sheet, tubular, hollow-fiber and spiral-wound configurations, of which the hollow-fiber type provides the largest packed membrane area per unit volume.

11.5.2.1 Flat-sheet Membranes

Figure 11.8 shows a schematic diagram of a plate-and-frame unit without feed spacer. Multiple units of these can be stacked in a filter press. The feed moves tangentially along the membrane surface. A portion passes through the membrane, enters the permeate channel and is withdrawn by a collection tube. The plate-and-frame module is one of the earliest designs developed for small-scale applications and is suitable for feeds of high viscosity. It is easy to dismantle for cleaning and the membranes are easy to remove for storage or replacement.

Another design of this type of module is the multi-flat-sheet membrane module, which has a much higher packing capacity. In this model, flat sheets of membrane are stacked and sealed together, with or without a separator screen (spacer) between the layers, into a package called a cassette (Figure 11.9).

11.5.2.2 Tubular Membrane Modules

Tubular membranes (Figure 11.10) are designed to operate in CF mode. In these membranes, the feed is introduced at one end and then flows through

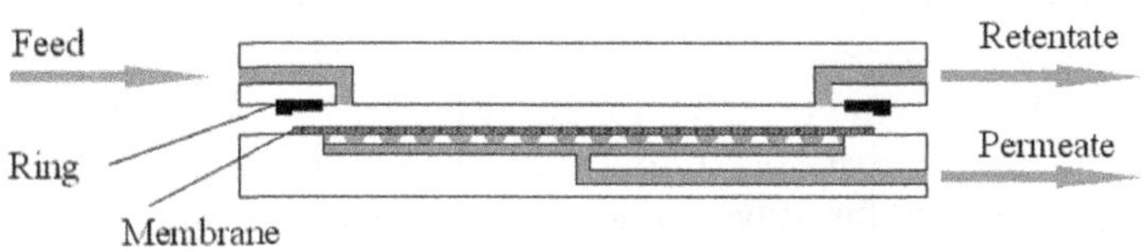

Figure 11.8 Schematic diagram of a plate-and-frame unit.

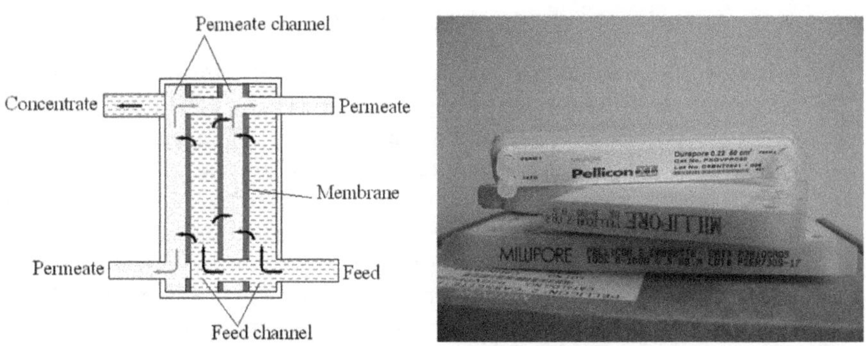

Figure 11.9 Millipore flat-sheet membrane cassettes that were cut for illustration. The advantage of the Millipore membrane cassette is the ability to scale it up linearly by adding other cassettes into the frame without using extra pipings and connections.

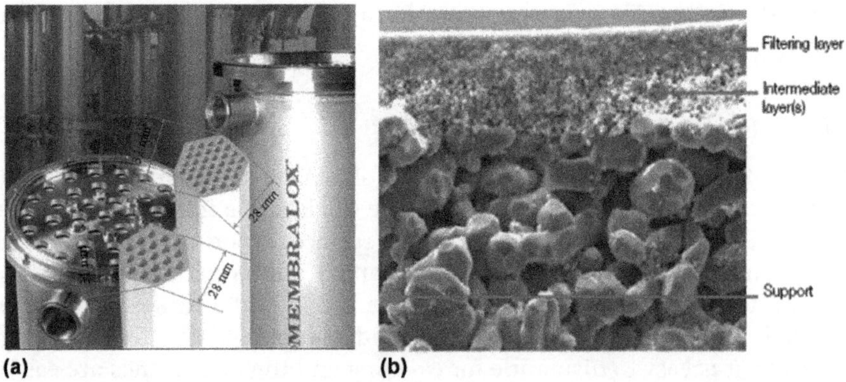

Figure 11.10 (a) Membralox® GP ceramic membranes and modules and (b) membrane cross-sectional view (×1010).
Image courtesy of Pall Corp., New York, NY, USA.

inside the support tube where the selective ceramic membrane layer is cast. Tubular membrane modules are suitable for feeds with a high suspended solids content, which could easily cause blockage of the feed channel in other modules. Some applications are in the clarification of juice and in fermentation broth and beer recovery. Feed prefiltration is not required in some cases. In addition, a tubular membrane allows the feed to be pumped with higher CFVs. This will help in minimizing the formation of CP and cake fouling, improve permeate flux and allow the retentate to be concentrated to a high level. Tubular membranes are also widely applied in MF of skim milk. The manufacturing costs of tubular and plate-and-frame modules seem to be the highest since they have the lowest membrane packing density for the same membrane area.

11.5.2.3 Spiral-wound Membrane Modules

In a spiral-wound membrane module, membrane-permeate spacer-membrane sandwiches are wrapped spirally around a central perforated tube that is designed to collect the permeate (Figure 11.11). The permeate moves spirally along its channel, which is inside the membrane sandwich, to the central tube, then exits through the header assembly. The retentate is pumped axially along the feed spacers from this to the other end of the module. Spiral-wound membrane modules are an option for low investment costs and operational costs (energy consumption) when a large membrane area is required.[48] The module accommodates plant space saving, since it allows the packing of large membrane area in a compact housing. Several of the modules can be connected and they are placed inside a tubular pressure vessel. Spiral-wound modules are the most common design applied in RO and UF.[49]

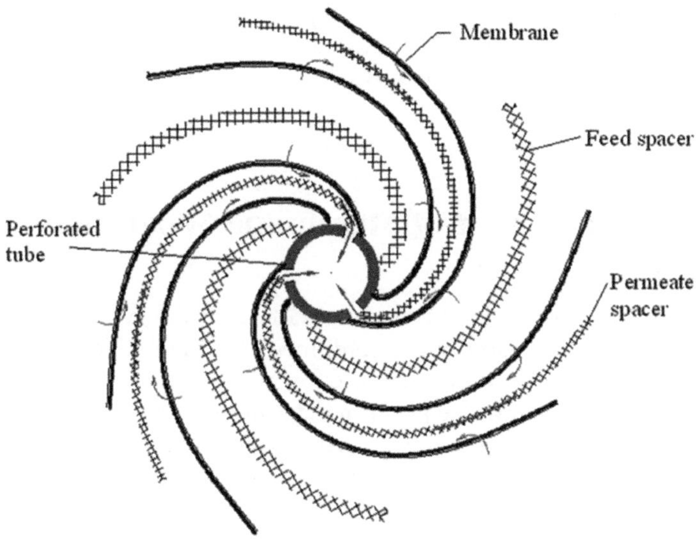

Figure 11.11 Schematic illustration of the cross-section of a multi-membrane spiral-wound module. Arrows show the permeate flow direction. The feed flows perpendicularly to the paper.

11.5.2.4 Hollow-fiber Membrane Modules

A hollow-fiber module consists of a bundle of small membrane tubes, with two possible configurations, namely outside-in and inside-out feeding fibers. In the former, the fibers are closed at one end or formed in a U shape where two ends are mounted on an end plate. The fibers are placed in a chamber into which the feed is introduced. The feed is forced from the outside of the fibers and permeate is collected from the open ends of the fibers. It is difficult to control CP and membrane fouling with outside-in feeding modules and the filtration mode is more like dead-end than CF filtration. This type of hollow fiber module is often applied in submerged filtration systems such as in wastewater treatment where the membrane lumens are submerged in the feed reservoir and permeate is pumped out at the open end of the fibers. Several techniques such as vibration or gas sparging are applied to improve the permeate flux.

The inside-out configuration is more commonly used for applications in the food industry. In this type, both ends of the tubes are open and molded on end plates, and these plates are fixed inside a tube. The feed is pumped into the fiber from one end and the retentate comes out at the other end. The filtrate permeates out of the porous tubes and is collected. The diameters of the hollow fibers range from 0.1 to 2.0 mm. The lower end of the range is for fibers engineered for outside-in type of feeding of the module where very high hydrostatic pressure can be applied, such as in RO and high-pressure gas separation. Fibers of 1.0–2.0 mm diameter are called capillary fibers and are normally found in inside-out feeding modules for UF or MF. Capillary membranes have the combined advantages of having a high packing capacity of spiral-wound membranes and the beneficial hydrodynamic behavior of tubular membranes.[50,51]

11.6 Enhancement of Membrane Separation Performance

The remarkable developments in membrane technology as applied in the food industry are the result of the combined efforts of all key stakeholders: the membrane manufacturers, scientists and end users.

Enhancement of membrane performance entails both the improvement of flux and preservation of membrane selectivity. In this section, up-to-date information on the setting of operational parameters, modification of membrane systems, selection of membrane materials and feed properties in relation to membrane performance is discussed.

11.6.1 Optimization of Operational Parameters

11.6.1.1 Crossflow Velocity (CFV)

Depending on the type of module and other manufacturing design (*e.g.* kind of turbulence promoter), membrane producers normally give a recommended range of operating velocities for their products. The normal range of CFV is 1–5 $m\,s^{-1}$. Tubular modules can be operated at higher velocities, *e.g.* 4–8 $m\,s^{-1}$.[52]

In CF filtration, a higher feed velocity will generally give a higher permeate flux at a certain TMP (see Figure 11.12B). When the average velocity increases, the turbulence increases, which then decreases the laminar boundary layer and in turn reduces the concentration gradient and deposition thickness on the membrane surface.[41] A decrease in CP also reduces the portion of TMP consumed to counter osmotic pressure.

At higher CFVs, more energy is required by the feed pump. It is then important to consider the energy consumption per unit of filtered product. (Very) high CF velocities cause a serious retentate pressure drop along the membrane modules, especially in modules with narrow feed channels such as spiral-wound or hollow-fiber modules or when turbulence promoters are used.[55] Owing to the pressure drop, TMP decreases along the modules, leading to non-uniform local fluxes. This causes inefficiencies in terms of energy consumption and overall membrane performance.[56]

Too high CFVs may cause food quality deterioration; for example, too high a shear rate resulting from too high a CFV may cause aggregation of proteins during protein concentration or disruption of yeast cells during MF of yeast fermentation broth.[57,58]

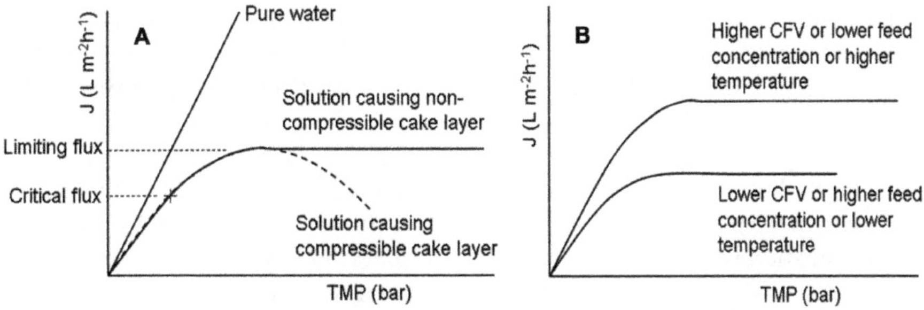

Figure 11.12 Profiles of permeate flux during filtration of polydisperse solutions. (A) Permeate flux versus TMP of feeds causing cake fouling that are non-compressible or compressible at high TMP values; (B) permeate flux behavior under operating conditions.[47,52–54]

11.6.1.2 Transmembrane Pressure

The relationship between flux and TMP in CF filtration can be seen in Figure 11.12. First, the flux increases linearly with increase in TMP and this stage is called the TMP-dependent period. The flux then increases more slowly and reaches a maximum value with further increase in TMP. From then on, the process is fouling layer controlled.

The maximum stationary flux obtained at a certain feed velocity is referred to as the *limiting flux.* Another concept that is also often used is *critical flux.* This is the flux value at which membrane fouling starts (*e.g.* conversion from CP to fouling) and is different in significance from limiting flux.[54,59,60] The TMP at the critical flux is called the critical TMP at the corresponding CFV.

The TMP at which the permeate flux starts to plateau increases with increase in CFV of the feed (Figure 11.12B). This TMP value also depends on the membrane morphology, membrane materials and modules and the feed properties, which are discussed in the following sections. Generally, the higher the permeate flux, the greater the extent of CP will be and also the higher is the tendency to form fouling cake. If a membrane is to be operated in the linear zone, which is called the subcritical regime, a higher membrane area will be required. Higher CFVs can also be considered to compensate for this. If the process is operated in the limiting-flux zone, the concentration of macromolecule components, which are rejected by the membrane, could exceed their solubility limitation. This leads to the precipitation or aggregation of these components on the membrane surface, forming a cake layer. For porous and non-compressible cake, the flux does not decrease with further increase in TMP. For compressible cake or in feeds where penetration and adsorption of smaller components into the cake reduce the porosity of the cake, the flux then decreases with excessive increase in TMP (Figure 11.12A).[61] This behavior between flux and TMP is often seen in MF of polydisperse suspensions, and this region is called the time-dependent flux zone.[62–65] This range of TMP values should be avoided in membrane processes.

The control of TMP is very important in MF. The optimum TMP for a separation process based on particle size is at the critical flux corresponding to each CFV. The critical flux was found to be two-thirds of the limiting flux from calculation using a model with a colloidal system.[54] The two-thirds rule to avoid deposits on the membrane surface is already applied in ED. Bacchin suggested that the transition from critical flux to limiting flux during an increase in TMP is from the local appearance of irreversible fouling to a condition such that a whole membrane is operated beyond the critical flux.[54] Above the critical point, membrane selectivity decreases with increase in TMP values. For MF of highly fouling feed such as milk, favorable operating conditions are high CFV, by use of tubular membrane modules, and low TMP to maintain a high flux and high transmission of membrane-permeable solutes.[66]

11.6.1.3 Control of Permeate Flux

In most studies carried out on CF separations, the feed velocity and pressure at the retentate side are controlled while the filtrate flow is unrestricted. As discussed previously, a higher flux causes a greater extent of CP, leading to faster formation of the cake layer, which in turn has a counteracting effect on the flux. For applications with a large pore size, such as MF and UF in multicomponent feeds, the permeate flux is initially very high but decreases very rapidly in a short time if the filtrate flow is not controlled. The filtration is then controlled by the fouling layer rather than the membrane. Therefore, restriction of permeate flow would show some advantages in terms of suppressing external membrane fouling[67] and perhaps, in some cases, improve productivity. A permeate valve or pump can be used to control the filtrate flow. If a volumetric pump is used, the operation will be in the mode of constant permeate flow. This mode is suitable when the membrane unit is coupled with a subsequent operational unit that requires a constant feed flow. Restriction of filtrate flow results in reduced TMP if the pressures of the feed and retentate sides remain the same. Retarded formation of the cake layer increases the exposure of the membrane surface to the feed components and, as a result, the internal fouling is greater.[68]

11.6.1.4 Operating Temperature and Time

A higher flux is obtained with higher temperatures because the latter reduce the fluid viscosity and increase the mass transfer coefficient, which improves the transfer of membrane-rejected components from the membrane surface back into the bulk stream.[3,10] For example, at the same feed velocities during CF MF of skim milk, the limiting flux obtained at a feed temperature of 55 °C was more than twice that obtained at 15 °C. Limiting flux was also reached at lower TMP values at lower operating temperatures.[52,459] For such reasons, food MF and UF in many cases are carried out at 40–50 °C. In UF of apple juice, an increase in temperature between 20 and 40 °C considerably enhanced the permeate flux.[460] However, further increases in temperature will impose a high risk of chemical deterioration of the product, such as denaturation of proteins and induction of oxidation and damage to membrane integrity.[461] For example, membrane swelling due to high temperatures may cause changes in membrane selectivity.[6,462] In another study, both the membrane performance and functionality of milk protein concentrates decreased significantly in UF at 50 °C compared with 15 and 30 °C.[463] Recently, studies by Baruckic *et al.* and Steinhauer *et al.* on UF of sweet whey indicated that both permeate flux and total resistance in the case of operating temperatures <20 °C are better than those at 50 °C.[464,465] The energy for pumping the feed is lower at higher feed temperatures. According to Méthot-Hains *et al.*, increase the UF temperature of skim milk from 10 to 50 °C reduced the pumping energy 2.3-fold.[459] However, energy is needed to warm the feed. Application at higher temperatures is more advantageous when the feed is already warm from a previous processing stage or there is excess heat in the factory that can be recycled.

Filtration processes cannot be continued for prolonged periods. Owing to the growth of microorganisms in either the retentate or permeate, or both, the organisms would accumulate in the retentate as concentration progresses. With time, biofilm fouling could occur on the membrane surface and adversely affect the membrane performance.[69] The growth of microorganisms can be reduced when the temperature is kept either fairly high or fairly low and the trend today is to work at low temperatures.[1] At lower temperatures, a larger membrane area or other additional techniques, as discussed later in Sections 11.6.4 and 11.6.5, can be employed to shorten the long recirculating filtration time due to low flux. Cold separation could minimize microbial fouling of the membrane and prevent the germination of thermophilic spores.[62] Low temperature also minimizes the risk of changes to the product quality.

11.6.1.5 Diafiltration

Diafiltration (DF) is applied to increase the purity of the macromolecule components rejected by the semipermeable membrane or as a means to increase the removal of membrane-permeable microsolutes. The DF process is carried out by diluting the retentate with a pure solvent, usually water or a buffer, and continuing the separation process until a desired purity of the membrane-rejected components is reached. There are several ways in which DF can be carried out, outlined in the following.

Batch DF consists of preconcentrating the feed to a certain VCR, diluting the retentate to the original volume with the DF solvent and then again concentrating the retentate. This process can be performed in several steps to maximize the concentration of the high molecular weight solutes. In *continuous DF* or *constant-volume DF* (CVD), the pure solvent is added continuously to the retentate at a rate equal to the permeate flow rate.

Addition of the solvent is stopped when a certain purity of the membrane-rejected components is obtained. The retentate is then concentrated to the desired value of solids content. Figure 11.13 presents the evolution of permeate flux during MF of skim milk in combination with batch or continuous DF.

Continuous DF requires more solvent than the batch DF to obtain the same level of purity of the retained components.[70] *Variable-volume diafiltration* (VVD), also known as *volume-decreasing diafiltration*, is a continuous diafiltration process in which pure solvent is continuously added at a flow rate lower than the permeate flow rate. By this means, concentration and diafiltration occur simultaneously.[71] This DF process is represented by $\alpha = V_D/V_P$, which is the ratio of the two volume values mentioned.

Using mathematical modeling and for the same separation level between macrosolutes and membrane-permeable microsolutes, less water is used in VVD than in CVD, if the latter is carried out at the original volume of the feed. If the CVD is performed at the final volume of the retentate or at an intermediate macrosolute concentration that minimizes the process time, it will consume less water than does VVD.[72,73] The drawback of DF, in general, is the increase in energy consumption and the generation of large volumes of diluted permeate, which impose difficulties for the recovery of valuable components in the permeate or increase disposal costs.[74] An approach for

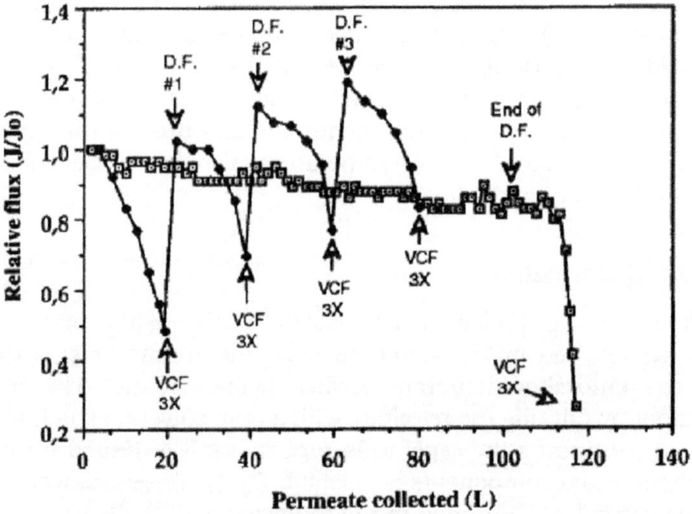

Figure 11.13 Evolution of relative flux (J/J_0) during MF of skim milk in combination with continuous or batch diafiltration with water. □, continuous; ●, batch.
Reprinted from *International Dairy Journal*, **6**, M. Pouliot, Y. Pouliot and M. Britten, On the conventional cross-flow microfiltration of skim milk for the production of native phosphocaseinate, 105–111,[70] Copyright (1996), with permission from Elsevier.

reducing wastewater is to use, for example, the filtrate of subsequent UF for diafiltration of the retentate of the preceding MF process.[75] In systems using an organic solvent, solvent recovery is an appropriate method to save on operating costs but leads to a low purity of the product. In a recent study, M. Schaepertoens *et al.*,[363] using a two-stage cascade combined with another membrane for integrated solvent recovery, solved the above problem to obtain products with high purity (~98.7%).

The combination of diafiltration and UF/NF has been widely applied in the dairy industry to increase protein concentration, isolate milk fat globules and enrich milk sugars. For example, Hozmüller and Kulozik[355] proved that isolation of purified milk fat globule membrane (MFGM) from buttermilk by multi-diafiltration steps leads to better quality than cream washing or the normal filtration method. Altmann *et al.*[356] combined NF and diafiltration on an industrial scale to increase significantly the content of milk oligosaccharides in retentate. Many studies[357–359] have demonstrated that feed components and diafiltration media play an important role in the separation efficiency and physicochemical properties of products. Not just limited to the dairy industry, diafiltration has been applied in various sectors such as refining sugarcane juice, purifying polyphenols in vegetable juice and recovering xanthan gum from palm oil-based broth.[360,361] In a study on betalain, a plant-based pigment, Merredy *et al.*[362] removed more than 95% of nitrates from a solution extract by using three diafiltration steps.

11.6.2 Effects of Feed Properties

11.6.2.1 Concentration of the Feed

The permeate flux of a membrane filtration declines faster with processing time when the feed concentration becomes high. Filtration of feed of lower concentration results in a higher steady-state flux and needs a longer time to reach this flux.[76,77] For a concentration process, the feed viscosity increases with increasing VCR.[78,79] Higher viscosities reduce the turbulence of the flow and hence increase the thickness of the laminar boundary layer. This results in a higher concentration gradient and a tendency to form a thicker external fouling layer on the membrane surface.[41] An increase in the wall concentration of feed components also increases the amount of TMP to counter osmotic pressure. Not only the flux but also the transmission of micromolecular components through membranes decreases with increasing feed concentration.[80]

The effect of feed concentration in relation to different TMP values is illustrated in Figure 11.12B. At a certain CFV, a lower critical flux and also a lower slope of the flux curve under the subcritical zone are obtained with higher feed concentration.[81] Therefore, in (batch) concentration processes in which TMP is kept constant, the operating conditions will possibly switch from subcritical to the transition zone or even above. When irreversible fouling occurs, diafiltration would hardly recover the original flux.

11.6.2.2 Composition of the Feed

The more complex the feed is, the more difficult it will be to suppress internal membrane fouling. "Complex" means having diverse components and different charges and sizes of the components. Examples of this complexity are the similarity in sizes of the components to be rejected by the membrane and those to be transmitted through the membrane, and the difference in the behavior of feed components with regard to their attraction to each other and to the membrane material. The viscosity of the feed is also affected by its composition. The complex interactions between feed components on the membrane surface can accelerate the growth of fouling. The study of Hausmann *et al.* on the concentration skim milk and whey by MOD revealed that fouling development depends on the type of protein present and the salt concentration in feed solution.[466]

The membrane performance in NF declines if the feed contains proteins that can form a gel layer on the membrane surface.[5,9,38,82] The use of UF to remove these proteins would improve the flux.[5,9,82] Some other components, such as sodium, calcium and lactose, also contribute to limit the flux in the MF of skim milk.[467] The flux during MF of a high-load feed of yeasts also decreased dramatically upon addition of bovine serum albumin (BSA), which can be adsorbed by the membrane.[83] For MF of colloidal feeds, particle size affects either positively or negatively the membrane performance depending on the membrane structure and the relative size of the feed particles compared with the membrane pore size.[76,84] In complicated colloidal feeds such as wines, many different fouling mechanisms can coexist.[468]

11.6.2.3 pH and Ionic Strength

The pH of the feed stream during membrane separation is an important factor. Unlike the feed composition, which is inherent to the product, the pH can be modified simply by adding an acid or a base, for example. Depending on the feed properties and the nature of the membrane, either an increase or a decrease in pH will improve the performance of the process.

Changing the pH will change the electrical charge or zeta potential of proteins, which will modify the electrostatic interaction among these molecules and between them and the membrane. In the UF and MF of skim milk, the flux declines more rapidly at lower pH owing to, *e.g.*, reduced electrostatic repulsion.[74,79,85] In the cited studies, the flux decreased by about 30% when the MF was carried out at a retentate pH of 6.0 compared with pH 6.6. The performance of MF of a silica suspension with a ceramic membrane was also reported to increase with increase in pH.[86] The pH had a more significant effect than temperature on membrane performance during MF of skim milk, which is a proteinaceous feed.[74] By combining optimum operating parameters with optimum electrostatics, Gautam Lal and Georges were able to increase the yield of IgG to be collected in the permeate during MF of transgenic goat milk from a low initial value of 0.7–95%.[87]

In separation processes (using MF) in which proteins are assumed to permeate through the membrane, operation near the pI of the proteins may increase pore blockage due to protein aggregates.[32] A study by Golbandi *et al.* on MF of whey solution revealed that the permeation flux is optimized at pH 7.0 and reaches the lowest value at pH 4.0 which is close to the isoelectric point of α-lactoglobulin, β-immunoglobulin and BSA.[469] The ionic strength influences the interaction within the cake layer and the packing density or porosity of the deposition.[28] Divalent cations, especially calcium, promote protein binding and thus intensify cake formation and fouling.[30,470] Higher concentrations of divalent cations can cause salt precipitation and protein denaturation in the bulk phase of the feed.[469] In addition, low pH increases the solubility of mineral salts and releases cations.

In UF and NF, two mechanisms often proposed to explain the change in flux and membrane selectivity are electroviscous effect/friction and membrane swelling. Both phenomena are affected by the pH of the feed and the latter by ionic strength, which increases in the order $CaCl_2 < NaCl < KCl$.[6,88,471–473] For example, in UF of sulfated pentasaccharides, the retention rate decreased by 50% on increasing the NaCl concentration to 0.5 mol L^{-1}.[474]

11.6.2.4 Pretreatment of Feed

Depending on the feeds and the streams (retentate or permeate) of interest, one can make use of several methods for the pretreatment of feed to reduce the flux loss and selectivity degradation. MF is normally performed as a prefiltration step for normal cheese whey to remove fouling-promoting species such as small fat globules, fat globule membrane fragments and bacteria before UF for making whey protein concentrate/isolate or before NF or RO for concentrating whey proteins. Traditional treatment methods for cheese whey include pH adjustment in combination with heat treatment and then removal of the residual fat precipitation by centrifugation or normal cloth filtration. Both MF prefiltration and conventional treatments increase the flux of subsequent UF several-fold.[5,89,90] Li *et al.* used MF to remove suspended particles from the extract of yellowfin tuna spleen before ultrafiltering it for recovery of proteases.[91,92] In the clarification of fruit juices, various pretreatment methods, such as enzymatic treatment, addition of fining agents, centrifugation and their various combinations, can be utilized before MF.[93]

Over recent decades, enzymatic treatment has always been the most popular method. The use of pectinolytic enzymes on feed not only helps to alleviate the viscosity and turbidity of juice but also contributes to increase the permeate flux and the concentration of high-value compounds in the product.[364–367] For example, studies by Echavarría *et al.*[365] and Campos *et al.*[366] on common fruit juices (grape, peach, pear, apple and mandarin) revealed that pectinase treatment combined with a ceramic membrane (CF filtration) can improve the permeate flux by 40% and enhance the quality and stability of the products. However, this method needs to be used with caution in the food industry since some enzymes can have negative effects on other compounds in feeds such as anthocyanins and flavanols.[367] Enzymatic treatment is also a good method to limit membrane fouling. Recently, Luo *et al.* improved the irreversible fouling removal process in the NF of monosaccharides by increasing the size of phenolic acid compounds through polymerization in the presence of laccase, a specific oxidase enzyme found in fungi.[368] In another study, the cheese whey recovery rate was improved and cake resistance decreased when protein transglutaminase was used to catalyze the whey protein crosslinking reaction before UF.[369] Enzymatic treatment also opens new routes in isolating bioactive compounds. UF of soy protein isolate assisted by phytase and phosphatase considerably improved the purity of the product and enhanced separation efficiency.[370] Similarly, a combination of UF and enzymatic hydrolysis of milk protein in buttermilk powder can enrich the phospholipid content eightfold compared with the starting substrate.[371]

The presence of insoluble suspended particles with diameters much larger than the membrane pore size has positive effects on membrane performance. Insoluble milled bean material (~25 μm) can cause scouring effects on the cake layer and improve the flux of UF concentration of aqueous extracts of defatted soy flour.[94] The mass flux of enzyme through a 0.2 μm MF membrane doubled when yeast cells, which can form a porous cake layer assisting the transmission of the enzyme, were added.[95]

11.6.3 Membrane Selection and Surface Modification

Correct selection of the membrane material plays an important role in minimizing the adsorption of feed components, especially for close membranes such as UF and NF membranes. In general, hydrophilic membranes are more resistant to fouling and hence offer better selectivity and higher flux than hydrophobic membranes. The fouling tendency of polysaccharides is lower than that of proteins. Both cause less fouling to hydrophilic membranes than to hydrophobic membranes. On the other hand, tannins (*e.g.* in MF of wine) are adsorbed more on membranes with higher polarity.[96] The hydrophilic macromolecular solutes can bind to less hydrophilic membrane surfaces via "surface dehydratation."[97] The performance recovery after microbiological fouling upon washing is higher for membranes with a chemically neutral surface than for charged membranes.[69] In MF of skim milk for the production of casein concentrate, ceramic membranes were found to be better than polymeric membranes in terms of flux and selectivity.[98]

Membrane pore size and pore size distribution affect the flux and separation efficiency. Membranes with larger pore size offer higher initial fluxes but, in some cases, the difference in steady-state fluxes is not great, depending of course on the feed properties and control of permeate flow.[93] Interactions between the membrane and the feed components are influenced by several forces, such as hydrophobicity and hydrophilicity, dipole–dipole, hydrogen and chemical bonds and van der Waals and electrical forces.[99] Many hydrophobic membranes, such as those made from PVDF, PP and PS, are rendered hydrophilic for better performance and better flux recovery after cleaning of fouled membranes.[100–104] In this method of "surface modification," membranes are immersed in an appropriate coating solution such as poly(ethylene glycol), poly(vinyl alcohol) or polydopamine to cover their surface partly *via* polymerization.[372–374] These new coating-based hydrophilic layers will contribute to limiting the hydrophobic interactions between organics/biofoulants and the membrane surface.[375,376] Further, nanoparticles such as silver, titania and alumina were attached on coated layer to improve permeate flux and anti-biofouling properties.[372,377,378,438] For example, in a study by Rahaman *et al.*,[378] over 95% of the activation of bacteria was inhibited on combining biocidal silver nanoparticles and a polyelectrolyte layer coated on the surface of a polyamide membrane. Recently, carbon nanotubes (CNTs) were considered an ideal material for fabrication on membranes. During synthesis, CNTs were capped with functional groups that provided them with a highly selective capacity towards various pollutants.[379] Incorporation of multiwalled CNTs with PA, PES and PS membranes led to excellent performance with respect to permeate flux and better antifouling properties.[380–382] Enzyme fabrication is also an advanced method for modifying membranes with the aim of controlling biofouling.[383,384] For gas separation, performance and selectivity enhancement can be achieved by combining an inorganic filler and metal–organic framework (MOF) material with mixed-matrix membranes.[385,386] Moreover, a reduction in membrane surface roughness decreases the extent of fouling.[37,69]

11.6.4 Modification of Membrane Module Configuration

11.6.4.1 Membrane Length

The retentate pressure decreases linearly from the membrane inlet to the outlet during CF filtration due to friction between the fluid and the membrane surface (Figure 11.14). Piry *et al.* reported a transmembrane pressure (TMP) decrease of more than fourfold (from 0.82×10^5 to 0.17×10^5 Pa) along a 1.2 m tubular membrane during MF of skim milk.[105] Because of this effect, membranes cannot be manufactured in very long modules, although increasing the length means an increase in membrane area, hence reducing the footprint area required for installation of the membrane unit. The average critical TMP determined for a whole membrane (Figure 11.12A) is not "critical" for all local sections of the membrane. The problem with a TMP decrease is successfully dealt with in uniform transmembrane pressure systems, which are described in the next section. When the pressure drop along the membrane is significant, membrane fouling in long membrane modules operated at high CFV is

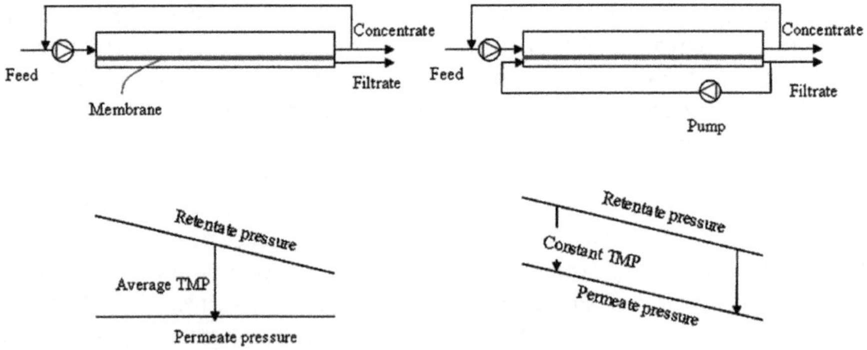

Figure 11.14 Module configuration and pressure profiles of TMP (left) and UTP (right) filtration modes.

more severe at the inlet section of the membrane than at the outlet section, hence permeation of small-molecule components may increase from the inlet to the outlet of the membrane.[26,105]

11.6.4.2 Uniform Transmembrane Pressure

As discussed previously, a high feed CFV in conventional CF filtration is essential to limit membrane fouling. However, a high CFV also leads to a high inlet TMP, which cause rapid membrane fouling. This issue can be resolved with filtration in uniform transmembrane pressure (UTP) mode, which was developed by Alfa-Laval in Sweden.[106] In this mode, the permeate side is pressurized by pumping the filtrate back to the membrane inlet at the permeate side and co-current flows of retentate and permeate are created in the module (Figure 11.14). The recirculating permeate flow is adjusted so that the pressure drop on the permeate side is equivalent to that on the retentate side. Thus a uniform TMP is created through the entire membrane module. Important features of the UTP mode are that the TMP can be adjusted independently with CFV and the whole system is easy to set under optimum conditions, *e.g.* low UTP at high CFVs or high membrane shear rate.[107] Under such conditions, the fouling layer, if it occurs, is expected to be less compact and durable than that in conventional CF filtration, which helps to reduce cleaning costs. Furthermore, the constant TMP would limit the decrease in flux and the variation of selectivity along the membrane.

The UTP mode is applied for tubular modules and is especially advantageous for membranes of length 1 m or greater compared with the normal separation mode. These systems with co-current permeate flow are used widely in MF of milk. In the fractionation of casein micelles from skim milk with a 0.2 μm nominal pore size, Vadi and Rizvi[108] reported that below VCR 4.0 the non-UTP mode performed better but above VCR 4.0 the UTP mode performed better, and beyond VCR 6.0 the non-UTP was not operable, whereas the UTP could reach VCR 8.0–10.0 depending on the set hydraulic parameters. The optimum conditions determined for this system were CFV of 7.1 ms^{-1} and a TMP of 241 ± 10 kPa.[108] Pafylias *et al.* reported an improvement in flux of more than 100% when the UTP mode was used compared with the conventional CF filtration during MF of skim milk to remove bacteria and spores.[107] The Bactocatch process, developed by Alfa-Laval, as a milder pasteurization method employs a MF membrane of 1.4 μm which is operated under UTP condition. Typical operation conditions are UTP of less than 1 bar and a very high CFV of 6–8 ms^{-1}.[45,47] Some studies on the applications of MF in UTP mode were reported by Saboya and Maubois,[109] Decloux *et al.*[110] and Huisman *et al.*[86]

The disadvantage of the UTP mode in comparison with the normal separation mode is the higher energy consumption due to the addition of an extra pump to bring back the permeate stream, especially when a very high CF velocity at low UTP is required. Extra tubing and pumps also add to the investment costs and increase the risk of contamination. To address these issues, several similar techniques have been developed that are sometimes denoted third-generation membranes. Pall (New York, NY, USA) introduced Membralox® GP, a ceramic graded-permeability module in which the mean porosity of the support increases progressively along the membrane in the direction of flow,[111] and Tami Industries (Nyons, France) designed Isoflux® membranes with decreasing thickness of the selective layer along the membrane length. In these membranes, an even profile of flux could be obtained. Nowadays, they are applied widely in the dairy industry for the MF of skim milk and whey proteins.[475–478]

11.6.4.3 Spacers and Turbulence Promoters

11.6.4.3.1 For Polymeric Membranes. In spiral-wound and multi-flat-sheet modules, screens or spacers are normally inserted between membranes to allow the flow of the streams (Figures 11.10 and 11.11). The spacer of the feed channel increases the flow turbulence and surface shear rate, which results in mixing in the boundary layer, and therefore a decrease in CP. This makes the process give higher fluxes at low CF rates of feed, hence saving energy. In RO membranes, the presence of feed spacer limited the development of biofouling and helped the system operate at low TMP.[479] Some methods, such as particle image velocimetry (PIV) or direct observation through the membrane (DOTM), can be used to evaluate the effect of spacers on the flux and fouling growth.[112,480] It was stated that the spacers are mainly effective at high CFV and if used inappropriately may increase fouling rather than reduce it. Incorporation of cylindrical promoters across the flow at a distance above and parallel to a flat UF membrane improved the flux 1.2–1.9-fold depending on the position of the promoters, whether straight or parabolic.[41] However, this type of design seems not to be industrially practical. Modification of feed spacers did not have a significant impact on biofouling control.[481]

11.6.4.3.2 For Tubular Membranes. The improvement of fluxes in MF, UF and NF separations using tubular membranes in combination with turbulence promoters is established and the extent of the improvement is different with different mixer configurations.[113–116]

The use of a Kenics™ static mixer with a short tubular membrane for MF of skim milk increased the permeate flux by 500–700% and reduced the specific energy consumption by 25–80% depending on the hydraulic conditions.[117,118] A similar effect was also found in MF of starch suspensions.[482] For energy-saving purposes, the use of the mixer should be combined with lower feed velocities and higher TMP values.[119] On the other hand, a normal tubular membrane with a helical insert can offer the same permeate flux at 4–5-fold lower feed CF rates.[115,120] This technique is therefore suitable for valuable feeds that are sensitive to high shear.[115,116] Helical inserts performed better than plain cylindrical inserts with a similar pressure drop.[115] In the filtration of yeast, the use of a rotating helical membrane can enhance the flux by 27% and consume less energy than a rotating flat membrane.[483] Experiments with UF of a dextran solution showed that inserting a rod with a linearly increasing radius central to a ceramic tubular membrane gave a greater improvement of flux than using a rod with a linearly decreasing radius. The former case gave a 90% flux enhancement compared with the empty tubular membrane.[121] Another kind of insert was investigated where a uniform-radius rod was wrapped with a wire at a varying angle along the rod.[122] CDF (computational fluid dynamics) modeling proved useful in investigating the flow behavior and mass transfer when filtration modules are incorporated with turbulence promoters.[115,123,124] More recently, Popovic *et al.* introduced a new turbulence promoter using semielliptical blades.[484] In MF of milk, this equipment can enhance the permeate flux by 500–650% while consuming less energy than a conventional system.

A disadvantage of adding a turbulence promoter is the increase in TMP drop along the membrane, which leads to an increase in dissipated power and energy consumption, especially with high feed velocities. The introduction of turbulence promoters then involves a trade-off between this drawback and the expected flux enhancement. Inserts are difficult to use with long tubular membranes. Another method is to use a microchannel turbulence promoter with micropores. Xie *et al.* found that micropores can reduce fouling, enhance the membrane effective area and prolong operating times.[485] It is also important for the inserts to be fixed in such a way that damage to the membrane due to vibrations of the insert is avoided during the process. When using turbulence promoters, the fouling is less resistant to cleaning, so it is easier to recover the original fluxes of the membranes.[118] Dead zones can be introduced with the use of inserts, which could probably create difficulties for cleaning and disinfecting the systems, *e.g.* in cleaning-in-place (CIP).[63] Several types of inserts for tubular membranes are shown in Table 11.1.

Table 11.1 Different types of static inserts for tubular membranes.

Insert	Ref.
Kenics® mixer 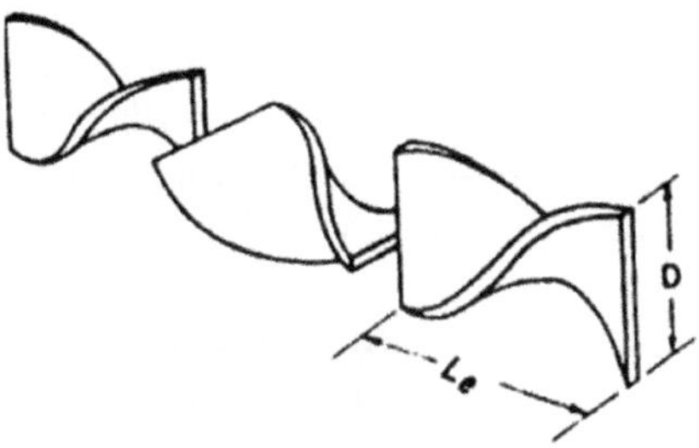 	118, 119, 125
Internal and external screw-threads 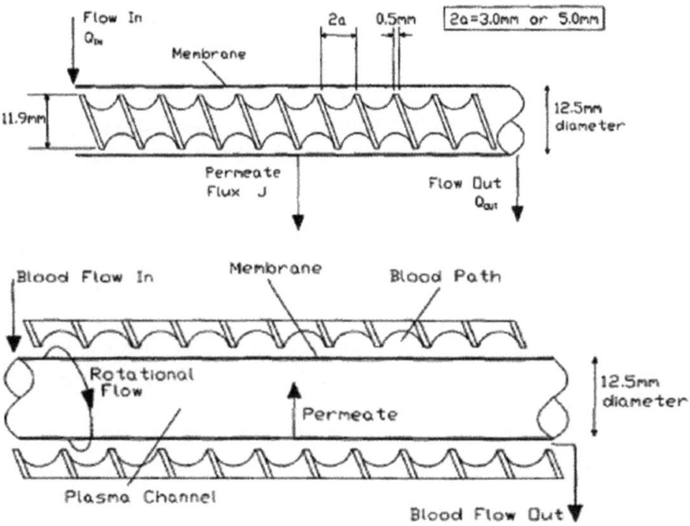 	115, 116
Wire-rod insert 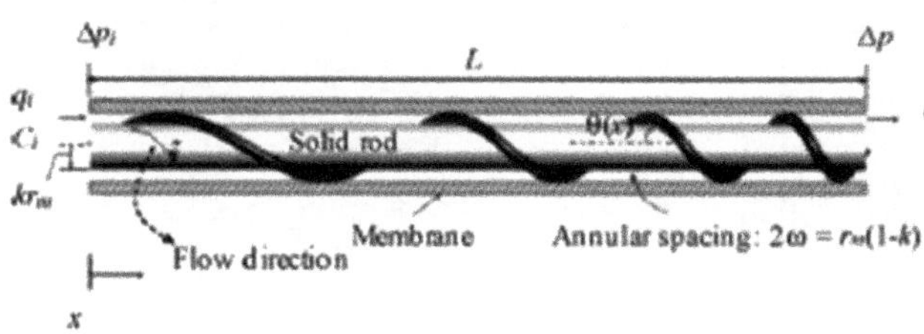 	122
Helical insert  	126

11.6.4.4 Dynamic Membrane Modules

Dynamic modules include systems composed of a rotating disk near the membrane and systems in which the membrane itself is rotated or vibrated during operation. These special designs can allow a very high wall shear rate, *e.g.* in the order of $(1–3)\times10^5\ s^{-1}$, which efficiently suppress CP and fouling formation and hence improve the permeate flux and membrane selectivity. Calculation of operating parameters such as shear rate and TMP for dynamic membrane modules can be also found in an extensive review.[127]

11.6.4.4.1 Rotating Disk Modules. A laboratory rotating disk module is shown in Figure 11.15. In the module, a solid disk is mounted on a shaft near stationary circular membranes and rotates at high speed.

With a requirement for a large membrane area and minimal recirculation of the feed, industrial systems normally have several disks placed alternately with circular membrane envelopes fixed on the wall of the chamber. Examples of this kind of design are the Dyno-Filter (Bokela, Karlsruhe, Germany) with available pore sizes ranging from 0.001 to 1.0 μm, and an improved version, the Kotobuki rotary filter, which was developed in Japan.[128] These systems are suitable for viscous and high suspension concentration feeds and were developed for use in the biotechnology and chemical industry. In comparison with CF filtration and dead-end filtration, the TMP in dynamic membrane modules is relatively uniform and higher VCR can be achieved.[486,487] In general, rotating disk modules assists CF filtration better than dead-end filtration.[488,489]

Numerous studies have been carried out on applying rotating disk modules in food processing,[130] especially work at the Technological University of

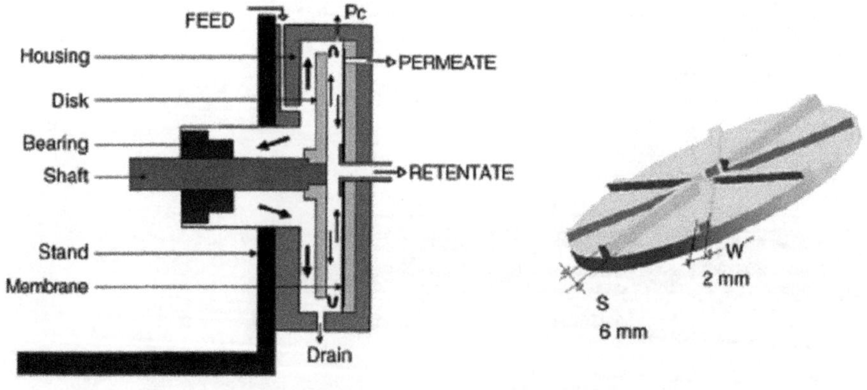

Figure 11.15 Schematic diagram of a simple rotating disk module. The disk can be smooth or with vanes as shown on the right.
Reprinted from *Journal of Membrane Science*, **282**, M. Frappart, M. Akoum, L. H. Ding and M. Y. Jaffrin, Treatment of dairy process waters modelled by diluted milk using dynamic nanofiltration with a rotating disk module, 465–472,[129] Copyright (2006), with permission from Elsevier.

Compiègne (France). Enhanced fluxes were observed, such as in UF for the concentration of skim milk,[129,131] MF of skim milk for protein fractionation,[132] MF for clarification of rough beer,[133] MF and UF of fruit and vegetable juices,[486,488,489] recovery of linseed oil from an oil-in-water emulsion,[134] MF in combination with UF for the recovery of polysaccharides from bacterial fermentation,[135] NF in the treatment of dairy process water[129] and RO in the concentration of milk.[136] Rotating disks fabricated with vanes (Figure 11.15) offer further improvement in flux compared with smooth disks.[131,137,138] The use of vane disks may result in lower specific energy consumption per cubic meter of permeate produced compared with the use of smooth disks.[134,135,139] The height and number of vanes influence the flux enhacement.[135] It was reported that the rotating disk module offered better performance than tubular membranes[131] or the vibrating system (VSEP) due to higher membrane shear rates of the former.[129] The effect of the rotation speed of the disk on the relation between flux and TMP is the same as that of CFV in standard CF filtration.[66]

11.6.4.4.2 Multi-shaft Disk Separators. Unlike rotating disk modules, where the membranes are stationary, in multi-shaft disk (MSD) separation systems the membranes are rotated. An MSD module is comprised of at least two shafts that are arranged in parallel or at an acute angle to one another.[137,140,141] On each shaft a stack of membrane disks (preferably ceramic sintered on a porous seamless substrate with hollow channels inside to collect permeate) are mounted and the whole system is enclosed in a housing. Permeate is collected through the bore of the shafts.

Several industrial designs in this group of modules are commercially available, *e.g.* those of Westfalia Separator[140] and Hitachi, and the Rotostream of Canzler (Düren, Germany). The main differences between the designs are in membrane diameter, number of membrane disks on a shaft, number of shafts and arrangement of the shafts. The stack of disks on one shaft is arranged alternately and with an overlap with the disks on the adjacent shaft(s). All shafts are rotated in the same direction as shown in Figure 11.16. The distance between two overlapping disks is of the order of a few millimeters.

With this configuration, the velocity directions of two facing disk sides are opposite. Multi-shaft modules with overlapping membrane disks eliminate the need for separate solid disks to create shear as in the single-shaft modules discussed in the previous section. Therefore, the packing capacity of MSD systems is higher than that of rotating disk systems. These systems can create very high shear rates and allow high retention times of feed in the housing, which is good for dewatering purposes, making it suitable for filtration in wastewater treatment and for the recovery of food components from diluted effluents. MF and UF modules are commercially available and membrane areas of up to 80 m^2 can be obtained in a Westfalia eight-shaft separator.[129]

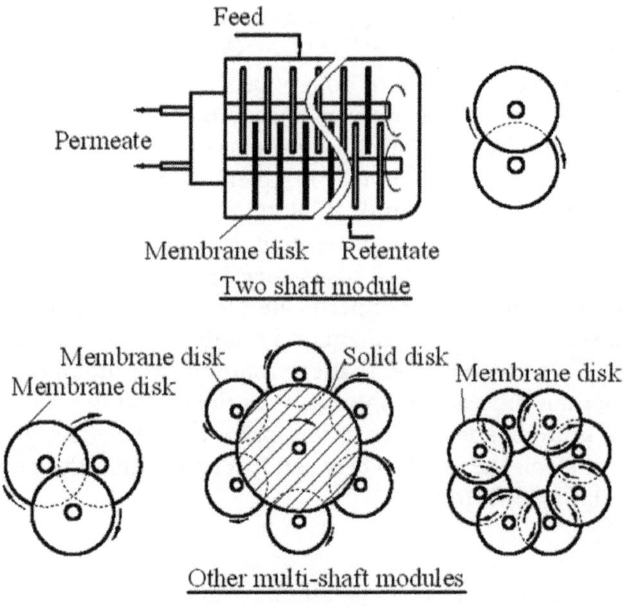

Figure 11.16 Schematic diagram of sample MSD modules. The systems can be designed horizontally or vertically.

Espina *et al.* stated that MSD with ceramic membranes can be an alternative to UTP tubular membranes in MF of skim milk for the separation of casein micelles and whey proteins.[132] In a test of MF of skim milk, the permeate fluxes at rotation speeds of 1492 and 1930 rpm were comparable to values reported in the literature with UTP systems at a CFV of 8 m s^{-1}.[66] The differential velocity at 1930 rpm in this module is 13.74 m s^{-1}.[130] Clearly, the rotation speed of the disk is the most important adjustable parameter influencing the permeate flux. The speed can be increased by increasing the rotation speed of the membrane and by increasing the disk radii.

Rotating membrane systems are not suitable for processing foods that are sensitive to high shear rates and temperature. Some energy is also lost due to heat dissipation. Bearings for rotating parts and sealings should not contaminate the products. NF and RO applications using dynamic membrane modules are rare, since few systems are able to withstand high pressures.

11.6.4.4.3 Vibrating Membrane Modules. The vibratory shear-enhanced processing (VSEP) system consists of a stack of circular plate-and-frame membranes. A bar acting as a torsion spring transfers the oscillation from a seismic mass driven by a motor with eccentric weight to the membrane module (Figure 11.17). The vertical module arrangement of VSEP results in a small installation footprint. Membranes of polymeric materials for the range of MF to RO are now available.

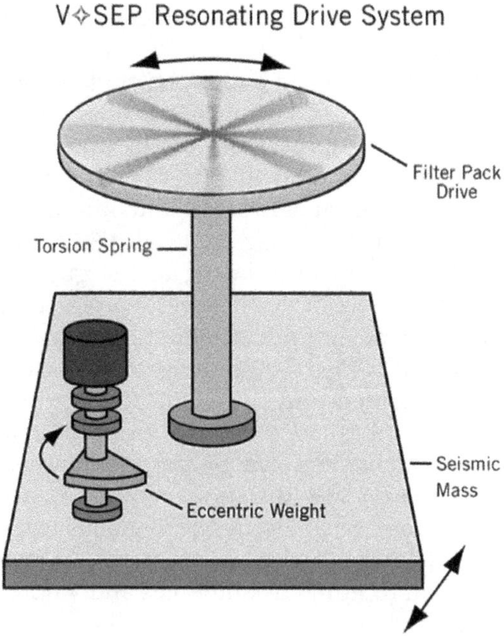

Figure 11.17 Schematic of a New Logic vibration system. Image courtesy of New Logic Research.

The vibration of the membrane back and forth at high speed creates the shear right at the membrane surface. Owing to inertia, the retentate moves at 180° out-of-phase with the membrane and varies sinusoidally with time. This cause the foulants to be removed from the membrane surface and remix with the bulk feed. The local permeate flux declines from the rim to the center of the membrane.[138,142] The applied feed velocities are normally low and they are not an important factor influencing the permeate flux.[143] In fouling reduction, transverse vibrations have better performance than longitudinal vibrations at low operating frequency.[490] To improve the performance of hollow-fiber membranes using vibrations, a small membrane looseness (1–2%) can be applied, but membrane cracking can occur at higher ratios.[491]

PallSep™ is another version of VSEP manufactured for industrial MF applications in biotechnology with an available membrane surface of 40 m^2 per module. Flows of permeate and retentate inside the membrane stacks have been described.[83] In vibrating membrane modules, the shear rate depends on the vibration frequency, vibration amplitude and membrane radii. These modules treat the products more gently than rotating modules. In UF of milk, large VSEP units, which are available up to 151 m^2 of membrane area per module, will consume significantly less energy per cubic meter of permeate than CF filtration.[144] The critical fluxes increase with increasing average membrane shear rate, $\bar{\gamma}_s$, in a power function, $J_{crit} = a\bar{\gamma}_s^b$ where a is a constant and b is an exponent ranging from 0.264 to 0.38 depending on the feed and feed concentration.[145,146]

Several studies have been carried out on the application of VSEP systems in MF of skim milk for the separation of micellar caseins,[143,147] and in UF of skim milk.[143,492] The improvement of flux with increasing frequency was more pronounced in UF than in MF. For MF with a 0.1 μm Teflon membrane, the critical flux at 60.75 Hz and initial concentration was 64 $L\,h^{-1}\,m^{-2}$.[143] Al-Akoum *et al.* suggested that VSEP can be a good alternative to UTP in MF of skim milk owing to the good performance and probably lower energy consumption.[143] Several production installations of vibrating systems are already in industrial use, such as for the MF of skim milk to obtain high VCR values for the removal of bacteria and spores.[63] The maximum VCR obtained and the permeate flux can be higher with larger gaps between membranes, since this provides a large core region that hosts feed components rejected by the membrane.[83] Rotating disk modules are better than VSEP filtration systems in terms of flux enhancement since the former can generate higher shear rates, especially when the metal disk is designed with radial vanes.[138] A good comparison of energy demand between rotating disk modules and VSEP can be found in a review by Zsirai *et al.*[493]

A laboratory module of hollow fibers vibrating axially was applied to the MF of yeast suspensions and the results showed that it was possible to filter at very low TMP (<0.1 bar) and a stable permeate flux was obtained when the flux was kept below the critical level.[145] Addition of a transverse vibration could double the flux.[148] In another study, the use of this technique improved significantly the separation efficiency of whey proteins and casein micelles in milk filtration.[494]

11.6.4.4.4 Conclusions. For the successful application of a technology in the food industry, it must satisfy several criteria: simplicity, economic advantage, scale-up feasibility, easy cleaning and food safety. Dynamic systems still have a small market share since they offer great improvements in process performance compared with normal CF systems. In normal CF, high velocity is used to suppress membrane fouling and narrow feed channels are necessary to reduce the thickness of the boundary layer, which cause a high pressure drop along the membrane. In dynamic systems, the shear rate is created by the movement of the disks or membranes so a high feed flow rate is not required, but energy is needed to create the movement. However, processing can be carried out at low feed velocity and low TMP, so less energy is consumed in maintaining the flow compared with normal CF systems. Dynamic membrane modules are more expensive and difficult to scale up, with problems with membrane cleaning and sanitization after use. Moreover, feed flow rates are slow and therefore it is possible to concentrate a solution to a target concentration with a single pass.

11.6.5 Flow Manipulation

11.6.5.1 Crossflushing

Crossflushing, also called forward flushing, is performed by pausing the permeate flow while maintaining the feed flow across the membrane. A shorter duration of application is required if it is performed more frequently.[149] It was reported that, during MF of yeast cells using a 0.45 μm PVDF membrane, crossflushing for 10–15 s is effective if it is applied after 0.5–1 min.[149] Moreover, the more frequent the application, the slower the flux decrease will be.[32,149] However, since no filtration occurs during forward flushing, an optimum time for crossflushing needs to be determined to balance the flux recovery and total permeate obtained. In NF for water production, crossflushing was shown to be more effective for capillary membranes than for flat-sheet membranes.[51] Crossflushing is simple and easy to apply. Although it is not effective against pore-block fouling, it can be applied when the fouling caused by pore blocking is minimal and the cake layer is not compact and is loosely attached to the membrane surface.

A pulsating crossflushing system that combines both simultaneous pulsation in TMP and CFV has been designed.[150] Periodically, the system is able to generate simultaneously negatives pulses on TMP and positive pulses on CFV. In this way, each filtration step is followed by membrane relaxation that promotes the disruption of the CP layer forming on its surface. There was a 250% performance improvement in UF of BSA aqueous solution compared with a standard UF unit. A pulse frequency profile can be constructed using artificial neural networks (ANNs) to obtain maximum maintenance of flux.[151] The use of volumetric pumps such as peristaltic, reciprocating and diaphragm pumps instead of centrifugal pumps creates pulsation of the feed flow and hence crossflushing effects. Dampers are normally connected after these volumetric pumps to damp the pulses.

11.6.5.2 Permeate Backflow Techniques

These techniques do not involve the modification of the modules such as discussed above, but involve supporting tools that are used to convert the TMP periodically during the operation and hence the backflow of the permeate lifts up the foulants, which are then flushed away by the CF of the retentate. Permeate backflow techniques include backwashing, backflushing, backpulsing and backshocking. The main differences in these methods are the applied frequency, pressure discrepancy and length of time for the permeate backflow.

11.6.5.2.1 Backwashing and Backflushing. Backwashing involves pausing the filtration process and pumping an external washing solution, commonly water, from the permeate side to the retentate side for a certain period of time. Backflushing is an *in situ* method for removal of membrane fouling by periodically pumping the permeate back to the retentate side. The terms backwashing and backflushing are sometimes used interchangeably. With backwashing, the flux decreases more slowly, allowing a prolonged process time before intensive chemical cleaning becomes necessary.

There is no common optimum backwashing or backflushing regime (frequency, flux rate and time) for all applications. Several studies have been carried out to optimize the backflush procedure in different applications[155–158] and model the effects of permeate backflow parameters.[158–160] Backwashing is often applied for 5–10 min after each 0.5–1 h of filtration, whereas backflushing with permeate can be carried out at shorter time interval from several seconds to minutes, after each filtration period of several minutes to an hour. The longer the backflushing interval is, the longer is the time needed for the reverse flow to remove all the reversible fouling. A typical flux profile for backwashing or backflushing with a long interval (low frequency) is illustrated in Figure 11.18.

The backwashing efficiency (flux recovery) increases with increasing frequency. It also increases with the duration of washing but levels off at longer durations. A way to control the backflush time, and thus minimize the water volume, is to measure the TMP automatically during washing. A controller can be used to stop the backwash when there are no changes of TMP in a set time, *e.g.* 5 s.[161] The efficiency depends on the membrane morphology, *e.g.* surface filtration *versus* depth filtration,[162,163] and the feed properties that determine the mechanisms, extent and compaction of membrane fouling.[28,30] The TMP difference during backwashing is another determining

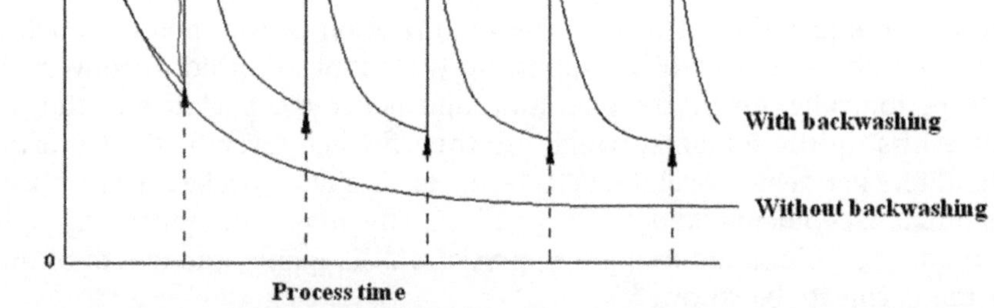

Figure 11.18 Typical flux profile for MF concentration with periodic application of backwashing and backflushing. A and B represent irreversible and reversible performance losses, respectively, in the first cycle of filtration. The cleaning efficiency β of this backwashing step is equal to $B/(A+B)$. Arrows show the places where the washing is applied.[28,152–154]

parameter.[158,164] Maximum backwashing TMP values are adapted for different module configurations and membranes. At certain higher values, the permeability of membranes is reduced owing to membrane compaction effects.[165]

The fouling reversibility is dependent not only on the backwashing parameters but also on the main filtration operating parameters. MF using a ceramic membrane in combination with backwashing using deionized water with sodium azide added showed that, at an operating TMP below an optimum value (critical TMP), irreversible fouling occurs mainly in the first cycle of filtration–backwashing, but at higher TMP the irreversible performance loss continues with succeeding cycles.[152,166] In some cases, the backflush efficiency decreases with increasing CF rates of the forward filtration,[158,167] since higher CF rates may produce stronger internal fouling.[167]

Backflushing can improve the steady-state permeate flux by 25–55% compared with a conventional CF filtration system.[32,155,157,168] Much greater flux enhancement can be achieved with more frequent backflushing.[169] In MF of milk protein, applying backflushing at 1 bar for 10 s every 5 min enhanced the permeate flux approximately 100% and decreased membrane resistance fourfold.[495] Meier *et al.* used backflushing to facilitate the separation of ultrafine particles for size classification using CF filtration.[170] CF separation incorporating backflushing is more effective than crossflushing in the filtration of BSA, yeasts or a combined solution of both,[32,154] and is also more effective than stirring dead-end filtration systems.[163] However, backflushing has a major effect on permeate flux while having less impact on the transmission of solid particles in the flux.[495] Gases such as compressed air, carbon dioxide[62] and nitrogen[171] are sometimes used as backwashing agents. A combination of permeate and CO_2 as backflushing medium can reduce the permeate loss by 40%.[172] In beer clarification, CO_2 backflushing can enhance the permeate flux 3–5-fold.[496] Moreover, the using of gas backwashing can limit significant anthocyanin loss in the clarification of red raspberry juice.[497]

Backwashing requires the incorporation of more tubing and pumps, which brings a higher risk of contamination. On the other hand, clean water for backwashing is expensive, at least in some parts of the world. A possible option is to use the filtrate of a closer membrane (UF compared with MF) as the washing solution for the separation employing a more open membrane. This is favorable when there are components in the first permeate that need to be passed back to the retentate. Backflushing uses the permeate as flushing solution, so it needs an optimum regime to limit permeate loss. Backwashing increases the non-productive time of membrane systems. Hence backwashing and backflushing do not remove the membrane pore blocking completely.

11.6.5.2.2 Backpulsing. Backpulsing has the same principle as backflushing but is characterized by very high frequencies, 0.1–1 s^{-1}, and is performed in a short time, generally less than 1 s. These characteristics require highly responding valves to pulse back the permeate flow. This technique was developed to overcome the disadvantage of backflushing because a short duration of the pulsation minimizes the permeate loss due to the reversed flow. This method is very efficient in maintaining a high flux for a long separation time. The cleaning effect of the method is due to the reverse flow of the permeate and, for flexible membranes, to the wave propagation (membrane vibration) caused by the pulse.[173] With the use of backpulsing, a high CFV is not required, as in standard CF to increase the transport of the components rejected by the membrane back to the bulk phase. This results in less shear stress, lower pressure drop along the membrane module and lower energy consumption.

As in backflushing, the backpulsing efficiency depends on the pulsing parameters, food properties and membrane material and modules. Longer duration, higher pressure and higher frequencies of backpulsation would lead to better cleaning effects but a greater loss of permeate volume. Therefore, there is a tradeoff between these parameters to achieve the maximum net flux.[32,173–177] A higher backpulsing frequency is necessary when the fouling formation rate is higher. The latter is in turn affected by the nature and morphology of the membrane, feed properties and operating conditions. Operation at low TMP values is most beneficial. Several models have been developed to predict the setting of these factors to maximize the permeate.[176,178–180] It is said that it is more important to choose an appropriate backpulse duration for a given optimum forward filtration time than *vice versa.*[176]

Backpulsing has been studied for applications in the food industry more extensively than backflushing owing to its good performance in membrane processes. A 100% increase in permeate flux was observed when backpulsing was applied in MF of skim milk to remove bacteria with ceramic membranes.[47] A huge flux increase was also observed by Beolchini *et al.*[181] in the same application. There was a decrease in permeate flux with process time. However, filtering for 13 h at a flux of 100 $\mathrm{L\,h^{-1}\,m^{-2}}$ was possible without cleaning and the skim milk was concentrated to a volume reduction factor (VRF) of 30.[47] The effect of backpulsing on the operating time (without cleaning) in a constant flux mode depends on the intensity of the flux. Arkell *et al.* concluded that backpulsing helped to prevent fouling in the MF of milk for 11 h at a flux of 350 $\mathrm{L\,h^{-1}\,m^{-2}}$ but the operating time decreased to 1.6 h when the flux reached 500 $\mathrm{L\,h^{-1}\,m^{-2}}$.[498] Gan studied beer clarification using MF and showed a permeate increase of up to 410%[177] compared with standard CF MF. MF of yeast broths using cellulose acetate membranes in combination with backpulsing showed a 2–4-fold increase in long-term (*e.g.* steady-state) flux.[57,174] However, reverse flow during the backpulsing was found to cause cell disruption.[57] MF applications in bacterial suspensions also showed significant enhancement of process performance, with an increase in net flux values of 2.6–10-fold.[100,176,179] In MF of oil-in-water emulsions, it was reported that backpulsing could maintain a long-term permeate flux at a level nearly three times that without membrane backpulsing.[158] For example, MF of a homogenized yeast suspension for recovery of soluble proteins showed a 5.4-fold higher permeability for proteins when backpulsing was applied.[80] Increases in permeability and selectivity also reduce the buffer volume required for diafiltration and shorten the process time,[173] which reduces the risk of chemical deterioration of sensitive components. When applying backpulses, the membrane surface is frequently exposed to the feed and it can be expected that until the end of the process, the internal fouling is more adhesive and tougher than in case of without backpulses.[100]

Results of economic analysis by Kuberkar *et al.* showed that CF MF with backpulsing has lower costs than centrifugation, rotary vacuum filtration and CF MF without backpulsing.[179] For example, in MF of milk, backpulsing can help to reduce energy consumption by 50%.[499] Most foods suitable for membrane processes are multicomponent systems and hence are highly susceptible to fouling. Backpulsing is a potential solution. However, these systems are difficult to scale up, because pressure pulses are damped and become less effective over long distances.[63]

11.6.5.2.3 Hybrid Backpulsing Techniques. *Low-power* or *low-pressure backpulsing* is also called infrasonic pulsing.[182] In backpulsing, the permeate pressure during pulsating is lower or slightly higher than the retentate pressure. The pulsing duration is in milliseconds, so the disruption of concentration and membrane fouling is greater because of membrane motion (vibration or relaxation) rather than the reversal of permeate flow.[173,182] A three-stage mechanism that explains the disruption of concentration and membrane fouling due to rapid vibration of the membrane has been reported.[182] In the same study, it was also shown that the technique helped increase the net flux 3–4-fold compared with normal CF MF of talc and yeast suspensions. A model to predict the net flux as a function of infrasonic frequency, duty cycle and amplitude was also developed.[182]

Low-power backpulsing is applicable for flexible membranes (*e.g.* polymeric flat sheets) that are configured in such a way that allows them to vibrate perpendicular to their surface. The vibration amplitude is not uniform for the entire membrane. The disturbance of CP and surface fouling is then not uniform. All these characteristics seem to make this technique difficult to apply in large-scale systems. The technique is effective in applications where the internal fouling is not critical and the attachment between foulants and membrane is not strong. Filtration with a PES microsieve of pore size 2 μm in combination with the low-power technique resulted in very high and stable fluxes, such as ≥6000 $L\,h^{-1}\,m^{-2}$ for BSA solution, 1600 $L\,h^{-1}\,m^{-2}$ for milk and 1900 $L\,h^{-1}\,m^{-2}$ for beer.[183] However, the pore size is too large for industrial applications. Backpulsing was also found to be superior to crossflushing, backflushing and air sparging at low air flows.[183]

Using *backshocking in combination with reverse asymmetric membranes*, excellent membrane performance was reportedly obtained in MF of beer.[184] This novel technique is often referred to as backshocking and has been the subject of several studies. Here, the feed is charged to the porous support side of the membrane, which has very high porosity, and the backpulse is applied from the active layer side. The open fouling layer formed inside the porous support layer can be easily removed by frequent backshocking.[47] For this, the role of the porous structure can be roughly understood as "particle holding and releasing".[177]

Complete (100%) transmission of proteins and milk components was found in MF of beer for retention of yeasts[184] and MF of skim milk for removal bacteria,[47] respectively. Both studies employed asymmetric polymeric hollow-fiber membranes. Reverse filtration with asymmetric membranes increases the permeability and so may reduce the membrane discrimination of hydrophilic and hydrophobic proteins. For example, a high concentration of chill haze (which indicates the concentration of hydrophilic proteins) due to its high transmission through the membrane into beer is undesirable.[172] Comparing different flux enhancement techniques for MF of beer with a ceramic membrane, the degree of flux enhancement increases in the following order: feed flow pulsing, introduction of helical insert, normal filtration with high-frequency backflush and reverse filtration in combination with the backflush.[177] A high porosity of the membrane surface where the feed is introduced reduces the shear stress at the pore entrance, which may reduce the tendency for protein denaturation and aggregation.[184] Without backshock, reverse asymmetric membranes would not work since the CF cannot remove the concentration layer and fouling inside the porous structure. The control of backshock parameters is very important. Reverse asymmetric membranes should be operated at low TMP. Selection of the membrane according to the nature of the feed is vital. There should also be control of the permeate flux since beyond a critical value a gel will form inside the porous layer by proteins, which could be difficult to remedy by backshock.[47]

11.6.5.2.4 Conclusions. When permeate backflow techniques are used, the membrane can be in service for longer before a chemical cleaning is required. This aids in reduction of cleaning costs and increased plant productivity and extends the lifespan of the membrane, and the fouling rate would be slower and membrane selectivity deterioration would decrease. Backward flow and forward filtration are similar in that a good distribution of permeate backflow is needed to have homogeneous washing for the entire membrane. Permeate backflow techniques are hardly used for multiple flat-sheet membranes (membrane cassettes) and spiral-wound membranes. Membrane installations utilizing these techniques are generally more expensive and difficult to scale up.

11.6.6 Applications of External-body Forces

11.6.6.1 Electrically Enhanced Membrane Separations

Constant or pulsed DC electric fields can be applied for the enhancement of the performance of membrane processes. Figure 11.19 is a schematic

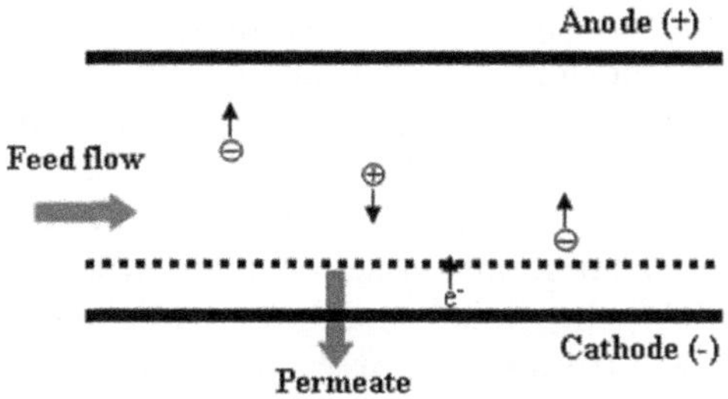

Figure 11.19 Illustration of a flat-sheet membrane with application of an electric field and electrophoretic migration of species.

diagram of an electro-membrane filtration unit where the cathode is placed at the permeate side and the anode at the retentate side. During the process, the negatively charged species move towards the membrane by convection and away from the membrane surface by back-diffusion and electrophoresis.[185] Therefore, the electric field plays a role as an additional driving force for the separation of species. Electric field pulses may dislodge the deposit layer on the membrane surface due to the electrophoretic mobility of charged foulants. The electric field not only affects the feed flux but also the membrane properties. Darestani *et al.* concluded that an intense electric field can reduce the surface roughness of many types of membranes using in MF, UF, NF and RO systems.[500]

Theories relevant to electrically enhanced membrane filtration have been reviewed.[186] Effects generated by an electric field are electrophoresis, electroosmosis, electrochemical reactions and Joule heating. The determining parameter for flux and selectivity improvement in this technique is the electric field strength. This can be defined in a flat-sheet membrane as the ratio of the applied voltage to the distance between two electrodes. For a tubular membrane, a certain modification of the equation is needed. The modes of application, namely continuous or pulsed electric field, and *e.g.* on-time/off-time ratio for the latter case are important. The process performance also depends on the operating conditions such as CFV and TMP. Feed properties such as pH, feed concentration and composition are significant since they can affect the mobility of charged species and also the flux in a way discussed earlier for conventional CF filtration. Many models have been developed to predict the permeate flux for electrically enhanced membrane filtration.[93,187–189] The models were developed and checked to have good agreement with practical data for specific applications. However, further investigation is still needed for generic application.

Selectivity improvement is obtained *via* two approaches: (i) reduction of membrane fouling, which preserves the permeability of the membrane, and (ii) electrophoresis, which facilitates the separation capacity. The separation empowered by electrophoresis allows the selection of membranes with pore sizes larger than normally accepted for standard CF filtration. This allows a further increase in permeate flux and shortens the processing time.[190]

Protein transmission through membranes under an electric field depends on the charges of the molecules, which can be modified by means of pH and ionic strength variations.[191] Hence these parameters influence the composition of permeate stream and also the retentate stream.[192]

Employing a metal net both as a retentate spacer and as an anode in UF concentration of a BSA solution, Oussedik and co-workers found that under a pulsed electric field, the global membrane resistance decreased significantly. This effect is attributed to three factors: the turbulence caused by the spacer, the turbulence created by the formation of oxygen bubbles at the membrane surface due to water electrolysis and the electrophoresis of the protein by the pulses.[193,194] Application of a DC electric field increased the permeate flux about twofold during UF of BSA aqueous solutions.[193,195,501] A pulsed electric field strength $E = 700\ \mathrm{V\,m^{-1}}$ applied at a frequency of 30 Hz even increased the flux fourfold compared with the standard UF process in which no electric field was applied.[193] The electric field exerts greater flux enhancement when the BSA concentration is low, while rejection is improved at both lower or higher concentrations.[190] Electric field washing, *e.g.* electric pulses applied periodically as in water backwashing, recovered 60–100% of the original flux during UF of BSA solutions[196,197] or a diluted bovine plasma (0.5 wt%, pH 7.8) with a ceramic tubular membrane.[198] The use of a magnetic field in UF of BSA improved significantly the water flux (150–200%) and BSA recovery (15–20%).[502] Making use of different pI values of lactoferrin (8.0–9.0) from β-lactoglobulin (5.13) and α-lactalbumin (4.2–4.5), the former can be separated from whey proteins employing by an electrically enhanced MF process.[191] MF of wine fermentation broth showed that a 20–85% decrease in membrane resistance was obtained, depending on the electric voltage.[192] There was an insignificant loss of permeate components due to membrane fouling.

Pectin may form a gel layer over the membrane surface during UF of fruit juice and is negatively charged at the natural pH of fruit juice. An external electrical field can be used to improve the UF process for depectinization in fruit juice processing.[93,199,503] A test with mosambi juice showed a 30% increase in steady-state flux at $500\ \mathrm{V\,m^{-1}}$ compared with no electric field.[93] At an electric field of $300\ \mathrm{V\,m^{-1}}$, a similar flux increase in UF of juice from the same fruit with 22% less total energy consumption per permeate unit volume was reported in another study.[188] The pulsed field is more advantageous with respect to energy consumption than a continuous electric field. With a suitable pulse ratio, the flux improvement in the former case can be similar to that in the latter.[188,200] A pulsed electric field is also used as a pretreatment step before UF in the purification of fruit juice. This combination brings better product quality and consumes less energy than the traditional method using heat treatment.[504–506]

For shear-sensitive feeds, such as blood or suspensions of mammalian cells, high shear operation should not be chosen. Therefore, an electric field provides a solution to increase process performance.[201,202] Several effects, such as electrolytic reactions and Joule heating under an electric field, are unavoidable. Water may dissociate and produces hydroxyl ions (OH^-) and hydrogen (H_2) at the cathode and hydronium ions (H_3O^+) and oxygen (O_2) at the anode. These effects are greater with higher electric field strength. Either alkalinization or acidification of the retentate and permeate may occur owing to water dissociation and depending on the arrangement of the electrodes.[191] Precipitation of proteins in these streams may create pumping difficulties. Localized heating, which increases with increase in conductivity of feeds, is also a cause of protein denaturation.[190] Moreover, a very high pulsed electric field intensity (20–30 kV cm^{-1}) may inactivate enzymatic activity,[203] but electrode oxidation and reduction could cause metal contamination in processed foods. Unfortunately, electrofiltration systems are difficult to scale up.

11.6.6.2 *Ultrasound*

Ultrasound uses sound waves at frequencies higher than the maximum value that human ears can hear (above 20 000 Hz). Figure 11.20 shows an example of a laboratory ultrasound-assisted filtration unit where the membrane module is immersed in an ultrasonic bath. In dead-end filtration, an ultrasound probe can be placed above a membrane. The theory and applications of ultrasound in membrane separation processes, not for food alone, have

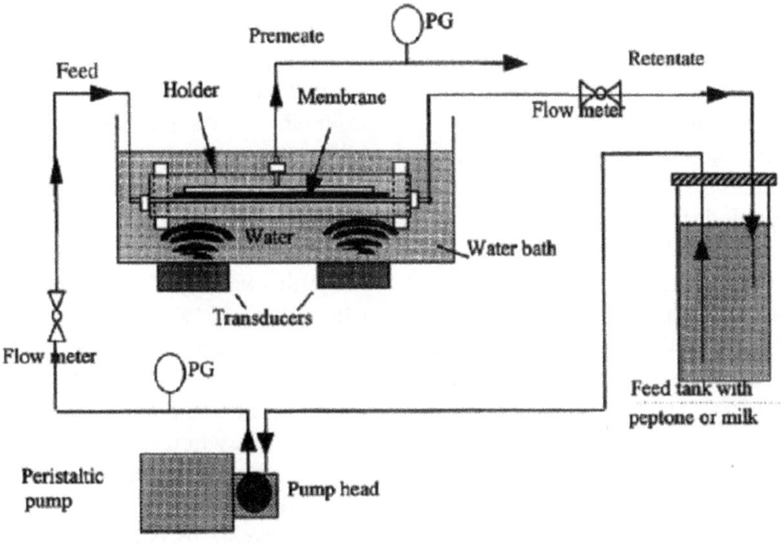

Figure 11.20 Experimental unit of ultrasonic-assisted CF filtration. PG, pressure gauge.
Reprinted from *Ultrasonics*, **41**, T. Kobayashi, T. Kobayashi, Y. Hosaka and N. Fujii, Ultrasound-enhanced membrane-cleaning processes applied water treatments: influence of sonic frequency on filtration treatments, 185–190,[212] Copyright (2003), with permission from Elsevier.

been reviewed.[84,204] Ultrasound can be integrated in membrane processes during on-line separation or during cleaning with water or chemicals. It does not change the inherent membrane selectivity, as does an electric force. The flux-increasing effects of ultrasound irradiation result from a combination of several mechanisms, such as cavitation, radiation pressure and acoustic streaming.[205] They increase the mass transfer across the CP layer, breaking and stirring up the fouling layer on the membrane surface, and the fragments are then carried away by the feed flow. A decrease in osmotic pressure at the membrane surface also increases the permeate flux.

The effectiveness of ultrasound-assisted membrane processes is dependent on the frequency, power intensity, mode of application (intermittent or continuous), apparatus configuration (*e.g.* ultrasound propagation direction; from the permeate side or from the retentate side[206,207]), distance between the membrane and the transducers,[208] membrane material, feed properties, and operating conditions such as TMP, CFV and feed temperature.[84] Ultrasonic power increases the flux,[207] with the lower frequencies improving cleaning efficiency.[207,209–212] Continuous waves offer slightly greater flux increases than intermittent waves.[208,209,213] However, the pulsed mode could be more beneficial when energy consumption and destructive effects on the membrane are considered.

Studies on applications of ultrasound have been performed with single flat-sheet and single tubular and hollow-fiber membranes. Fiber and tubular membranes can be immersed in an ultrasonic bath where they are directly exposed to the ultrasonic irradiation, increasing the ultrasonic effects. Low-frequency ultrasound increased the flux 1.2–1.7-fold during UF of whey.[209,214] In combination with a turbulence promoter, a double increase in flux was observed. Similar flux improvement was observed in the separation of a binary mixture of proteins.[206] In MF of fresh milk, ultrasound power of 40 W enhanced the flux by 460%.[507] A flux increase of up to 4–6-fold during MF accompanied by ultrasound of a suspension of baker's yeast and BSA was reported.[215] Ultrasound can also be applied as a flux enhancement technique for RO, since it effectively removes both inorganic and organic foulants on the membrane surface,[216] and an increase in flux of 50–264% was reported.

Filtration coupled with ultrasound results in a more labile gel layer and less compressive cake growth, which is readily removed with less washing.[209,212,214] Hence this method can reduce the cleaning frequency. An increase of 5–10% in cleaning efficiency was observed when a whey-fouled UF membrane was cleaned with a surfactant under the influence of ultrasound irradiation.[217,508] UF and MF membranes fouled with peptone and milk, respectively, have been efficiently cleaned with water in combination with ultrasound.[212,218] In whey UF, preheating of feed can assist ultrasound significantly in controlling fouling growth.[509] Ultrasound is especially effective in assisting the clarification of fruit juice.[393–396,510] This technique not only increases the permeate flux but also affects to the quality of the product. Rojas *et al.*[511] observed an increase of antioxidant compounds in cactus pear juice after ultrasound treatment, whereas Aghdam *et al.*[512] found a reduction of total anthocyanin content in pomegranate juice. Ultrasound also reduced effectively the fouling resistance during UF of dextran solution and cherry juice.[207,513] Recently, some studies have suggested that a combination of ultrasound with MF or UF is an appropriate method to inhibit and control some microorganism in fruit juice.[514,515]

Generally, ultrasound treatment can be applied to all kinds of feeds, which are not highly fouling, but not to all kinds of membrane materials. Many authors have reported the occurrence of membrane damage or erosion due to ultrasound irradiation.[211,213] The membrane degradation increases with ultrasound intensity.[213] PVDF and hydrophilic PAN are more tolerant to ultrasound than PES, cellulose nitrate–cellulose acetate and nylon 6 membranes.[219,220] Membrane cracking can even occur with alumina membranes at high ultrasound power (3.8 W cm^{-2}) and when the membrane is placed within the cavitation region.[208] In another study, no damage to a ceramic membrane from the same material was observed after prolonged exposure to ultrasonic power of 20 W cm^{-2}.[205] Increasing the CFV may reduce the physical erosion caused by repeated oscillations of cavitation bubbles trapped at a point on the membrane surface.[217]

For a design such as presented in Figure 11.20, the loss of acoustic energy in the ultrasonic bath is important. There are still several issues with this technique that need to be addressed before it can be widely applied in industry. An apparatus still needs to be developed to facilitate large-scale applications with effective use of energy through even distribution and minimum energy waste. Thin transducers can be incorporated near the membrane to reduce energy loss.[213] Membrane damage during long-term membrane processes still needs to be investigated. It must be assured that the ultrasound does not adversely affect the quality of feeds. Cavitation may cause denaturation of proteins.[221] Suitable selection of parameters is essential. Favorable conditions are low frequency, a sufficient power level, but not high enough to damage the membrane, and a low TMP. Stainless-steel reflection plates with flat or round shapes can be placed at the opposite side to the transducer to increase the flux enhancement of the ultrasound.[222]

11.6.7 Other Techniques

Several other techniques that can be used to reduce the deterioration of membrane performance due to fouling include the use of scouring particles and gas sparging . Addition of scouring particles finds limited use in food processing. In this method, the particles can scour the cake layer but they should be recollected, washed and reused, which makes membrane processes more laborious. The addition may also exert a greater wearing effect on the pumps and damage the membrane. Several tests on adding scouring material during membrane cleaning have been performed.[63]

Gas sparging involves injecting gas (*e.g.* air) into the feed stream *via* a Y-tubular piece and a two-phase gas–liquid flow is created in the membrane module. The method can be employed during on-line filtration or during membrane cleaning. Depending on the gas injection factor *r*, defined as the ratio of the superficial flow velocity of gas to the sum of the superficial velocities of gas and feed, the two-phase flow pattern in tubular membranes or hollow fibers changes from bubble flow to slug flow and finally to annular flow. Gas bubbles disrupt the CP and slug flow may erode the cake layer, hence enhancing mass transfer at the surface. Cui *et al.*[223] gave an extensive review of the acting mechanisms and applications of air sparging in all fields of membrane applications, but mostly in environmental engineering.

Air sparging is widely applied in submerged membrane bioreactors for the treatment of sewage or industrial wastewater, where air bubbles are introduced externally to the lumen of the hollow fibers with a twofold function: aeration for microbial activities and control of fouling. In food research applications, a significant improvement in flux has been confirmed in several tests. Gas sparging increases the critical and limiting fluxes during membrane filtration.[92,224] MF for removal of suspended particles from tuna spleen extract[91] showed a flux increase of 300% when nitrogen gas was introduced at $r = 0.38$. Introduction of gas was not effective when *r* was higher than certain values, which means switching to a different flow pattern.[91,92] In MF of a yeast suspension, and depending on the membrane characteristics, cake formation is more important than internal fouling at higher feed concentrations and *vice versa*. This then dictates which method, air sparging or backflushing, should be used to improve the steady-state flux.[225] Gas entrained in an *Escherichia coli* broth was observed to improve the flux by 80% during concentration of the bacteria with a ceramic MF membrane.[65]

When using gas sparging, deterioration of food quality should be taken into account. Proteins could be denatured (structure changed) during contact with air. Oxygen in injected air is an oxidizing agent, especially at high temperatures. Shear-induced denaturation of enzyme activities was observed at high *r* values.[91] Gas sparging improves the permeate flux but either increases or decreases the selectivity depending on the flow patterns, *e.g.* slug or bubble flow.[92,95] Success in employing this technique is highly dependent on the feed properties. Introduction of air can cause foaming with some kinds of foods, such as milk.[183]

11.6.8 Selection of the Techniques

All factors, including membrane material and geometry, feed properties and operating conditions, determine which fouling mechanisms will be dominant in a filtration process. These factors can also dictate the choice of techniques for the enhancement of the membrane separation processes. For instance, backpulsing is least dependent on feed properties compared with backflushing, crossflushing and air sparging.[183] The extent and mechanisms of fouling are premises for selecting a suitable technique since a certain treatment may be effective for one type of fouling but not for other types.

Combination of several methods in one application is a trend in resolving membrane fouling issues. Using backpulsing in a uniform transmembrane pressure (UTP) system offers significant improvements in performance compared with the use of UTP alone in MF of transgenic goat milk.[45] However, the energy consumption efficiency for this particular study of a pilot system has not yet been determined. Several studies in environmental research showed that an electric field and ultrasound when applied simultaneously can have a synergistic effect in suppressing fouling.[226] However, this strategy seems to have no potential for food applications since both of these techniques are harsh treatments. Ultrasound may increase the possibility of oxidation caused by an electric field.[227] Ultrasound can be used in conjunction with crossflushing during cleaning of the membrane to remove surface fouling effectively.[228] Since ultrasound is not very effective at removing in-pore fouling,[214,229] its use in conjunction with techniques involving permeate backflow is a promising strategy. Air sparging in combination with backflushing was found to increase the flux tremendously in the treatment of wastewater using a membrane bioreactor.[230] A more easily controllable approach is to use air sparging together with static mixers. This combination creates strong shear on the surface and increases the flux due to reduction of material deposition on the membrane surface.[55,231,232]

In food processing, it is not only high productivity that is essential but also the quality of the processed products. Hence maintaining the selectivity of the membrane during the filtration process is important. All additional techniques discussed have been proven to improve the permeate flux and, at the same time, maintain membrane selectivity. However, the addition of more techniques makes the system more difficult to control and to clean, increases the risk of microbial contamination, decreases the ease of scaling up, increases the footprint of the system and increases the investment costs. The energy consumption is considered per volume of product filtered and the possible adverse effects of the process on product quality must be taken into account. Considering all these factors, a suitable method or combination of several methods can be selected. An experimental approach can then be used to validate the option and set the optimum processing conditions. Available realistic models from the literature will assist in focusing the experimental work.

11.7 Membrane Cleaning and Sanitation

Membrane cleaning is just as important in determining the success of a membrane process as the other factors discussed. Cleaning is usually performed when the flux falls below a predetermined level. Generally, membrane cleaning is carried out in three steps: flushing with water, extensive cleaning with chemicals and/or enzymes and finally rinsing with clean water. In first step, the permeate flow is first closed for a period of time and then opened again. This will prevent loosely attached foulants on the membrane surface from entering the membrane pores. After the cleaning, a sanitizing solution can be applied to the system followed by rinsing with clean water before putting it to use.

The cleaning efficiency or flux recovery (J_r, %) is determined using eqn (11.8), where J_c is the water flux after cleaning and J_w is the water flux measured before the fouling took place. Both are measured under the same operating conditions.

$$J_r(\%) = \frac{J_c}{J_w} \times 100 \tag{11.8}$$

When a system must be stopped for cleaning, the production also stops. Hence the working principle should be to try to diminish as much as possible the formation of (irreversible) fouling rather than let it form and clean later. A production line normally possesses several membrane filtration units. To guarantee the continuity of production, one unit at a time can be cleaned while the rest are in operation.

An effective cleaning process is one that causes the least damage to the membrane and the installation, achieves maximum recovery of the original flux, is performed in the minimum time with the least consumption of the cleaning materials, is easy to perform and meets the sanitary requirements of the whole process. Selection of the types and concentrations of chemicals and time of chemical cleaning is very important since they may cause membrane damage (swelling, cracking), which then shortens the lifespan of a membrane and increases the costs of membrane replacement.[3,233,234,516] Flux recovery is normally lower than 100% compared with the flux of the original membrane.[235] If it is higher than 100%, the membrane integrity is affected by cleaning. A combination of an acid and a base (two steps) is often used to clean membranes, but not for organic membranes such as PVDF and polyamide that are sensitive to NaOH.[3,233] Recently, Filloux *et al.* suggested a new method using only acidified nitrite solution (one step) to remove biofouling in RO membranes.[517] In a plant-scale MOD system, the combination of oxalic acid and citric acid can be used to remove fouling layers.[518] Inorganic membranes are more resistant to chemicals, and also to other cleaning parameters such as temperature and pH, than organic membranes. Ceramic membranes can withstand CIP chemicals (*e.g.* in dairy processing), which gives them an advantage for expanding applications of this type of membrane. In general, there is no common cleaning regime that is effective for all membrane materials or all types of foulants. There have been numerous reports on cleaning and sanitation of different membrane materials in processing applications of a variety of food types. Membrane suppliers normally deliver, along with their products, a list of applicable chemicals for cleaning and sanitation and also the recommended cleaning procedure.

Cleaning solutions are normally circulated for a period of time. A high TMP has a detrimental effect on the cleaning efficiency since it causes deeper penetration of the foulants into the membrane during cleaning. That is why, during cleaning, the TMP is normally set lower during filtration.[236,237] The CFV has, to some extent, a positive effect. There are optimum cleaning times and temperatures over which the flux recovery is reduced.[3,236] There is also sometimes an optimum cleaner concentration.[233] The flux recovery also depends on the pH of the cleaning solution.[238,239]

In addition to acids and bases, other components such as complexing agents (*e.g.* EDTA), detergents and surfactants (*e.g.* sodium dodecyl sulfate, Triton-X series), can be used to increase the cleaning efficiency.[233,240] An extra cleaning step using enzymes, for example proteases and amylases, is also helpful.[241,242] Enzymes and enzyme–detergent mixtures in some cases offer better flux recovery than do chemicals.[243–245] Reuse of enzymes for consecutive cleaning steps is possible, but a certain percentage of enzymatic activity is lost after each cleaning step.[244] It is important to optimize the cleaning parameters and concentration of enzymes to prevent the deposition of the enzymes in the membranes.[244,246] Enzyme cleaning is normally performed at neutral pH. It is less destructive to the membrane and causes less impact on the environment than chemical cleaning. In the dairy industry, rhamnolipid, a low-cost and environmentally friendly biosurfactant, has been suggested as an appropriate solution for replacing chemical surfactants.[519] For a long time, non-ionic surfactants were rarely used in membrane cleaning because of their irreversible adsorption on the membrane surface. However, recent study of Kingma showed that some alkyl alcohol-based non-ionic surfactants have a low affinity with the membrane surface while showing excellent cleaning performance and are environmentally friendly.[520] Other physical means of supporting membrane cleaning, such as sonication, electric field and scouring particles, have also been studied.[210,217,229,247] As discussed previously, introduction of these methods is difficult for large-scale membrane systems.

The selection of suitable sanitizers should be considered, since some of them may cause damage to the membrane or membrane-supporting facilities.[248–250]

11.8 Comparison Between Membrane Separations and Corresponding Traditional Technologies

11.8.1 General Applications and Technological Advantages of Membrane Separations

The number of applications of membrane technology and the extent of the areas in which it is applicable have been steadily increasing since its introduction to the food industry about four decades ago. Membrane processes are being used to modify conventional processing methods, as an alternative to traditional steps in a processing line or as a new technique in developing new ingredients with desired nutritional and technological characteristics. Applications of pressure-driven membrane processes in food industry can be summarized as follows:

- Concentration using, *e.g.*, UF and RO as alternatives for thermal evaporation or distillation. If NF is used for this purpose, both concentration and partial demineralization are accomplished.
- Separation in combination with concentration using, *e.g.*, MF and UF as alternatives to centrifugation, gravity separation or chemical extraction.
- Cold pasteurization and sterilization for microbial stability as alternatives for heat treatments or use of preservatives, such as in MF of skim milk and in UF of fruit juices.
- Rejection of impurities such as in MF clarification of beer and wine as alternatives for conventional filtration methods using filter aids.
- Purification of water for processing or treatment of wastewater. Depending on the membrane processes used, treated water can be reused for certain purposes in the food processing plant such as for cleaning and boiler makeup.
- Recovery of valuable components from by-products or effluents of processing activities.

Concentration using membrane technologies does not involve a phase change of a liquid, hence it consumes less energy than thermal evaporation. For complex products with high viscosity and concentration, a high VCR is hardly obtained with membrane separations and then it can be used to replace the initial stages of conventional concentration methods. Membrane separation eliminates the need for coagulants to separate proteins from a solution containing other smaller molecules such as salt, non-protein nitrogen (NPN) and sugars. MF, with appropriate pore sizes, may offer a similar food quality-preservation function to heat pasteurization but, additionally, MF also reduces the heat-induced destruction of nutritionally and technologically important functional properties of components (*e.g.* vitamins and proteins) and also preserves the natural food flavor (*e.g.* does not exert a cooking flavor). Other advantages relating to the use of the technology and the resulting food quality will be discussed in the subsequent sections with reference to specific foods.

However, vigorous filtration may have some damaging effects on food quality. Denaturation of proteins may occur at high shear rates and the tendency for fat oxidation increases in the presence of oxygen at elevated temperatures. Therefore, the selection of suitable membrane modules, operating conditions and permeate flux-enhanced techniques is important for the processing of feeds that are sensitive to the effects of those factors. Proper design of the equipment/instruments will also help. For example, restricting small-diameter tubing for control of pressure instead of using valves can reduce shear.[251]

11.8.2 Economic Aspects of Membrane Processing Applications

The economic feasibility of a membrane process is affected by the initial investment costs and operational costs. The higher the permeate flux, the lower is the membrane surface area required for any level of productivity, which leads to lower investment costs. The installation cost is also dependent on the membrane material, membrane modules and membrane suppliers. Operational costs are determined by the energy and time consumption per unit of production, automation of the installation, periodical replacement of membranes and washing chemicals. Space requirements of the installation in the processing plant are also a factor.

To illustrate, economic analyses for several plants employing membrane technologies are presented. In the concentration of skim milk from 8.8 to 45% solids content, the combination of RO and a multistage evaporator required the same capital and operational costs as a multistage evaporator alone. The combined processes, however, consumed much less energy per unit of the final concentrated product.[252] The operating cost (energy) for NF was as low as 0.04 times the evaporation cost of dehydration of diluted feed solutions containing organics.[253,254] The permeated water from RO and NF can then be recycled. Clarified fruit juices produced after partial enzymatic liquefaction using MF with a 0.2 μm pore size was highly competitive in economic costs and could offer a higher production yield than other established processes. In addition, the MF retentate can be used for the production of pulpy juices.[255] Computer modeling showed that a ceramic MF membrane under optimum conditions for clarifying 113.6 $m^3 h^{-1}$ of corn starch hydrolyzate for the production of dextrose could save over $1.3 million per year compared with the conventional method. The capital cost for the membrane system could be paid back within 1.6 years.[53] Using a non-linear programming optimization model for evaluating the net revenues and potential profitability of cheese making in a 3-year period (1998–2000), the use of MF prior to cheese making for the separation of soluble components from a bacterial fermentation broth (see Section 11.9.1.1) produced higher net revenues, relative to conventional cheese technology, most of the time, for both Cheddar and low-moisture, part-skim mozzarella, leading to an appreciable increase in net revenue for both cheeses.[256] The annual cost of a backpulsing-supported MF was found to be 25 and 50% lower than that of a centrifugation system and a rotary vacuum filtration system, respectively.[179]

High investment costs and membrane fouling are issues often cited as hindering the widespread uptake of membrane separations. In the example mentioned above,[53] the total capital cost for the installation of such a membrane plant in 1998 was $2.08 million, with membrane modules accounting for 85% of the cost.[53] This high investment is a real challenge for food manufacturers, especially in small- and medium-scale factories. Organic membranes generally have a short service life whereas ceramic membranes can be used for 5–10 years, but they are much more expensive and not as compact.[53,257] Research on membrane materials is aimed at producing membranes that are more resistant to chemicals, heat and pH. Such membranes would need less frequent replacement, thus increasing the economic attraction of the process. With current advanced materials and techniques, membrane modules with an active skin layer as thin as 0.1 μm or less can be fabricated. With thin membranes, a smaller in-pore filtering effect occurs, which leads to internal fouling. The technique is more energy efficient when operated in a continuous or semi-continuous mode than in a batch process.[258] Integration of membrane processes into existing processing lines rather than separate implementation may increase energy efficiency use and plant space utilization. For example, excess heat from other stages can be used to heat the feed for the filtration. The use of cleaning water and chemicals can also be reduced.

A brief review of applications of membrane separation in the processing of several common types of food is presented in the following section.

11.9 Applications of Membrane Separations in the Food Industry

11.9.1 Membrane Processes in the Dairy Industry

Milk is a complex and polydisperse system with particles having different charges and a wide range of sizes, making membrane separation feasible but dependent on the membrane material, pH, temperature and other operating parameters. In MF of skim milk, casein micelles are responsible for the initiation of irreversible fouling whereas whey proteins participate in irreversible fouling by adsorption phenomena and/or by accumulation of the proteins in the micelle deposit. Milk Ca^{2+} ions are assumed to create bonds in the deposit between the membrane and the micelles and between micelles themselves.[259]

Many food membrane processing applications come from the dairy industry.[44] In the early 1970s, dairy applications mainly used UF for the separation of whey protein from cheese and RO for the concentration of whey. UF was later used for milk protein standardization in the production of cheese.[99] More recently, research on MF applications flourished owing to the development of ceramic membranes and the new UTP techniques for suppressing membrane fouling. The trend in applications went from the use of membrane processing steps to the use of a membrane process in a single processing step and to the use of several membrane types in several steps in integrated production lines. Membrane processes can replace conventional processing methods and be implemented as innovative methods for the production of custom ingredients in the development of new food products, or improvement of existing food products.[1] A downstream processing line for milk with the corresponding membrane processes involved at each level is shown in Figure 11.21.

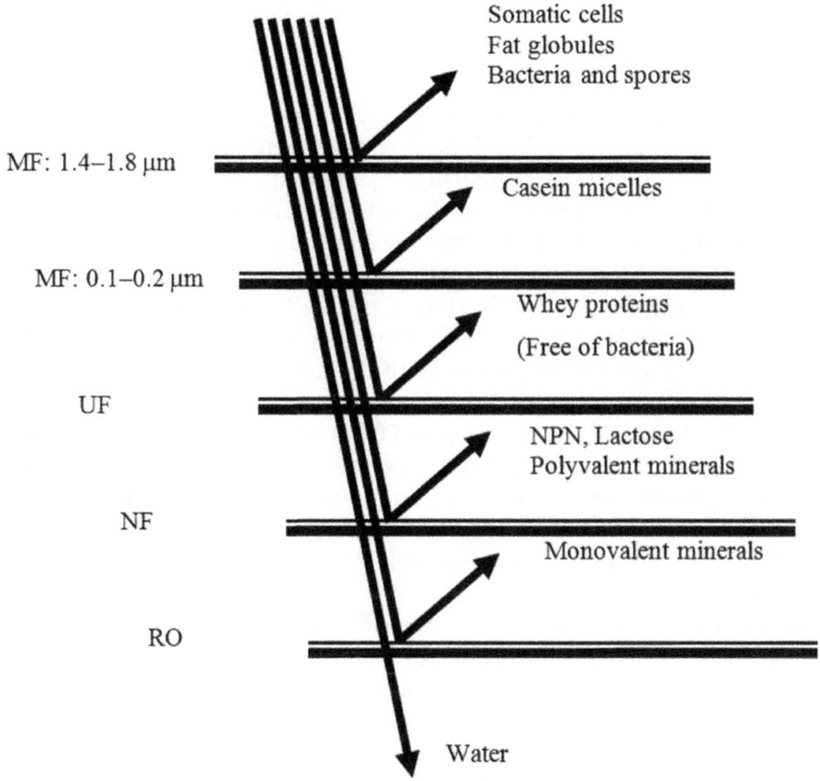

Figure 11.21 Membrane processes involved in the processing of milk to its purified groups.

The main applications of membrane processing in the dairy industry and their advantages relative to corresponding conventional technologies are briefly discussed in the following section. For a detailed discussion, readers are referred to several reviews.[4,7,63,109,260]

11.9.1.1 *Microfiltration*

Removal of bacteria and spores from skim milk or the so-called cold pasteurization (*e.g.* Bactocach process) for the production of milk[47,107,261] and cheese262 is one of the main applications of MF. The process preserves the milk flavor and other milk components. It reduces β-lactoglobulin/κ-casein complexation and, possibly, the needs of preservatives in cheese making. MF membranes with a pore size of 0.1–0.2 μm are used to fractionate skimmed milk into (1) a fraction enriched with micellar caseins for vatless production of cheeses (semi-hard and hard)[74,75,256,263–266] and for the production of casein-enriched ingredients without using coagulating and neutralizing agents,[70,132,267] and (2) a sterile permeate enriched with native serum proteins. MF manufacture of cheese requires smaller quantities of rennet and additives, reduces the total production time and increases plant capacity. It also allows modification of the casein-to-whey protein ratio and casein-to-fat ratio to achieve optimum structural and sensorial properties of the final products. In addition, the sweet whey permeate contains the major whey proteins in a highly functional state without rennet, bacteria and starter culture. It has a consistent composition and is extremely low in fat.[268] Using UF, this fraction is suitable for the direct manufacture of functionality-enhanced whey protein concentrate (WPC) and whey protein isolate (WPI) with improved oxidation stability. MF can be applied to separate fat globules from whole milk for standardization of the milk fat content.[99]

MF can potentially replace clarification processes such as pH adjustment or salt addition followed by gravity settling or centrifugation for the removal of cheese-making remnants (fat, starter culture, *etc.*) in the manufacture of WPC and WPI.[269,270] Bacteria and fat removal from cheese brine is another use of MF.[109] Many attempts have been made to use 0.1–0.2 μm MF membranes for the isolation of milk fat globules membrane fragments from dairy processing co-products such as buttermilk or butter serum, for the production of innovative ingredients with nutritional and functional properties.[271]

MF can be employed as a more efficient method than gravity separation and centrifugation to fractionate fat globules according to their diameters to improve the textural and sensorial properties of dairy products.[272,273] MF with a membrane pore size of >10 μm can remove somatic cells from whole milk.[274] Extraction of lactoferrin from whey using electro-MF is under investigation and could potentially replace more expensive techniques such as ion-exchange chromatography.[191]

11.9.1.2 Ultrafiltration and Nanofiltration

Ultrafiltered milk, used for the continuous production of cheese (*e.g.* cottage cheese, soft cheese varieties, some hard cheeses such as Cheddar and Parmesan), increases the cheese yield by retaining whey proteins and other milk constituents in the cheese matrix. It also reduces the total processing time due to a shorter clotting time and elimination of the cutting and whey drainage steps, and reduces the use of salt, starter culture, rennet and calcium chloride.[275–287] UF is used in the production of delactosed milk and in protein standardization in dairy processing. Other uses include the production of milk protein concentrate and milk protein isolate and in the development of new products with desired texture and functional properties such as fermented milk and ice cream.[288–292] UF and RO of milk on dairy farms would lower transportation costs.[275,293] UF is widely used to concentrate cheese whey for the production of whey protein concentrate and whey protein isolate with reduced concentrations of lactose and minerals,[294–299] which is incorporated into cheese.[300] Recently, UF was applied to produce and enhance bioactive compounds in the dairy industry such as lactic acid from fermented whey and phospholipids from buttermilk.[387,388]

NF is an alternative method to ED and ion exchange, achieving significant desalination and simultaneous concentration of cheese whey and acid casein whey for the recovery of whey proteins.[5,8,9,11,38,82,301–303] Various milk components such as lactose, lactate and milk oligosaccharides can also be recovered from whey or milk ultrafiltrate by NF.[389–392] The nanofiltrate can be discharged directly into the sewer, hence reducing the amount of wastewater required to be treated at the plant.[10] However, the greatest problem when applying UF and NF in the dairy industry is the formation of biofoulants on the membrane surface.[397] Therefore, these methods always need the assistance of methods such as ultrasound and electric fields,[393–396] which will contribute to increased investment costs.

11.9.1.3 *Reverse Osmosis*

The main use of RO is in the preconcentration of milk and other dairy liquids (*e.g.* whey) and requires less energy than conventional evaporation.[304,305] Permeate from single RO or NF + RO operations could be reused as heating, cooling, cleaning and boiler feed water to reduce the amount of waste effluents from dairy processing.[14] Purified water complying with drinking water criteria could be achieved by a two-stage RO + RO process for the treatment of dairy effluents.[136,301,306,307]

11.9.2 Membrane Processes in the Brewing Industry

Beer is a multicomponent beverage containing carbohydrates (mono- and oligosaccharides, dextrin (α-glucans), β-glucans, pentosans, *etc.*), proteins and polypeptides, phenolic compounds, alcohols, minerals and vitamins. It is the second most consumed beverage in the world after tea and consumption is growing.[42] There is currently an increasing demand for cleaner and more economical clarification techniques in beer production. Applications of membrane separation in the brewery industry have been reviewed by Moraru and Schrader[257] and more recently by Ambrosi *et al.*[444] Major areas of applications are clarification of rough beer (main filtration after fermentation and aging), final filtration for microbial stability before bottling/canning/kegging and recovery of beer from yeast slurry in tank bottoms.

Beer clarification is still carried out mostly by dead-end filter presses with the aid of diatomaceous earth (DE). This material is potentially harmful to human health and disposal of the used DE slurry is an environmental concern. Replacing this step with CF MF will eliminate the contamination of filtered beer with foreign materials and, in addition, will minimize space requirements in processing plants. The challenge in MF clarification of beer is to achieve a balance between the retention of particulates (yeasts, haze flocs, *etc.*) and the transmission of soluble macromolecules contributing to the quality of beer.[172] Carbohydrates, protein–polyphenol complexes and yeast cells (in the case of rough beer) are key components contributing to membrane fouling.[42,308,445] Internal and external fouling are important factors. Membrane fouling reduces the flux and affects the consistency of the final products. However, Gan *et al.* reported that it is possible to obtain a permeate with a quality better than that set by European Brewing Convention (EBC) standards, over a period of 24 h.[172] The necessity to perform the clarification at low temperatures (0–4 °C) also causes a low permeate flux, thus requiring a larger membrane surface area. Permeate fluxes of 10–100 $L\,h^{-1}\,m^{-2}$ in beer membrane clarification are normally cited in the literature, as required in the industry.[183] Maximizing the utility of previous investments in equipment is another factor limiting the switching from DE filtration to MF, although several membrane suppliers, such as Pall/GEA, Alfa Laval/Sartorius and Norit (The Netherlands), already have industrial systems for beer filtration. The availability of ceramic membranes with decreasing prices, the development of microsieves[183] and optimization of performance-enhanced techniques such as backflushing or backpulsing are expected to advance the application of membrane separation in beer filtration.

Final filtration of the clarified beer using MF with 0.45 μm pore size for the removal of spoilage microorganisms before bottling can be used instead of thermal pasteurization. This will preserve the freshness and avoid unwanted changes in taste of the final product. This type of product is denoted cold-filtered or non-pasteurized beer. MF for recovery of beer from tank bottoms (yeast slurry) at the end of fermentation of lager-type beers can replace the conventional methods such as centrifugation and pressing. Commercial systems for beer recovery are available (*e.g.* Pall's Keraflux beer recovery system). Water is an essential factor deciding the "real beer" quality of beer and RO is more advantageous than ion exchange in the purification of brewing water since it also remove hardness, impurities and other organic materials (substances, microorganisms, *etc.*). In the past, alcohol removal from fermented beer was conducted through thermal evaporation, resulting in the loss of sensorial components at high temperatures. Isolating these components through composite membranes before blending them to final products is a good alternative method. Studies by Olmo *et al.*[446] and Catarino and Mendes[447] showed that low-alcohol beer produced using this method preserved most of the aroma of the original product. Dialysis is also a popular method in the production of high-quality alcohol-reduced beer. Recently, membrane osmotic distillation has been widely used to eliminate alcohol from beer.[448,449] In comparison with other methods, MOD is simpler and faster while still retaining the nutrients and organoleptic properties of the product.

11.9.3 Membrane Processes in the Winemaking Industry

As in breweries, major applications of membrane separation in winemaking are wine clarification (pore size >0.65 μm), replacing DE filtration, and final filtration (pore size 0.2–0.65 μm) before bottling. The use of MF in wine clarification dates back to the early 1970s. The application was controversial since the rejection of colorants and aromatic and polyphenolic compounds by the membranes may adversely affect the quality of the wine produced. Nowadays, these applications are fairly popular in industrial enology. In contrast to beer filtration, wine filtration is normally performed at room temperature (20 °C), which promotes a higher flux than chilling temperatures.

Selection of the membrane material according to the type of wine being processed is very important. Decreases in concentration of 12 of the 100 aromatic compounds analyzed in Cabernet Sauvignon wine (red wine) were observed after MF in a study by Arriagada-Carrazana *et al.*[309] The polyphenolic profile was also affected, probably owing to the adsorption of these components on the membrane. Polyphenols and polysaccharides are adsorbed strongly by PES but only marginally by PP membranes.[97] Adsorption of polysaccharides increases with decreasing hydrophilicity of the membrane while adsorption of tannins has the opposite trend.[96]

Optimization of the filtration parameters is also necessary.[310] About 1000 KMS WINEFILTER™ (Koch Membrane Systems, Wilmington, MA, USA) systems have been installed worldwide for wine filtration. These CF MF systems hold 6–12 cartridges comprising hollow fibers developed specifically for wine filtration and operated with the assistance of a backpulse system using gas. Membrana-Charlotte (USA) has commercialized Liqui-Flux® systems that utilize hollow fibers made from PP for the purpose of wine clarification.

NF and RO can be employed for the concentration of grape must to increase the concentration of sugars and other components before fermentation.[311,312] It can also be used for the concentration of valuable red wine components while maintaining the alcohol content of the wine, since water and alcohol readily pass through the NF membrane.[313,314] Lowing the alcohol content is a current trend in the beverage industry to limit negative impacts on consumers' health.[449] Membrane processes are mild treatments and the use of membrane processes aimed at reducing the alcohol content will preserve the aromatic compounds better than the use of heat. Similarly as in the brewing industry, membrane osmotic distillation is a promising technology to reduce alcohol levels in the winemaking industry.[420,421] In a study by Liguori *et al.*, there was no significant change in volatile acidity, color or polyphenol and organic acids content when partial dealcoholizing Aglianico red wine to about 2 vol% using the MOD method.[420] Tartaric acid stabilization in wine is an interesting application of ED to wines. ED removes certain ions such as potassium and calcium and tartaric acid from wine, hence the undesired precipitation of potassium hydrogentartrate at the bottom of bottles of wine, which consumers find objectionable, can be avoided.[20,315] "Cold pasteurization" using nano/micro-tubular cellulose, a cheap material, to remove spoilage microorganisms in wine such as *Saccharomyces cerevisiae* and *Acetobacter pasteurianus* at low temperature replaces traditional methods that can destroy bioactive compounds in wine at high temperatures.[450] Other applications of NF and RO have been reported.[13]

11.9.4 Membrane Processes in the Production of Fruit and Vegetable Juices

In the production of fruit and vegetable juices, the main area of application is in the clarification step, using MF and UF processes.[93,316,415–417] Pretreatments with enzymes, depectinization and flocculation are some of the treatment methods used to increase the membrane separation performance. Ultrasound is used to reduce membrane fouling after membrane clarification.[439] UF-clarified juice is microbially stable without heat treatment or addition of preservatives.[93] MOD is a novel approach for the production of fruit juice concentrates.[398–400] In this method, two sides of a hydrophobic porous membrane are contacted with fruit juice and a stripping solution (salt/sugar solution) to create a water flux based on the water activity difference. The organoleptic characteristics and nutrient bioavailability of the products after dilution resemble those of the fresh fruit drink.[401,402] Separation and purification of bioactive compounds in fruit/vegetable and their by-products also play an important role in the study of membrane technology. This approach not only helps enhance the value of agro-products but also reduces environmental costs.[403–405] Crossflow filtration has been applied to recover carotenoids in gac fruit and phenolic compounds in orange juice.[406,407]

In the past, NF was considered an alternative method to ion exchange for demineralizing (NO_3^-, NO_2^-, *etc.*) vegetable juice. More recently, NF has been widely used to enhance the concentration of phenolic compounds in various types of fruit.[408–411] Some fruit juices are too acidic for consumption. ED with bipolar membranes has been found to be a useful method to deacidify fruit juices, such as orange, pineapple, cranberry and passion fruit juices, with only slight organoleptic changes.[17,18,412] The traditional techniques for this purpose are calcium salt precipitation, which involves the addition of chemical reagents, and the use of ion-exchange resins, which modify drastically the aroma profile of the juices and generate polluting effluents during the regeneration step. In addition to deacidification, ED can be combined with UF to enrich antioxidant compounds in juices.[413,414]

11.9.5 Membrane Processes in the Sugar Industry

UF of clarified sugar juice allows the production of a less colored sugar in the industrial production of cane sugar.[317,318] UF can also be used for the treatment of effluent derived from the regeneration of anionic resins used to separate colorings and salt. This salt solution can then be reused in the regeneration step to save water and salt and to reduce the amount of wastewater.[319] In a preliminary investigation, a combination with RO could save 33% energy consumption in the concentration of dilute sugar juice, compared with thermal evaporation alone.[15] A combination of MF and UF in the purification and decoloarization of raw sugar beet juice was observed to be give better results than lime-carbonated carbonation treatment in terms of color, turbidity and hardness of the clarified juice.[320] UF and NF can thus be used in the refiningery of sugar syrups.[321,322] RO is a feasible method for the treatment of press water for reuse in the extraction of sugar beet cassettes. The retentate can be subjected to crystallization of low-quality sugar.[323]

11.9.6 Membrane Processes in the Production of Soy Ingredients and Products

The use of UF to concentrate aqueous soy extract for the production of soy products would reduce transportation costs, improve the production of soy ingredients and standardize tofu and soy cheese production.[441,442] UF of defatted soybean flour extract in the production of soy protein concentrate eliminates the use of chemicals and provides higher yields with superior functionalities, to of the products obtained because of the incorporation of whey proteins in the UF soy protein concentrate by conventional soy technology.[94,324] ED eliminates the use of HCl and NaOH in the conventional production of protein isolate. However, the loss of proteins which that are not precipitated by at low pH still occurs. Traditional Asian traditional products such as soy sauce contain very high contents of NaCl (18–20%) and consumption at high doses may pose a risk to the health. NF is a realistic method for the partial removal of NaCl while maintaining the product flavor.[325] MF of soy sauce has a dual functions: clarification and removal of microorganisms.[326,443] The Soy protein fraction can be separated into different fractions by using isoelectric precipitation and MF.[327] Soy protein isolate fractions with different technologically functional properties are potential ingredients for the improvement of certain food products and, or for the development of new food products.

11.9.7 Membrane Processes in Tea and Coffee Production

Tea and coffee are traditional drinks over throughout the world and play an important role in the beverage industry. Recently, the rapid growth of ready-to-drink products has brought many chances for membrane application. The traditional method using thermal treatment to evaporate water often results in decreases in the concentration of bioactive compounds and their antioxidant capacity, which are very important in these products.[451] The use of membrane separation as an alternative green method can help solve these issues. MF and UF were applied to improve significantly the concentration and purity of polyphenols in tea such as theaflavins[452] and epigallocatechin gallate.[453] Similarly, the use NF and UF helped isolate aroma compounds, caffeine, polysaccharides and chlorogenic acid from coffee extracts, which can be added to the final products to improve the quality.[454–458] Most of these approaches are easy to scale up in the food industry.

11.9.8 Membrane Emulsification

After 30 years of development, membrane technology ultimately became an important processing tool in the food industry. In comparison with traditional methods, membrane emulsification consumes less energy and reduces negative effect on the structure of the feed materials.[422,423] Two popular systems in membrane emulsification are rotating membranes and crossflow membranes.[424] Both techniques have been applied to create various food emulsion systems, including water-in-oil, oil-in-water and multiple emulsions.[425–429] Membrane emulsification is especially excellent in controlling the size uniformity of the droplets, resulting in higher stability of foams and Pickering emulsions (emulsions that are stabilized by solid particles) and higher encapsulation efficiency for double emulsions.[429–434] Shirasu porous glass is the most popular material in membrane emulsification[427,431,433] but it is not suitable for application in the food industry because of the low mechanical strength and high contaminant risk. Hence recent studies have focused on using other membrane materials such as ceramics, stainless steel and aluminum.[435–437]

11.9.9 Other Applications in the Food Industry

The application of MF in vinegar making is an established industrial process.[328] UF fractionation of protein hydrolyzates to collect bioactive and functional peptides from plant such as alfalfa and soy[329,330] or animal sources such as milk[331–335] and fish[336,337] has attracted many researchers. This application is inexpensive and easier to scale up than size-exclusion chromatography.

Another research trend is the implementation of membrane separation in recovering valuable components from co-products of various processing activities. This strategy has a twofold benefit for food manufacturers: increasing the profit from recovered components and reducing the cost of wastewater treatment. Protein recovery is carried out by UF,[338–340] whereas lipids (oil-in-water emulsion) can be recovered by MF with a pore size smaller than the emulsion droplet size.[134,138,440] Food processing effluents normally have very low concentrations of components to be recovered, so thermal evaporation is not economically feasible. Dynamic MF systems seem to be a good option since they can operate at high shear rates, leading to high fluxes, and there is a low tendency for solute modification owing to the low concentration, *e.g.* at the initial stage of filtration.

Membrane separation is readily used for water, food processing and wastewater treatment. More economical wastewater treatment methods are being investigated. Effluents (*e.g.* with a low oligomer content) treated with UF in many cases are already suitable to be discharged to sewers or recycled and combined with feed water in prescribed ratios for washing purposes.[339] The fraction (retentate) requiring treatment in which valuable components can still be recovered will then be small. Food processing lines that have a high water consumption (*e.g.* fish processing) also have a great need for wastewater treatment. Reusing the treated effluents will be beneficial, especially where water resources are limited.[340–344]

Although the investment necessary with UF is high, its lifecycle costs are comparable to those of conventional treatments such as screening, diatomaceous earth filtration, ozonation, and the use of chlorine dioxide in the treatment of poultry processing wastewater and its environmental footprint is significant smaller.[339]

Owing to the limited scope of this chapter, it is not possible to cover all industrially viable membrane separation applications. Generally, applications deal with liquid streams. Common applications of MF are for clarification (*e.g.* removal of colloidal particles) as an alternative to traditional filtration with filter aids and centrifugation, and for yeast and bacteria removal as an alternative to heat treatment. Applications of UF are based on its capacity to separate soluble macromolecules such as proteins and polysaccharides and their hydrolyzed products (oligosaccharides, oligopeptides, amino acids, *etc.*) and other molecules such as salts. A suitable selection of UF pore size in combination with appropriate feed and process conditions can even facilitate the separation of proteins with different molecular weights. The main application of RO in food processing is in concentration. NF can play the same role and at the same time offer partial desalination.

11.10 Conclusions and Perspectives

Research to develop new membrane materials and new module design configurations to address membrane fouling and to improve separation efficiency is very active. A more comprehensive understanding of membrane technology is being gained and, from this, additional techniques for increasing membrane process performance are being developed and optimized. Moreover, membrane producers and scientists are now searching for cheaper and simpler methods of manufacturing membranes. All these factors are increasing the scope of membrane applications in the food industry.

The number of applications and also the membrane surface area have been increasing very rapidly since the introduction of membrane technology into the food processing industry about four decades ago. The significance of membrane processes in industrial food processing, such as in dairy processing, beer and wine production, and fruit and vegetable juice processing, is now being realized in developed countries. In developing countries, where food processing is predominantly carried out in large factories, the role of membrane separation in the production of traditional products, such as soy sauce and rice wine, is becoming increasingly important.

On the macro-scale, the increasing costs of energy caused by the intensive use of natural resources and the scarcity of clean water are encouraging producers to replace conventional processing methods with greener and more efficient technologies. The recovery of valuable components, which is not feasible with methods other than membrane separation, reduces the losses of processing materials and provides more benefits to the producers. Membrane processing permits the development of new ingredients and food products, especially nutritional and functionally enhanced food products. Diversification of food products is a means for industry to keep pace with the competitive market. Safer and healthier products manufactured with green technology have a positive influence on consumers' purchasing decisions.

In many cases, the cost of investing in membrane separation units is currently higher than for conventional processes. However, the long-term application of green technology will be more beneficial when industry profits, consumer health and impacts on the environment are considered and realized.

Acknowledgements

The completion of this chapter was financially supported by a BOF scholarship of Ghent University (BOF07/DOS/047) and a 911 scholarship of Vietnam government. The authors would like to thank Ms Angeli D. Cabaltica and Ms Van Thao Nguyen for proofreading the manuscript.

References

1. P. Walstra, J. T. M. Wouters and T. J. Geurts, *Dairy Science and Technology*, CRC Press, Boca Raton, FL, 2nd edn, 2006.
2. J.-L. Maubois, *Aust. J. Dairy Technol.*, 2002, **57**, 92–96.
3. A. Makardij, X. D. Chen and M. M. Farid, *Food Bioprod. Process.*, 1999, **77**, 107–113.
4. P. A. Marcelo and S. S. H. Rizvi, in *Handbook of Membrane Separations: Chemical, Pharmaceutical, Food, and Biotechnological Applications*, ed. A. K. Pabby, S. S. H. Rizvi and A. M. Sastre, Taylor & Francis Group, Boca Raton, FL, 2nd edn, 2009.
5. J. Kelly and P. Kelly, *Int. Dairy J.*, 1995, **5**, 291–303.
6. M. Nilsson, G. Trägårdh and K. Östergren, *J. Membr. Sci.*, 2008, **312**, 97–106.
7. M. Rosenberg, *Trends Food Sci. Technol.*, 1995, **6**, 12–19.
8. A. Román, J. Wang, J. Csanádi, C. Hodúr and G. Vatai, *Desalination*, 2009, **241**, 288–295.
9. E. Suárez, A. Lobo, S. Álvarez, F. A. Riera and R. Álvarez, *Desalination*, 2006, **198**, 274–281.
10. R. Atra, G. Vatai, E. Bekassy-Molnar and A. Balint, *J. Food Eng.*, 2005, **67**, 325–332.
11. B. Cuartas-Uribe, M. I. Alcaina-Miranda, E. Soriano-Costa, J. A. Mendoza-Roca, M. I. Iborra-Clar and J. Lora-García, *Desalination*, 2009, **241**, 244–255.
12. V. M. Matta, R. H. Moretti and L. M. C. Cabral, *J. Food Eng.*, 2004, **61**, 477–482.

13. A. Massot, M. Mietton-Peuchot, C. Peuchot and V. Milisic, *Desalination*, 2008, **231**, 283–289.
14. M. S. Yorgun, I. A. Balcioglu and O. Saygin, *Desalination*, 2008, **229**, 204–216.
15. S. S. Madaeni and S. Zereshki, *Chem. Eng. Process. Process Intensif.*, 2008, **47**, 1075–1080.
16. G. Pourcelly and L. Bazinet, in *Handbook of Membrane Separations: Chemical, Pharmaceutical, Food, and Biotechnological Applications*, ed. A. K. Pabby, S. S. H. Rizvi and A. M. Sastre, Taylor & Francis Group, Boca Raton, 2nd edn, 2009.
17. L. Bazinet, F. Lamarche and D. Ippersiel, *Trends Food Sci. Technol.*, 1998, **9**, 107–113.
18. E. Vera, J. Sandeaux, F. Persin, G. Pourcelly, M. Dornier and J. Ruales, *J. Food Eng.*, 2009, **90**, 67–73.
19. A. L. Quoc, M. Mondor, F. Lamarche, D. Ippersiel, L. Bazinet and J. Makhlouf, *Food Res. Int.*, 2006, **39**, 755–760.
20. F. Gonçalves, C. Fernandes, P. Cameira dos Santos and M. N. de Pinho, *J. Food Eng.*, 2003, **59**, 229–235.
21. M. Greiter, S. Novalin, M. Wendland, K.-D. Kulbe and J. Fischer, *J. Membr. Sci.*, 2002, **210**, 91–102.
22. M. Mondor, D. Ippersiel, F. Lamarche and J. I. Boye, *J. Membr. Sci.*, 2004, **231**, 169–179.
23. S. Nene, G. Patil and K. S. M. S. Raghavarao, in *Handbook of Membrane Separations: Chemical, Pharmaceutical, Food, and Biotechnological Applications*, ed. A. K. Pabby, S. S. H. Rizvi and A. M. Sastre, Taylor & Francis Group, Boca Raton, 2nd edn, 2009.
24. S. Mahesh Kumar and S. Roy, *Desalination*, 2008, **229**, 348–361.
25. H. Susanto, H. Arafat, E. M. L. Janssen and M. Ulbricht, *Sep. Purif. Technol.*, 2008, **63**, 558–565.
26. G. Gésan, G. Daufin, U. Merin, J. P. Labbé and A. Quémerais, *J. Membr. Sci.*, 1993, **80**, 131–145.
27. L. Song, *J. Membr. Sci.*, 1998, **139**, 183–200.
28. S. Hong, P. Krishna, C. Hobbs, D. Kim and J. Cho, *Desalination*, 2005, **173**, 257–268.
29. L. Fillaudeau and M. Lalande, *Food Bioprod. Process.*, 1998, **76**, 217–223.
30. K. Katsoufidou, S. G. Yiantsios and A. J. Karabelas, *Desalination*, 2008, **220**, 214–227.
31. A. Vernhet and M. Moutounet, *J. Membr. Sci.*, 2002, **201**, 103–122.
32. M. Girones, R. G. H. Lammertink and M. Wessling, *J. Membr. Sci.*, 2006, **273**, 68–76.
33. P. S. Tong, D. M. Barbano and W. K. Jordan, *J. Dairy Sci.*, 1989, **72**, 1435–1442.
34. J. Li, R. D. Sanderson, G. Y. Chai and D. K. Hallbauer, *J. Colloid Interface Sci.*, 2005, **284**, 228–238.
35. M. O. Lamminen, H. W. Walker and L. K. Weavers, *Sep. Sci. Technol.*, 2006, **41**, 3569–3584.

36. D. N. Lee and R. L. Merson, *J. Dairy Sci.*, 1975, **58**, 1423–1432.
37. Q. Li, Z. Xu and I. Pinnau, *J. Membr. Sci.*, 2007, **290**, 173–181.
38. G. Rice, A. Barber, A. O'Connor, G. Stevens and S. Kentish, *J. Membr. Sci.*, 2009, **330**, 117–126.
39. J. Li, J. Liu, T. Yang and C. Xiao, *Water Res.*, 2007, **41**, 4595–4610.
40. R. Sanderson, J. Li, L. J. Koen and L. Lorenzen, *J. Membr. Sci.*, 2002, **207**, 105–117.
41. S. Pal, S. Ambastha, T. B. Ghosh, S. De and S. DasGupta, *J. Membr. Sci.*, 2008, **315**, 58–66.
42. L. Fillaudeau and H. Carrere, *J. Membr. Sci.*, 2002, **196**, 39–57.
43. M. Cheryan, *Ultrafiltration and Microfiltration Handbook*, Taylor & Francis Routledge, Pennsylvania, 1998.
44. D. R. Henning, R. J. Baer, A. N. Hassan and R. Dave, *J. Dairy Sci.*, 2006, **89**, 1179–1188.
45. G. L. Baruah, A. Nayak and G. Belfort, *J. Membr. Sci.*, 2006, **274**, 56–63.
46. J. Zulewska, M. Newbold and D. M. Barbano, *J. Dairy Sci.*, 2009, **92**, 1361–1377.
47. A. Guerra, G. Jonsson, A. Rasmussen, E. W. Nielsen and D. Edelsten, *Int. Dairy J.*, 1997, **7**, 849–861.
48. M. Cheryan and K. P. Kuo, *J. Dairy Sci.*, 1984, **67**, 1406–1413.
49. R. W. Baker, *Membrane Technology and Applications*, John Wiley & Sons Ltd, West Sussex, 2004.
50. M. Frank, G. Bargeman, A. Zwijnenburg and M. Wessling, *Sep. Purif. Technol.*, 2001, **22–23**, 499–506.
51. B. Van der Bruggen, I. Hawrijk, E. Cornelissen and C. Vandecasteele, *Sep. Purif. Technol.*, 2003, **31**, 193–201.
52. G. Samuelsson, P. Dejmek, G. Trägårdh and M. Paulsson, *Int. Dairy J.*, 1997, **7**, 237–242.
53. N. Singh and M. Cheryan, *J. Food Eng.*, 1998, **38**, 57–67.
54. P. Bacchin, *J. Membr. Sci.*, 2004, **228**, 237–241.
55. G. N. Vatai, D. M. Krstic, W. Höflinger, A. K. Koris and M. N. Tekic, *Desalination*, 2007, **204**, 255–264.
56. S.-H. Yoon, S. Lee and I.-T. Yeom, *J. Membr. Sci.*, 2008, **310**, 7–12.
57. W. D. Mores and R. H. Davis, *Ind. Eng. Chem. Res.*, 2003, **42**, 130–139.
58. W. D. Mores and R. H. Davis, *Ind. Eng. Chem. Res.*, 2002, **42**, 130–139.
59. R. W. Field, D. Wu, J. A. Howell and B. B. Gupta, *J. Membr. Sci.*, 1995, 259–272.
60. W. Youravong, M. J. Lewis and A. S. Grandison, *Food Bioprod. Process.*, 2003, **81**, 303–308.
61. H. K. Vyas, A. J. Mawson, R. J. Bennett and A. D. Marshall, *J. Membr. Sci.*, 2000, **176**, 113–119.
62. J. Fritsch and C. I. Moraru, *J. Dairy Sci.*, 2008, **91**, 3744–3760.
63. G. Brans, C. G. P. H. Schroën, R. G. M. van der Sman and R. M. Boom, *J. Membr. Sci.*, 2004, **243**, 263–272.

64. M. Mercier-Bonin, C. Fonade and G. Gésan-Guiziou, *Chem. Eng. Sci.*, 2004, **59**, 2333–2341.
65. S.-L. Li, K.-S. Chou, J.-Y. Lin, H.-W. Yen and I. M. Chu, *J. Membr. Sci.*, 1996, **110**, 203–210.
66. V. S. Espina, M. Y. Jaffrin, M. Frappart and L.-H. Ding, *J. Membr. Sci.*, 2008, **325**, 872–879.
67. H. Carrère, F. Blaszkowa and H. Roux de Balmann, *Desalination*, 2002, **145**, 201–206.
68. H. K. Vyas, R. J. Bennett and A. D. Marshall, *Int. Dairy J.*, 2002, **12**, 473–479.
69. V. M. Kochkodan, N. Hilal, V. V. Goncharuk, L. Al-Khatib and T. I. Levadna, *Colloid J.*, 2006, **68**, 267–273.
70. M. Pouliot, Y. Pouliot and M. Britten, *Int. Dairy J.*, 1996, **6**, 105–111.
71. M. Y. Jaffrin and J. P. Charrier, *J. Membr. Sci.*, 1994, **97**, 71–81.
72. G. Foley, *Desalination*, 2006, **196**, 160–163.
73. D. M. Krstic, M. N. Tekic, Z. Z. Zavargo, M. S. Djuric and G. M. Ciric, *Desalination*, 2004, **165**, 283–288.
74. R. L. Brandsma and S. S. H. Rizvi, *J. Dairy Sci.*, 1999, **82**, 2063–2069.
75. B. K. Nelson and D. M. Barbano, *J. Dairy Sci.*, 2005, **88**, 1891–1900.
76. M. Zhang and L. Song, *Environ. Sci. Technol.*, 2000, **34**, 3767–3773.
77. H. K. Vyas, R. J. Bennett and A. D. Marshall, *Int. Dairy J.*, 2000, **10**, 855–861.
78. M. H. A. Elsalam and N. Shahein, *J. Dairy Res.*, 1989, **56**, 147–149.
79. M. Caric, M. N. Tekic, G. N. Vatai and S. Milanovic, *Acta Aliment.*, 1995, **24**, 241–250.
80. J. A. Levesley and M. Hoare, *J. Membr. Sci.*, 1999, **158**, 29–39.
81. M. D. Caric, S. D. Milanovic, D. M. Krstic and M. N. Tekic, *J. Membr. Sci.*, 2000, **165**, 83–88.
82. E. Suárez, A. Lobo, S. Alvarez, F. A. Riera and R. Álvarez, *Desalination*, 2009, **241**, 272–280.
83. J. Postlethwaite, S. R. Lamping, G. C. Leach, M. F. Hurwitz and G. J. Lye, *J. Membr. Sci.*, 2004, **228**, 89–101.
84. H. M. Kyllonen, P. Pirkonen and M. Nystrom, *Desalination*, 2005, **181**, 319–335.
85. M. C. Almécija, R. Ibáñez, A. Guadix and E. M. Guadix, *J. Membr. Sci.*, 2007, **288**, 28–35.
86. I. H. Huisman, D. Johansson, G. Trägårdh and C. Trägårdh, *Chem. Eng. Res. Des.*, 1997, **75**, 508–512.
87. B. Gautam Lal and B. Georges, *Biotechnol. Bioeng.*, 2004, **87**, 274–285.
88. Y. Xu and R. E. Lebrun, *J. Membr. Sci.*, 1999, **158**, 93–104.
89. D. A. Musale and S. S. Kulkarni, *J. Agric. Food Chem.*, 1998, **46**, 4717–4722.
90. S. Bhattacharjee, C. Bhattacharjee and S. Datta, *J. Membr. Sci.*, 2006, **275**, 141–150.
91. Z. Li, A. H-Kittikun and W. Youravong, *Biochem. Eng. J.*, 2008, **38**, 226–233.

92. Z.-Y. Li, A. H-Kittikun and W. Youravong, *J. Membr. Sci.*, 2008, **311**, 104–111.
93. P. Rai and S. De, *Curr. Sci.*, 2009, **96**, 1361–1371.
94. T. R. Noordman, K. Kooiker, W. Bel, M. Dekker and J. A. Wesselingh, *J. Food Eng.*, 2003, **58**, 135–141.
95. M. Mercier-Bonin and C. Fonade, *Desalination*, 2002, **148**, 171–176.
96. A. Vernhet, M. N. BellonFontaine, J. M. Brillouet, E. Roesink and M. Moutounet, *J. Membr. Sci.*, 1997, **128**, 163–174.
97. M. Ulbricht, W. Ansorge, I. Danielzik, M. Konig and O. Schuster, *Sep. Purif. Technol.*, 2009, **68**, 335–342.
98. N. D. Lawrence, S. E. Kentish, A. J. O'Connor, A. R. Barber and G. W. Stevens, *Sep. Purif. Technol.*, 2008, **60**, 237–244.
99. U. Merin and G. Daufin, *Lait*, 1990, **70**, 281–291.
100. H. M. Ma, C. N. Bowman and R. H. Davis, *J. Membr. Sci.*, 2000, **173**, 191–200.
101. M. Girones, L. Bolhuis-Versteeg, R. Lammertink and M. Wessling, *J. Colloid Interface Sci.*, 2006, **299**, 831–840.
102. M.-G. Yan, L.-Q. Liu, Z.-Q. Tang, L. Huang, W. Li, J. Zhou, J.-S. Gu, X.-W. Wei and H.-Y. Yu, *Chem. Eng. J.*, 2008, **145**, 218–224.
103. J. Gilron, S. Belfer, P. Väisänen and M. Nyström, *Desalination*, 2001, **140**, 167–179.
104. K. S. Kim, K. H. Lee, K. Cho and C. E. Park, *J. Membr. Sci.*, 2002, **199**, 135–145.
105. A. Piry, W. Kühnl, T. Grein, A. Tolkach, S. Ripperger and U. Kulozik, *J. Membr. Sci.*, 2008, **325**, 887–894.
106. R. M. Sandblom, *US Pat.*, US4105547, 1978.
107. I. Pafylias, M. Cheryan, M. A. Mehaia and N. Saglam, *Food Res. Int.*, 1996, **29**, 141–146.
108. P. K. Vadi and S. S. H. Rizvi, *J. Membr. Sci.*, 2001, **189**, 69–82.
109. L. V. Saboya and J. L. Maubois, *Lait*, 2000, **80**, 541–553.
110. M. Decloux, M.-L. Lameloise, A. Brocard, E. Bisson, M. Parmentier and A. Spiraers, *Process Biochem.*, 2007, **42**, 693–699.
111. D. Garcera and E. Toujas, *US Pat.*, US6375014, 2002.
112. M. Gimmelshtein and R. Semiat, *J. Membr. Sci.*, 2005, **264**, 137–150.
113. V. Mavrov, N. D. Nikolov, M. A. Islam and J. D. Nikolova, *J. Membr. Sci.*, 1992, **75**, 197–201.
114. H. M. Yeh and K. T. Chen, *J. Membr. Sci.*, 2000, **178**, 43–53.
115. B. J. Bellhouse, G. Costigan, K. Abhinava and A. Merry, *Sep. Purif. Technol.*, 2001, **22–23**, 89–113.
116. H. R. Millward, B. J. Bellhouse and G. Walker, *J. Membr. Sci.*, 1995, **106**, 269–279.
117. D. M. Krstic, M. N. Tekic, M. D. Caric and S. D. Milanovic, *Desalination*, 2004, **163**, 297–309.
118. D. M. Krstic, M. N. Tekic, M. D. Caric and S. D. Milanovic, *J. Membr. Sci.*, 2002, **208**, 303–314.

119. D. M. Krstic, W. Höflinger, A. K. Koris and G. N. Vatai, *Sep. Purif. Technol.*, 2007, **57**, 134–139.
120. D. M. Krstics, M. N. Tekic, M. D. Caric and S. D. Milanovic, *Desalination*, 2004, **163**, 297–309.
121. H. M. Yeh and P. C. Huang, *J. Membr. Sci.*, 2004, **243**, 357–364.
122. H. M. Yeh and Y. F. Chen, *J. Membr. Sci.*, 2006, **278**, 205–211.
123. Y. Liu, G. He, X. Liu, G. Xiao and B. Li, *Sep. Purif. Technol.*, 2009, **67**, 14–20.
124. S. Pal, R. Bharihoke, S. Chakraborty, S. K. Ghatak, S. De and S. DasGupta, *Sep. Purif. Technol.*, 2008, **62**, 659–667.
125. D. M. Krstic, M. G. Antov, D. M. Pericin, W. Höflinger and M. N. Tekic, *Biochem. Eng. J.*, 2007, **33**, 10–15.
126. X.-h. Zhen, S.-l. Yu, B.-f. Wang and H.-f. Zheng, *J. Environ. Sci.*, 2006, **18**, 1077–1081.
127. M. Y. Jaffrin, *J. Membr. Sci.*, 2008, **324**, 7–25.
128. M. Shirato, *Filtr. Sep.*, 2006, **43**, 40.
129. M. Frappart, O. Akoum, L. H. Ding and M. Y. Jaffrin, *J. Membr. Sci.*, 2006, **282**, 465–472.
130. L. H. Ding, M. Y. Jaffrin, M. Mellal and G. He, *J. Membr. Sci.*, 2006, **276**, 232–240.
131. L. Ding, O. Al-Akoum, A. Abraham and M. Y. Jaffrin, *Desalination*, 2002, **144**, 307–311.
132. V. S. Espina, M. Y. Jaffrin and L. H. Ding, *Desalination*, 2009, **245**, 714–722.
133. L. Fillaudeau, B. Boissier, A. Moreau, P. Blanpain-Avet, S. Ermolaev, N. Jitariouk and A. Gourdon, *J. Food Eng.*, 2007, **80**, 206–217.
134. L. Li, L. Ding, Z. Tu, Y. Wan, D. Clausse and J.-L. Lanoisellé, *J. Membr. Sci.*, 2009, **342**, 70–79.
135. A. Brou, M. Y. Jaffrin, L. H. Ding and J. Courtois, , 2003, **82**, 429–437.
136. M. Frappart, M. Jaffrin and L. H. Ding, *Sep. Purif. Technol.*, 2008, **60**, 321–329.
137. M. Y. Jaffrin, G. He, L. H. Ding and P. Paullier, *Desalination*, 2006, **200**, 269–271.
138. N. Moulai-Mostefa, O. Akoum, M. Nedjihoui, L. Ding and M. Y. Jaffrin, *Desalination*, 2007, **206**, 494–498.
139. A. Brou, L. Ding, P. Boulnois and M. Y. Jaffrin, *J. Membr. Sci.*, 2002, **197**, 269–282.
140. *Germany Pat.*, 2006.
141. G. He, L. H. Ding, P. Paullier and M. Y. Jaffrin, *J. Membr. Sci.*, 2007, **300**, 63–70.
142. O. A. Akoum, M. Y. Jaffrin, L. Ding, P. Paullier and C. Vanhoutte, *J. Membr. Sci.*, 2002, **197**, 37–52.
143. O. Al-Akoum, L. H. Ding and M. Y. Jaffrin, *Sep. Purif. Technol.*, 2002, **28**, 219–234.
144. O. Akoum, M. Y. Jaffrin and L.-H. Ding, *J. Membr. Sci.*, 2005, **247**, 211–220.

145. S. P. Beier and G. Jonsson, *Desalination*, 2006, **199**, 499–500.
146. S. P. Beier and G. Jonsson, *Sep. Purif. Technol.*, 2007, **53**, 111–118.
147. O. Al-Akoum, L. Ding, R. Chotard-Ghodsnia, M. Y. Jaffrin and G. Gésan-Guiziou, *Desalination*, 2002, **144**, 325–330.
148. G. Genkin, T. D. Waite, A. G. Fane and S. Chang, *J. Membr. Sci.*, 2006, **281**, 726–734.
149. J. N. Kuruzovich and P. R. Piergiovanni, *J. Membr. Sci.*, 1996, **112**, 241–247.
150. S. Curcio, V. Calabrò and G. Iorio, *Desalination*, 2002, **145**, 217–222.
151. S. Curcio, G. Scilingo, V. Calabrò and G. Iorio, *J. Membr. Sci.*, 2005, **246**, 235–247.
152. S. A. Mourouzidis-Mourouzis and A. J. Karabelas, *J. Membr. Sci.*, 2008, **323**, 17–27.
153. K. Katsoufidou, S. G. Yiantsios and A. J. Karabelas, *J. Membr. Sci.*, 2007, **300**, 137–146.
154. V. T. Kuberkar and R. H. Davis, *J. Membr. Sci.*, 2001, **183**, 1–14.
155. E. Gabrus and D. Szaniawska, *Desalination*, 2009, **240**, 46–53.
156. H. G. Kim, C. Park, J. Yang, B. Lee, S. S. Kim and S. Kim, *Desalination*, 2007, **202**, 150–155.
157. B. G. Park, W. G. Lee, Y. K. Chang and H. N. Chang, *Bioprocess Eng.*, 1997, **16**, 253–256.
158. J. Cakl, I. Bauer, P. Dolecek and P. Mikulasek, *Desalination*, 2000, **127**, 189–198.
159. N. Mugnier, J. A. Howell and M. Ruf, *J. Membr. Sci.*, 2000, **175**, 149–161.
160. C. M. Silva, D. W. Reeve, H. Husain, H. R. Rabie and K. A. Woodhouse, *J. Membr. Sci.*, 2000, **173**, 87–98.
161. P. J. Smith, S. Vigneswaran, H. H. Ngo, R. Ben-Aim and H. Nguyen, *J. Membr. Sci.*, 2005, **255**, 99–106.
162. T. J. Phelps, A. V. Palumbo, B. L. Bischoff, C. J. Miller, L. A. Fagan, M. S. McNeilly and R. R. Judkins, *J. Microbiol. Methods*, 2008, **74**, 10–16.
163. A. Nabe, E. Staude and G. Belfort, *J. Membr. Sci.*, 1997, **133**, 57–72.
164. K. Matsumoto, K. Kitabayashi and H. Ohya, *Hakkokogaku Kaishi-J. Soc. Ferment. Technol.*, 1991, **69**, 477–481.
165. K. Katsoufidou, S. G. Yiantsios and A. J. Karabelas, *J. Membr. Sci.*, 2005, **266**, 40–50.
166. S. A. Mourouzidis-Mourouzis and A. J. Karabelas, *J. Membr. Sci.*, 2006, **282**, 124–132.
167. H. K. Vyas, R. J. Bennett and A. D. Marshall, *Int. Dairy J.*, 2000, **10**, 477–487.
168. R. B. Amar, B. B. Gupta and M. Y. Jaffrin, *J. Dairy Sci.*, 1990, **55**, 1620–1625.
169. G. R. Sanjeev and H. D. Robert, *AIChE J.*, 1995, **41**, 501–508.
170. J. Meier, G. M. Klein and V. Kottke, *Sep. Purif. Technol.*, 2002, **26**, 43–50.
171. G. Naja, B. Volesky and A. Schnell, *Biotechnol. Bioeng.*, 2006, **95**, 584–598.

172. Q. Gan, J. A. Howell, R. W. Field, R. England, M. R. Bird, C. L. O'Shaughnessy and M. T. MeKechinie, *J. Membr. Sci.*, 2001, **194**, 185–196.
173. F. Meacle, A. Aunins, R. Thornton and A. Lee, *J. Membr. Sci.*, 1999, **161**, 171–184.
174. W. D. Mores and R. H. Davis, *Desalination*, 2002, **146**, 135–140.
175. E. Wen, L. D. Cinelli, D. Murray, R. J. Lander, S. L. Sagar and A. L. Lee, *J. Membr. Sci.*, 2005, **258**, 23–34.
176. W. D. Mores, C. N. Bowman and R. H. Davis, *J. Membr. Sci.*, 2000, **165**, 225–236.
177. Q. Gan, *Chem. Eng. Process.*, 2001, **40**, 413–419.
178. S. Redkar, V. Kuberkar and R. H. Davis, *J. Membr. Sci.*, 1996, **121**, 229–242.
179. V. Kuberkar, P. Czekaj and R. Davis, *Biotechnol. Bioeng.*, 1998, **60**, 77–87.
180. H. Mallubhotla and G. Belfort, *Ind. Eng. Chem. Res.*, 1996, **35**, 2920–2928.
181. F. Beolchini, F. Vegho and D. Barba, *Desalination*, 2004, **161**, 251–258.
182. P. Czekaj, W. Mores, R. H. Davis and C. Güell, *J. Membr. Sci.*, 2000, **180**, 157–169.
183. M. Gironès i Nogué, I. J. Akbarsyah, L. A. M. Bolhuis-Versteeg, R. G. H. Lammertink and M. Wessling, *J. Membr. Sci.*, 2006, **285**, 323–333.
184. I. G. Wenten, *Filtr. Sep.*, 1995, **32**, 252–253.
185. B. Sarkar, S. Pal, T. B. Ghosh, S. De and S. DasGupta, *J. Membr. Sci.*, 2008, **311**, 112–120.
186. H. M. Huotari, G. Tragardh and I. H. Huisman, *Chem. Eng. Res. Des.*, 1999, **77**, 461–468.
187. B. Sarkar, S. DasGupta and S. De, *J. Membr. Sci.*, 2009, **341**, 11–20.
188. B. Sarkar, S. DasGupta and S. De, *J. Membr. Sci.*, 2009, **331**, 75–83.
189. B. Sarkar, A. Sengupta, S. De and S. DasGupta, *Sep. Purif. Technol.*, 2009, **65**, 260–268.
190. R. J. Wakeman, *Food Bioprod. Process.*, 1998, **76**, 53–59.
191. G. Brisson, M. Britten and Y. Pouliot, *J. Membr. Sci.*, 2007, **297**, 206–216.
192. Y. G. Park, *Desalination*, 2006, **191**, 411–416.
193. S. Oussedik, D. Belhocine, H. Grib, H. Lounici, D. L. Piron and N. Mameri, *Desalination*, 2000, **127**, 59–68.
194. N. Mameri, S. Oussedik, R. Yeddou, D. L. Piron, D. Belhocine, H. Lounici and H. Grib, *Sep. Purif. Technol.*, 1999, **17**, 203–211.
195. B. Sarkar, S. DasGupta and S. De, *J. Colloid Interface Sci.*, 2008, **319**, 236–246.
196. C. W. Robinson, M. H. Siegel, A. Condemine, C. Fee, T. Z. Fahidy and B. R. Glick, *J. Membr. Sci.*, 1993, **80**, 209–220.
197. C. C. Tarazaga, M. E. Campderros and A. P. Padilla, *J. Membr. Sci.*, 2006, **283**, 339–345.

198. C. C. Tarazaga, M. E. Campderros and A. P. Padilla, *J. Membr. Sci.*, 2006, **278**, 219–224.
199. B. Sarkar, S. DasGupta and S. De, *J. Membr. Sci.*, 2008, **307**, 268–276.
200. B. Sarkar, S. De and S. DasGupta, *Sep. Purif. Technol.*, 2008, **63**, 582–591.
201. J. Jurado and B. J. Bellhouse, *Filtr. Sep.*, 1994, **31**, 273–281.
202. P. vonZumbusch, R. Portner, B. Brun, G. Brunner and H. Markl, *Anim. Cell Technol.*, 1997, 215–220.
203. S. Bendicho, G. V. Barbosa-Canovas and O. Martin, *J. Dairy Sci.*, 2003, **86**, 697–703.
204. S. Muthukumaran, S. E. Kentish, G. W. Stevens and M. Ashokkumar, *Rev. Chem. Eng.*, 2006, **22**, 155–194.
205. M. O. Lamminen, H. W. Walker and L. K. Weavers, *J. Membr. Sci.*, 2004, **237**, 213–223.
206. M.-Y. Teng, S.-H. Lin, C.-Y. Wu and R.-S. Juang, *J. Membr. Sci.*, 2006, **281**, 103–110.
207. T. Kobayashi, X. Chai and N. Fujii, *Sep. Purif. Technol.*, 1999, **17**, 31–40.
208. D. Chen, L. K. Weavers and H. W. Walker, *Ultrsons. Sonochem.*, 2006, **13**, 379–387.
209. S. Muthukumaran, S. E. Kentish, G. W. Stevens, M. Ashokkumar and R. Mawson, *J. Food Eng.*, 2007, **81**, 364–373.
210. A. Maskooki, T. Kobayashi, S. A. Mortazavi and A. Maskooki, *Sep. Purif. Technol.*, 2008, **59**, 67–73.
211. H. Kyllonen, P. Pirkonen, M. Nystrom, J. Nuortila-Jokinen and A. Gronroos, *Ultrason. Sonochem.*, 2006, **13**, 295–302.
212. T. Kobayashi, T. Kobayashi, Y. Hosaka and N. Fujii, *Ultrasonics*, 2003, **41**, 185–190.
213. M. O. Lamminen, H. W. Walker and L. K. Weavers, *J. Membr. Sci.*, 2006, **283**, 225–232.
214. S. Muthukumaran, S. E. Kentish, M. Ashokkumar and G. W. Stevens, *J. Membr. Sci.*, 2005, **258**, 106–114.
215. Y. Matsumoto, T. Miwa, S. Nakao and S. Kimura, *J. Chem. Eng. Jpn.*, 1996, **29**, 561–567.
216. D. Feng, J. S. J. van Deventer and C. Aldrich, *Sep. Purif. Technol.*, 2006, **50**, 318–323.
217. S. Muthukumaran, S. Kentish, S. Lalchandani, M. Ashokkumar, R. Mawson, G. W. Stevens and F. Grieser, *Ultrsons. Sonochem.*, 2005, **12**, 29–35.
218. X. J. Chai, T. Kobayashi and N. Fujii, *Sep. Purif. Technol.*, 1999, **15**, 139–146.
219. I. Masselin, X. Chasseray, L. Durand-Bourlier, J.-M. Lainé, P.-Y. Syzaret and D. Lemordant, *J. Membr. Sci.*, 2001, **181**, 213–220.
220. X. H. Wang, X. F. Li, X. Q. Fu, C. A. Rui and B. Gao, *Desalination*, 2005, **175**, 187–196.
221. M. Villamiel and P. de Jong, *J. Agric. Food Chem.*, 2000, **48**, 3068.
222. T. Kobayashi and Y. Hosaka, *Jpn. J. Appl. Phys., Part 1*, 2003, **42**, 2954–2955.

223. Z. F. Cui, S. Chang and A. G. Fane, *J. Membr. Sci.*, 2003, **221**, 1–35.
224. M. Mercier-Bonin, G. Gésan-Guiziou and C. Fonade, *J. Membr. Sci.*, 2003, **218**, 93–105.
225. H. Fadaei, S. R. Tabaei and R. Roostaazad, *Desalination*, 2007, **217**, 93–99.
226. R. J. Wakeman and C. J. Williams, *Sep. Purif. Technol.*, 2002, **26**, 3–18.
227. B. G. Pollet, J. P. Lorimer, J. Y. Hihn, S. S. Phull, T. J. Mason and D. J. Walton, *Ultrsons. Sonochem.*, 2002, **9**, 267–274.
228. J. X. Li, R. D. Sanderson and E. P. Jacobs, *J. Membr. Sci.*, 2002, **205**, 247–257.
229. A. L. Lim and R. Bai, *J. Membr. Sci.*, 2003, **216**, 279–290.
230. C. Psoch and S. Schiewer, *J. Membr. Sci.*, 2006, **283**, 273–280.
231. A. F. Derradji, S. Taha and G. Dorange, *Desalination*, 2005, **184**, 377–384.
232. A. F. Derradji, A. Bernabeu-Madico, S. Taha and G. Dorange, *Desalination*, 2000, **128**, 223–230.
233. S. S. Madaeni and Y. Mansourpanah, *Desalination*, 2004, **161**, 13–24.
234. V. Gitis, R. C. Haught, R. M. Clark, J. Gun and O. Lev, *J. Membr. Sci.*, 2006, **276**, 185–192.
235. P. Blanpain-Avet, J. F. Migdal and T. Bénézech, *Food Bioprod. Process.*, 2004, **82**, 231–243.
236. P. Blanpain-Avet, J. F. Migdal and T. Bénézech, *J. Membr. Sci.*, 2009, **337**, 153–174.
237. M. L. Cabero, F. A. Riera and R. Álvarez, *J. Membr. Sci.*, 1999, **154**, 239–250.
238. O. O. Ogunbiyi, N. J. Miles and N. Hilal, *Desalination*, 2008, **220**, 273–289.
239. M. Kazemimoghadam and T. Mohammadi, *Desalination*, 2007, **204**, 213–218.
240. S. Siavash Madaeni, T. Mohamamdi and M. Kazemi Moghadam, *Desalination*, 2001, **134**, 77–82.
241. K. E. Smith and R. L. Bradley, Jr., *J. Dairy Sci.*, 1987, **70**, 1168–1177.
242. K. E. Smith and R. L. Bradley, Jr., *J. Dairy Sci.*, 1987, **70**, 243–251.
243. A. Maartens, P. Swart and E. P. Jacobs, *J. Membr. Sci.*, 1996, **119**, 9–16.
244. M. A. Argüello, S. Álvarez, F. A. Riera and R. Álvarez, *J. Membr. Sci.*, 2003, **216**, 121–134.
245. T. Coolbear, C. Monk, K. Peek, H. W. Morgan and R. M. Daniel, *J. Membr. Sci.*, 1992, **67**, 93–101.
246. H. B. Petrus, H. Li, V. Chen and N. Norazman, *J. Membr. Sci.*, 2008, **325**, 783–792.
247. W. Jin, W. Guo, X. P. Lu, P. F. Han and Y. R. Wang, *Chin. J. Chem. Eng.*, 2008, **16**, 801–804.
248. H. F. Bohner and R. L. Bradley, *J. Dairy Sci.*, 1990, **73**, 2309–2317.
249. H. F. Bohner and R. L. Bradley, *J. Dairy Sci.*, 1991, **74**, 3348–3352.
250. K. E. Smith and R. L. Bradley, Jr., *J. Dairy Sci.*, 1986, **69**, 1232–1240.

251. D. M. Barbano, D. G. Bynum and G. F. Senyk, *J. Dairy Sci.*, 1983, **66**, 2447–2451.
252. R. L. Stabile, *J. Dairy Sci.*, 1983, **66**, 1765–1772.
253. G. S. Murthy, S. Sridhar, M. S. Sunder, B. Shankaraiah and M. Ramakrishna, *Sep. Purif. Technol.*, 2005, **44**, 205–211.
254. G. S. Murthy, S. Sridhar, M. Shyam Sunder, B. Shankaraiah and M. Ramakrishna, *Sep. Purif. Technol.*, 2005, **44**, 221–228.
255. F. Vaillant, A. Millan, M. Dornier, M. Decloux and M. Reynes, *J. Food Eng.*, 2001, **48**, 83–90.
256. A. Papadatos, M. Neocleous, A. M. Berger and D. M. Barbano, *J. Dairy Sci.*, 2003, **86**, 1564–1577.
257. C. I. Moraru and E. U. Schrader, in *Handbook of Membrane Separations: Chemical, Pharmaceutical, Food, and Biotechnological Applications*, ed. A. K. Pabby, S. S. H. Rizvi and A. M. Sastre, Taylor & Francis Group, Boca Raton, 2nd edn, 2009.
258. S. Mirza, *Desalination*, 2008, **224**, 132–142.
259. A. J. E. Jimenez-Lopez, N. Leconte, O. Dehainault, C. Geneste, L. Fromont and G. Gésan-Guiziou, *Sep. Purif. Technol.*, 2008, **61**, 33–43.
260. Y. Pouliot, *Int. Dairy J.*, 2008, **18**, 735–740.
261. M. W. Elwell and D. M. Barbano, *J. Dairy Sci.*, 2006, **89**, E20–30.
262. Y. Amornkul and D. R. Henning, *J. Dairy Sci.*, 2007, **90**, 4988–5000.
263. S. Thomann, P. Schenkel and J. Hinrichs, *LWT – Food Sci. Technol.*, 2008, **41**, 826–835.
264. A. V. Ardisson-Korat and S. S. H. Rizvi, *J. Dairy Sci.*, 2004, **87**, 3601–3613.
265. S. Govindasamy-Lucey, J. J. Jaeggi, M. E. Johnson, T. Wang and J. A. Lucey, *J. Dairy Sci.*, 2007, **90**, 4552–4568.
266. M. Neocleous, D. M. Barbano and M. A. Rudan, *J. Dairy Sci.*, 2002, **85**, 2415–2424.
267. N. Lawrence, S. Kentish, A. O'Connor, G. Stevens and A. Barber, *Desalination*, 2006, **200**, 305–306.
268. M. Britten and Y. Pouliot, *Lait*, 1996, **76**, 255–265.
269. C. D. Pereira, O. Diaz and A. Cobos, *Int. Dairy J.*, 2002, **12**, 773–783.
270. A. Rektor and G. Vatai, *Desalination*, 2004, **162**, 279–286.
271. K. Dewettinck, R. Rombaut, N. Thienpont, T. T. Le, K. Messens and J. V. Camp, *Int. Dairy J.*, 2008, **18**, 436–457.
272. H. Goudédranche, J. Fauquant and J. L. Maubois, *Lait*, 2000, **80**, 93–98.
273. M. C. Michalski, N. Leconte, V. Briard-Bion, J. Fauquant, J. L. Maubois and H. Goudedranche, *J. Dairy Sci.*, 2006, **89**, 3778–3790.
274. J. C. Lesqueren and G. Canteri, *Trends Food Sci. Technol.*, 1997, **8**, 207.
275. F. V. Kosikowski, *J. Dairy Sci.*, 1985, **68**, 2403–2410.
276. H. R. Covacevich and F. V. Kosikowski, *J. Dairy Sci.*, 1978, **61**, 529–535.
277. J. L. Maubois and G. Mocquot, *J. Dairy Sci.*, 1975, **58**, 1001–1007.
278. J. L. Maubois and F. V. Kosikowski, *J. Dairy Sci.*, 1978, **61**, 881–884.
279. H. R. Covacevich and F. V. Kosikowski, *J. Dairy Sci.*, 1978, **61**, 701–709.
280. A. Fernandez and F. V. Kosikowski, *J. Dairy Sci.*, 1986, **69**, 2011–2017.
281. M. M. Furtado and J. A. Partridge, *J. Dairy Sci.*, 1988, **71**, 2877–2884.

282. C. A. Ernstrom, B. J. Sutherland and G. W. Jameson, *J. Dairy Sci.*, 1980, **63**, 228–234.
283. Y. Kemal Erdem, *J. Food Eng.*, 2005, **71**, 366–372.
284. M. A. Mehaia, *Food Chem.*, 2002, **79**, 445–452.
285. M. A. Mehaia, *J. Food Technol.*, 2006, **4**, 206–212.
286. J. U. McGregor and C. H. White, *J. Dairy Sci.*, 1990, **73**, 314–318.
287. S. Govindasamy-Lucey, J. J. Jaeggi, A. L. Bostley, M. E. Johnson and J. A. Lucey, *J. Dairy Sci.*, 2004, **87**, 2789–2799.
288. F. Y. Lee and C. H. White, *J. Dairy Sci.*, 1991, **74**, 1170–1180.
289. N. Rajagopalan and M. Cheryan, *J. Dairy Sci.*, 1991, **74**, 2435–2439.
290. R. J. Premaratne and M. A. Cousin, *J. Dairy Sci.*, 1991, **74**, 788–795.
291. H. K. Vyas and P. S. Tong, *J. Dairy Sci.*, 2003, **86**, 2761–2766.
292. F. V. Kosikowski, *J. Dairy Sci.*, 1979, **62**, 41–46.
293. M. Howie, *Feedstuffs*, 1999, **71**, 33.
294. F. E. McDonough, W. A. Mattingly and J. H. Vestal, *J. Dairy Sci.*, 1971, **54**, 1406–1409.
295. F. E. McDonough, R. E. Hargrove, W. A. Mattingly, L. P. Posati and J. A. Alford, *J. Dairy Sci.*, 1974, **57**, 1438–1443.
296. P. D. Veith and E. C. Reynolds, *J. Dairy Sci.*, 2004, **87**, 831–840.
297. M. O. Nigam, B. Bansal and X. D. Chen, *Desalination*, 2008, **218**, 313–322.
298. K. W. K. Yee, D. E. Wiley and J. Bao, *J. Membr. Sci.*, 2007, **290**, 125–137.
299. L. Fachin and W. H. Viotto, *Int. Dairy J.*, 2005, **15**, 325–332.
300. R. J. Brown and C. A. Ernstrom, *J. Dairy Sci.*, 1982, **65**, 2391–2395.
301. B. Balannec, M. Vourch, M. Rabiller-Baudry and B. Chaufer, *Sep. Purif. Technol.*, 2005, **42**, 195–200.
302. B. Cuartas-Uribe, M. I. Alcaina-Miranda, E. Soriano-Costa and A. Bes-Pia, *J. Dairy Sci.*, 2007, **90**, 1094–1101.
303. H. C. van der Horst, J. M. K. Timmer, T. Robbertsen and J. Leenders, *J. Membr. Sci.*, 1995, **104**, 205–218.
304. P. L. Spangler and C. H. Amundson, *J. Dairy Sci.*, 1986, **69**, 1498–1509.
305. J. Hiddink, R. de Boer and P. F. C. Nooy, *J. Dairy Sci.*, 1980, **63**, 204–214.
306. M. Vourch, B. Balannec, B. Chaufer and G. Dorange, *Desalination*, 2005, **172**, 245–256.
307. H. Chmiel, V. Mavrov and E. Bélières, *Filtr. Sep.*, 2000, **37**, 24–27.
308. M. Taylor, D. B. F. Faraday, C. L. O'Shaughnessy, B. O. Underwood and R. J. R. Reed, *Sep. Purif. Technol.*, 2001, **22–23**, 133–142.
309. J. P. Arriagada-Carrazana, C. Sáez-Navarrete and E. Bordeu, *J. Food Eng.*, 2005, **68**, 363–368.
310. S. Gergely, E. Bekassy-Molnar and G. Vatai, *J. Food Eng.*, 2003, **58**, 311–316.
311. A. Versari, R. Ferrarini, G. P. Parpinello and S. Galassi, *Food Bioprod. Process.*, 2003, **81**, 275–278.
312. A. Rektor and G. Vatai, *Desalination*, 2004, **162**, 271–277.

313. S. Banvolgyi, I. Kiss, E. Bekassy-Molnar and G. Vatai, *Desalination*, 2006, **198**, 8–15.
314. J. Labanda, S. Vichi, J. Llorens and E. López-Tamames, *LWT – Food Sci. Technol.*, 2009, **42**, 1390–1395.
315. J. Gómez Benítez, V. M. Palacios Macías, P. Szekely Gorostiaga, R. Veas López and L. Pérez Rodríguez, *J. Food Eng.*, 2003, **58**, 373–378.
316. G. Daufin, J. P. Escudier, H. Carrère, S. Bérot, L. Fillaudeau and M. Decloux, *Food Bioprod. Process.*, 2001, **79**, 89–102.
317. A. M. Ghosh and M. Balakrishnan, *J. Food Eng.*, 2003, **58**, 143–150.
318. M. Balakrishnan, M. Dua and J. J. Bhagat, *Sep. Purif. Technol.*, 2000, **19**, 209–220.
319. S. Cartier, M. A. Theoleyre and M. Decloux, *Desalination*, 1997, **113**, 7–17.
320. V. Hakimzadeh, S. M. A. Razavi, M. K. Piroozifard and M. Shahidi, *Desalination*, 2006, **200**, 520–522.
321. M. Djuric, J. Gyura and Z. Zavargo, *Desalination*, 2004, **169**, 167–183.
322. J. Gyura, Z. Seres, G. Vatai and E. B. Molnár, *Desalination*, 2002, **148**, 49–56.
323. M. Bogliolo, A. Bottino, G. Capannelli, M. De Petro, A. Servida, G. Pezzi and G. Vallini, *Desalination*, 1997, **108**, 261–271.
324. N. S. Krishna Kumar, M. K. Yea and M. Cheryan, *J. Membr. Sci.*, 2004, **244**, 235–242.
325. J. Luo, L. Ding, X. Chen and Y. Wan, *Sep. Purif. Technol.*, 2009, **66**, 429–437.
326. M. Li, Y. Zhao, S. Zhou, W. Xing and F.-S. Wong, *J. Membr. Sci.*, 2007, **299**, 122–129.
327. B. E. Chove, A. S. Grandison and M. J. Lewis, *Food Hydrocoll.*, 2007, **21**, 1379–1388.
328. F. López, P. Pescador, C. Güell, M. L. Morales, M. C. García-Parrilla and A. M. Troncoso, *J. Food Eng.*, 2005, **68**, 133–136.
329. L. Firdaous, P. Dhulster, J. Amiot, A. Gaudreau, D. Lecouturier, R. Kapel, F. Lutin, L.-P. Vézina and L. Bazinet, *J. Membr. Sci.*, 2009, **329**, 60–67.
330. Y. Popineau, B. Huchet, C. Larré and S. Bérot, *J. Cereal Sci.*, 2002, **35**, 327–335.
331. J.-F. Poulin, J. Amiot and L. Bazinet, *J. Membr. Sci.*, 2007, **299**, 83–90.
332. F. Lebrun, A. Bazus, P. Dhulster and D. Guillochon, *J. Membr. Sci.*, 1998, **146**, 113–124.
333. J.-F. Lapointe, S. F. Gauthier, Y. Pouliot and C. Bouchard, *J. Membr. Sci.*, 2005, **253**, 89–102.
334. S. Butylina, S. Luque and M. Nyström, *J. Membr. Sci.*, 2006, **280**, 418–426.
335. Y. Pouliot, S. F. Gauthier and C. Bard, *J. Membr. Sci.*, 1993, **80**, 257–264.
336. A. Chabeaud, L. Vandanjon, P. Bourseau, P. Jaouen, M. Chaplain-Derouiniot and F. Guerard, *Sep. Purif. Technol.*, 2009, **66**, 463–471.

337. L. Vandanjon, M. Grignon, E. Courois, P. Bourseau and P. Jaouen, *J. Food Eng.*, 2009, **95**, 36–44.
338. Y. M. Lo, D. Cao, S. Argin-Soysal, J. Wang and T.-S. Hahm, *Bioresour. Technol.*, 2005, **96**, 687–698.
339. R. Y. Avula, H. M. Nelson and R. K. Singh, *Innovative Food Sci. Emerging Technol.*, 2009, **10**, 1–8.
340. M. D. Afonso and R. Bórquez, *Desalination*, 2003, **151**, 131–138.
341. M. Kuca and D. Szaniawska, *Desalination*, 2009, **241**, 227–235.
342. E. Ellouze, R. B. Amar and A. H. Ben Salah, *Desalination*, 2005, **177**, 229–240.
343. K. Walha, R. B. Amar, F. Quemeneur and P. Jaouen, *Desalination*, 2008, **219**, 231–239.
344. E. Ferjani, E. Ellouze and R. Ben Amar, *Desalination*, 2005, **177**, 43–49.
345. F. Perreault, A. F. Faria and M. Elimelech, *Chem. Soc. Rev.*, 2015, **44**, 5861–5896.
346. R. K. Joshi, P. Carbone, F. C. Wang, V. G. Kravets, Y. Su, I. V. Grigorieva, H. A. Wu, A. K. Geim and R. R. Nair, *Science*, 2014, **343**, 752–754.
347. R. R. Nair, H. A. Wu, P. N. Jayaram, I. V. Grigorieva and A. K. Geim, *Science*, 2012, **335**, 442–444.
348. D. An, L. Yang, T. Wang and B. Liu, *Ind. Eng. Chem. Res.*, 2016, **55**, 4803–4810.
349. P. Sun, K. Wang and H. Zhu, *Adv. Mater.*, 2016, **28**, 2287–2310.
350. C. Zhao, X. Xu, J. Chen and F. Yang, *J. Env. Chem. Eng.*, 2013, **1**, 349–354.
351. M. Hu and B. Mi, *Environ. Sci. Technol.*, 2013, **47**, 3715–3723.
352. S. Liu, M. Hu, T. H. Zeng, R. Wu, R. Jiang, J. Wei, L. Wang, J. Kong and Y. Chen, *Langmuir*, 2012, **28**, 12364–12372.
353. O. Akhavan and E. Ghaderi, *ACS Nano*, 2010, **4**, 5731–5736.
354. S. Gurunathan, J. W. Han, A. A. Dayem, V. Eppakayala, M. R. Park, D. N. Kwon and J. H. Kim, *J. Ind. Eng. Chem.*, 2013, **19**, 1280–1288.
355. W. Holzmüller and U. Kulozik, *Int. Dairy J.*, 2016, **63**, 88–91.
356. K. Altmann, A. Wutkowski, S. Kampfer, M. Klempt, P. C. Lorenzen and I. C. Redecker, *Eur. Food Res. Technol.*, 2015, **241**, 803–815.
357. M. Ferrer, M. Alexander and M. Corredig, *Food Sci. Technol.*, 2014, **59**, 173–180.
358. A. Jukkola, R. Partanen, O. J. Rojas and A. Heino, *J. Dairy Sci.*, 2016, **99**, 1–11.
359. H. Li, Y. C. Hsu, Z. Zhang, N. Dharsana, Y. Ye and V. Chen, *Sep. Sci. Technol.*, 2017, **2**, 381–391.
360. J. Luo, X. Hang, W. Zhai, B. Qui, W. Song, X. Chen and Y. Wan, *J. Membr. Sci.*, 2016, **509**, 105–115.
361. M. S. So'aib, J. Krishnan and M. V. P. S. Veluri, *Chem. Biochem. Eng. Q.*, 2014, **28**, 357–365.
362. R. Mereddy, A. Chan, K. Fanning, N. Nirmal and Y. Sultanbawa, *Food Chem.*, 2017, **215**, 311–317.

363. M. Schaepertoens, C. Didaskalou, J. F. Kim, A. G. Livingston and G. Szekely, *J. Membr. Sci.*, 2016, **514**, 646–658.
364. S. Maktouf, M. Neifar, S. J. Drira, S. Baklouti, M. Fendri and S. E. Châabouni, *Food Bioprod. Process.*, 2014, **92**, 14–19.
365. A. P. Echavarría, V. Falguera, C. Torras, C. Berdún, J. Pagán and A. Ibarz, *Food Sci. Technol.*, 2012, **46**, 189–195.
366. P. R. F. Campos, A. N. Modenes, F. R. E. Quinõnes, D. E. G. Trigueros, S. T. D. Barros and N. C. Pereira, *Ann. Brazil Acad. Sci.*, 2016, **88**, 423–436.
367. N. Pap, M. Mahosenaho, E. Pongracz, H. Mikkonen, M. Jaakkola, V. Virtanen, L. Myllykosky, Z. Hovorka, C. Hodur, G. Vatai and R. L. Keiski, *Food Bioprocess Technol.*, 2012, **5**, 921–928.
368. J. Luo, B. Zeuner, S. T. Morthensen, A. S. Meyer and M. Pinelo, *J. Membr. Sci.*, 2015, **482**, 83–91.
369. W.-Q. Wang, L.-W. Zhang, X. Han and Y. Lu, *Food Chem.*, 2017, **215**, 31–40.
370. J. Yang, J. Guo, X. Q. Yang, N. N. Wu, J. B. Zhang, J. J. Huo, Y. Y. Zhang and W. K. Xiao, *J. Food Eng.*, 2014, **143**, 25–32.
371. K. M. Barry, T. G. Dinan and P. M. Kelly, *Int. Dairy J.*, 2017, **68**, 80–87.
372. L. Shao, Z. X. Wang, Y. L. Zhang, Z. X. Jiang and Y. Y. Liu, *J. Membr. Sci.*, 2014, **461**, 10–21.
373. Y. Hu, K. Lu, F. Yan, Y. Shi, P. Yu, S. Yu, S. Li and C. Gao, *J. Membr. Sci.*, 2016, **501**, 209–219.
374. S. R. V. Castrillón, X. Lu, D. L. Shaffer and M. Elimelech, *J. Membr. Sci.*, 2014, **450**, 331–339.
375. G. R. Xu, J. N. Wang and C. J. Li, *Desalination*, 2013, **328**, 83–100.
376. V. Kochkodan, D. J. Johnson and N. Hilal, *Adv. Colloid Interface Sci.*, 2014, **216**, 116–140.
377. T. A. Saleh and V. K. Gupta, *Sep. Purif. Technol.*, 2012, **89**, 245–251.
378. M. S. Rahaman, H. T. Aubin, M. B. Sasson, C. K. Ober, M. Nielsen and M. Elimelech, *J. Mater. Chem. B*, 2014, **2**, 1724–1732.
379. R. Das, M. E. Ali, S. B. A. Hamid, S. Ramakrishna and Z. Z. Chowdhury, *Desalination*, 2014, **336**, 97–109.
380. H. Zhao, S. Qiu, L. Wu, L. Zhang, H. Chen and C. Gao, *J. Membr. Sci.*, 2014, **450**, 249–256.
381. P. Daraei, S. S. Madaeni, N. Ghaemi, M. A. Khadivi, B. Astinchap and R. Moradian, *J. Membr. Sci.*, 2013, **444**, 184–191.
382. V. Vatanpour, M. Esmaeili and M. H. D. A. Farahani, *J. Membr. Sci.*, 2014, **446**, 70–81.
383. D. Y. K. Imer, N. Dizge and Y. Koyuncu, *Colloids Surf., B*, 2012, **92**, 334–339.
384. W. Z. Qiu, Q. Z. Zhong, Y. Du, Y. Lv and Z. K. Xu, *Green Chem.*, 2016, **18**, 6205–6208.
385. N. N. R. Ahmad, H. Mukhtar, D. F. Mohshim, R. Nasir and Z. Man, *Rev. Chem. Eng.*, 2016, **32**, 181–200.

386. R. Lin, L. Ge, L. Hou, E. Strounina, V. Rudolph and Z. Zhu, *Appl. Mater. Interfaces*, 2014, **6**, 5609–5618.
387. B. Wojtynyak, J. Kolodziejczyk and D. Szaniawska, *Chem. Eng. J.*, 2016, **305**, 28–36.
388. G. Konrad, T. Klenschmidt and C. Lorenz, *Int. Dairy J.*, 2013, **30**, 39–44.
389. J. Chandrapala, G. Q. Chen, K. Kezia, E. G. Bowman, T. Vasiljevic and S. E. Kentish, *J. Food Eng.*, 2016, **177**, 59–64.
390. K. Altmann, I. C. Radecker, W. Hoffmann and P. C. Lorenzen, *Food Bioprocess Technol.*, 2016, **9**, 1924–1936.
391. J. Cao, W. Zhang, S. Wu, C. Liu, Y. Li, H. Li and L. Zhang, *J. Dairy Sci.*, 2015, **98**, 100–105.
392. M. Bédas, G. Tanguy, A. Dolivet, S. Méjean, F. Gaucheron, G. Garric, G. Senard, R. Jeantet and P. Schuck, *Food Sci. Technol.*, 2017, **79**, 355–360.
393. M. J. C. Báguena, S. A. Blanco and M. C. V. Vela, *Sep. Purif. Technol.*, 2014, **125**, 1–10.
394. M. J. L. Facundo, J. A. M. Roca, B. C. Uribe and S. A. Blanco, *Ultrason. Sonochem.*, 2016, **33**, 18–25.
395. Y. Jin, N. Hengl, S. Baup, F. Pignon, N. Gondrexon, M. Sztucki, G. G. Guiziou, A. Magnin, M. Abyan, M. Karrouch and D. Blésès, *J. Membr. Sci.*, 2014, **470**, 205–218.
396. M. Abel, Z. L. Kiss, S. Beszedés, C. Hodúr, G. K. Szabó and Z. László, *J. Food Process. Eng.*, 2015, **38**, 467–473.
397. J. Luo, W. Cao, L. Ding, Z. Zhu, Y. Wan and M. Y. Jaffrin, *Sep. Purif. Technol.*, 2012, **96**, 194–203.
398. P. Onsekizoglu, *Crit. Rev. Food Sci. Nutr.*, 2015, **55**, 1098–1113.
399. J. Kujawa, E. G. Burrieza, H. A. Arafat, M. Kurzawa, A. Wolan and W. Kujawski, *Food Bioprocess Technol.*, 2015, **8**, 2146–2158.
400. S. Ilame and S. Singh, *Crit. Rev. Food Sci. Nutr.*, 2015, **55**, 964–987.
401. W. Kujawski, A. Sobolewska, K. Jarzynka, C. Guell, M. Ferrando and J. Warczok, *J. Food Eng.*, 2013, **116**, 801–808.
402. C. Zambra, J. Romero, L. Pino, A. Saavedra and J. Sanchez, *J. Food Eng.*, 2015, **144**, 58–65.
403. R. C. Munoz, J. Y. Fernandez and V. Fila, *Food Chem.*, 2016, **213**, 753–762.
404. M. Servili, S. Esposto, G. Veneziani, S. Urbani, A. Taticchi, I. D. Maio, R. Selvaggini, B. Sordini and G. F. Montedoro, *Food Chem.*, 2011, **124**, 1308–1315.
405. H. Kiai, M. C. G. Payo, A. Hafidi and M. Khayet, *Chem. Eng. Process.*, 2014, **86**, 153–161.
406. F. Destani, A. Cassano, A. Fazio, J. P. Vincken and V. Gabriele, *J. Food Eng.*, 2013, **117**,, 263–271.
407. H. C. Mai, V. Truong and F. Debaste, *J. Food Sci.*, 2014, **79**, 2222–2231.
408. O. Acosta, F. Vaillant, A. M. Perez and M. Dornier, *J. Food Process. Eng.*, 2017, **40**, 12343–12349.

409. G. D. Arend, W. T. Adorno, K. Rezzadori, M. D. Luccio, V. C. Chaves, F. H. Reginatto and J. C. C. Petrus, *J. Food Eng.*, 2017, **201**, 36–41.
410. C. Conidi and A. Cassano, *Desalin. Water Treat.*, 2015, **56**, 3510–3518.
411. N. A. Arriola, G. D. Santos, E. S. Prudencio, L. Vitali, J. C. C. Petrus and R. D. M. C. Amboni, *Int. J. Food Sci. Technol.*, 2014, **49**, 2052–2060.
412. E. Serre, E. Rozoy, K. Pedneault, S. Lacour and L. Bazinet, *Sep. Purif. Technol.*, 2016, **163**, 228–237.
413. E. Husson, M. A. Farias, Y. Desjardins and L. Bazinet, *Innovative Food Sci. Emerging Technol.*, 2013, **17**, 153–162.
414. L. Bazinet, S. Brianceau, P. Dubé and Y. Desjardins, *Sep. Purif. Technol.*, 2012, **87**, 31–39.
415. R. C. Oliveira, R. C. Docê and S. T. D. Barros, *J. Food Eng.*, 2012, **111**, 432–439.
416. S. Baklouti, R. E. Ghorbel, A. Mokni and S. Chaabouni, *Fruits*, 2012, **67**, 215–225.
417. R. C. C. Domingues, A. A. Ramos, V. L. Cardoso and M. H. M. Reis, *J. Food Eng.*, 2014, **121**, 73–79.
418. M. Purwasasmita, D. Kurnia, F. C. Mandias, Khoiruddin and I. G. Wenten, *Food Bioprod. Process.*, 2015, **94**, 180–186.
419. G. D. Francesco, G. Freeman, E. Lee, O. Marconi and G. Perretti, *J. Agric. Food Chem.*, 2014, **62**, 3279–3286.
420. L. Liguori, P. Russo, D. Albanese and M. D. Matteo, *Food Bioprocess Technol.*, 2013, **6**, 2514–2524.
421. M. T. Lisanti, A. Gambuti, A. Genovese, P. Piombino and L. Moio, *Food Bioprocess Technol.*, 2013, **6**, 2289–2305.
422. F. Spyropoulos, D. M. Lloyld, R. D. Hancocks and A. K. Pawlik, *J. Sci. Food Agric.*, 2014, **94**, 613–627.
423. K. Muijlwijk, C. B. Carabin and K. Schroen, *Trends Food Sci. Technol.*, 2016, **49**, 51–63.
424. R. D. Hancocks, F. Spyropoulos and I. T. Norton, *J. Food Eng.*, 2013, **116**, 382–389.
425. G. T. Vladisavljevic, B. Wang, M. M. Dragosavac and R. G. Holdich, *Colloids Surf., A*, 2014, **458**, 78–84.
426. V. Eisinaite, D. Juraite, K. Schroen and D. Leskauskaite, *Food Chem.*, 2016, **206**, 59–66.
427. R. Berendsen, C. Guell, O. Henry and M. Ferrando, *Food Hydrocolloids*, 2014, **38**, 1–10.
428. S. Morelli, R. G. Holdich and M. M. Dragosavac, *Chem. Eng. J.*, 2016, **288**, 451–460.
429. S. Ramakrishnan, M. Ferrando, L. A. Munoz, S. D. L. Castellvi and C. Guell, *Food Bioprocess Technol.*, 2013, **6**, 3088–3101.
430. M. S. Manga, O. J. Cayre, R. A. Williams, S. Biggs and D. W. York, *Soft Matter*, 2012, **8**, 1532–1538.
431. G. Sun, F. Qi, J. Wu, G. Ma and T. Ngai, *Langmuir*, 2014, **30**, 7052–7056.
432. B. Ahmad, E. Stride and M. Edirisinghe, *Food Bioprocess Technol.*, 2012, **5**, 2848–2857.

433. M. Matos, G. Gutierrez, O. Iglesias, J. Coca and C. Pazos, *J. Food Eng.*, 2015, **166**, 212–220.
434. R. Berendsen, C. Guell and M. Ferrando, *Food Hydrocolloids*, 2015, **43**, 636–648.
435. S. Gehrmann and H. Bunjes, *Eur. J. Pharm. Biopharm.*, 2016, DOI: 10.1016/j.ejpb.2016.11.006.
436. R. D. Hancocks, F. Spyropoulos and I. T. Norton, *J. Membr. Sci.*, 2016, **497**, 29–35.
437. M. Matos, M. A. Suarez, G. Gutierrez, J. Coca and C. Pazos, *J. Membr. Sci.*, 2013, **444**, 345–358.
438. I. Sawada, R. Fachrul, T. Ito, Y. Ohmukai, T. Maruyama and H. Matsuyama, *J. Membr. Sci.*, 2012, **387–388**, 1–6.
439. M. Aghdam, H. Mirassedghazi, M. Aboonajmi and M. H. Kianmehr, *Int. J. Food Sci. Technol.*, 2015, **50**, 892–898.
440. S. Chakraborty, E. Drioli and L. Giorno, *Biomass Bioenergy*, 2012, **46**, 574–583.
441. S. K. Giri and S. Mangaraj, *Int. J. Food Sci. Technol.*, 2014, **49**, 666–672.
442. H. J. Kim and B. K. Kim, *Food Sci. Biotechnol.*, 2015, **24**, 67–73.
443. K. H. Viswanathan, K. Govindaraju, V. Singh and R. Subramanian, *J. Food Sci.*, 2011, **76**, 158–164.
444. A. Ambrosi, N. S. M. Cardozo and I. C. Tessaro, *Food Bioprocess Technol.*, 2014, **7**, 921–936.
445. R. G. M. Sman, H. M. Vollebregt, A. Mepschen and T. R. Noordman, *J. Membr. Sci.*, 2012, **396**, 22–31.
446. A. Olmo, C. A. Blanco, L. Palacio, P. Pradanos and A. Hernandes, *J. Food Eng.*, 2014, **133**, 1–8.
447. M. Catarino and A. Mendes, *Sep. Purif. Technol.*, 2011, **79**, 342–351.
448. M. Purwasasmita, D. Kurnia, F. C. Mandias, Khoiruddin and I. G. Wenten, *Food Bioprod. Process.*, 2015, **94**, 180–186.
449. L. M. Schmidtke, J. W. Blackman and S. O. Agboola, *J. Food Sci.*, 2012, **71**, 25–41.
450. A. I. Gialelli, M. Kallis, A. Bekatorou, M. Kanellaki and A. A. Akoutinas, *Food Bioprocess Technol.*, 2015, **8**, 539–547.
451. R. Subramanian, C. S. Kumar and P. Sharma, *Crit. Rev. Food Sci. Nutr.*, 2014, **54**, 1151–1157.
452. C. S. Kumar, L. J. Rao and R. Subramanian, *Food Bioprocess Technol.*, 2013, **6**, 1926–1943.
453. A. Kumar, B. K. Thakur and S. De, *Food Bioprocess Technol.*, 2012, **5**, 2568–2577.
454. Q. Liu, Y. Zhao, J. Pan, B. V. d. Bruggen and J. Shen, *Sep. Purif. Technol.*, 2016, **164**, 70–80.
455. Y. Wang, Z. Yang and X. Wei, *Int. J. Biol. Macromol.*, 2012, **50**, 558–564.
456. B. Pan, P. Yan, L. Zhu and X. Li, *Desalination*, 2013, **317**, 127–131.
457. C. Brazinha, M. Cadima and J. G. Crespo, *J. Food Eng.*, 2015, **149**, 123.
458. T. A. Weschenfelder, P. Lantin, M. C. Viegas, F. D. Castilhos and A. D. P. Scheer, *J. Food Eng.*, 2015, **159**, 57–65.

459. S. Méthot-Hains, S. Benoit, C. Bouchard, A. Doyen, L. Bazinet and Y. Pouliot, *J. Dairy Sci.*, 2016, **99**, 8655–8664.
460. K. S. Bahceci, *Int. J. Food Sci. Technol.*, 2012, **47**, 315–324.
461. C. E. Jorgensen, R. K. Abrahamsen, E. O. Rukke, A. G. Johansen, R. B. Schuller and S. B. Skeie, *J. Dairy Sci.*, 2016, **99**, 6164–6179.
462. M. C. Y. Wong, K. Martinez, G. Z. Ramon and E. M. V. Hoek, *Desalination*, 2012, **287**, 340–349.
463. X. Luo, L. Ramchandran and T. Vasiljevic, *Dairy Sci. Technol.*, 2015, **95**, 15–31.
464. I. Baruckic, R. Bozanic and U. Kulozik, *Int. Dairy J.*, 2015, **51**, 1–7.
465. T. Steinhauer, S. Hanely, K. Bogendorfer and U. Kulozik, *J. Membr. Sci.*, 2015, **492**, 364–370.
466. A. Hausmann, P. Sanciolo, T. Vasiljevic, M. Weeks, K. Schroen, S. Gray and M. Duke, *J. Membr. Sci.*, 2013, **442**, 149–159.
467. M. C. Adams, E. E. Hurt and D. M. Barbano, *J. Dairy Sci.*, 2015, **98**, 7483–7497.
468. Y. E. Rayess, C. Albasi, P. Bacchin, P. Tailandier, M. M. Peuchot and A. Devatine, *Innovative Food Sci. Emerging Technol.*, 2012, **16**, 398–408.
469. R. Golbandi, M. A. Abdi, A. A. Babaluo, A. B. Khoshfetrat and T. Mohammadlou, *J. Membr. Sci.*, 2013, **448**, 135–142.
470. J. Lee, S. Jeong, Y. Ye, V. Chen, S. Vigneswaran, T. Leiknes and Z. Liu, *Sep. Purif. Technol.*, 2017, **176**, 323–334.
471. H. H. Himstedt, H. Du, K. M. Marshall, S. R. Wickramasinghe and X. Qian, *Ind. Eng. Chem. Res.*, 2013, **52**, 9259–9269.
472. J. L. Cohen, D. Barile, Y. Liu and J. M. L. N. Bell, *Int. Dairy J.*, 2017, **66**, 68–75.
473. C. Roblet, A. Doyen, J. Amiot and L. Bazinet, *J. Membr. Sci.*, 2013, **435**, 207–217.
474. C. Delcroix, J. P. Bonnet, M. Etienne and P. Moulin, *Carbohydr. Polym.*, 2015, **116**, 243–248.
475. M. C. Adams, E. E. Hurt and D. M. Barbano, *J. Dairy Sci.*, 2015, **98**, 7527–7543.
476. M. C. Adams and D. M. Barbano, *J. Dairy Sci.*, 2013, **96**, 2020–2034.
477. J. Zulewska and D. M. Barbano, *J. Dairy Sci.*, 2014, **97**, 2619–2632.
478. L. F. Garcia and F. A. R. Rodriguez, *Chem. Eng. Commun.*, 2015, **202**, 1455–1462.
479. S. R. Suwarno, X. Chen, T. H. Chong, V. I. Puspitasari, D. McDougald, Y. Cohen, S. A. Rice and A. G. Fane, *J. Membr. Sci.*, 2012, **405–406**, 219–232.
480. H. J. Tanudjaja, W. Pee, A. G. Fane and J. W. Chew, *J. Membr. Sci.*, 2016, **513**, 101–107.
481. P. A. Araujo, J. C. Kruithof, M. C. M. V. Loosdrecht and J. S. Vrouwenvelder, *J. Membr. Sci.*, 2012, **403–404**, 58–70.
482. B. Ikonic, Z. Zavargo, Z. Seres, J. Gyura and A. Jokic, *Sep. Sci. Technol.*, 2012, **47**, 972–982.
483. L. Liu, B. Gao, J. Liu and F. Yang, *Chem. Eng. J.*, 2012, **181–182**, 486–493.

484. S. Popovic, D. Jovicevic, M. Muhadinovic, S. Milanovic and M. N. Tekic, *J. Membr. Sci.*, 2013, **425–426**, 113–120.
485. F. Xie, W. Chen, J. Wang and J. Liu, *J. Membr. Sci.*, 2015, **495**, 361–371.
486. Z. Zhu, J. Luo, L. Ding, O. Bals, M. Y. Jaffrins and E. Vorobiev, *J. Food Eng.*, 2013, **115**, 264–271.
487. P. Meyer, A. Mayer and U. Kulozik, *Int. Dairy Sci.*, 2015, **51**, 75–83.
488. W. Zhang, N. Grimi, M. Y. Jaffrins and L. Ding, *J. Membr. Sci.*, 2015, **489**, 183–193.
489. Z. Zhu, H. Mhemdi, W. Zhang, L. Ding, O. Bals, M. Y. Jaffrins, N. Grimi and E. Vorobiev, *Food Bioprocess Technol.*, 2016, **9**, 493–500.
490. T. Li, A. W. K. Law and A. G. Fane, *J. Membr. Sci.*, 2014, **455**, 83–91.
491. T. Li, A. W. K. Law, M. Cetin and A. G. Fane, *J. Membr. Sci.*, 2013, **427**, 230–239.
492. S. Kertesz, A. Veszpremi, Z. Laszlo, G. K. Szabo and C. Hodur, *Desalin. Water Treat.*, 2015, **55**, 2836–2842.
493. T. Zsirai, H. Qiblawey, M. J. A. Marray and S. Judd, *J. Membr. Sci.*, 2016, **516**, 56–63.
494. M. Chai, Y. Ye and V. Chen, *J. Membr. Sci.*, 2017, **524**, 305–314.
495. L. E. Head and M. R. Bird, *J. Food Process. Eng.*, 2013, **36**, 228–240.
496. A. Cimini and M. Moresi, *Food Bioprocess Technol.*, 2014, **7**, 2694–2710.
497. G. T. Vladisavljevic, P. Vukosavljevic and M. S. Veljovic, *Food Bioprod. Process.*, 2013, **91**, 473–480.
498. A. Arkell, F. Vrgoc, O. Wallberg and A. S. Jonsson, *Int. Dairy J.*, 2015, **41**, 23–25.
499. A. Arkell, F. Vrgoc and A. S. Jonsson, *Int. Dairy J.*, 2014, **35**, 1–5.
500. M. T. Darestani, T. C. Chilcott and H. G. L. Coster, *J. Membr. Sci.*, 2014, **449**, 158–168.
501. H. Dong, K. Xiao, W. Liu and S. Guo, *Desalin. Water Treat.*, 2012, **49**, 234–239.
502. R. Vardanega, M. V. Tres, M. A. Mazutti, H. Treichel, D. Oliveira, M. D. Luccio and J. V. Oliveira, *Bioprocess Biosyst. Eng.*, 2013, **36**, 1087–1093.
503. B. Sarkar, *J. Food Process. Preserv.*, 2015, **39**, 1372–1384.
504. F. Almohammed, H. Mhemdi and E. Vorobiev, *Sep. Purif. Technol.*, 2017, **173**, 156–164.
505. H. Mhemdi, O. Bals, N. Grimi and E. Vorobiev, *Food Bioprocess Technol.*, 2014, **7**, 795–805.
506. Z. Zhu, O. Bals, N. Grimi, L. Ding and E. Vorobiev, *Ind. Crops Prod.*, 2013, **46**, 8–14.
507. A. Mirzaie and T. Mohammadi, *J. Food Eng.*, 2012, **108**, 77–86.
508. M. J. L. Facundo, J. A. M. Roca, B. C. Uribe and S. A. Blanco, *Food Bioprod. Process.*, 2016, **100**, 36–46.
509. L. L. A. Koh, H. T. H. Nguyen, J. Chandrapala, B. Zisu, M. Ashokkumar and S. E. Kentish, *J. Membr. Sci.*, 2014, **453**, 230–239.
510. M. A. Aghdam, H. Mirsaeedghazi, M. Aboonajmi and M. H. Kianmehr, *Innovative Food Sci. Emerging Technol.*, 2015, **30**, 127–131.

511. Q. Y. Z. Rojas, N. C. Cansino, E. R. Moreno, L. D. Olivares, J. V. Sanchez and E. A. Garcia, *Ultrason. Sonochem.*, 2013, **20**, 1283–1288.
512. M. A. Aghdam, H. Mirsaeedghazi, M. Aboonajmi and M. H. Kianmehr, *Int. J. Food Sci. Technol.*, 2015, **50**, 892–898.
513. M. H. Shahraki, A. Maskooki and A. Faezian, *J. Food Process. Preserv.*, 2017, **41**, e12795.
514. D. M. Sango, D. Abela, A. McElhatton and V. P. Valdramidis, *J. Appl. Microbiol.*, 2014, **5**, 1067–1083.
515. A. Bevilacqua, M. Sinigagliaand and M. R. Corbo, *Food Bioprocess Technol.*, 2013, **6**, 1153–1163.
516. A. Simon, J. A. McDonald, S. J. Khan, W. E. Price and L. D. Nghiem, *J. Membr. Sci.*, 2013, **447**, 153–162.
517. E. Filloux, J. Wang, M. Pidou, W. Gernjak and Z. Yuan, *J. Membr. Sci.*, 2015, **495**, 276–283.
518. E. G. Buriezza, A. R. Aguirre, G. Zaragoza and H. A. Arafat, *J. Membr. Sci.*, 2014, **468**, 360–372.
519. X. Long, Q. Meng and G. Zhang, *J. Membr. Sci.*, 2014, **457**, 113–119.
520. A. J. Kingma, *Food Bioprod. Process.*, 2015, **93**, 304–309.

Subject Index